WITHDRAWN

North Lincs College
1002656

Library of Congress Cataloging-in-Publication Data

Viru, A.
Biomechanical monitoring of sport training / Atko Viru and Mehis Viru.
p. cm.
Includes bibliographical references and index.
ISBN: 0-7360-0348-7
1. Sports–Physiological aspects. 2. Exercise–Physiological aspects. 3. Biochemistry.
4. Clinical biochemistry. I. Viru, Mehis, 1963- II. Title.
RC1235 .V57 2001
612'.044–dc21 2001024107

ISBN: 0-7360-0348-7

Acquisitions Editor: Mike Bahrke; **Developmental Editor:** Renee T. Thomas; **Assistant Editors:** Amanda S. Ewing, J. Gordon Wilson, Sandra Merz-Bott; **Copyeditor:** Judi Wolken; **Proofreader:** Erin T. Cler; **Indexer:** Betty Frizzéll; **Permission Manager:** Dalene Reeder; **Graphic Designer:** Fred Starbird; **Graphic Artist:** Jody Boles; **Cover Designer:** Jack W. Davis; **Art Manager:** Craig Newsom; **Illustrator:** Sharon Smith; **Printer:** Sheridan Books

Permission notices for material reprinted in this book from other sources can be found on pages xi–xii.

Printed in the United States of America 10 9 8 7 6 5 4 3 2 1

Human Kinetics
Web site: www.humankinetics.com

United States: Human Kinetics, P.O. Box 5076, Champaign, IL 61825-5076
800-747-4457
e-mail: humank@hkusa.com

Canada: Human Kinetics, 475 Devonshire Road Unit 100, Windsor, ON N8Y 2L5
800-465-7301 (in Canada only)
e-mail: orders@hkcanada.com

Europe: Human Kinetics, Units C2/C3 Wira Business Park, West Park Ring Road, Leeds LS16 6EB, United Kingdom
+44 (0) 113 278 1708
e-mail: hk@hkeurope.com

Australia: Human Kinetics, 57A Price Avenue, Lower Mitcham, South Australia 5062
08 8277 1555
e-mail: liahka@senet.com.au

New Zealand: Human Kinetics, P.O. Box 105-231, Auckland Central
09-523-3462
e-mail: hkp@ihug.co.nz

Biochemical Monitoring of Sport Training

Atko Viru, PhD, DSc
University of Tartu, Estonia

Mehis Viru, PhD
University of Tartu, Estonia

Human Kinetics

Dedicated to the athletes who intend to become a *Homo olympicus*
and to the coaches and scientists who help the athletes
on their way to superior performance

Contents

Preface

From a biological point of view, sport training represents the body's adaptation to conditions of increased muscular activity. Such adaptation can be achieved as the result of several changes in the body, which extend from the level of cellular structures and metabolic processes to the integral level of functional activities, their control, and the build of their body. The changes include the molecular mechanisms of metabolic processes and functional capacities of cellular structures. Together, all these changes ensure the increase of physical work capacity and sport performance, assist in the optimal development of a child or adolescent, guarantee health promotion, and help to maintain the quality of life in elderly people. However, the appearance of these expected developments depends on the quality and organization of the training. The training effects are specifically related to certain characteristics of performed exercises, their intensity and duration, and work/rest ratio both during the exercise session and the training microcycle (usually 4 to 7 days). Accordingly, the purpose of sport training is to intentionally change the body with the aid of appropriate exercises and training methods.

The background for this approach to training originates from the following results of physiological and biochemical studies:

- A number of changes and peculiarities in the body distinguish the top athlete *(homo olympicus)* from the sedentary person (*homo sedentarius*).
- Training experiments and cross-sectional studies confirm that systematic exercise may induce the changes necessary to improve physical abilities.
- The nature, intensity, and duration of training exercises and the peculiarities in the involvement of various muscles and motor units are the determinants of adaptive changes in the body.
- Specific changes in the body depend on performing specific exercises. Adaptive protein synthesis is the foundation for this dependence. It has been hypothesized that hormonal changes and metabolites accumulated during and after exercising are inductors of specific syntheses of proteins. The evoked adaptive protein syntheses produce increases in the most active cellular structures and augment the number of enzyme molecules, catalyzing the most responsive metabolic pathways.

Thus, training exercises result in specific changes in the body that are necessary to obtain the goal of training. For instance, improved endurance requires enhanced oxidative potential of muscle fibers, which is founded on increased numbers of mitochondria and elevated activity of oxidative enzymes. Collectively, the changes caused by various exercises produce improved performance levels.

The main advantage of biochemical monitoring is that each exercise is performed to achieve a specific change in the body and that the resulting change makes it possible to check the effectiveness of each exercise or a group of exercises. In this way, training becomes a well-controlled process. In turn, the changes in the body serve as a means for the expeditious and specific feedback control of the effectiveness of training.

For training arrangement, particularly for current corrections of the previously established plan, feedback information is required to know what happens in the body of the athlete; to know how training sessions, regimes, and stages are influenced (e.g., stage of hard training or taper stage); and to know what the main outcomes of training are. In many cases, hormonal and metabolic studies are necessary to obtain the required information.

Biochemical methods are increasingly used in athletes' training. A number of athletes and coaches recognize the value of biochemical indexes and use them as guides to training. However, cases still remain in which the results of biochemical studies are not understood and thereby useless. Cases in which biochemical methods are used incorrectly or inconsistently also exist.

The purpose of *Biochemical Monitoring of Sport Training* is to supply qualified coaches, sport physicians, researchers, and postgraduate students with the knowledge of scientific principles for the use of biochemical methods in monitoring training. In this way, we hope to ensure the propagation of information for choosing the appropriate biochemical methods for the desired results, to avoid overestimation of results of biochemical studies, and to determine whether a method is meaningful or inconsistent in obtaining the needed information. Therefore, this book deals with background and methodological problems of biochemical monitoring in sport training.

When and how to use biochemical monitoring in sport training is the focus of this book. The methodology of biochemical field studies and their validity, limitations, and eventual sources of errors are considered. The main task is to outline the relevant information and the scientifically founded background for using the various methods of monitoring. Our idea is to show that any tool is only good if used consistently and correctly. The analysis of various methods will be extended to an evaluation of obtained results, taking into consideration the concrete goal of training monitoring.

The book consists of three parts. The first part concerns the purpose, necessity, and tasks of biochemical monitoring of training. The second part deals with a general analysis of the means used in monitoring. The main focus is on metabolites and substrates and hormones as tools for monitoring. The third part analyzes the use of various methods for assessment of training-induced changes, evaluation of the load of the training session (including the evaluation of the trainable effect), control over the influence of training microcycles, assessment of peak performance, monitoring of the body's ability to adapt, and timely establishment of what constitutes overtraining.

The reader can benefit from this book by acquiring knowledge of how to use biochemical methods and understand the results obtained. We hope that *Biochemical Monitoring of Sport Training* will help solidify the basis for the management of athletes' training, which will, in turn, allow the training to be more efficacious.

We hope that researchers who specialize in training monitoring will find information in *Biochemical Monitoring of Sport Training* that stimulates new ideas, shows how to make monitoring more effective, and explains how to elaborate on new methods. Our desire is to help those who use biochemical monitoring in training to better understand its results.

This book is a sequel to *Adaptation in Sports Training* (Viru 1995). An excellent and comprehensive review by Saltin and Gollnick (1983) in *Handbook of Physiology—Skeletal Muscle* and monographs by Yakovlev (1977), Hollmann and Hettinger (1976), Hargreaves (1995), and Lehmann et al. (1999a, 1999b), as well as textbooks by Åstrand and Rodahl (1986), Brooks et al. (1996), and Garret and Kirkendall (2000) provided the background information for both *Adaptation in Sports Training* and *Biochemical Monitoring of Sport Training*. In addition, several review articles provided information pertaining to metabolic adaptation in muscular activity.

Acknowledgments

The authors would like to express their gratitude to Aino Luik, Kaire Paat, and Eve Luik for valuable technical assistance. The book was supported by the Estonian Academy of Sciences and by grants from the Estonian Science Foundation (3897, 3962, 4461). They would also like to express their gratitude to professors Manfred Lehmann and Antonio Hockney for their constructive criticism and valuable suggestions.

Credits

Figure 3.2 Reprinted, by permission, from H. Itoh, Y. Yamazaki, and Y. Sato, 1995, "Salivary and blood lactate after supramaximal exercise in sprinters and long-distance runners," *Scandinavian Journal of Medicina and Sciences of Sports* 5:285-290.

Figure 3.5 Reprinted, by permission, from A. Viru, 1987, "Mobilization of structural proteins during exercise," *Sports Medicine* 4:95-128.

Figure 3.6 Adapted, by permissions, from A. Viru, 1987, "Mobilization of structural proteins during exercise," *Sports Medicine* 4:95-128.

Figure 3.7, 3.8 Reprinted, by permission, from A. Viru et al., 1995, "Variability in blood glucose change during a 2-hour exercise," *Sports Medicine, Training and Rehabilitation* 6:127-137.

Figure 4.5 Reprinted, by permission, from K. Toode et al., 1993, "Growth hormone action on blood glucose, lipids, and insulin during exercise," *Biology of Sport* 10(2):99-106.

Figure 4.6, 5.2, 5.5 Reprinted, by permission, from A. Viru, K. Karelson, and T. Smirnova, 1992, "Stability and variability in hormone responses to prolonged exercise," *International Journal of Sports Medicine* 13:230-235. Georg Thieme Verlag

Figure 4.7 Reprinted, by permission, from A. Viru et al., 1990, "Changes of ß endorphin level in blood during exercise," *Endocrinologica Experimentalis* 24:63-68.

Figure 5.6 Adapted, by permission, from J. Wahren et al., 1975, Metabolism of glucose, free fatty acids and amino acids during prolonged exercise in man. In *Metabolic adaptation to prolonged physical exercise,* edited by H. Howald and J.R. Poortmans (Basel: Birkhäuser Verlag), 146, 147.

Figure 6.3 Adapted, by permission, from *European Journal of Applies Physiology*, Influence of prolonged physical exercise on the erythropoietin concentration in blood, J. Schwandt et al., vol. 63, pp. 463-466, 1991, © Springer-Verlag Gmbh & CO.KG.

Figure 6.6 Reprinted, by permission, from M.N. Sawka et al., 2000, "Blood volume: importance and adaptation to exercise training, environmental stresses, and trauma/sickness," *Medicine and Science of Sports & Exercise* 32:332-348.

Figure 6.10 Adapted, by permission, from T. Clausen and M.E. Everts, 1991, "K^+-induced inhibition of contractile force in rat skeletal muscle: role of active Na^+-K^+ transport," *American Journal of Physiology* 30:C791-C807.

Table 7.2 Adapted, by permission, from E. Hultman and R.C. Harris, 1988, Carbohydrate metabolism. In *Principles of exercise biochemistry* (Basel, Switzerland: S. Karger), 78-119. Adapted, by permission, form E. Hultman et al., 1990, Energy metabolism and fatigue. In *Biochemistry of Exercise VII* (Champaign, IL: Human Kinetics), 74.

Table 7.3 Reprinted, by permission, from B. Saltin, 1990, Anaerobic capacity: part, present and prospective. In *Biochemistry of Exercise VII* (Champaign, IL: Human Kinetics), 406.

Figure 7.4, 9.8, 9.14 Reprinted, by permission, from W. Kindermann, 1986, "Ausdruck einer vegatativen Fehlsteuerung," *Deutche Zeitschrift für Sportmedizin* 37:238-245.

Figure 7.5 Reprinted, by permission, A. Urhausen, et al., 1993, "Individual anaerobic threshold and maximum lactate steady state," *International Journal of Sports Medicine 14:*134-139.

Figure 7.7 Reprinted, by permission, from E. Horton and R.L. Terjung, 1988, Dietary intake prior to and after exercise. In *Exercise, nutrition, and energy metabolism* (New York: McGraw-Hill Companies), 134.

Table 7.4 Adapted, by permission, from G. Neumann, 1992, Cycling. In *Endurance in sport*, edited by R.J. Shephard and P.-O. Åstrand (Oxford, UK: Blackwell Sciences Pulbication), 582-596.

Table 7.7 Adapted, by permission, from H. Liesen, 1985, "Trainingsteigerung im Hochleistungssport: einge Aspekte und Beispiele," *Deutsche Zeitschrift für die Sportmedizin* 1:8-18. Adapted, by permission, from E. Hultman et al., 1990, Energy metabolism and fatigue. In *Biochemistry of Exercise VII* (Champaign, IL: Human Kinetics), 74.

Table 7.8 Adapted, by permission, A. Urhausen, A.B. Coen, and W. Kindermann, 2000, "Individual assessment of the aerobic-anaerobic transition by measurements of blood lactate." In *Exercise and sport science,* ed. W.E. Garrett and D.T. Kirkendall (Philadelphia: Lippincott, Williams and Wilkens), 267-275.

Figure 7.13, 8.5 Reprinted, by permission, from G. Neumann, 1992, Cycling. In *Endurance in sport*, edited by R.J. Shephard and P.-O. Åstrand (Oxford, UK: Blackwell Sciences Publication), 582-596.

Figure 7.16 Reprinted, by permission, from *European Journal of Applies Physiology*, Steroid and pituitary hormone responses to rowing exercising: relative significance of exercise intensity and duration and performance, V. Snegovskaya and A. Viru, vol. 67, pp. 59-65, 1993, © Springer-Verlag Gmbh & CO.KG.

Figure 7.17, 9.11 Reprinted, by permission, from Lehmann et al., 1989, *Sur Bedeutung von Katecholamin – und Adrenorezeptorverhalten für Leistungsdiagnostik and Trainingsbegleitung* (Münster, Germany: Philippka-Sportverlag), 15, 19.

Figure 8.2 Reprinted, by permission, from A. Viru et al., 1992, "3-methylhistidine excretion in training for improved power and strength," *Sports Medicine Training and Rehabilitation* 3:183-193.

Figure 8.7 Reprinted, by permission, V. Oopik and A. Viru, 1992, "Changes of protein metabolism in skeletal muscle in response to endurance training," *Sports Medicine, Training and Rehabilitation* 3(1):55-64.

Figure 8.8 Reprinted, by permission, from R.H.T. Edwards, 1983, Biochemical bases of fatigue in exercise performance: catastrophy theory of muscular fatigue. In *Biochemistry of Exercise* (Champaign, IL: Human Kinetics), 6.

Figure 9.6 Reprinted, by permission, from V. Snegovskaya and A. Viru, 1993, "Elevation of cortisol and growth hormone levels in the course of further improvement of performance capacity in trained rowers," *International Journal of Sports Medicine* 14:202-206. Georg Thieme Verlag

Figure 9.9 Adapted, by permission, from B.B. Pershin et al., 1988, "Reserve potential of immunity," *Sports Training Medicine and Rehabilitation* 1(1):55.

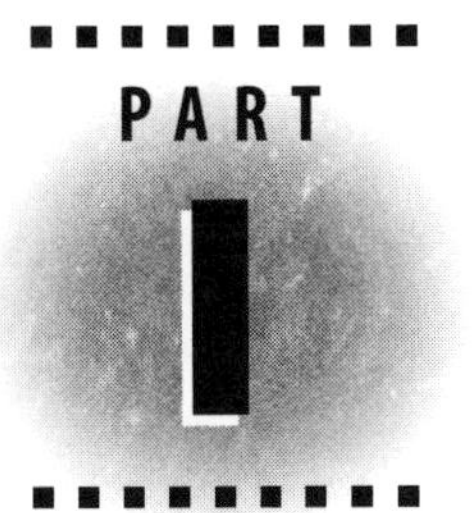

Purpose of Biochemical Monitoring of Training

The aim of this book is to discuss the problems of monitoring athletes' training with the aid of metabolic and hormonal studies. In the first part of chapter 1, the reader will find a brief historical overview on metabolic studies in humans performing physical exercises and participating in sport activities. Chapter 1 also discusses the development of research methods and the accumulation of specific knowledge that will allow researchers to use the results of blood and urine analyses to make recommendations about improving the training design and workloads.

If we want to discuss a special branch of research work, we have to establish what is known in this area, what principles guide the process, and what the special traits of design-related studies are. Accordingly, it is necessary to establish what the biochemical monitoring of training is. The second part of chapter 1 deals with this problem.

The monitoring process is effective if it has profound scientific foundations. Therefore, the researchers dealing with the biochemical monitoring of training have to be familiar with the results and conceptions of the metabolic adaptation to muscular activity. These problems also have to be understandable for those who will use the outcomes of biochemical monitoring in the practical guidance of training. Therefore, chapter 2 provides a short overview on metabolic adaptation in training.

CHAPTER

1

Introduction: Necessity and Opportunity

In 1992, Urhausen and Kindermann published the review article "Biochemical Monitoring of Training." The authors gave an overview of how several metabolic and hormonal parameters, recorded in athletes during their usual training, may be useful to get information about changes in the body. The paper did not contain the definition of the term "biochemical monitoring of training." However, the essence of this concept was made understandable with the discussion of the results. Urhausen and Kindermann (1992a) wrote: "A training stimulus can be effective only if intensity and duration of the training workload correspond to the present individual workload capacity. On this narrow strait between training below an effective threshold on the one hand and overtraining on the other, sports medicine has different blood parameters at its disposal. . . ." *Accordingly, biochemical monitoring can be considered a complex of means for effective guidance of training with the aid of information founded on biochemical analysis.*

Urhausen and Kindermann (1992a) dealt with blood parameters (substrates, enzymes, hormones, and immunological parameters). Undoubtedly, these parameters constitute the main tools for biochemical monitoring of training. Biochemical monitoring may also involve analysis of urine, sweat, and muscle tissue (by use of the needle biopsy method).

Historical Remarks

In the 19th century, studies appeared that considered metabolic processes in men during muscular activity. In 1866, Peterkofer and Voit indicated that in muscular activity, protein, unlike carbohydrate and fat, did not play a major role as a substrate for the energy turnover of contracting muscles. According to Fick and Wislicenus (1866), a person ascending a mountain (1,956 m) on a protein-free diet had urinary excretion of only 6 g of nitrogen. Consequently, the degradation process amounted to 37.6 g of protein, which could release only 635 kJ of energy (i.e., several times less than the actual energy expenditure). These studies seem to be the first (or at least among the first) biochemical studies on humans performing muscular work. The main conclusion of these studies was confirmed with the aid of calculations founded on changes in respiratory coefficient (Zunts 1901; Krogh and Lindhard 1920). However, exercise-induced protein degradation was confirmed in several papers. Exercise did cause increased urinary nitrogen excretion or an increased amount of urea in urine and blood (Rakestraw 1921; Levine et al. 1924; Cathcart 1925). Thus, the exercise-induced protein breakdown was established. This fact was later used to form the opinion that acute exertion induces catabolism, which is substituted by anabolism during postexercise recovery.

When the marathon race was included in the program of athletic competitions (first in Athens' Olympic Games in 1896), the possibility of hypoglycemia became an interesting research task. Hypoglycemia after marathon races has been documented in several articles (Barach 1910; Levine et al. 1924; Gordon et al. 1925).

Investigations of muscle energetics performed by Meyerhof (for an overview, see Meyerhof 1930) and Hill (for an overview, see Hill 1925) directed research interest to blood lactate changes. Several studies of exercise action on lactate level were soon published (e.g., Schenk and Craemer 1929; Margaria et al. 1933; Bang 1936). In this regard, interest has been extended to changes in blood pH (Dill et al. 1930) and blood buffer systems assessed by the blood alkaline reserve (Herxheimer 1933).

These studies were not directly aimed at supplying coaches with information on changes in the body to improve training. Biochemical studies performed for this purpose appeared in the 1950s and 1960s. The articles of Yakovlev and his team were among the first of these studies. To find out the optimal training workload, lactate levels after training sessions and competitions were compared among the athletes of sport games (Yakovlev et al. 1952) and rowers (Yakovlev et al. 1954). Data on lactate changes in sport exercises were analyzed by Donath et al. (1969b). Haralambie (1962) analyzed the possibilities of alterations in acid-base balance to assess the adaptation to muscular exercises. The problem with metabolic acidosis in severe sport exercises has been widely studied by Kindermann and Keul (1977). The material presented by Kindermann and Keul (1977) may serve as a standard response for the use of acid-base balance in assessing exercise intensity.

In 1959, Yakovlev and coworkers compared the improvement of strength, speed, and endurance in athletes in various events with lactate responses to test exercises. Makarova (1958) recorded the accumulation of protein metabolites after weight exercises with bars corresponding to 25%, 50%, and 75% of the athlete's maximal strength. The 75% exercise caused a larger accumulation of nonprotein nitrogen in the blood than exercise with bars corresponding to 25% or 50%. Further longitudinal study showed that training was most effective in improving strength when the bar corresponded to 75% of the athlete's ability. This material indicated that the accumulation of protein metabolites could be used in assessing the trainable effect of training sessions. To exploit protein metabolites, Haralambie (1964a) showed that blood urea accumulation depends on the intensity and duration of sport exercises. He also suggested that urea responses might be used as diagnostic tests for overtraining (Haralambie and Berg 1976). Plausibly, Eric Hultman was the first to use the biopsy method for training monitoring.

The term "biochemical monitoring of training" was not used in these articles. Instead, the terms "biochemical diagnostics" (Yakovlev 1962), "biochemical control" (Yakovlev 1970; Volkov 1977), or "biochemical criteria" (Yakovlev 1970, 1972; Volkov 1974), as well as "means for sports medical functional diagnostics" (Donath et al. 1969b) or "control means for assessing adaptation to physical exercises" (Haralambie 1962) was used. The term "training monitoring" became popular in the 1990s. Therefore, when biochemical methods are used for monitoring, the general term may be specified as "biochemical monitoring of training," as Urhausen and Kindermann (1992a) did. Compared with the terms "biochemical diagnostics" or "biochemical control," the term "biochemical monitoring" points out the purpose of the process. According to *Chambers Twentieth Century Dictionary* (Macdonald 1972), the monitor is the one who admonishes and advises, and the monitorship means to give admonition or warning.

By analyzing the tasks of biochemical diagnostics in sport, Yakovlev enumerated optimal relief intervals, evaluation of training effects and reaching peak performance (Yakovlev 1962), recording of characteristics of the influences of diverse training sessions, prediction of the actual performance level, diagnostics of prepathological states, and analysis of nutritional states (Yakovlev 1970). Therefore, biochemical criteria should be recorded at rest and after standard test exercises, training sessions, and competitions (Yakovlev 1970). In 1972, Yakovlev recommended the use of molar ratios of glucose/lactate, lactate/pyruvate, lactate/free fatty acids (FFA), pyruvate/FFA, and glucose/FFA. According to the results of his research team, the mobilization and use of FFA occur in highly trained athletes when the levels of glucose and lactate in the blood are higher (Krasnova et al. 1972).

Biochemical testing became popular in national-team athletes of the former Soviet Union, German Democratic Republic, and several other countries. However, most of the results from these tests remained unknown because they were considered secret. Only a limited number of articles were published due to the stringent requirements that had to be met for publication. Usually, these articles reported biochemical studies without describing the practical benefit of the information for training.

In the 1970s, lactate and urea studies of athletes became popular all over the world. Lactate studies led to the assessment of the anaerobic threshold (Wasserman and McIlroy, 1964; Mader et al. 1976). Lactate studies were also used for the assessment of anaerobic (glycogenolytic) capacity or power. Lactate was determined by the Wingate test or after similar supramaximal exercises (Szogy and Cherebetiu 1974; Jacobs et al. 1983). Volkov (1963) proposed the assessment of anaerobic glycogenolytic capacity by use of the blood lactate response in running the 400 m four times. The same possibility appears when pH decline is recorded in supramaximal exercises (Hermansen and Osnes 1972; Sahlin et al. 1978).

Special studies and reviews considered exercise-induced changes in levels of urea and other protein degradation products (Chailley-Bert et al. 1961; Gontzea et al. 1961; Haralambie 1964a; Gorokhov et al. 1973; Refsum and Strömme 1974; Haralambie and Berg 1976; Lorenz and Gerber 1979), including amino acids (see Holz et al. 1979; Viru 1987). The investigations of Chailley-Bert and coworkers provided data on electrolyte metabolism during prolonged exercise (Chailley-Bert et al. 1961).

Urhausen and Kindermann (1992a) showed that the biochemical monitoring of training should not be limited by lactate and urea determinations. They pointed out that valuable information can be obtained with the aid of hormonal studies. French researchers used corticosteroid excretion for fatigue diagnostics during prolonged exercise (Rivoire et al. 1953; Bugard et al. 1961). The main result of these studies was that in exercise-induced fatigue, an increase of corticosteroid excretion is substituted by decreased excretion, pointing out suppressed adrenocortical activity.

Hormonal studies in usual sport activities on top athletes were initiated in the former Soviet Union. During the first period, the urinary excretion of catecholamines, corticosteroids, and their precursors and metabolites was assessed (for an overview, see Kassil et al. 1978; Viru 1977). Later these studies were extended to hormone determinations in blood (e.g., Keibel 1974). Hormonal studies were mentioned in a popular book about Lasse Virén (Saari 1979), the four-time winner of the Olympic Games (1972 and 1976) in the 5,000 and 10,000 m. The responses of blood hormones to training sessions were used to determine the necessity of increasing the workload. In the 1980s, blood hormone responses were extensively investigated in top athletes of the former Soviet Union (Kostina et al. 1986).

Urhausen and Kindermann (1992a) also pointed out the usefulness of immunological studies and assessment of the enzyme activity in blood plasma for monitoring training. In 1970, Donath discussed the significance of the activity of enzymes in plasma in the evaluation of influence of training sessions and competitions.

Yakovlev (1977) emphasized that in biochemical control of training, success depends on the design of the test, choice of methods, and biochemical parameters. These three components must adequately correspond to the specific nature of the sport event and to the task tested.

In summary, metabolic changes in the human body related to muscular activity have been studied for more than a century. During the last half of the 20th century, attempts were made to provide information for trainers on metabolic changes during training sessions, competition, and several training stages. The value of this information was founded on the established links between the metabolic changes concerned and training effectiveness. Thus, elaboration of approaches for improved guidance of training became possible. In the future, extended studies are necessary to create informative monitoring systems specifically related to performance capacity in diverse sport events.

Principles and Design of Training Monitoring

Many measurements have been made in athletes. Special issues require various tests or measurements in athletes. However, just because something is measured in athletes does not mean that the measurement constitutes training monitoring. The following five principles comprise training monitoring:

1. It is a process performed for the purpose of increasing the effectiveness of training.
2. It is based on recording changes in an athlete during various stages of training or under the influence of the main elements of sport activities (training session, competition, microcycle of training).
3. It is a highly specific process, depending on the sport event, performance level of the athlete, and age/gender differences. Therefore, the methods for training monitoring have to be chosen specifically for the event and the athlete's characteristics.
4. Any method or measurement makes sense in training monitoring if it provides reliable information related to the task being monitored.
5. The information obtained from measurements has to be understandable; that is, it must be scientifically sound so that necessary corrective changes in training design can be made.

The main principle of the design of training monitoring is *minimum testing—maximum reliable information.* The alternative principle, *more testing—more information,* is not acceptable because monitoring is not a thing unto itself but a means to help coaches and athletes. The testing must be tailored to the training and should not overload the athlete. Every test or measurement is chosen as the most appropriate among a number of possibilities.

First, the tasks for training monitoring have to be established. Methods, tests, and necessary recordings (parameters) have to be chosen with the task and the specific event in mind. In turn, the account of the specific event means that valuable information is related to the specific foundations for performance in the event. The more direct the link between a parameter and a specific performance, the more value the testing has. For example, cross-country skiers require not only endurance but also muscle strength. However, testing for maximal strength will not provide the necessary information because the performance of cross-country skiers depends on the capability to maintain the optimal level of power output in each movement cycle. Maximal strength is only indirectly related to local muscle endurance, which limits the performance level considerably.

The successful choice of methods, tests, and parameters is the main condition for minimizing testing and maximizing information. Recording several parameters providing the same information should be avoided. Parameters that are related to the performance are preferable. In these cases, the link between performance and recorded parameter is important.

From an ethical point of view

- athletes should be hurt as little as possible during the study;
- athletes should be completely free from constraint;
- athletes should give informed consent for the use of any procedure, manipulation, or method;
- participation should not cause negative emotions in an athlete; and
- athletes should be informed about who will know about the results and have the right to require limitations on the distribution of information.

According to his experience, Yakovlev (1977) affirmed that besides the ethical aspect, constraint versus voluntariness might change results of biochemical studies in athletes.

From a medical point of view, any long-lasting aftereffect (soreness, inflammation caused by infections during the biopsy or blood sampling, etc.) has to be completely avoided. The possibility for infection arises when biopsy sampling or catheterization of blood vessels is done in field conditions. Even a prick on the fingertip for blood sampling before a game may harm basketball or volleyball players' performance because of soreness, and it can open the way for infection during intensive playing activities. In field studies, blood sampling should be substituted with urinalysis if possible. However, in most cases several cautions arise that substantially reduce the value of information obtained from the urinalysis compared with blood analysis. These cautions arise because a urine sample obtained does not reflect the metabolic situation at that moment but during the time from the last emptying of the urinary bladder to the sampling time. Accordingly, the metabolites and hormones in the urine sample represent mean values of the renal excretion (inflow into the urine) during this period. However, the renal excretion is related to the urine production by the kidneys, as well as to the conditions for renal clearance of the studied compound (see chapter 4, pp. 62-63).

Biopsies, arterial puncture, assessment of arteriovenous differences, and administration of isotopes can only be done in clinical laboratory conditions. Also, in these cases, the time elapsed from the previous testing has to be sufficient to avoid any harmful influence.

Biochemical analysis of microsamples obtained with biopsy or fingertip blood sampling and the precise determination of hormones and several other compounds require complicated and expensive tests. In studies on muscle energetics, the invasive biochemical methods may be substituted by nuclear magnetic resonance. However, the necessary apparatus is much more expensive than the use of biochemical methods. Moreover, nuclear magnetic resonance is only available in laboratory conditions and with a limited number of exercises.

With certain limitations, biopsy studies for evaluation of structural changes in muscle fibers may be substituted with muscle tomography. This equipment is also expensive.

Many articles have been published that indicate the possibility of substituting invasive lactate determinations for anaerobic threshold assessment with assessment of breaking points in the pattern of ventilatory and gas exchange indexes or heart rate during incremental exercise testing. However, several questions must be answered to be certain the precision and specificity of the proposed indirect methods are preserved, because direct tests with determination of blood lactate are preferred for anaerobic threshold studies. The determination of lactate is not expensive; the necessary amount of blood can be obtained by means of an earlobe prick. Lactate determinations are necessary for the assessment of the power of anaerobic glycogenolysis and for other purposes.

The time for testing is when a problem is found. Several tasks of training monitoring require that the actual changes appearing during the training or competition be recorded. In this case, the study design has to ensure that the most typical changes are recorded at their maximum. Therefore, the timing of the sampling/recording and the time between the actual maximum of change and sampling are critical. In other cases, the aim is to assess the cumulative effects of exercise, the pattern of developing training effects. In these cases, fatigue from previous activities of the day of (or day before) testing and an incomplete recovery process may alter the picture.

To ensure standard conditions for testing, Bulgarian athletes used a special microcycle for testing. The microcycle consisted of two stages. The first stage was aimed at standardization of the condition. Therefore, during the training year, the workload in the first 2 or 3 days (the first stage) has to be the same intensity, volume, exercises, and rest intervals. The second stage (2 days) was for testing, and the testing was performed by means of a strict schema. For an athlete, the testing meant not only recordings but also a repetition of exercises up to the maximum by a ring-training method. Thus, the idea was to use testing exercises so that they resulted in a trainable workload as a result of the sum of their actions. On the first day, the tests were performed for maximal power output. On the second day, tests were repeated at 75% individual maximum power for the highest possible number of repetitions (Matveyev 1980). This example has to be considered one of many, not a single possible solution.

The purpose of training monitoring is founded on the necessity of

- having feedback information on the actual training effects;
- knowing that the training design is adequate at a specific stage for that athlete; and
- recognizing the pattern of adaptational possibilities of the athlete.

The assessment of the pattern of developing training effects has to provide the opportunity for evaluating the link between performed exercise and resultant specific changes in the body. The analysis of training design in training tactics requires evaluation of the workload of training sessions (both the intensity and volume of the workload) and of training microcycles. The most essential is to ascertain whether the training session exerted the expected trainable effect. For evaluation of the microcycles, information on the recovery processes is necessary. Analysis of recovery processes may also be essential for establishment of optimal rest intervals between exercises during a training session. Evaluation of training sessions and microcycles is strongly related to fatigue diagnostics.

From the aspect of training strategy, it is important to know the dynamics of performance capacity. However, the account of results of competition does not provide sufficient information for several reasons. Therefore, more general information about the state of the body and particularly about the foundation of the specific performance capacity is necessary. In this regard, the prediction of the peak performance is more important, as it is usually recognized. Reaching the peak performance level exhausts the body's adaptivity to a great extent. Thus, a play at the borderline between effective training and overtraining begins. Consequently, monitoring training has to supply coaches and athletes with information about the body's adaptivity, including the diagnostics of earlier manifestations of overtraining.

The best information on training effectiveness and on adequate training design will be obtained if the adaptation processes and their structural, metabolic, and functional manifestations are recorded on the cellular level. In most cases, it is technically possible. However, these studies use complicated and expensive methods. Several ethical and medical considerations were noted earlier that constrict the use of biochemical methods in training monitoring. Even in laboratories, the most informative methods can be used only in a small number of persons. Nevertheless, biochemical methods are indispensable to fulfill several tasks of training monitoring. They are particularly necessary for evaluation of training effects on the power and capacity of energy production (adenosine triphosphate resynthesis pathways) and accomplishment of metabolic possibilities and metabolic control, for assessment of the trainable effect of training sessions, for analyses of the recovery processes and the design of the microcycle, for prediction of peak performance, for diagnostics of fatigue and overtraining, and for monitoring adaptivity changes.

Biochemical monitoring has to be considered a part of training monitoring, which makes it possible to get more profound information and information on changes with the aid of other methods. On one hand, when suitable information is obtainable without biochemical methods, biochemical monitoring may not be necessary. On the other hand, even in these cases, more information may be obtained with the aid of biochemical methods. For example, in several cases in which the effectiveness of training disappears, the studies may provide the cause. The idea is get into the fundamentals of training-induced adaptation by use of the metabolic and hormonal characteristics of adaptation.

The following points determine the essence of the biochemical monitoring of training:

- Metabolic adaptation constitutes the background for improved specific performance in the athlete's main sport event. These adaptations have to be characterized qualitatively and quantitatively to complete the training guide.
- In training, metabolic adaptation is also essential for improving general and event-specific motor fitness. The adaptations may have significance for exploring the effectiveness of the training.
- The effectiveness of training management during short periods may be evaluated by the metabolic and functional changes, which are known to occur as a result of certain exercises and training methods.
- The foundation for effective training is cellular enzymatic-structural adaptation evoked by metabolic and hormonal changes during and after training sessions. Recording these parameters opens a way for assessment of the trainable effect of training sessions.
- The erroneous training management causing the wrong direction in the metabolic adaptation or the hazardous drop in adaptivity and body resources may be detected with the aid of metabolic and hormonal studies.
- In monitoring training, metabolic and hormonal studies are useful if they provide results in which the information is significantly

greater than the information obtainable with the aid of more simple and less expensive physiological methods or specific testing of physical abilities and performance capacities.

The choice of tools and methods for biochemical monitoring has been founded on knowledge of the specific nature of training-induced metabolic adaptation. Accordingly, the organizers of training monitoring have to know what changes in the body make an ordinary teenager or a young adult into a homo olympicus, capable of competing in the Olympic Games or international championships for medals. The task is to establish the pathway for getting information on achieving the necessary characteristics, discriminating tasks for training, and significance of genotypical peculiarities (figure 1.1). This kind of information is necessary to make corrections in the training management and to objectively record the accumulated experience.

The necessary recorded characteristics have to provide valid and event-specific information on developmental processes over several years. In many cases, these years include prepubertal age, pubertal period, postpubertal development, and young adulthood. A problem exists as to whether the informative value of recorded parameters remains the same during the ontogenetic development and maturation. Although it is an important problem, we are not going to discuss it because the studies are still too few for generalizations.

Another consideration is that biochemical monitoring should become more accomplished over time; that is, the higher the performance level, the deeper should be the information. In an advanced level of performance, more frequently than previously, special investigations

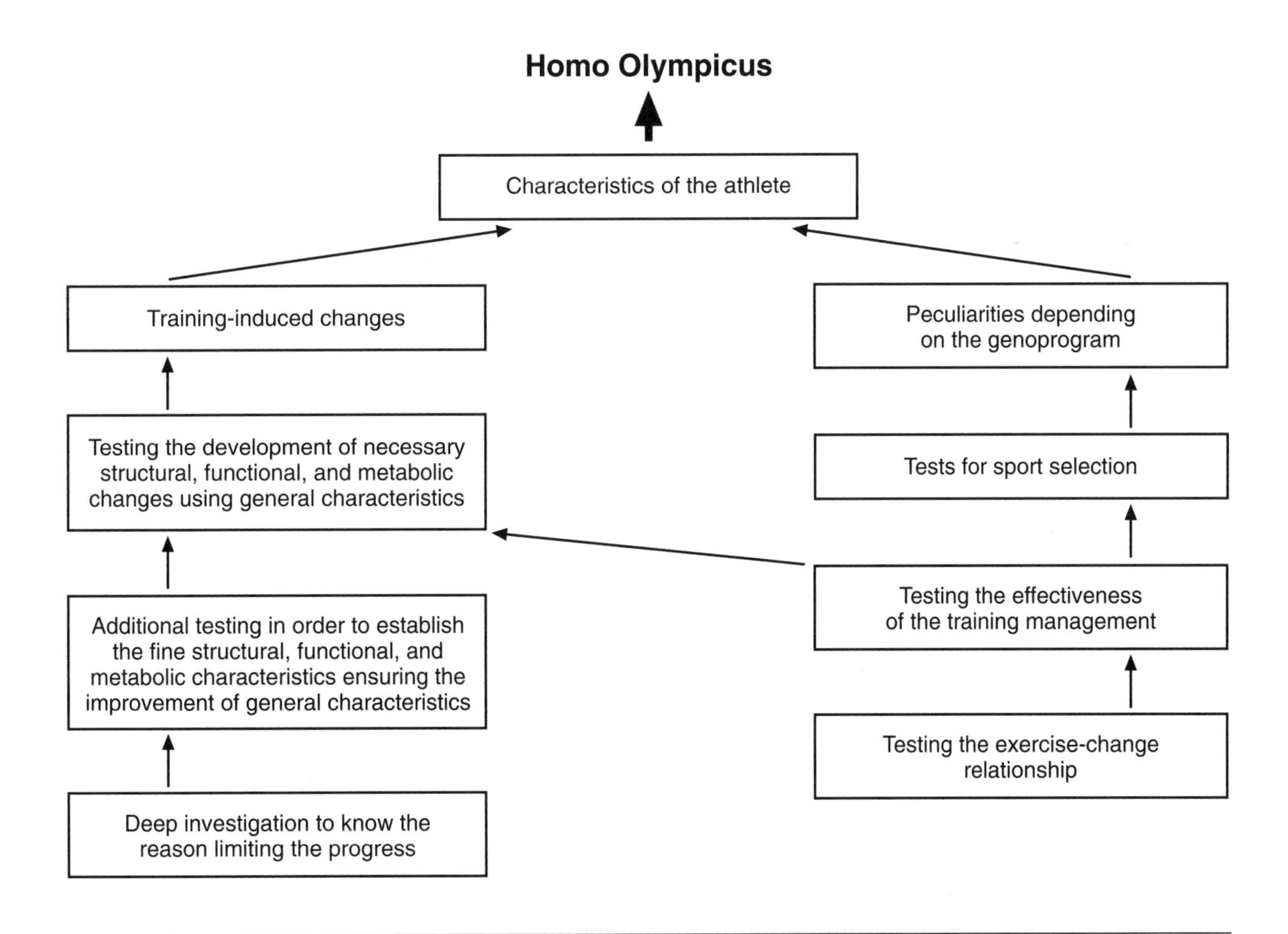

Figure 1.1 Schema for testing the training effects.

are necessary to understand the reason(s) limiting the rate of or eliminating further progress in adaptibility. Training practitioners suggest that every athlete has his or her inherent potential for improvement in training. When this potential is reached, an athlete can maintain his or her performance level but not increase it any further. The practical experience of athletes shows that after reaching a certain performance level, further increases in training volume and/or intensity do not improve the results. Unexpectedly, they may result in overtraining. The problem arises as to how to assess the potential of an athlete. At present, we are not able to quantify that potential.

In the historical overview (pp. 4), we mentioned that Yakovlev (1970) considered the diagnostics of prepathological states and the analysis of nutritional states among the tasks for biochemical diagnostics in sport. These tasks are not directly related to the evaluation of training design and are specific. The biochemical diagnostics of prepathological states in athletes still require serious research work.

Summary

Monitoring training is a complex means for studying the effectiveness of training. It is a purposeful and event-specific process. Only those investigations that result can promote the effectiveness of training guidance and may be considered a means of training monitoring. These investigations have provided information that is scientifically reliable and understandable to make necessary corrections in training design. The purpose for training monitoring is founded on the necessity of having feedback information on training effects, adequacy of training design, and pattern of adaptational possibilities of an athlete.

Biochemical monitoring is a part of training monitoring. It consists of recording metabolic and/or hormonal parameters that make it possible to get more profound information on adaptive processes in the body of an athlete, which are useful for the solution of tasks of training monitoring.

CHAPTER

2

Metabolic Adaptation in Training

To understand the body's adaptation in sport training, we have to penetrate deeply into the interior of the athlete's body. Increased performance levels and improved motor abilities are reflections of these adaptations. They indicate the achievement of the goal of the training and its effectiveness. However, performance indexes speak little about the adaptation processes inside the body. One way of visualizing the situation in a general way is by placing a "black box" on the cognitive pathway between training exercises and improved performance (figure 2.1). We know the entrance into the box (exercises) and the outcome from the box (improved performance), but we do not know how exercising improves performance and what the secrets of adaptation are.

This chapter presents a brief summary of the events in the "black box" during training in metabolic processes. Attention will be paid to adaptive protein synthesis as the foundation for cellular morphofunctional improvements. A control system, called metabolic control, exists for metabolic adjustments during exercise performance. This system ensures the effective use of cellular capacities and whole-body resources. The improvement of this control is also an essential consequence of training, together with increased ending reserves.

Role of Cellular Adaptation in Training-Induced Changes

The body's adaptation processes involve several organ systems and their control mechanisms. It is not difficult to understand that in endurance training the heart's improvement in functional capacity provides improved blood

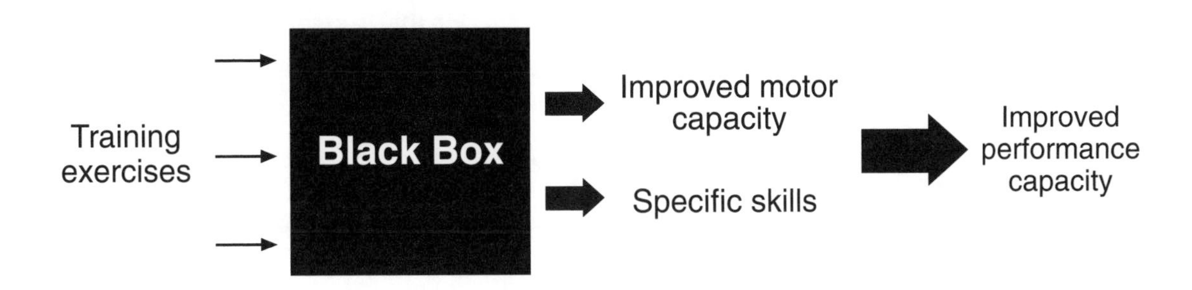

Figure 2.1 "Black box" on the cognitive pathway to understanding the essence of training.

supply to the working muscles. This change is an unavoidable link connecting training results and performance in endurance exercises. Almost all kinds of training inevitably lead to changes in skeletal muscles. First, muscle volume may increase. In heavy resistance training, this change is plainly visible. However, in endurance training, athletes do not see well-developed muscles. Inside the muscles one can see that the adaptation is expressed differently in the volume of muscle fibers of various types. Resistance training causes hypertrophy of muscle fibers of all types, with predominance of hypertrophy in fast-twitch fibers (figure 2.2) (Dons et al. 1978; Costill et al. 1979). A study indicated that the area of muscle occupied by fast-twitch fibers (type II) increased by 90%, despite retaining a fiber-type composition within the normal range (Tesch and Karlsson 1985). Speed or power training results in a selective hypertrophy of fast-twitch glycolytic (type IIb) or fast-twitch oxidative-glycolytic (type IIa) fibers (Saltin et al. 1976; Tihanyi et al. 1982). In resistance or power training and partly in speed training, changes appear in myofibrils, actualizing the muscle contraction. The increase of myofibrillar size is related to the augmentation of myofibrillar proteins, which are related to the act of contraction (Yakovlev 1978). These changes are necessary for the improvement of muscle strength and power.

Increases in the number and volume of mitochondria of muscle fibers, mainly of type I fibers (figure 2.3) (Gollnick and King 1969; Hoppeler et al. 1985), are typical for endurance training. The mitochondria are essential in producing more energy at the expense of the oxidative process because substrate oxidation together with formation of adenosine triphosphate (the main intracellular donor of energy) takes place in mitochondria. The developmental change at the level of mitochondria is associated with increased endurance capacity (Hoppeler et al. 1985).

The changes in cellular structures provide possibilities for improved performance both for the cell and whole organs, as well as for the whole body. However, these possibilities for improvement depend on the available energy and other conditions necessary for muscular contraction and other functional manifestations. Enzymes that catalyze biochemical processes, which make function possible or ensure the energy release for that function, and resynthesis of used energy-rich compounds are decisive. The increased activity of those enzymes is a typical training-induced response, which seems to depend on the nature of the exercises used (for a review, see Saltin and Gollnick 1983; Viru 1995).

The constant composition of ions in the cellular compartment and their fast effectual shifts are essential conditions for normal life activities. Ionic shifts between extracellular and intracellular fluids initiate the functional act of a cell. Each functional cycle has to be terminated with ionic shifts in the opposite direction. These shifts, on one hand, depend on the difference in ionic concentrations in intracellular and extracellular compartments. On the other hand, the ionic shifts, which restore the basal conditions, have to be carried out against the ionic gradient (from lower concentration to higher concentration). Therefore, it is an energy-consuming process. This process is actualized with the aid of ionic pumps in

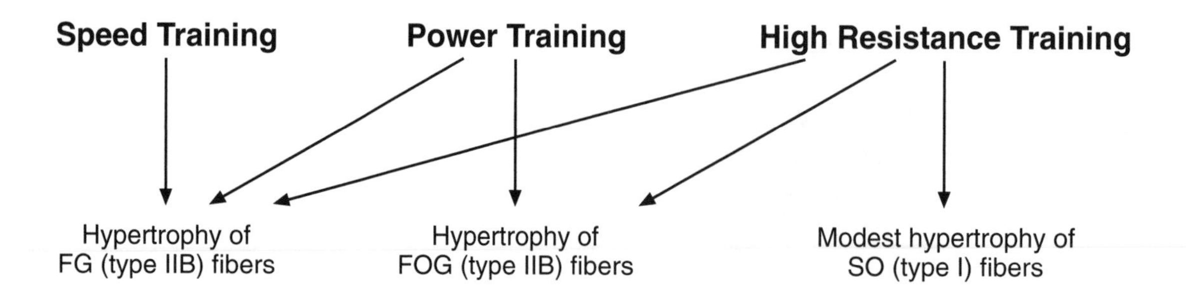

Figure 2.2 Effects of various types of training on the hypertrophy of muscle fibers.

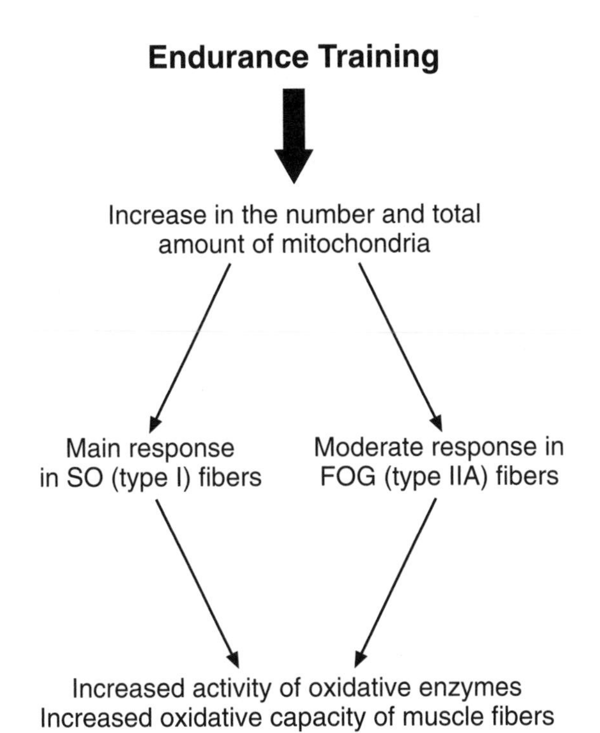

Figure 2.3 Specific effect of endurance training on the mitochondria of muscle fibers.

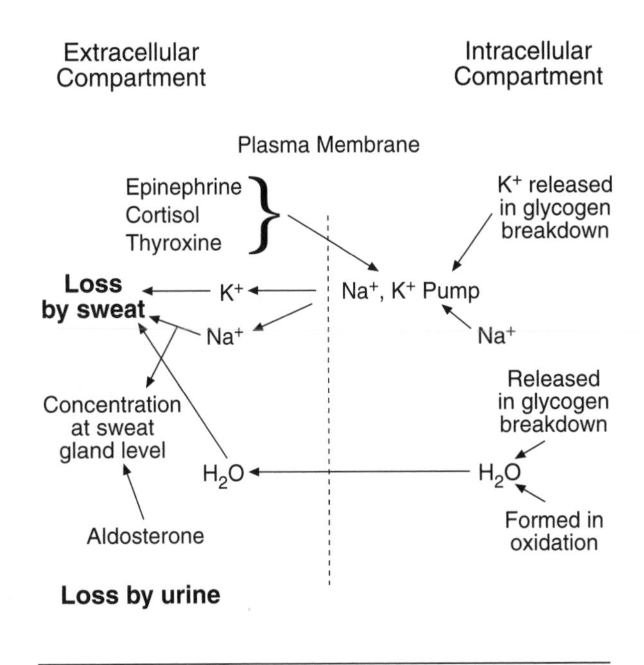

Figure 2.4 Ionic shift initiating and terminating the functional act of a cell.

cellular membranes (figure 2.4). Training increases the number of these pumps. This fact is demonstrated in humans with the aid of biopsy studies (Klitgaard and Clausen 1989; McKenna et al. 1993).

Increased intensity or duration of cellular activities makes outside help necessary. Increased amounts of oxygen have to be provided. Also, an additional inflow of substrates for oxidation has to reach into the cell. The blood gets oxygen from lung alveoli and substrates from liver, adipose, or some other tissues. In the arterial end of each capillary, the blood plasma is filtered into the interstitial fluid (extracellular fluid in tissues). This makes oxygen and substrates available for cells. In the venous end of a capillary, blood plasma is reabsorbed into the vessel. In this way, waste products of cellular activity are removed and transported into the specific organs for further metabolic changes or elimination of end products of metabolic degradation from the body.

Improvement in athletes' performance is founded first on changes in structures and metabolic capacities of skeletal muscle fibers. Improved metabolism in muscle fibers needs collaboration with several organs. Functional capacities of all concerned organs have to improve. Improvement also has to take place in integral coordination with the body's activity and in the control of activities of contributing systems, organs, tissues, and cells. Therefore, training-induced cellular adaptation involves myocardial, hepatic, renal, neuronal, endocrine, and other cells.

Adaptive Protein Synthesis

An intracellular mechanism exists that links the cellular function with the activity of the cellular genetic apparatus (Meerson 1965). Through this mechanism, an intensive functioning of cellular structures increases the synthesis of proteins specially related to the functional manifestation (e.g., contraction of muscle synthesis and secretion of hormones). These proteins are (1) "building material" for renewing and increasing protein structures that performed the functional activity, and (2) enzyme proteins that catalyzed the most important metabolic pathways, which make the functional activity possible. As a result, (1) involved cellular structures enlarge, and (2) enzyme activities are enhanced because the number of molecules of enzymes increased. Thus, the concerned synthesis ensures the adaptive effect. Therefore, this process is generalized by the term "adaptive protein synthesis."

It has been hypothesized (Viru 1984, 1994b) that training exercises cause an accumulation of metabolites that will specifically induce the adaptive synthesis of structure and enzyme proteins related to the most active cellular structures and metabolic pathways. Among hormonal changes induced by the training session are those that amplify the inductor effect of metabolites. The hormonal influence is probably necessary to increase the rate of protein synthesis more than is needed for the usual renewal of structure and enzyme proteins. Therefore, the adaptive effect will be achieved—structures will enlarge and the quantities of enzyme molecules will be augmented (figure 2.5).

A number of results have confirmed the increased rate of protein synthesis in muscles during hypertrophy (Hamosh et al. 1967; Goldberg 1968; for a review, see Poortmans 1975). Simultaneously, an increase in RNA content takes place (Millward et al. 1973). The increased genome activity has been indicated by elevated activities of DNA-dependent polymerase (Sobel and Kaufman 1970; Rogozkin and Feldkoren 1979) and amino-acyl-RNA-synthase (Rogozkin 1976). This result was obtained not only in strength training but also in endurance training. The main result of increased genome activity is the production of specific mRNA in response to the inductor action (the transcription phase of the protein synthesis). mRNA contains information on the structure of the protein that has to be synthesized. Production of various species of mRNA has been found after training exercises (Wong and Booth 1990a, 1990b) and during training (Marone et al. 1994; for a review, see Essig 1996; Carson 1997).

Actually, the activity-induced increase in protein synthesis is controlled not only at the level of transcription but also at the levels of translation and posttranslational control (figure 2.6) (Booth and Thomason 1991). Transcriptional control has been evidenced by an increase of α-actin in RNA and translational control by an increase of total RNA. The posttranslational control was suggested by a lesser increase of protein content compared with an increase in mRNA.

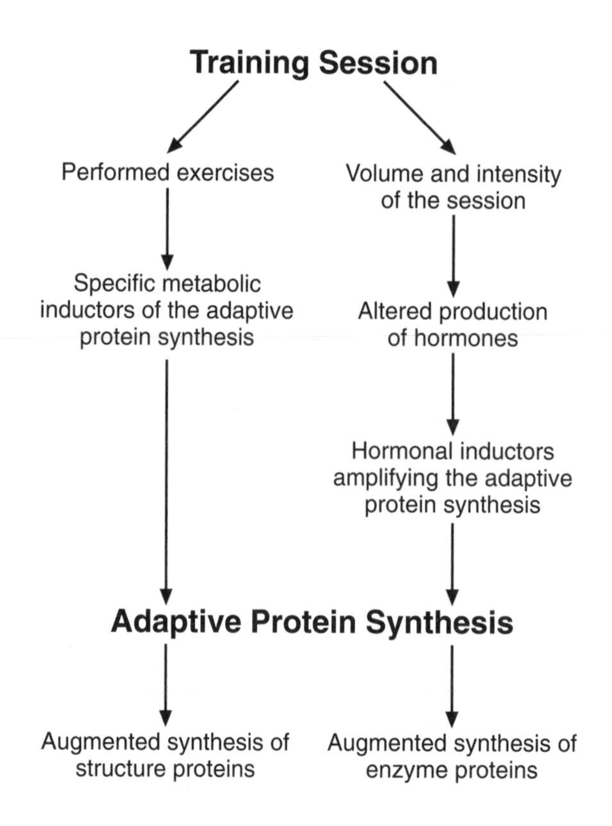

Figure 2.5 Adaptive protein synthesis triggered by training exercises.

The contribution of posttranslational control indicates the significance of protein degradation during training-induced adaptation. It has been asserted that the rapid growth of skeletal muscle coincides with rapid protein breakdown (Waterlow 1984). The degradation process, together with synthesis, constitutes the protein turnover. When a weight was attached to the wing of a chicken, a 140% increase in the protein content occurred in the slow-twitch anterior latissimus dorsi muscle (Laurent et al. 1978). It was calculated that only 20% of the increase in the protein synthesis rate accounts for net muscle growth, whereas 80% of the increase contributes to an increased turnover of proteins. In fast-twitch muscle hypertrophy, even a larger proportion of increase in protein synthesis contributes to the normal replacement of outworn protein structures (up to 91%). The actual muscle growth was warranted only by 9% of the increase in protein synthesis (Millward 1980). Thus, most of the newly synthesized proteins are used for renewal of cellular protein structures or degradated by the post-

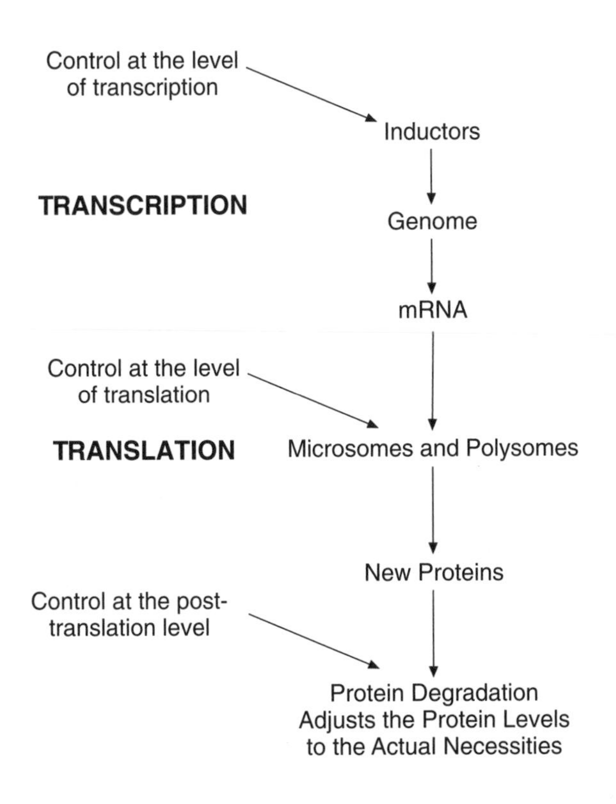

Figure 2.6 Three levels of control of adaptive protein synthesis.

translational control. Consequently, the muscle hypertrophy and the increase of cellular structures within a fiber require a cumulative effect of influences of several training sets.

Metabolites and hormones contribute to the control of protein synthesis at all three levels. The adjustment of transcriptional activity may be brought about by the "housekeeping" genes to meet the cell's protein demands. These genes are activated if protein-specific low–molecular weight compounds (specific fragments of protein subunits) are produced by protein degradation (Mader 1988). However, other metabolites act as inductors. Among protein metabolites, the contribution of creatine (Ingwall et al. 1974) and some amino acids, particularly leucine (Fulks et al. 1975), exerts an inductor effect. The effect of creatine has been demonstrated in preparations of skeletal (Ingwall et al. 1972) and heart muscles (Ingwall and Wildenthal 1976). However, creatine administration in vivo did not increase the synthesis rate of sarcoplasmic and myofibrillar proteins in muscle fibers of various types (Ööpik et al. 1993). The leucine action on protein synthesis consists of the stimulation of transcription (Fulks et al. 1975). Administration of leucine in vivo failed to convincingly confirm the stimulatory action on muscle protein synthesis (Sugden and Fuller 1991). Nevertheless, when leucine, valine, and isoleucine were administered in combination in vivo, the result was a stimulated protein synthesis in myocardium, diaphragm, and soleus muscle (Hedden and Buse 1982). When isoleucine and valine were added in vitro without leucine, the labeled amino acid incorporation was augmented but to a lesser extent than after administration of leucine alone (Buse 1981). In skeletal muscles, raised intracellular glutamine concentrations increase protein synthesis and inhibit overall, but not myofibrillar, protein degradation in vitro (Sugden and Fuller 1991).

According to the results of Yakovlev (1979), muscular activity causes subsequent activation of arginase, ornithine-decarboxylase, and ornithine-α-ketoglutarate transaminase in active muscles. The final result is an augmented formation of the polyamines spermidine and spermine, which are the inductors of protein synthesis (Tabor and Tabor 1976).

Stretching can induce increased protein synthesis in either cultured or whole muscle (Buresova et al. 1969, Vandenburgh and Kaufman 1981). Three possibilities are suggested: protein synthesis may be influenced through the function of the Na^+-K^+ pump, Ca shifts, or prostaglandin synthesis. The stretch-stimulated increase in protein synthesis may be caused by a chain of events: activation of sarcolemma phospholipases, release of arachidonic acid, and a consequent increase in prostaglandin synthesis. The arachidonic acid is considered to be a signal for protein synthesis (Smith et al. 1983). Of the metabolites of arachidonic acid, prostaglandin $F_{2\alpha}$ is effective (Rodemann and Goldberg 1972). The addition of this compound increased the effect of muscle contraction on protein synthesis (Palmer et al. 1983). Prostaglandins failed to stimulate the synthesis of proteins in calcium-deprived muscle preparations (Hatfaludy et al. 1989).

Little is known about the contribution of various tissue growth factors in training-induced adaptive protein synthesis. Growth factors are produced by the liver (e.g., insulin-like growth factors). They are known as metabolic modulators and stimuli for tissue growth (see chapter 5, p. 95). No background excludes their possible significance (see Adams 1998). In skeletal muscles undergoing hypertrophy, expression of insulin-like growth factor I has been found (Devol et al. 1990). This compound stimulates muscle protein synthesis but only in the presence of insulin and amino acids (Jacob et al. 1996). Independent of these growth factors, a family of myogenic factors has been discovered that is expressed in skeletal muscle cells and influences the transcription process in genomes (see Booth 1988; Carson and Booth, 2000).

Various experiments have provided much information on the inductor action of hormones and on the contribution of hormones to translation control. The role of hormones in the actualization of training effects becomes plausible when these results are plotted with the wide spectrum of exercise-induced changes in hormone levels. It is likely for one to assume that there are two separate mechanisms for the regulation of control of genome activity and gene expression: metabolic factors and hormonal influences.

The control of metabolic changes has to be highly specific and determines the choice of proteins for their adaptive synthesis. Therefore, metabolites accumulated in response to training exercise should be the main factor for selecting proteins for their adaptive synthesis. The hormonal control mechanism may be less specific and more dependent on the overall influence of training sessions on endocrine function. Accordingly, the training session has to be sufficient in its intensity and duration to activate the general adaptation mechanism and thereby induce the changes in endocrine functions. The related intensities and duration of training workloads depend on the ratio of performed exercises and the previous adaptation to related forms of muscular activity.

The two main hormones participating in the induction of the adaptive protein synthesis in postexercise periods are testosterone and thyroxin/triiodothyronine. The testosterone induction of protein synthesis responsible for training-induced hypertrophy was evidenced by the lack of hypertrophy after blockade of the androgen receptors (figure 2.7) (Inoue et al. 1994). In humans, confirming results were provided by Urban et al. (1995). Evidence indicates a further augmentation of protein synthesis in the muscles if training is accompanied by administration of an androgen preparation of anabolic action (Rogozkin 1979). The prohibited use of these preparations by athletes (the doping effect) made their training more effective and improved their results in strength and power events (Wilson 1988; Lamb 1989). Here we have a theoretically striking example of the amplifying action of a hormone on the metabolic effect of training.

Thyroid hormones are known to exert a stimulatory influence on the biogenesis of mitochondria. More than 30 years ago, it was demonstrated that treatment with triiodothyronine enhanced the training-induced increase in the activities of glycerol-*p*-dehydrogenase and succinate dehydrogenase (Kraus and Kinne 1970). The role of thyroid hormones in the adaptive synthesis of mitochondrial proteins in endurance training was supported by the results, indicating that the specific increase in the synthesis of mitochondrial proteins and the increase of total protein synthesis in oxidative-glycolytic muscle fibers after an endurance exercise (30-min run at 35 m/s) does not appear in hypothyroid rats (figure 2.8) (Konovalova et al. 1997).

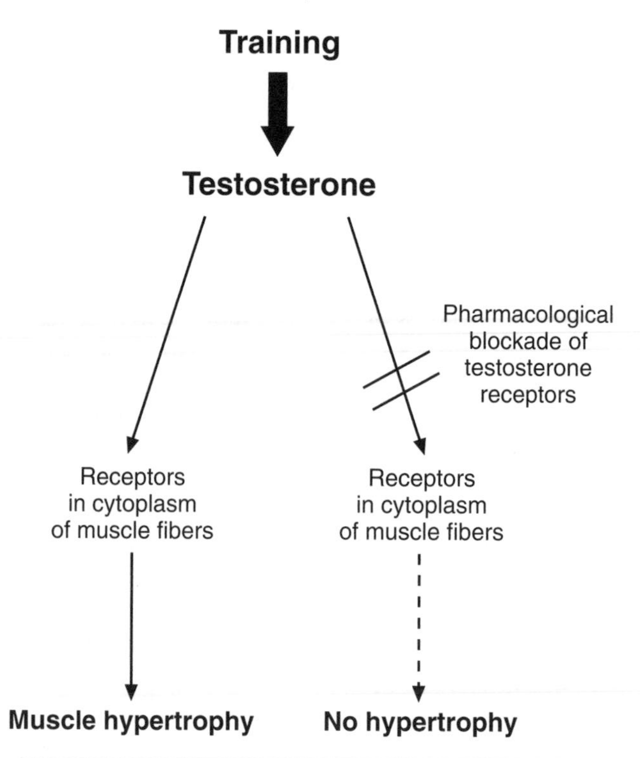

Figure 2.7 Significance of testosterone in skeletal muscle hypertrophy.

Thyroid hormones and androgens seem to be important in separately amplifying the two main directions of training effects on muscle proteins. However, this does not mean that thyroid hormones do not contribute to the synthesis of myofibrillar proteins or testosterone to the synthesis of other proteins. Nevertheless, certain specificity remains (for a review, see Caiozzo and Haddad 1996).

In translation control, certain significance belongs to charging tRNA by specific amino acids. The charging releases the protein synthesis system from tonic inhibition (Vaughan and Hansen 1973). The control of muscle protein synthesis could be modulated at the translational level by the cytoplasmic redox potential (Poortmans 1988). Essential contributions to translational control belong to growth hormone (Goldberg and Goodman 1969; Fryburg et al. 1991) and insulin (Wool and Cavicchi 1966; Balon et al. 1990). When the normal adrenocortical responses were blocked during training exercises, swimming training did not increase the endurance in rats (Viru 1976b). This fact may be related to the two regulatory effects of glucocorticoids. First, the glucocorticoid catabolic effect is necessary for the mobilization of precursors for protein synthesis. On the other hand, the same catabolic effect may be involved to warrant an intensive protein turnover and, thereby, an effective renewal of the most responsible protein structures. The contribution of glucocorticoids in the posttranslational control seems plausible.

In adrenalectomized rats, cardiac hypertrophy was more pronounced after training than in normal rats (Viru and Seene 1982). This fact suggests that glucocorticoids may contribute to the determination of the optimal magnitude of structural changes of the myocardial cells. However, apart from these results, glucocorticoid treatment increased the training-induced cardiac enlargement in intact female rats (Kurowski et al. 1984). High unphysiological doses of glucocorticoids induce muscle atrophy. This fact has to be considered if glucocorticoids are used as therapeutic means. Rat experiments showed that training can avoid or reduce this harmful result of high-dose glucocorticoid administration (Hickson and Davis 1981; Seene and Viru 1982). These results suggest that muscular activity may change the glucocorticoid effect on protein metabolism (see also chapter 5, p. 86).

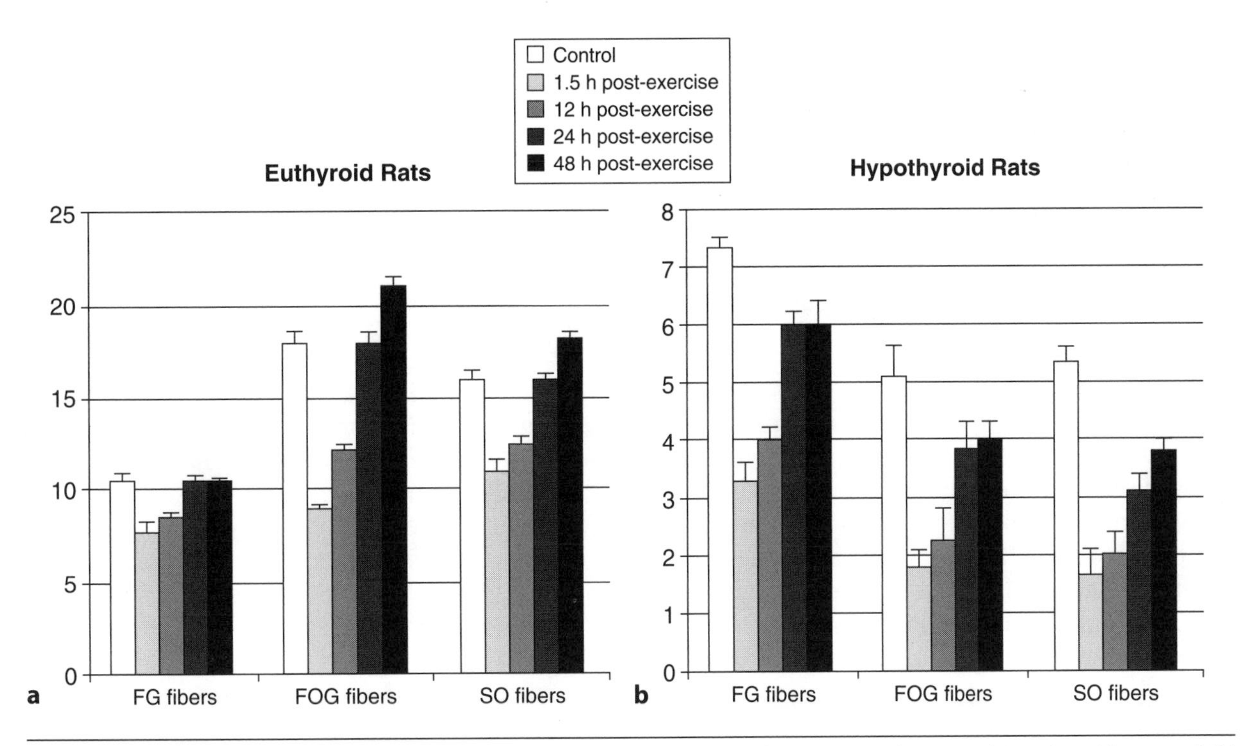

Figure 2.8 Significance of thyroid functions in postexercise increase of protein synthesis on *(a)* euthyroid rats and *(b)* hypothyroid rats. Columns indicate the rate of amino acid incorporation into protein of muscle fibers.

By results of Konovalova et al. 1997.

Other factors created by exercising also seem to be connected with posttranslational control. One of these may be testosterone; administration of its synthetic analogue, nandrolone, diminished the decrease of myosin Mg^{2+}-ATPase activity in slow-twitch muscles caused by forced training in rats (Viru and Kõrge 1979).

To adjust the final suitable amount of new proteins, stimulation of protein degradation by leucine transamination product α-ketoisocaproate (Tischler et al. 1982), arachidonic acid, prostaglandin E_2 (Rodemann and Goldberg 1972), or a low level of glutamine (MacLennan et al. 1988) may be essential. Thus, control of adaptive protein synthesis at the posttranslational level is actualized with the aid of hormones and metabolites.

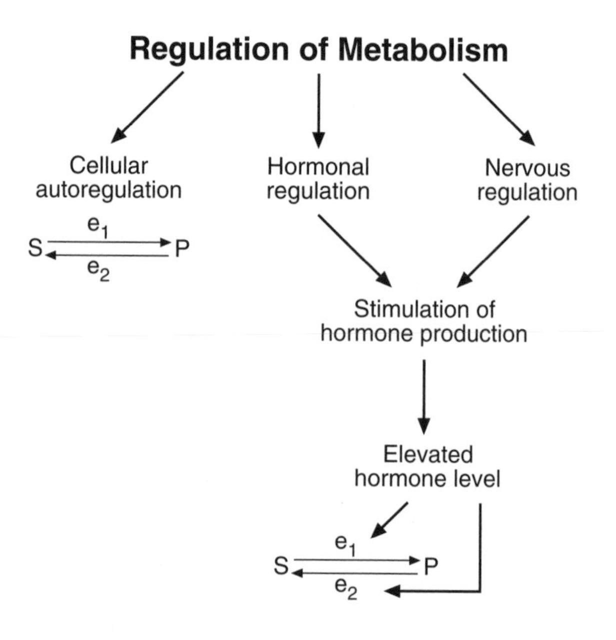

Figure 2.9 Three levels of regulation of metabolism.

Metabolic Control

Metabolic control is the tool for adjusting metabolic processes in various tissues to the demands of the body's activities. Metabolic control is actualized by an influence on the enzyme activities. As a result, the direction of biochemical reactions, founding the metabolic processes, and the rate of metabolic cycles change. Exercises can be performed if metabolic processes are adjusted to the demands of increased energy expenditure. On this order, the energy reserves and other resources (e.g., protein resources) have to be made available for their use during exercise. This is the task of metabolic control. Metabolic control is actualized on three levels: cellular autoregulation, hormonal regulation, and nervous regulation (figure 2.9).

Cellular Autoregulation

The main principle of metabolic control is that the substrate/product ratio determines the activity of enzymes catalyzing, respectively, the conversion of a substrate (S) into a certain product (P) or the reaction in the opposite direction.

The increase of the substrate and the decrease of the product stimulate enzyme e_1 activity (catalyzes the conversion of the substrate S into the product P) and inhibit enzyme e_2 (catalyzes the opposite process) activity. The substrate can be converted into the product if the activity of enzyme e_1 surpasses the activity of enzyme e_2. The opposite situation emerges with the decrease of the substrate and the increase of the product. Then, an inhibition of enzyme e_1 and a stimulation of enzyme e_2 occur. As a result, the reaction stops, and it is replaced by the opposite reaction.

Actually, cellular autoregulation is more complicated than this. In most cases, the regulation is actualized in the course of the whole metabolic pathway or metabolic cycle.

If the rate of glycogenolysis surpasses the intensity of the oxidation processes, a portion of pyruvate will be converted into lactate. The latter, a product of the metabolic pathway, inhibits a number of glycogenolytic enzymes; the inhibition is also due to the concomitant accumulation of hydrogen ions (decline of pH).

In muscle fibers, Ca^{2+} and inorganic phosphate have an essential contribution to cellular autoregulation. Ca^{2+} has the primary role in the coordination of skeletal muscle contraction (cross-bridge formation in myofibrils) and the activation of glycolysis and several mitochondrial enzymes. Inorganic phosphate accumulation arises from adenosine triphosphate (ATP) and, more importantly, from phosphocreatine hydrolysis during muscle contraction. It plays a key role in the regulation of glycogenolytic activity of phosphorylase a. Thus, inorganic phosphate links ATP turnover associated with contraction and the rate of substrate mobilization.

At the same time, the ATP turnover rate also influences the oxidation rate in mitochondria. The oxidation rate augments the increase in adenosine diphosphate and adenosine monophosphate (products of ATP degradation) with the decrease in ATP and phosphocreatine (for a review, see Greenhaff and Timmons 1998).

Cellular autoregulation is directed toward the immediate satisfaction of cellular needs and the exclusion of pronounced changes in the cellular compartment (table 2.1). It is important for the resting state. However, it is not suitable for an overall mobilization of cellular and bodily resources. For example, metabolic events concomitant with the initiation of contraction stimulate glycogenolysis only for a short time in skeletal muscle.

Hormonal Metabolic Control

The overall mobilization of cellular and bodily resources requires the interference of hormonal regulation with cellular autoregulation (table 2.1). The main aim of hormonal regulation is to adjust the metabolic processes to the level corresponding to the actual needs of life activities, despite the opposite effects of cellular autoregulation. This aim is achieved through the action of hormones on the activity of enzymes.

The hormone effects on enzyme activities are realized in two ways. First, in a number of cases, the structure of the enzyme molecule changes under the influence of a hormone. As a result, enzyme activity increases or decreases. In many cases, the corresponding change consists of either phosphorylation or dephosphorylation of the enzyme molecule. The second possibility is the change in the number of enzyme molecules. A number of hormones can induce or inhibit the synthesis of enzyme proteins, resulting in an increase or decrease in the number of enzyme molecules. In some cases, hormones are able to either intensify or suppress the degradation of enzyme proteins.

The epinephrine effect on the activity of glycogen phosphorylase in contracting muscle has been convincingly demonstrated (Richter et al. 1982; Arnall et al. 1986; Spriet et al. 1988). Obviously, this is the mechanism that enables athletes to perform short-term competitive exercises. Through the action of hormones on enzyme activities, the mobilization of hepatic glycogen stores, lipids, and protein resources is also actualized during prolonged exercises.

Hormonal regulation is also necessary for performing the tasks of homeostatic regulation during muscular activity. Through hormonal actions, constant levels of ions and water are maintained in intracellular and extracellular compartments. A constant level of glucose in the blood is also maintained with the aid of hormonal regulation.

Table 2.1
Cellular Autoregulation and Hormonal Regulation in Metabolic Control

Cellular autoregulation	Hormonal regulation
Substrate/product ratio determines the activities of concerned enzymes	Hormones result in • Conversion of inactive enzymes to active or • Increased/decreased rate of synthesis of enzyme molecules
Immediate satisfaction of cellular needs Exclusion of pronounced changes in the cellular compartment	To adjust the metabolic processes to the level of corresponding actual needs of life activities
Important for the resting state	Important for an overall mobilization of cellular and bodily resources

Nervous Metabolic Control

The production of hormones by endocrine glands is primarily regulated with the aid of feedback systems: a high level of hormone suppresses and a low level stimulates the activity of corresponding endocrine systems. However, rapid adjustments of metabolic processes require quicker interference of hormones with cellular autoregulation than occurs as a result of feedback regulation of hormone production. The necessary rapid changes in hormone levels are evoked by nervous regulation of endocrine function. Functional nerves—the excitement of which causes changes in hormone secretion—directly supply some of the endocrine glands. For instance, excitement of the n. splanchnicus results in rapid epinephrine secretion by the adrenal medulla at the beginning of exercise. The activities of other endocrine glands are altered by a two-step system: (1) hypothalamic neurosecretory cells produce neurohormones (liberins or statins) that stimulate or inhibit the release of pituitary trophic hormones, and (2) pituitary trophic hormones stimulate the activity of peripheral endocrine glands.

Influence acting on the constant parameters of the internal environment

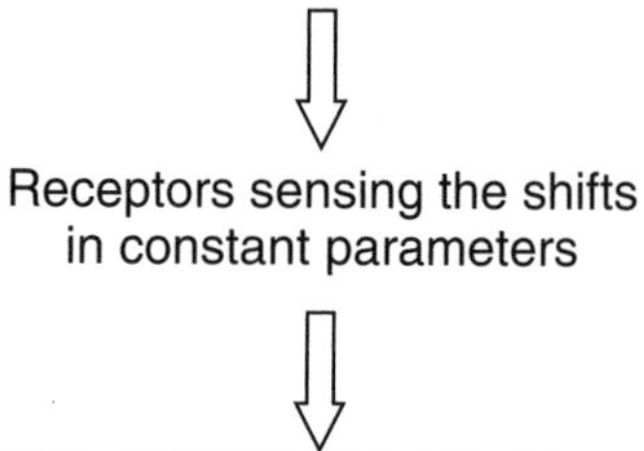

Coordinated changes in various functions, secretions of hormones, and metabolic processes to compensate the influence and/or restore the constant level of temperature, pH, ionic constant, osmotic pressure, pO_2, water content, glucose level

To ensure the optimal activity of enzymes and to avoid metabolic disturbances

Figure 2.10 Homeostatic regulation.

Acute and Long-Term Adaptation

Exercises performed by athletes during training sessions or competitions evoke adjustments belonging to the group of *acute* adaptation processes. These processes include homeostatic regulatory responses (figure 2.10) and activation of oxygen transport and use of energy reserves. Every exercise results in an increased oxygen demand and the necessity to eliminate the produced CO_2. Accordingly, the activity of cardiovascular and respiratory systems has to increase. The greater the role of anaerobic glycogenolysis in ATP resynthesis, the greater the need for homeostatic means to avoid the increase in the concentration of H^-. Augmented energy metabolism results in increased heat production. Adjustments at the level of thermoregulation have to follow. Increased perspiration alters water and electrolyte balances. Again, necessities arise for related homeostatic responses. Homeostatic responses are also necessary to maintain euglycemia.

When exercise intensity or duration increases over certain threshold values, an overall mobilization of energy and protein resources (activation of the mechanism of general adaptation) will take place (figure 2.11). High activity of the mechanism of general adaptation creates conditions for transition from acute to long-term adaptation. The accumulation of inductors of protein synthesis and increased free amino acid pool during training exercises are essential.

Structural and functional changes developing in an athlete during prolonged periods of training express *long-term* adaptation. It is founded on adaptive protein synthesis. Adaptive protein synthesis requires

- creation of inductors acting on the cellular genetic apparatus and calling forth the specifically related synthesis of the concerned proteins;
- supply of synthesis processes by "building materials" (amino acids and precursors for synthesis of ribonucleic acids);
- destruction of old, physiologically exhausted cellular elements; and
- supply of synthesis processes by energy.

Therefore, the accumulation of inductors of protein synthesis and increased free amino acid pool are essential. These changes happen during training exercise. However, to evoke conditions for the adaptive protein synthesis, the load of training sessions has to be sufficiently high to activate the mechanism of general adaptation, including the pronounced alterations in endocrine functions.

During the recovery period after training sessions or competitions, the body's energy reserves and protein resources can be extensively used for the adaptive synthesis of enzyme and structural proteins to restore the functional capacity of cellular structures (for a review, see Viru 1996). The enlargement of active cellular structures and, thus, an improvement of the functional capacity are actualized as a result of postexercise synthetic processes.

Training Effects on Metabolic Resources

Energy releases immediately for muscle contraction and for any cellular process in the hydrolysis of ATP. Resources of ATP are maintained at the expense of rapid degradation of phosphocreatine and less rapid degradation of glycogen. The most voluminous but slowest pathway for ATP resynthesis arises at the expense of energy released in oxidation (oxidative phosphorylation). Rats trained by repeated short-term intensive exercise had an increased phosphocreatine content, but not ATP content, in skeletal muscles. The effect of continuous exercise on phosphocreatine was only modest (Yakovlev 1977). Some of the human biopsy studies confirm the increase of phosphocreatine stores and

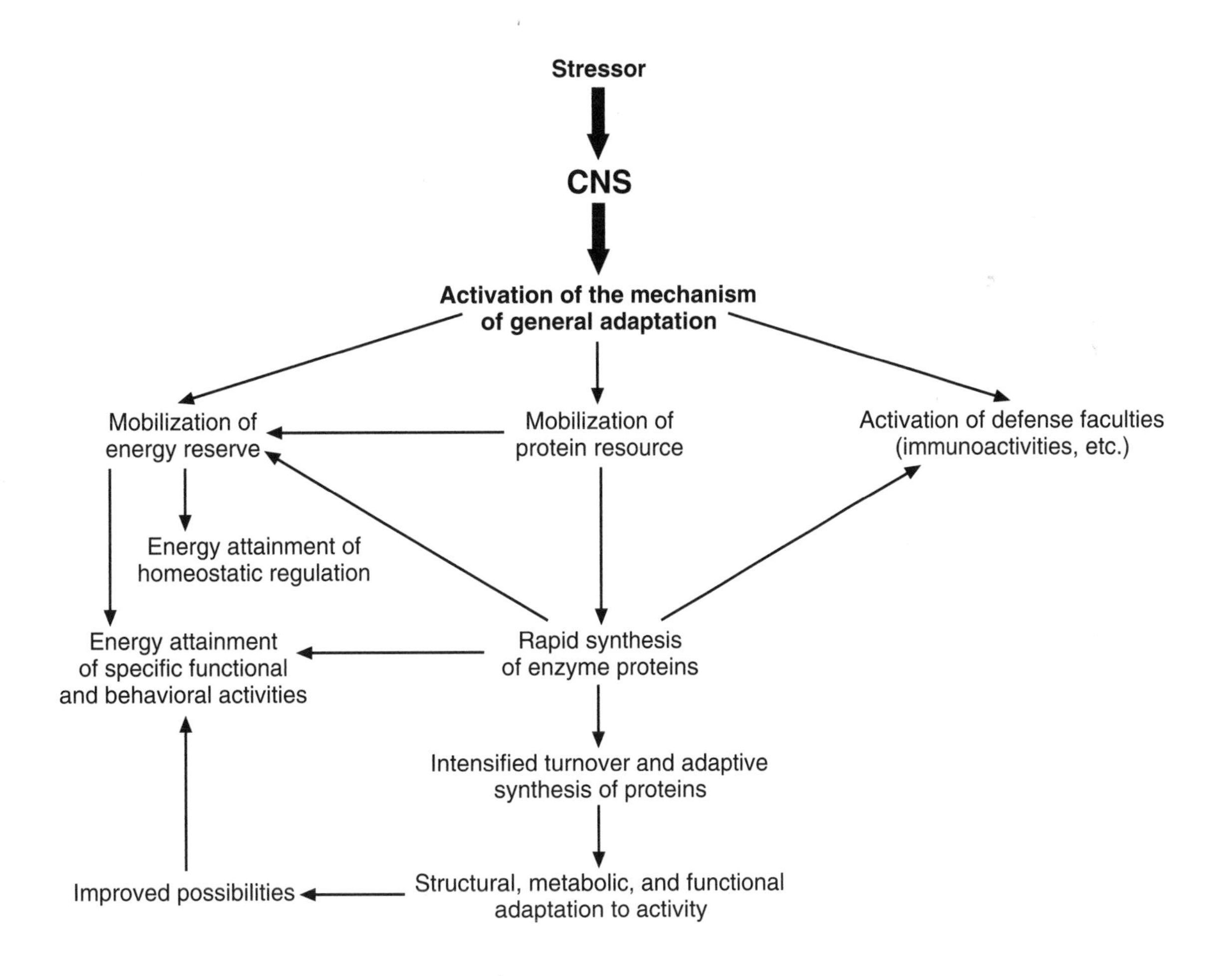

Figure 2.11 Mechanism of general adaptation.

also a minimal increase of ATP content in limb muscles after short-term endurance (Karlsson et al. 1972) or heavy resistance training MacDougall et al. 1977). The training effect on phosphocreatine content should be expected in sprint training. However, after 8 weeks of training with 30-s cycling bouts, phosphocreatine concentration increased by only 9%, and no change in ATP content occurred (Boobis et al. 1983); 8 weeks of training with 30-s dashes did not change phosphocreatine and did not significantly increase ATP content (Nevill et al. 1989). A cross-sectional comparison of 68 men and 11 women did not show any significant differences in ATP and phosphocreatine content in the quadriceps femoris muscle of cyclists, weightlifters, sprinters, long-distance runners, and sedentary people (Rehunen 1989).

The muscle glycogen store is greater in trained athletes than in sedentary men (Hultman 1967). Longitudinal and cross-sectional studies confirm that this training effect appeared in subjects undergoing strength-, sprint-, or endurance-training programs (Karlsson et al. 1972; Gollnick et al. 1973b; Piehl et al. 1974; MacDougall et al. 1977). In rats, sprint training did not alter the glycogen store of either slow-twitch or fast-twitch fibers. Aerobic training with continuous running or swimming was highly effective. In slow-twitch fibers, the effect of aerobic endurance training was more pronounced than the response to anaerobic interval or strength training. Aerobic endurance training, anaerobic interval, and strength training produced approximately the same changes in fast-twitch fibers (Viru M 1994).

Substrates for oxidation are obtained as a result of glycogenolysis, degradation of blood-borne glucose, or release of free fatty acids. Free fatty acids are released in adipose tissue or muscles (as a result of degradation of the local triglycerol store). In humans, the muscle triglycerol content increases after endurance training (Morgan et al. 1969; Bylund-Fellenius et al. 1977).

Experiments with rats indicate that endurance training increases the glycogen store in the myocardium (Poland and Blount 1968; Scheuer et al. 1970), but no changes were found in the contents of phosphocreatine and ATP (Scheuer et al. 1970). Myocardial triglycerols did not alter with training (Watt et al. 1972).

Endurance training increases the glycogen content in the livers of rats (Yakovlev 1977; Ööpik and Viru 1992). Increases up to 50% have been found (Yakovlev 1977). Thus, the training effect on body energy stores consists of increased glycogen content in skeletal and heart muscles and in the liver. The training effect on phosphocreatine content is questionable. Training does not significantly change the content of the main energy donor—ATP.

Training Effects on Enzyme Activities

The rate of biochemical processes depends on the catalytic activity of the concerned enzymes. As was mentioned previously, a typical reflection of activity-induced adaptation is the increase or decrease in the number of enzyme molecules. The significance of this enzyme adaptation appears, first of all, in cellular autoregulation. Gollnick and Saltin (1982) demonstrated that with a high-enzyme concentration it is possible to attain high rates of substrate fluxes at low substrate levels. Thus, the high-enzyme concentration favors more extensive substrate use. Moreover, when enzyme concentrations are doubled, the velocity of the biochemical reaction will be doubled at any substrate concentration. An elevation in the enzyme concentration would increase the enzyme activity and the sensitivity for control, particularly at low substrate levels. Thus, with a higher enzyme concentration, it would be easier to regulate the metabolic pathway on the basis of substrate and cofactor regulation of the enzymes.

Another adaptation of enzyme activities is the alteration in sensitivity of the enzyme toward stimulating or inhibiting factors. This variant of adaptation is particularly significant for enzymes of anaerobic glycolysis. In this metabolic pathway, the flux of substrate is controlled by glycogen phosphorylase and phosphofructokinase. The concentration of these enzymes is high in most skeletal muscles, particularly in the fast-twitch fibers without training (Saltin and Gollnick 1983). It has been estimated that a 5% activation of phosphorylase could account for the maximal lactate production in skeletal muscle of glycogenolysis, depending on phosphorylase (Fischer et al. 1971). Further augmentation of the enzyme effect seems to be useless.

Therefore, increasing the total amount of enzyme available would not increase the precision for regulating substrate flux through the metabolic pathway (Saltin and Gollnick 1983).

The decrease of phosphofructokinase activity in rats occurred after a 10-week interval anaerobic or continuous aerobic training both in slow- and fast-twitch fibers. Sprint training caused this change in slow-twitch but not in fast-twitch fibers (figure 2.12). The muscle samples were obtained 48 h after the final training session. Obviously, the elapsed time was enough to suggest that a rapid enzyme turnover had eliminated the increased enzyme activity (Viru M 1994). Molecules of phosphofructokinase and other glycolytic enzymes are characterized by a short-term life span (Pette and Dölken 1975). Therefore, when 2 or 3 days have passed after the last training session, the training effect on these enzymes may not appear. However, an important result of the mentioned study was the fact that 4 min of intensive running (at 60 m/min) changed the muscle phosphofructokinase activity differently, depending on the training regimen used. The test exercise induced a decrease in the enzyme activity in the oxidative muscle of untrained control rats; the activity did not change in glycolytic muscle. Instead, a twofold or threefold increase was found in muscles of rats trained by either interval or continuous running after the test exercise. After the test exercise, the enzyme activity was not only greater than the resting levels in these trained rats but also greater than the resting levels in sedentary control rats. The effect of sprint training was different: in the test exercise muscle enzyme activities decreased in both types of fibers, but in slow-twitch fibers the change was greater than in the control group (Viru M 1994).

It may be speculated that training sensitizes the enzyme activity to stimulatory factors. This explanation implies that the enzyme sensitivity to inhibitory factors is also increased. Therefore, the activity was reduced in resting conditions. The final conclusion of the results points to an enhanced effectiveness of the regulation of phosphofructokinase activity in the organism trained by interval or continuous exercises.

For highly qualified athletes rest intervals of 2 or 3 days between sessions is an unusual situation, particularly in the period of intensive or extensive training. Therefore, the increased ac-

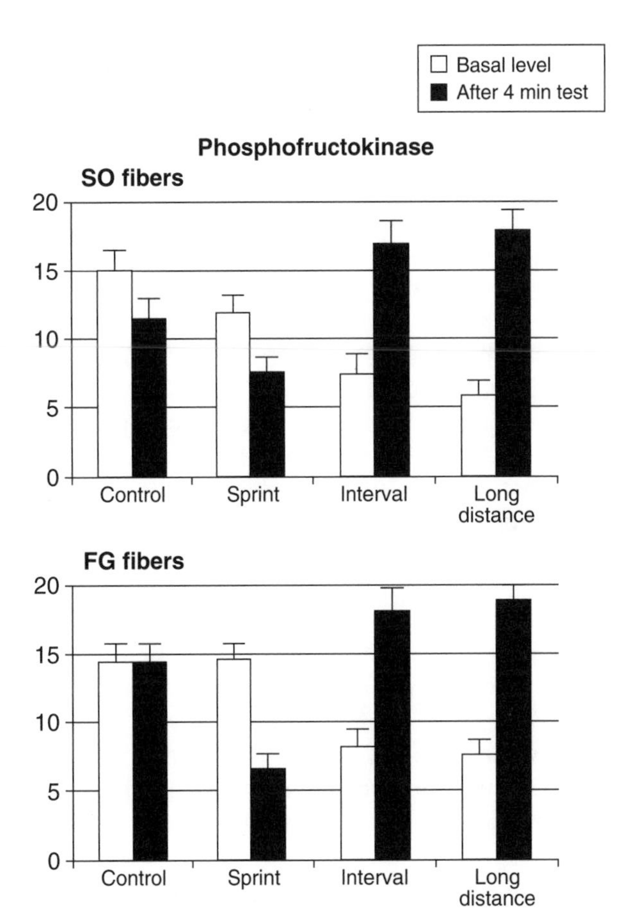

Figure 2.12 Changes in the activity of phosphofructokinase in slow-twitch (SO) and fast-twitch fibers (FG).
By results of Viru M 1994.

tivity of phosphofructokinase found in the biopsy studies of athletes (Gollnick et al. 1973a; Costill et al. 1976a) is not in disagreement with the results obtained in rats. However, it should be a result of enzyme up-regulation but not of an increased number of enzyme molecules. Newsholme (1986) believes, that one reason for the remarkable performance of elite athletes is the fact that their metabolic control mechanisms are so well developed that they provide maximum sensitivity when required in the control of energy-producing pathways in muscle.

There is one more possibility of the training effect on enzymes: training changes the spectrum of isozymes. For example, muscles of sprinters and jumpers contain a relatively high percentage of lactate dehydrogenase isozymes (LDH_{4-5}), whereas muscles from endurance athletes have a high percentage of LDH_{1-2} (Sjödin et al. 1976). Marathon training induced increased activity of LDH_{1-2} and decreased the activity of

$LDH_{3\text{-}5}$ (Apple and Rogers 1986). Training may selectively influence the content of creatine kinase MB isozyme in muscle fibers (Jansson and Sylven 1985; Apple and Tesch 1989). In this way, the specific adaptation to anaerobic exercises may concern the formation of isozymes less sensitive to lowered pH values. This question has been investigated only in one study in regard to hexokinase (Goldberg 1985). An interval-training program (training sessions caused a pronounced increase in the blood lactate concentration and a decrease in the blood pH from 6.98 to 6.90) resulted in an increase in hexokinase activity in the medium pH range of 8.0 to 6.5. This change was found in both slow-twitch and fast-twitch fibers and in brain tissue. The enzyme activity in fast-twitch fibers increased most at pH 6.5. After training with continuous aerobic exercises, the increase in the enzyme activity was also found at various medium pHs, but at pH 6.5 extensive increase was not observed. In association with these events, augmentation of muscle type II hexokinase isozyme occurred in muscles and brain tissue.

Improved Metabolic Control

Three main outcomes from training-induced improvement of metabolic control are:

1. The rapid and stable mobilization of the body's resources
2. More economical use of the body's resources
3. Increased lability of metabolic control

An expression of the first outcome is the previously discussed catecholamine effect on anaerobic glycogenolysis and thereby on anaerobic working capacity. Other expressions of the first point are the faster adjustment of VO_2 during exercise (Hickson et al. 1978), and the significance of maintaining an adequate level of glucocorticoids in blood for performance in long-lasting exercises.

The former studies, founded on investigation of excretion of hormone metabolites, showed that prolonged exercises, lasting 2 h or more, led to decreased excretion of corticosteroids. The period of decreased excretion was preceded by the period of increased excretion. Decreased excretion appeared first in untrained persons (for a review, see Viru 1985a). Animal experiments showed the same pattern in blood and adrenal glucocorticoid levels (Viru and Äkke 1969). The low blood level of glucocorticoids is associated with shifts of water and sodium into the intracellular compartment in skeletal and heart muscles (Kõrge et al. 1974a, 1974b), decreased activity of Na, K-ATPase (Kõrge et al. 1974a, 1974b), and increased rate of decrease of hepatic glycogen store (Viru M et al. 1994). However, neither administration of adrenocorticotropic hormone nor blockade of adrenocortical response to exercise with a previous dexamethasone treatment altered maximal aerobic power in either trained or untrained persons (Viru and Smirnova 1982).

The training-induced increase in the mechanical efficiency of muscular work is reflected in reduced oxygen uptake when endurance-trained athletes were compared with sedentary persons during exercise at the same rate of power output (Costill et al. 1973b; Conley and Krahenbuhl 1980). In addition, training modifies the rates at which fuels are used (for a review, see Coggan and Williams 1995). Compared with untrained persons, endurance-trained athletes oxidize less carbohydrate and more fat during exercise performed at the same power output (Henriksson 1977; Coggan et al. 1993; Martin et al. 1993; Karlsson et al. 1974). This is reflected in the marked muscle glycogen-sparing effect of endurance training (Karlsson et al. 1974).

The interaction between insulin and lipolytic hormone levels is highly significant for the portion of fat used during exercise. Insulin blocks the effect of lipolytic hormones on adipose tissue and, thereby, avoids the release of free fatty acids and glycerol. The lipolysis becomes effective after the blood insulin level is decreased (a 10- to 15-min lag period precedes from the onset of exercise to the decrease of insulin level in the blood). Epinephrine and other lipolytic hormones become effective, and skeletal muscles are supplied by free fatty acids. Hypoinsulinemia does not exclude glucose transport to muscle cells. However, hypoinsulinemia is a decisive factor in the mobilization of lipid resources; therefore, increased portions of blood glucose will be reserved for fuel for neurons (for a review, see Viru 1995).

The economical use of body resources also appears in trained organisms regarding the hormonal responses in exercise. The exercise intensity has to surpass a certain threshold to activate endocrine functions significantly. As a result of training, the intensity threshold is shifted to more strenuous exercises (see Galbo 1983; Viru 1985a, 1985b, 1992).

Athletes are capable of performing an exercise with rather high power output without significant changes in blood hormone levels. Consequently, they reserve the hormonal responses for metabolic control at very high exercise intensities. The most pronounced hormone responses are found mainly in highly qualified athletes during and after exercise that requires mobilization of body capacities close to maximum, as is usual in competitions (Adlercreutz et al. 1976; Weicker et al. 1981; Farrell et al. 1987; Bullen et al. 1984; Kjaer and Galbo 1988; Petraglia et al. 1988; Snegovskaya and Viru 1993a).

The increased catecholamine responses to supramaximal exercises (Bullen et al. 1984; Kjaer et al. 1986; Nevill et al. 1989) may be an essential pathway for increased anaerobic power output in trained athletes. Training also influences hormone receptors in tissues, including muscle tissue. This is essential for rapid mobilization of the body's reserves and for economical use of the resources of the endocrine glands.

Adrenaline administration in trained rats caused an exaggerated increase in the activity of adenylate cyclase in muscle, liver, and adipose tissues (Yakovlev 1975). This result emphasizes that the elevated sensitivity to catecholamines in trained rats is related to the processes that are initiated by interaction of adrenaline and β-adrenoreceptors (for a review, see Yakovlev and Viru 1985). However, training also increases the activity of 3', 5'-AMP-phosphodiesterase (Yakovlev 1975). Although adenylate cyclase catalyzes the formation of cAMP, which is responsible for the actualization of adrenaline metabolic effects, 3', 5'-AMP-phosphodiesterase catalyzes the degradation of cAMP. Consequently, training makes the metabolic control more labile: a fast increase of the cAMP production and an intensive degradation of the mediator compound are possible.

Summary

In exercise training, cellular adaptation is expressed by increases of muscle fiber structures specifically related to the nature of the training exercises used (either myofibrils or mitochondria). These changes are accompanied by alterations in enzyme activities and increases in energy stores in muscle fibers (mainly glycogen). Improvement also takes place in the supply of muscles with oxygen and extramuscular energy substrates. The adaptive protein synthesis, induced by metabolic and hormonal changes during and after training exercises and actualized during the recovery period, is the foundation for cellular morphofunctional improvements. Summation of cellular improvements warrants the increase of whole-body capacities.

Metabolic control becomes more effective in aspects of mobilization of metabolic and functional capacities and more economical in the use of metabolic resources. The accomplishment of metabolic control is related to increased enzyme activities, elevated sensitivity of muscle and other tissues for influencing controlling agents, and increasing possibilities of endocrine glands for augmented and stable responses during exercises requiring mobilization of maximum body functions. The main reflections of accomplished metabolic control are rapid and more stable mobilization of the body's resources, more economical use of body reserves, and increased lability of metabolic control.

The training-induced metabolic changes constitute the background for improved performance. Consequently, the methods used for the evaluation of training effectiveness also have to involve indexes providing information on cellular adaptation.

Tools for Biochemical Monitoring of Training

Part II of this volume considers the methodological problems of various means of assessment for biochemical monitoring of training.

Chapter 3 begins with an analysis of the use of muscle biopsy sampling for training monitoring. The information obtained from several metabolites and substrates in blood is discussed. In chapter 4, general methodological considerations of hormonal studies are discussed to help avoid methodological errors and misinterpretation of results. Chapter 5 adds specific information about the function of various hormones in metabolic control, particularly in exercise. A summary is provided regarding the effect of exercise on blood levels of the hormones discussed. Chapter 6 addresses the question of whether the hematological parameters and immunological indexes and characteristics of water-electrolyte balance are useful in training monitoring.

CHAPTER

3

Metabolites and Substrates

In training monitoring, evaluation of the body's metabolic state is usually done by assessment of several metabolites and substrates found in blood, urine, saliva, or sweat. The results obtained characterize what happens in working muscles. For direct information, biopsy sampling is necessary. In the investigation of muscle metabolism, the value of the information obtained from each method decreases in the following order when metabolites or substrates are determined:

Muscle biopsy → Arteriovenous difference → Venous blood → Capillary blood → Urine and saliva → Sweat

However, the feasibility of each method increases in this same order with muscle biopsy as the least feasible and sweat as the most feasible. Later, several options will be discussed regarding the use of biopsy sampling. In field conditions, arterial puncture and, therefore, arteriovenous difference assessment are not usable.

Thus, the researcher has to choose the "golden mean," the method with the most feasibility in the particular circumstances of the activity and that provides sufficient information for the evaluation of the function to be measured. The metabolites and substrates chosen for measurement have to be related to the task of monitoring. However, interpreting the information obtained depends on the knowledge of the metabolic pathway resulting in the formation of the metabolite, the further metabolic fate of the substance, and the production/use of the substrate.

Muscle Biopsy

The most valuable information on energy reserves and metabolic processes can be achieved by muscle biopsy, following determinations of metabolites/substrates in muscle tissue. This method has provided valuable results for establishing the principal characteristics of energy metabolism in human muscles and is advisable for training experiments in laboratory conditions.

However, the method is limited in its use. Hultman reported that he had used this method in investigating soccer players. The method is based on the use of a needle for percutaneous biopsy sampling of muscle tissue. The needle was introduced by Bergström (1962). The method was first used in clinical conditions for the determination of muscle electrolytes in humans (Bergström 1962). Today, it is a widely used method in human exercise physiology and sports medicine. The studies of Hultman (1967, 1971) contributed a great deal to the promotion of the expanded use of this method for biochemical studies on humans during exercise.

Early Studies

From 1966 forward, publications appeared describing the results obtained by the biopsy method on muscle glycogen (Bergström and Hultman 1966a; Ahlborg et al. 1967; Hultman 1967), electrolyte (Bergström and Hultman 1966a; Ahlborg et al. 1967), and energy-rich phosphate (Hultman 1967; Karlsson 1971) changes in humans during exercise. Results showed that an exercise-induced drop in muscle glycogen content stimulates the activity of glycogen synthesis (Bergström and Hultman 1966b). It became clear that (1) muscle glycogen content is a factor limiting performance capacity for prolonged severe exercises and (2) postexercise glycogen supercompensation can be enhanced by a combination of exhaustive exercise and altered carbohydrate diet (Bergström et al. 1967; Hultman 1967, 1971). The training effect on human muscle glycogen content was confirmed in several studies (see Saltin and Gollnick 1983), including the training effect in ischemic conditions (Viru and Sundberg 1994).

Methods for assessing the activity of oxidative (Björntorp et al. 1970; Moesch and Howald 1975) and other enzymes—for example, phosphorylase (Taylor et al. 1972), glycogen synthetase (Piehl et al. 1974), and myofibrillar adenosine triphosphatase (ATPase) (Ingjer 1979), and for determination of lactate and other metabolites (Hultman 1971; Karlsson 1971; Karlsson et al. 1971; Harris et al. 1974) or muscle lipids (Carlson et al. 1971)—were adapted for biopsy samples and used in studies concerned with the effects of exercise. An extensive study was performed on changes of muscle intracellular and extracellular pH (Sahlin 1978). By use of human skeletal muscle biopsy samples, an increase in mitochondrial ATP production has been demonstrated as a result of endurance training (Wibom and Hultman 1990).

Applying Morphological Studies

Muscle biopsy was used for determination of the individual distribution of muscle fibers of different types in athletes in various sport events (Gollnick et al. 1973a; Costill et al. 1976a; Thornstensson et al. 1977). It became possible to discriminate the acute exercise effects on metabolic changes between muscle fibers of various types (Costill et al. 1973a;Gollnick et al. 1973b; Edgerton et al. 1975). Results began to accumulate on the training-induced changes of the cross-sectional area and metabolic characteristics of muscle, depending on their type and the exercises used (Gollnick et al. 1973a; Karlsson et al. 1975; Thorstensson et al. 1977; Henriksson 1977, Henriksson and Reitman 1977; Costill et al. 1979). These studies were extended to teenaged athletes (Eriksson et al. 1973) and for the evaluation of gender differences (Costill et al. 1976b; Komi and Karlsson 1978; Nygaard 1982). The essential role of a genetic factor in the interindividual distribution of muscle fibers of various types has been established (Komi et al. 1977). Biopsy material also provided a background for the content of fiber-type transformation in training, at least transformation of one subgroup into another (Andersen and Henriksson 1977; Jansson et al. 1978).

To solve the problem of hypertrophy versus hyperplasia in trained skeletal muscles, a calculation of the total number of fibers was performed from the measurements of the total cross-sectional area of muscle (computed tomography) and the area of individual fibers (biopsy samples). The results obtained demonstrated the existence of considerable variations in the number of fibers in different subjects. However, no evidence of a systemic difference was found between sedentary and trained persons. The greater total cross-sectional area in athletes was attributable to the larger cross-sectional area of individual fibers (Nygaard 1980).

With the aid of the muscle biopsy technique, increased muscle capillary supply and mitochondrial contents have been demonstrated in humans as a result of endurance training (Ingjer 1979).

Combining Biopsy With Other Methods

The serial muscle biopsy samples taken during electrically induced repeated muscle contractions afforded an opportunity for precise studies on the metabolic background of muscle fatigue (Chasiotis 1983; Hultman and Sjöholm 1983; Hultman and Spriet 1988; Spriet et al. 1988). A prospective trend is to use the muscle biopsy sampling in combination with the estimation of arteriovenous differences at multiple sites and measurements of leg blood flow. For example, the latter approach during and after electrically induced muscle work allowed the detection of lactate and potassium fluxes from human skeletal muscle during and after intense knee extensor exercise (Juel et al. 1990), anaerobic energy production, and the O_2-debt relationship during exhaustive exercise (Bangsbo et al. 1990), as well as lactate/glycogen relationships during the postexercise recovery period (Bangsbo et al. 1991).

Information on metabolic processes also increases when the biopsy method is combined with isotope studies (see p. 32).

Separating Individual Muscle Fibers From Biopsy Samples

The trend of separating individual muscle fibers from biopsy samples presented the opportunity for a highly specific comparison of the peculiarities of fiber types I and II in metabolic studies. New analytical procedures were necessary for these studies. A new luminometric method was elaborated, which enabled the determination of ATP and phosphocreatine (PCr) in single muscle fibers (Wibom et al. 1991). It was found that the anaerobic ATP turnover rate was nearly three times greater in type II compared with type I fibers. The loss of PCr was especially fast in type II fibers. After stimulation, the resynthesis of ATP was not complete in type II fibers after 15 min of oxidative recovery (Söderlund 1991). During electrical stimulation, rapid glycogenolysis occurred in type II fibers, with hardly any detectable glycogenolysis in type I fibers. Epinephrine infusion caused a 10-fold increase in the rate of type I fibers but did not enhance the rate in type II fibers (Greenhaff et al. 1991).

Studies of Protein Metabolism

Combinations of muscle biopsy and isotope methods made it possible to study protein metabolism in muscular activity. Biopsy sampling, blood analyses, and administration of isotopes were used in combination to assess the amino acid metabolism, urea kinetics, and synthesis in humans (Wolfe et al. 1982, 1984). Intravenously infused amino acids labeled with stable isotopes (^{13}C-leucine, ^{15}C-glycine) enabled the study of the dynamics of protein synthesis and protein turnover during and after exercise (Millward et al. 1982; Nair et al. 1988). It was demonstrated that delayed phases of postexercise recovery are characterized not only by substitution of protein catabolism by anabolism but also by persisting high rates of protein turnover (Millward et al. 1982; Carraro et al. 1990). In resistance training, the fractional rate of muscle protein synthesis increased to a comparable rate in young and elderly subjects (Yarasheski et al. 1993).

Sarcoplasmic membrane Na^+-K^+ pumps are essential protein structures that are investigated with the aid of muscle biopsy. Their density on the membrane can be estimated by the number of ouabain-binding sites. Microsamples of a skeletal muscle will be incubated in buffer solution containing radioactive ouabain. The specific activity of [^3H]-ouabain in the incubation medium and the amount of [^3H]-ouabain taken up by the muscle tissue and retained in the specimen allow a researcher to calculate the pump concentration (Nøgaard et al. 1983). In the vastus lateralis muscle, the density of the Na^+-K^+ pump was greater in swimming-, running-, and strength-trained persons compared with untrained age-matched subjects (Klitgaard and Clausen 1989).

Recently, biopsy of the vastus lateralis muscle was used for determination of the training effect on heat-shock proteins (HSP) also called stress proteins. In humans, the predominant HSP is HSP70, with a molecular weight of 72. Animal experiments have shown that exercise induces HSP70 production (Salo et al. 1991; Skidmore et al. 1995). Liu et al. (1999) tested trained male rowers during a 4-week training stage for the World Championship. In accordance with augmented endurance capacity, the content of HSP70 increased in the thigh muscle several times in relation to the training workload. The positive significance of the HSP70 response might be in the capacity of HSP70 to protect vital function and prevent damage (Hutter et al. 1994) and in HSP influence on protein synthesis and folding (Beckmann et al. 1990).

Studies of Hormone Receptors

Human open biopsy muscle samples taken at surgery were used for the investigation of steroid hormone cytoplasmic receptors. The binding sites were specific for androgens and glucocorticoids. In most human muscles analyzed, both receptors were detectable (Snochowski et al. 1981).

In this research, the amount of muscle tissue sample was comparatively great, 2 to 4 g. It is impossible to obtain this amount of tissue with the usual needle biopsy technique. When the amount of tissue is approximately 100 times less, special methods for receptor studies have to be developed.

Biopsy Sampling of Other Tissues

In a limited number of persons, biopsy samples of hepatic tissue were obtained before and after heavy exercise. Glycogen content decreased from 15 to 8 g (in resting state) to levels as low as 2 g/kg body weight (Hultman and Nilsson 1973). However, liver biopsy samples can be taken only in clinical conditions. This procedure is ethically approved only for diagnostics and prognostics.

Subcutaneous adipose tissue biopsy samples present a wide area for study. This approach showed that endurance training increased the lipolytic effect of epinephrine (Despres et al. 1984; Crampes et al. 1986) and the lipogenetic effect of insulin (Savard et al. 1985). In former athletes, the elevated sensitivity to the lipolytic action of epinephrine seemed to disappear, and sensitivity to the lipogenetic action of insulin was reduced (Viru et al. 1992b).

Methodological Considerations

The methodological aspects of skeletal muscle biopsy have been addressed in several overviews (Edwards et al. 1983; Jansson 1994, Viru 1994). Therefore, a short summary will be sufficient here.

Bergström's (1962) needle is the most commonly used needle in human exercise physiology. Using suction with Bergström's needle makes it possible to increase the sample size (Evans et al. 1982). A modified needle has been designated for small muscles (Bylund et al. 1981). Another percutaneous technique is the conchotome technique ("semi-open" biopsies), in which an alligator-type forceps, or conchotome, is used instead of Bergström's needle or other similar needles (Henriksson 1979). The advantage of the conchotome technique compared with Bergström's needle is questionable. No convincing method-specific differences have been found in the results.

Before the needle is inserted, the area has to be washed, shaved, and anesthetized. Usually 1 to 2 ml of 1% lidocaine is administered into the skin, subcutaneous tissue, and muscle fascia. Caution must be exercised so that the anesthetic will not reach the muscle tissue. The needle, or conchotome, is inserted through a 5-mm incision made with a pointed scalpel blade through the skin and fascia. It is a fast procedure that leaves a nearly invisible scar. The subjects do not have to restrict their activity after the biopsy (Jansson 1994). The possibility of hurting the athlete's performance or causing effects persisting for a long time after the biopsy is slight.

The usual amount of muscle tissue obtained with Bergström's needle is 25 to 50 mg. When suction is added or the conchotome is used, the amount of tissue may be increased up to 75 to 150 mg.

In exercise physiology, most biopsy specimens are taken from the vastus lateralis muscle. This muscle is involved in cycling and running. Other muscles such as the lateral portion of the gastrocnemius and the soleus, tibial anterior, deltoideus, biceps brachii, and triceps brachii have also been investigated. Various muscles of the human body differ in fiber composition. Some muscles have a predominance of either slow-twitch or fast-twitch fibers (for a review, see Saltin and Gollnick 1983). However, the results from autopsy studies (persons who died suddenly without having a known muscle disease) (Johnson et al. 1973) or from studies in which several muscles in the same subjects were biopsied (Sjøgaard 1979) showed that the person with a higher percentage of slow-twitch fibers in the soleus muscle also had a higher percentage of these fibers in the other three (the biopsy study) or five (the autopsy study) muscles, including the vastus lateralis muscle, compared with other persons. Accordingly, male and female athletes have a higher percentage of fast-twitch fibers in the vastus lateralis, gastrocnemius, deltoid, biceps brachii, and triceps brachii muscles compared with the same muscles in sedentary persons (Nygaard 1981).

A variation in the relative occurrence of fiber types does exceed 5% to 15% when calculated for adjacent regions of the muscle (Saltin and Gollnick 1983). The variations within a certain muscle have been more precisely mapped out in detail on "whole muscle" sections obtained from individuals who died suddenly: regional differences exist in the relative number of different fiber types (Lexell et al. 1983). The methodological error was about 12% when the percentage of slow-twitch fibers was estimated from one single biopsy (Glenmark et al. 1992). When two biopsy samples were obtained, the error decreased to approximately 8% (Blomstrand and Ekblom 1982).

When multiple biopsy samples were taken from the vastus lateralis muscle, the coefficient of variation for the relative occurrence of a fiber type was 5% to 15%. It was 5% for the size of the corresponding fiber type (Thorstensson 1976; Halkjaer-Kristensen and Ingemann-Hansen 1981).

In the vastus lateralis muscle, a clear tendency for a higher percentage of slow-twitch fibers was found in deeper parts of the muscle (Lexell et al. 1983). This difference was not confirmed in other studies (Elder et al. 1982; Nygaard and Sanchez 1982). However, a standardized biopsy procedure is preferred, taking into consideration the depth of the biopsy site, left or right leg, the direction of the needle insertion, and the position on the muscle belly (Jansson 1994).

In several cases, it is better to analyze fresh material without freezing (i.e., membrane-bound enzymes are more sensitive to freezing than cytoplasmic enzymes). The mitochondrial cytochrome-c oxidase activity may dramatically change after freezing the tissue (Bylund-Fellenius et al. 1982). If isolated mitochondria are to be studied, fresh tissue again must be used. On the other hand, citrate synthase located in the mitochondrial matrix does not change its activity after freezing or after freeze-drying (Henriksson et al. 1986). However, if the enzyme is not altered by freezing, it should be used to avoid interassay differences. When freeze-dried samples are used, it is possible to remove blood, fat, and connective tissue from the muscle tissue. In this way analysis error is reduced.

Freezing the tissue immediately after exercise is recommended for analyses of ATP and PCr. Even resting levels of PCr were elevated when freezing was delayed for 1 to 6 min (Söderlund and Hultman 1986). When freezing was delayed for 10 s, phosphorylase a fraction decreased by 12% (Ren and Hultman 1988).

When histochemical analysis is to be performed on cross-cut sections from frozen tissue, isopentane should be used, with approximately a 2-min delay after being removed from the muscle belly (Larsson and Skogsberg 1988). Delaying freezing is important for the measurement of muscle fiber areas because removal of tissue and/or immediate freezing may induce muscle contraction, which leads to an increased cross-sectional area of fibers (Jansson 1994).

Usually, the biopsy sample is taken only from one muscle. When biopsy sampling is used for evaluation of metabolic changes during an exercise, it is necessary to take into consideration the function of the muscle in exercise performance: correct experimental designs have to be prearranged for the most active muscle for sampling. An additional problem arises in regard to possible differences in activity of various parts of the muscle (superficial or deep portion, distal or proximal portion) because of the location of muscle fibers belonging to the recruited motor units. When training effects are studied, it is necessary to choose the muscle for biopsy according to the exercise used and the specific nature of training effects. In these studies, the following must be determined: what adaptive changes in the most active muscles are determined by the exercises used and how the training design satisfies the need for improvement in a specific performance level regarding the sport event. The choice of exercise determines which muscles are the most active. However, it does not follow that the most active muscles are ultimately those whose adaptation is decisive in the improvement of performance. This is an essential problem when biopsy sampling is used for practical evaluation of training efficiency in athletes.

Conclusion

Comprehensive evidence has been obtained on the benefits of skeletal muscle biopsy in metabolic studies related to muscular activity. The ease of obtaining human muscle samples by use of the needle biopsy technique and the availability of valid histochemical and biochemical methods for estimating fiber composition, fiber cross-section area, enzyme activities, and energy stores made this approach a valuable experimental tool. Much of the information regarding adaptation to varying types of training has come from studies in which muscle samples were obtained by means of a needle biopsy.

However, biopsy sampling is not a method for field studies in the practice of biochemical monitoring of training. Biopsy studies have to be done in laboratories that adhere to clinical standards for biopsy sampling. The biopsy is necessary to evaluate the fiber composition of skeletal muscles. Therefore, biopsy seems to be essential in the selection of athletes for various events and for other specific tasks (e.g., testing of depletion or supercompensation of energy stores, measuring muscle enzyme activities, changes of regulatory proteins, muscle buffer capacity, or its antioxidant systems). This is all possible if the researcher is able to overcome psychological limitations (fear/repulsion of biopsy sampling).

According to Gollnick and coworkers (1980), the muscle biopsy is mostly a research tool. Needle biopsy shows relationships between some characteristics of skeletal muscle fibers and athletic performance, but its significance cannot be overestimated; moreover, the results of muscle biopsy can be used as predictors of athletic success only in combination with methods ensuring additional information about other factors.

Blood Metabolites

The effective use of metabolites for training monitoring presupposes specific knowledge. First, to understand the information provided by changes of a metabolite, it is necessary to know the position of the metabolite in metabolism. It means knowing the metabolic pathway or pathways that lead to the formation of the metabolite and of the further metabolic fate of this metabolite. In several cases, it is essential to imagine how this metabolite is used in any synthetic process, degraded further, and what the elimination rate from the body fluid studied is (e.g., from the blood plasma). At the same time, it is necessary to know the significance of applicable metabolic processes. It is also necessary to be familiar with the main outcome of physiological exercise studies concerning the dynamics of the metabolite during various exercises and training. Of course, the researcher has to be acquainted with methodological considerations to avoid errors in metabolite assessment.

The aim of this chapter is not to list the metabolites useful for training monitoring but to provide a brief overview of the knowledge necessary for metabolic studies and analysis of results obtained. Today, some metabolites are considered, even though their significance in training monitoring is modest or nonexistent. The reason they are considered is that they were measured in training studies and their significance in training monitoring may be established in the future.

Blood Lactate

Measurement of blood lactate is frequently used to determine the contribution of anaerobic glycogenolysis in energy production during exercise. Lactate is the end product of anaerobic degradation of glycogen, or glucose. Nevertheless, lactate level provides only one means to measure energy metabolism. It is only a semiquantitative estimation of the contribution of anaerobic glycogenolysis in energy formation.

Lactate is formed from pyruvate produced by glycogenolysis, or glucose degradation. A certain part of pyruvate is always oxidized. In addition, a part of pyruvate may be used for the synthesis of alanine (figure 3.1). The latter consists of adding an amino group to the pyruvate molecule (see p. 45).

When exercise intensity is low or moderate, the rate of pyruvate formation is in equilibrium with the rate of pyruvate oxidation. Therefore, the part of pyruvate that is transformed to lactate remains constant. At the same time, a certain amount of branched-chain amino acids is oxidized. Another situation occurs when the exercise intensity increases over the level of the anaerobic threshold. Then, the rate of pyruvate formation surpasses the maximal rate of pyruvate oxidation. Thus, the ratio between pyruvate oxidation and pyruvate transformation to lactate is shifted to the predominance of the latter. The part of pyruvate used for alanine synthesis remains dependent on the oxidation of branched-chain amino acids. Even in conditions of "pure" short-term anaerobic exercises, the accumulated lactate is not in precise quantitative relationship to the amount of energy produced by anaerobic glycogenolysis. This is so because the oxidation of a limited amount of lactate, at least by oxidative fibers in the same muscle, is not excluded (see Brooks 2000). The rate of anaerobic glycogenolysis is greater in fast glycolytic (FG) fibers, but oxidation capacity is greater in slow oxidative (SO) fibers. As a result, FG fibers produce more lactate, and SO fibers can oxidize more lactate. On the whole, oxidative ATP resynthesis signifi-

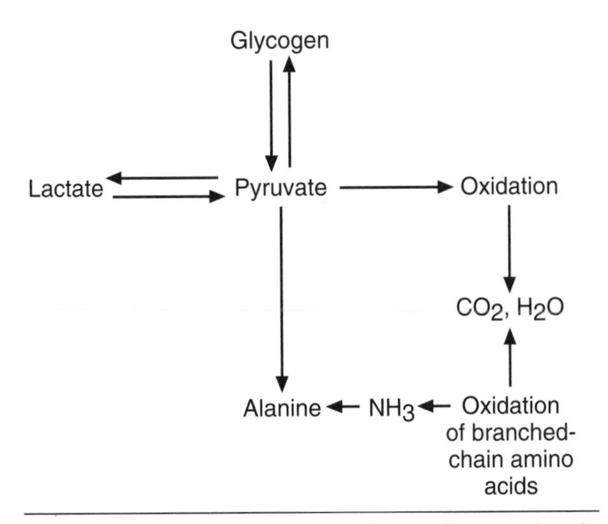

Figure 3.1 Fate of pyruvate originating from the breakdown of glycogen (or glucose).

cantly contributes, even in short-term supramaximal exercise bouts (Greenhaff and Timmons 1998). For instance, at the end of a 30-s bout of exercise performed at the highest possible rate, the oxidation rate is increased this much to provide most of the energy for ATP resynthesis (Trump et al. 1996).

On the other hand, at the beginning of intensive exercise, the energy for ATP resynthesis is obtained from PCr degradation. ATP resynthesis at the expense of combining two molecules of ADP may also contribute to the anaerobic energy processes. Therefore, the lactate accumulation in blood is characteristic of only the anaerobic glycolysis but not anaerobic energy production as a whole.

The blood level of lactate actually expresses the ratio between lactate inflow from active muscles and lactate outflow from blood to sites of its metabolic use in the oxidation process (mainly in SO fibers of resting muscles and myocardium), glycogen resynthesis (in resting muscles), or gluconeogenesis (in liver). Radioisotope studies have proved that the ratio between lactate appearance and disappearance in blood remains constant up to a certain exercise intensity. This intensity constitutes the anaerobic threshold. In greater exercise intensities, the lactate appearance exceeds the lactate disappearance (Brooks 1985). Consequently, blood lactate can be used for characterizing the contribution of anaerobic energy production in working muscles, but the caution regarding quantitative calculations has to be taken into consideration.

Furthermore, when an exercise is performed by glycogen-depleted muscles because of previous prolonged exercise or a carbohydrate-poor diet, lactate concentrations are reduced at the identical submaximal workload, whereas maximal performance and lactate production are diminished (Yoshida 1989). This leads to the overestimation of aerobic endurance capacity when estimated by the calculation of the anaerobic threshold on the basis of fixed lactate values (4 mmol/L) or to the underestimation of exercise intensity in the monitoring of training (Urhausen and Kindermann 1992a). In heavy-resistance exercise, a protein-carbohydrate nutritional supplement significantly reduced the blood lactate response (Kraemer et al. 1998).

Saliva Lactate

The prospect of using lactate in saliva has been pointed out by Ohkuwa and colleagues (1995). Salivary and blood lactate showed parallel dynamics after running 400 m and 3000 m (figure 3.2). The authors of the study believe that salivary lactate may serve as a relevant indicator in determining the contribution of anaerobic glycogenolysis (Ohkuwa et al. 1995). However, the concentrations of saliva consistently depend on the saliva secretion rate, which alters under the influence of autonomic balance (ratio between sympathetic and parasympathetic actions). Physical exercise increases the sympathetic nervous activity and the blood level of epinephrine, inhibiting the saliva secretion rate. Therefore, precise evaluation of the salivary lactate response requires an account of the saliva secretion rate.

Methodological Considerations

Lactate is usually determined in samples of arterial, arterialized capillary, and venous blood. In field studies, puncture of the arteries is out of the question. Determination of lactate in venous blood is not the best choice because there is a problem regarding lactate shifts between plasma and erythrocytes. For the precise determination of lactate in plasma, the distribution of lactate between plasma and blood cells and the kinetics of the exchange have to be taken into account (Foxdal et al. 1990; Smith et al. 1997).

Lormes et al. (1998) listed cautions for lactate determinations in venous blood:

- Blood has to be immediately cooled to approximately 4° C.
- Blood should not be treated with any stabilizing agent.
- The blood sample has to be centrifuged immediately at 4° C.
- The supernatant (plasma) of this centrifugation is used for subsequent analysis.

When a venous catheter is used to obtain blood samples, the catheter has to be kept free flowing for long periods by displacing the blood trapped in the catheter with saline or heparinized saline. In this situation, the catheter flush solution may contaminate the blood samples and artificially lower the lactate level (Bishop and Martino 1993). The treatment of blood with either potassium fluoride or heparin causes immediate volume shifts of erythrocytes, influencing plasma volume and thereby plasma lactate concentration (Lormes et al. 1998).

In training monitoring, the common sampling site for lactate is the earlobe or the fingertip. In the case of free blood flow, particularly after warming of the sampling site, the sample yielded arterialized capillary blood (Bishop and Martino 1993). The sample site can be warmed with clean hot water. "Milking" the tissue to move more blood into the area is questionable. It has been thought that this procedure dilutes the blood sample by introducing extravascular fluid into the sample. Godsen et al. (1991) found that this is not the case.

Care should be taken to avoid contamination of the sample by sweat because the sweat lactate level is higher than the blood lactate concentration. It is important to dry the area thoroughly before sampling.

Several analytical procedures prescribe prompt mixing of the sample with trichloride acid, which precludes any concern for clotting or continued glycolysis in erythrocytes (adding a new portion of lactate into the sample) but raises new concerns for dilution errors. This is also a problem with the use of automated analyzers. If the procedure requires lysing of red cells, dilution errors can occur (Bishop and Martino 1993). Lormes and coauthors (1998) affirmed that it has not been proved that plasma lactate offers advantages compared with lactate values from hemolyzed whole blood samples.

In exercise studies, the main goal is to obtain the peak recovery lactate. According to Bishop and Martino (1993), the times of peak lactate vary from 1 to 10 min. In cases of prolonged moderate exercise, the peak lactate may appear

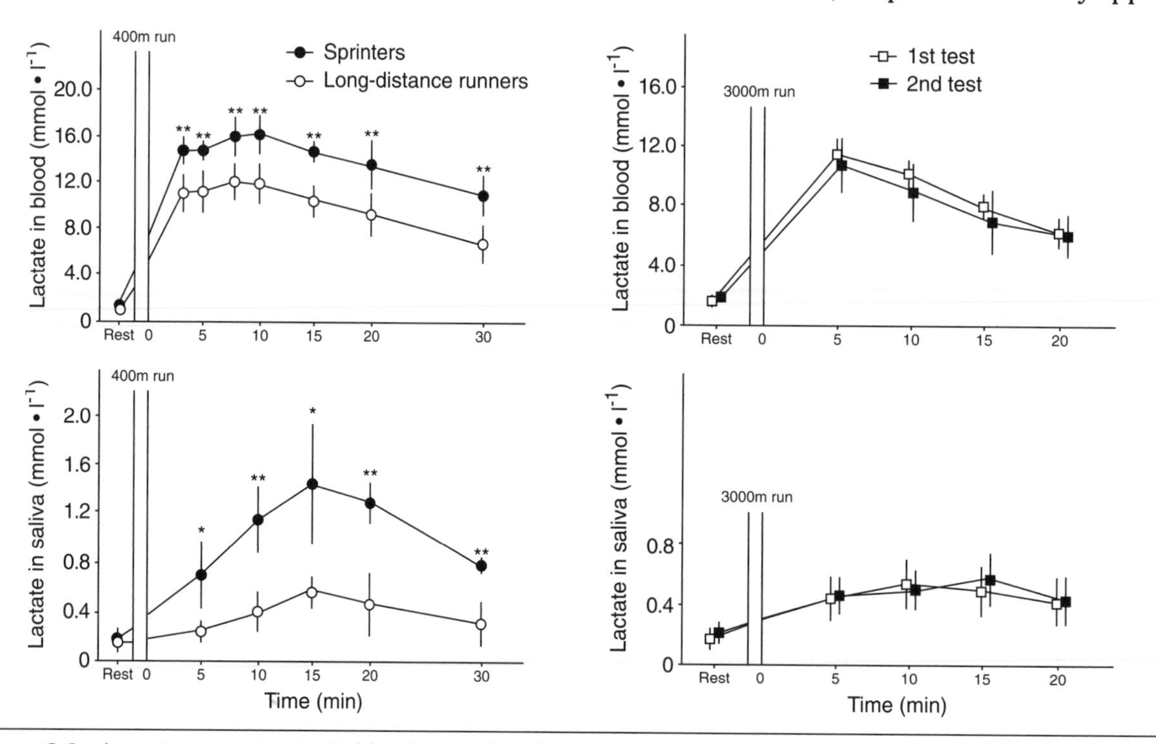

Figure 3.2 Lactate concentration in blood and saliva after running 400 m and 3000 m.

Reprinted from Itoh et al. 1995.

immediately after the exercise. In an incremental exercise test for the assessment of $\dot{V}O_2max$, the most common time of peak lactate was approximately 3 min.

Ammonia

At the beginning of intensive exercise, ATP resynthesis is founded on PCr breakdown, as was mentioned earlier. If most of the PCr is used up, the level of ADP formed during muscle contraction rises. This leads to increased formation of adenosine monophosphate (AMP) by way of the myokinase reaction:

$$\overset{\textit{Myokinase}}{2\ ADP \rightarrow AMP + ATP}$$

ADP is an effective activator of adenylate deaminase. This enzyme catalyzes the removal of AMP:

$$\overset{\textit{AMP deaminase}}{AMP + H_2O \rightarrow IMP + NH_3}$$

As a result, inosine monophosphate (IMP) and ammonia (NH_3) accumulate. The dephosphorylation of IMP leads to the formation of inosine that converts into hypoxanthine and uric acid. IMP and ammonia serve as activators of anaerobic glycogenolysis. IMP contributes to the activation of phosphorylase b, whereas ammonia is an activator of phosphofructokinase. Both enzymes are also activated by AMP.

In addition to adenine nucleotide degradation, the deamination of amino acids also contributes to ammonia accumulation in the skeletal muscles. The accumulation is minimized by the formation of glutamate and glutamine. Glutamine and alanine carry amino groups to the liver and kidney for disposal (see pp. 44-46).

Increased ammonia production is specifically linked to FG fibers. Therefore, a measurement of ammonia, taken during training, can provide clues to the fiber types predominantly recruited (Meyer and Terjung 1979; Dudley et al. 1983). In rats, ammonia and IMP increased in muscle fibers during running, depending on running velocity. However, accumulation of AMP deamination products was found only in fast-twitch but not in slow-twitch fibers (Meyer et al. 1980). In athletes, ammonia accumulation in blood plasma depends on the intensity of sprint exercises (Itoh and Ohkuwa 1991). After a 75-m sprint, blood ammonia increased more than after a 1000-m run. The response was greater in sprinters than in middle-distance runners. On the basis of these results, the ratio of ammonia concentration after 75-m and 1000-m distances was proposed for talent selection in runners. A high response is typical for sprinters, whereas a low index is particular to middle-distance runners (Hageloch et al. 1990).

According to the results of Katz et al. (1986), submaximal cycle exercise does not increase NH_3 levels either in blood plasma or in muscle tissue. During exercise at 97% $\dot{V}O_2max$ to exhaustion, NH_3 increased significantly in muscle in accordance with the decrease in the total adenine nucleotide pool (ATP + ADP + AMP). These changes are accompanied by an increase of lactate concentration in muscles up to 104 ± 5 mmol/kg · dm. (dry muscle), as are NH_3 increases in blood plasma.

Ammonia is also excreted in expired air. In incremental exercise, the respiratory ammonia excretion increases exponentially with the workload. However, the exponential increase of ammonia in expired air was less pronounced than the increase in blood ammonia level (Ament et al. 1999).

Hypoxanthine

IMP may be reaminated to AMP. If the IMP accumulation is excessive, IMP may be converted to inosine. The latter catabolizes to hypoxanthine by means of the purine nucleoside phosphorylase. During prolonged cycling (at 74% $\dot{V}O_2max$) until fatigue (average duration of exercise 79 ± 8 min), plasma concentrations of hypoxanthine increased eightfold parallel to the decrease in muscle total adenine nucleotide (ATP + ADP + AMP) content. After 60 min of exercise, the plasma hypoxanthine level correlated with the concentrations of plasma ammonia and blood lactate. Endurance time was inversely correlated with plasma hypoxanthine but not with plasma ammonia and blood lactate (Sahlin et al. 1999). The potential use of plasma hypoxanthine to assess training status deserves attention but needs further study.

Uric Acid

In relation to the degradation of adenonucleotides, the blood level of uric acid increases in prolonged exercises (Chailley-Bert et al. 1961; Cerny 1975; Rougier and Babin 1975). Its urinary excretion also increases (Nichols et al. 1951; Rougier and Babin 1975). Uric acid has been found in sweat collected during exercise (Cerny 1975). A special necessity for uric acid determination in training monitoring has not been established.

Urea

The main end product of protein metabolism (degradation of amino acids) is urea. The main site of urea formation is the liver. The possibility for urea formation has also been found in muscles (Pardridge et al. 1982) and in kidney. Urea synthesis is related to the deamination of amino acids. Whereas in muscles, deamination of branched-chain amino acids supplies the synthesis of alanine by amino groups, in liver, deamination of alanine and other amino acids occurring with gluconeogenesis is concordant with the synthesis of urea. Urea is produced in the process of formation of ornithine from arginine after the introduction of NH_3 groups into the urea cycle. The NH_3 groups come into the urea cycle from ammonia after the formation of carbonyl phosphate or aspartate. A part of the urea formed is related to the ammonia release in the degradation of AMP.

Prolonged exercises have been shown to cause increased urea concentration in the blood, liver, skeletal muscles, urine, and sweat (for a review, see Lorenz and Gerber 1979; Lemon and Nagle 1981; Poortmans 1988; Viru 1987). This is considered to reflect an augmented urea production. Conversely, the determination of urea production rate in humans after the administration of stable isotopes failed to demonstrate any increase during prolonged exercise (Wolfe et al. 1982, 1984; Carraro et al. 1993). However, the exercises used did not increase urea concentration in blood as well (Wolfe et al. 1984; Carraro et al. 1993). Therefore, these results did not correspond to the metabolic situation occurring frequently in athletes performing long-lasting exercise and reflecting the increased urea concentrations in blood. In experiments on Wistar rats, the urea production rate was assessed by the ^{14}C-urea content in liver tissue after the administration of $NaH^{14}CO_3$. In intact rats, swimming caused increases in ^{14}C-urea content in liver (by 35% after 3h of swimming and by 103% after swimming for 10 h). In adrenalectomized rats, swimming for 3 h resulted in a decrease in liver ^{14}C-urea (by 24%). Thus, the results confirmed the exercise-induced increase in urea production and indicated the essential role of adrenal hormones for this response (Litvinova and Viru 1995a). The possibility for increased urea production in the liver and skeletal muscles has been confirmed by the exercise-induced activation of arginase, an enzyme contributing to the synthesis of urea (Yakovlev 1977, 1979). In rats, adrenalectomy suppressed liver arginase activity at rest and after swimming (Viru A et al. 1994).

Urea accumulation is most frequently used as a measure of protein catabolism. On the basis of measured urinary urea excretion, it was calculated that protein breakdown was 2.5 to 11.0 g/h during 70- and 90-km cross-country ski races (Refsum and Strömme 1974) and 3.8 g/h during a 100-km run (Decombaz et al. 1979). Taking into account the urea nitrogen excretion in sweat (Lemon and Mullin 1980), the calculation indicated protein breakdown of 13.7 g/h in carbohydrate-loaded subjects after a 26-km run, lasting 128 ± 6 min (Lemon and Nagle 1981). The study of urea excretion from urine and sweat led to the suggestion that the protein use threshold may exist between 42% and 55% $\dot{V}O_2max$ (Lemon and Nagle 1981).

The exercise-induced protein catabolism and, thereby, urea accumulation and excretion depend on carbohydrate availability (see Lemon and Nagle 1981). In agreement, the rate of protein catabolism during exercise depends on the initial muscle glycogen level (Lemon and Mullin 1980).

A widespread tendency exists with the use of blood urea for the evaluation of the load of a training session and of the recovery processes (Lorenz and Gerber 1979; Voznesenskij et al. 1979; Urhausen and Kindermann 1992a). It is believed that a pronounced increase in the urea concentration indicates strong influence of a training session, whereas normalization of the urea level in blood is an index of time to perform subsequent strenuous training sessions.

The activity of enzymes concerned with urea synthesis depends on diet (Schimke 1962). In training, the intake of creatine increases blood urea concentration in resting and exercised rats (Ööpik et al. 1993). Hence, possible nutritional effects have to be considered when urea is used for monitoring training. Low ambient temperature increases urea nitrogen excretion in exercise (Dolny and Lemon 1988). The latter result has been explained by the intensified protein degradation. Thus, at low temperature, urea levels form the combined metabolic adjustment to cold and exercise. Exercise-induced urea responses also change during the acclimatization to a medium altitude (Matsin et al. 1997).

A limitation for the use of urea levels is the suppression of urea production in exercises inducing high lactate levels (figure 3.3) (Litvinova and Viru 1997). In animal experiments, it has been demonstrated that at pH 7.1 urea synthesis is reduced by 40% in the liver (Saheki and Katunuma 1975). Accordingly, the change in blood pH caused by lactate accumulation might suppress urea synthesis during intense exercise. This suggestion is supported by the fact that urea concentration failed to increase when lactate concentration rose to 10 to 17 mmol/L (Litvinova and Viru 1997). Gorski et al. (1985) failed to observe an increased blood urea concentration or altered renal urea excretion in short-term exercises that caused blood lactate to rise to 15 to 17 mmol/L. In agreement, Poortmans (1988) reported that when short-term intensive exercises are performed, blood urea remains relatively stable. In athletes, the usual dependence of the blood urea increase on the volume of training workload disappeared when highly intensive exercises were included in a training session (Voznesenskij et al. 1979).

Gorski et al. (1985) found a high rate of urea excretion by sweat in short-term intensive exercise (blood lactate arouse to 15 to 17 mmol/L). This result shows that besides suppressed urea synthesis, elevated elimination by sweat may also concern the blood urea response.

It has been believed that altered urea elimination during and after an exercise may influence the urea level in blood (Zerbes et al. 1983). Thus, renal retention of urea excretion may cause or facilitate the exercise-induced increase in blood urea concentration. However, in rats, the postexercise period is characterized by an increased renal clearance of urea after swimming for various durations. Only after 10 h of swimming did the lag period of up to 12 h precede the increased urea excretion and the elevated renal clearance rate. The postexercise increase in urea renal elimination is glucocorticoid dependent; it is absent in adrenalectomized rats (Litvinova et al. 1989).

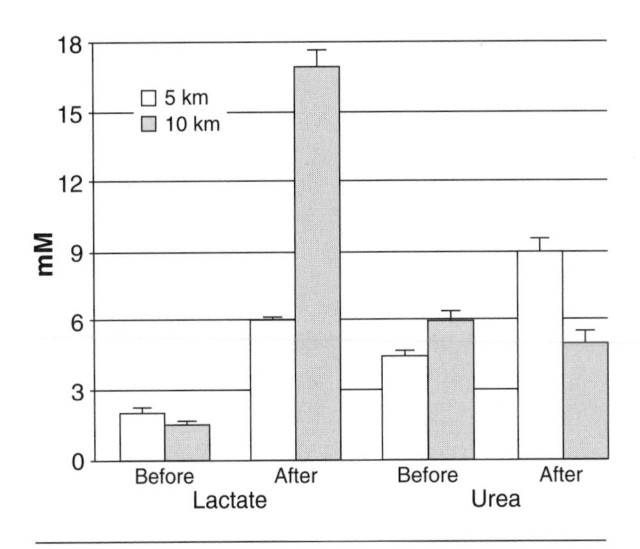

Figure 3.3 Lactate and urea levels before and after skiing for 5 km and 10 km in qualified male skiers.

By results of Litvanova and Viru 1997.

Creatine and Creatinine

Creatine is a specific constituent of muscle tissue; 95% to 98% of its total pool can be found in skeletal muscles (Waterlow et al. 1972). According to biopsy material, total creatine content in skeletal muscles is 115 to 140 mmol/kg · dm. The energy-rich PCr constitutes 60% to 65% of the total creatine content in human skeletal muscles.

Creatine is obtained in the digestive tract from food constituents or has been synthesized in liver from the amino acids arginine, glycerine, and methionine. Creatine production by the kidney has also been found. From the sites of creatine production or absorption, the compound is transported by the blood flow to muscle tissue. In muscles, the main function of creatine is related to the PCr metabolism. The free creatine is simultaneously the substrate for PCr synthesis and the product of PCr breakdown.

The synthesis of PCr consists of the formation of a high-energy bond that links creatine and phosphate in an energy-consuming process. The degradation of PCr releases energy, which is used for ADP rephosphorylation (for the ATP resynthesis). During muscle contraction, ATP hydrolyzes under the catalytic activity of myosin to ADP. Rapid rephosphorylation of ADP takes place, with PCR serving as the donor of energy-rich phosphate. As a result, free creatine that will be rephosphorylated under the catalytic action of mitochondrial creatine kinase by the use of ATP, which has been formed in mitochondria at the expense of oxidation energy, releases. Thus, PCr is extremely important for the maintenance of ATP provision during high-intensity exercise. At the same time, the maintenance of optimal creatine content is necessary to maintain PCr stores (Spriet 1995).

Of the total body creatine, 1.5% to 2% undergoes dehydration, yielding creatinine that is subsequently excreted by urine (Crim et al. 1975). In the kidneys, creatinine is filtered from blood plasma completely without further reabsorption. Therefore, creatinine excretion is used for the evaluation of the renal filtration process.

Although creatine is not solely a by-product of muscle metabolism, creatinine is excreted functionally to total muscle mass. However, creatinine excretion is influenced by diet, exercise, emotional strain, menstrual cycle, and certain disease states in addition to normal day-to-day variation (Heymsfield et al. 1983). Because of these influences, it is questionable whether creatinine excretion may be used for the assessment of training effects on muscle mass. The results are noticeable if urine is collected during a 24-h period, diet is standardized 1 or 2 days before the exercise, and emotional strain is avoided. It is difficult to practically fulfill these demands. Moreover, the training effects on muscle creatine/PCr metabolism and on myofibrillar mass may dissociate. Instead of the assessment of creatine excretion, the easiest way to calculate the approximate muscle mass is the assessment of the body's specific gravity with the aid of hydrostatic weighing. The most precise noninvasive method for the evaluation of muscle development is founded on the ultrasound principle (see Pollock et al. 1995).

It has been suggested that the creatine accumulation in blood might be used for semiquantitative assessment of the magnitude of use of the PCr mechanism in exercise, similar to the evaluation of anaerobic glycogenolysis by lactate accumulation (Tchareyeva 1986a). The suggestion is founded on the assumption that during short-term intensive exercise, creatine released during PCr breakdown may penetrate into the blood in a quantitative relationship to the drop in PCr content. However, this quantitative relationship is not established. Normally, free creatine is bound by the PCr kinase on the outer membrane of mitochondria. To accept the aforementioned suggestion, the binding capacity of the enzyme has to be established, and it must be significantly lower than the use of PCr during short-term intensive exercises.

In the 1930s, results began to accumulate that exercise results in changes in the blood level of creatine and creatinine (Kacl 1932) and their urinary excretion (Margaria and Foa 1929/1930). However, it has been found that well-trained persons may fail to show this response (Castenfors et al. 1967).

A 2-h exercise resulted in suppressed urinary excretion of creatinine (Cerny 1975). This change probably expressed the decreased renal blood supply during exercise (see Rowell 1986). The combination of changes at the kidney level and of transforming creatine released from muscle fibers into the creatinine might be the reason for the creatinine increase in blood serum during the Boston Marathon in postcoronary patients (Shephard and Kavanagh 1975). Neumann et al. (1980) found increased creatinine but not creatine levels in blood serum after a marathon race. After a 90-km skiing hike, creatine excretion remained increased (Refsum and Strömme 1974).

Inorganic Phosphate

Similar to the creatine accumulation in blood serum as an indirect characteristic of PCr/ATP metabolism, attention also turned to the increase of inorganic phosphate in the blood serum during exercise (Gerber and Roth 1969; Tchareyeva 1986a). It was thought that inorganic phosphate releases into the bloodstream when the rate of breakdown of ATP and PCr is greater than their resynthesis. However, the dependence of the inorganic phosphate level in blood plasma during exercise on the breakdown of ATP and PCr has not been established. Therefore, the use of inorganic phosphate for assessment of the degradation/resynthesis ratio of energy-rich phosphates is not justified.

Free Amino Acids

The main site of free amino acid deposition is skeletal muscle, accounting for 50% to 80% of the total amount of free amino acid in the whole body (Smith and Rennie 1990). Blood plasma contains only 0.2% to 6% of the total for individual amino acids. The exercise effect on plasma-free amino acids is variable with respect to the results of different studies and with respect to changes in the concentration of various individual amino acids. Studies of arteriovenous differences in amino acids demonstrated the release of amino acids from working skeletal muscles (Felig and Wahren 1971; Keul et al. 1972) and myocardium (Carlsten et al. 1961; Keul et al. 1964). As a result, the total amount of amino acids may increase in blood plasma (Carlsten et al. 1962; Felig and Wahren 1971; Poortmans et al. 1974). However, during prolonged exercise, the decrease of the total pool of amino acids in plasma is typical (Haralambie and Berg 1976; Decombaz et al. 1979; Holz et al. 1979). Cerny (1975) observed an increase in the concentration of total amino acids during the first 80 min of exercise at 60% to 75% of $\dot{V}O_2max$. At the end of 120 min of exercise, the concentration was less than the initial values. The decrease of total free amino acid content is associated with increased renal excretion (Decombaz et al. 1979) and with enhanced amino acid catabolism, indicated by a negative correlation between the increase in urea level and the decrease in amino acid content (Haralambie and Berg 1976).

Rat experiments showed that prolonged swimming sets induced a significant drop of the total content of free amino acids in skeletal muscles and liver. In the myocardium, a decrease of free amino acid content was found after swimming for 1.5 h with a load of 6% body weight but not after swimming for 12 h without an additional load (Eller and Viru 1983).

The changes in the total pool of amino acids integrate into a whole diversity of shifts in the concentrations of individual amino acids. Therefore, the total pool of free amino acids in blood plasma only has a limited significance, if at all, for the evaluation of the metabolic state. With respect to training monitoring, instead of the total free amino acid pool, attention should be paid to alanine, glutamine, leucine, and the other branched-chain amino acids, tyrosine and tryptophan, keeping an eye on the physiological meaning of their changes in muscular activity. These amino acids provide information on ammonia release and transport from muscle fibers and on catabolic changes in muscles and conditions of synthesis of several hormones and neurotransmitters.

Tyrosine

In biochemical experiments, the liberation of tyrosine is used as an index of protein catabolism in muscles. Muscle cannot metabolize this amino acid. During exercise, increased protein catabolism is associated with the augmented free tyrosine content in rat muscles (Dohm et al. 1981; Eller and Viru 1983) and with tyrosine release into the bloodstream in humans (Ahlborg et al. 1974). As a result, exercise induces the increased level of free tyrosine in serum (Haralambie and Berg 1976), urine, and sweat (Haralambie 1964b). After the Boston Marathon, the concentration of free tyrosine was increased in runners (Conlay et al. 1989). A rat experiment showed that free tyrosine peaked in the quadriceps femoris muscle 2 h in blood 6 h after 10 h of swimming. In the muscle, the high level of tyrosine persisted for 24 h after swimming (figure 3.4) (Viru et al. 1984).

3-Methylhistidine

3-Methylhistidine excretion is a specific index of the catabolism of muscle contractile proteins. In the last stages of synthesis of actin and myosin molecules, histidine is methylated. When myosin and actin are degraded, 3-methylhistidine is released. This amino acid is not further re-used and is excreted by urine. Therefore, 3-methylhistidine excretion is a quantitative measure of the degradation of contractile proteins (Young and Munro 1978). A substantial limitation is caused because of the effect of protein intake on 3-methylhistidine excretion; 3-methylhistidine releases in the digestion of meat products. This quantity of 3-methylhistidine is absorbed into blood and excreted by urine. Therefore, 3-methylhistidine can be used for biochemical monitoring either in the case of a meat-free diet or when the amount of consumed meat products is recorded for the calculation of exogenous inflow of 3-methylhistidine. The actual excretion of 3-methylhistidine must be corrected by subtraction of exogenous 3-methylhistidine from the total value. A special experiment showed that a training session for improved strength resulted in changes in corrected 3-methylhistidine

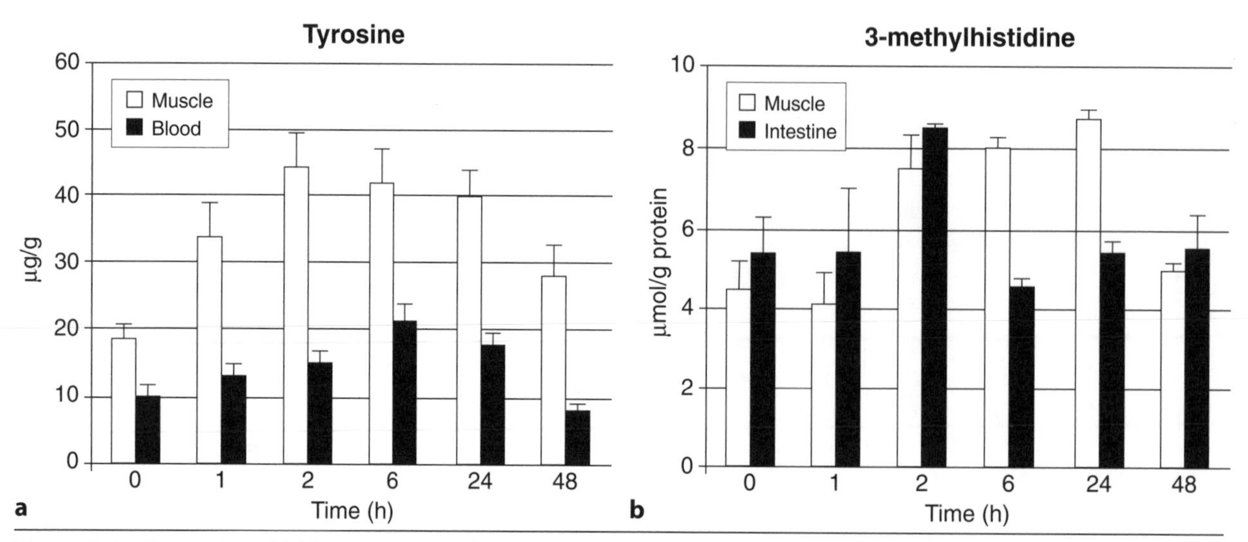

Figure 3.4 Dynamics of *(a)* free tyrosine in blood and muscle, and *(b)* 3-methylhistidine in skeletal and smooth muscles in rats after swimming for 12 h.
By results of Viru et al. 1984.

excretion very close to that obtained after the same training session performed by persons living 5 days on a meat-free diet (Viru and Seli 1992).

Muscular exercises may influence 3-methylhistidine excretion in persons living on a meat-free diet and when excretion was corrected for exogenous inflow. Despite the great variability in the changes that occur in 3-methylhistidine excretion during exercise, increased excretion is a typical postexercise phenomenon (for a review, see Dohm 1986; Viru 1987). The postexercise increase in 3-methylhistidine liberation also is seen in the blood 3-methylhistidine response (Dohm et al. 1985) and the accumulation of this metabolite in skeletal muscles (Varrik et al. 1992).

Another limitation for the use of 3-methylhistidine excretion is the release of this amino acid besides skeletal muscles from other tissues also containing myosin and actin (smooth muscles of digestive tract, skin, and myocardium). The calculations showed that, at least in rats, the skeletomuscular pool of 3-methylhistidine contributes only 50% of the 3-methylhistidine excretion. The intestinal pool contributes 20%, and other tissues contribute 30% (Millward and et al. 1980).

To assess the significance of the intestinal pool in elevated 3-methylhistidine excretion during exercise, the 3-methylhistidine content in intestine, skeletal muscle, and urine was simultaneously measured in rats after 10 h of swimming. The skeletomuscular 3-methylhistidine content was increased from 2 to 24 h after exercise. This was accompanied by a significant increase in 3-methylhistidine excretion during the second day of the recovery period. The 3-methylhistidine level in the intestine was increased only during the first hours of recovery (figure 3.4). Obviously, this did not contribute to the delayed increase in the excretion of 3-methylhistidine (Viru et al. 1984).

Measurable amounts of 3-methylhistidine have not been found in sweat during exercise (Dohm et al. 1982).

In several studies, 3-methylhistidine excretion has been expressed as a ratio to creatinine excretion. The idea is to avoid the exercise-induced changes in the production of urine. However, exercises may also influence the creatinine excretion (see p. 40). Therefore, in muscular activity, a methodological error may interfere with the results when the ratio for 3-methylhistidine/creatine is used.

A rat experiment indicated that the accumulation of free tyrosine and 3-methylhistidine in skeletal muscle is different in various types of muscle fibers. During 10 h of swimming, the contents of both metabolites increased in the white portion of the quadriceps muscles (containing FG fibers). In the red portion of this muscle (containing fast oxidative glycolytic [FOG] fibers), tyrosine did not change, and the 3-methylhistidine content increased less than in the white portion (figure 3.5). Judging by the drop in glycogen, FOG fibers were more active than FG fibers. Thus, the activity sup-

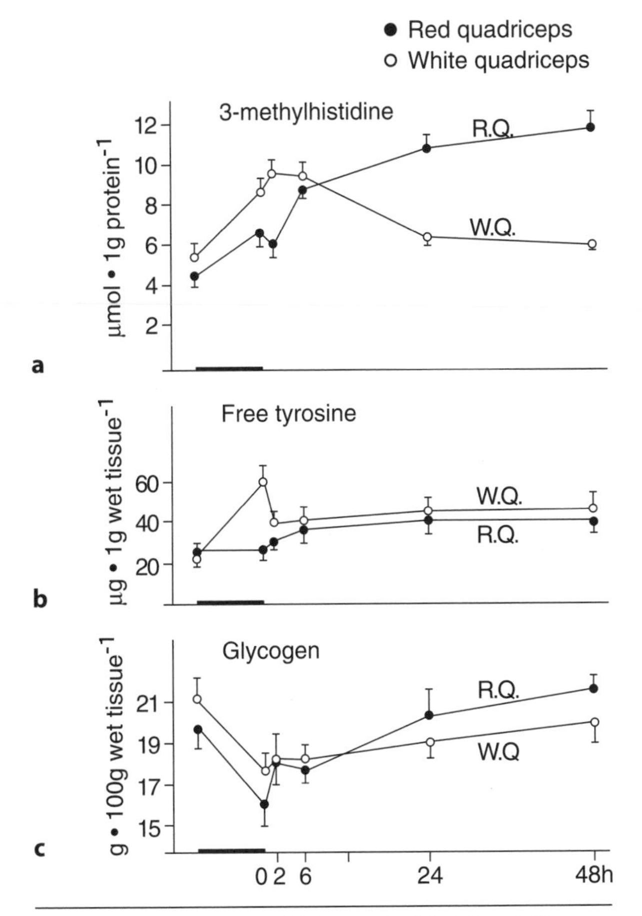

Figure 3.5 Dynamics of *(a)* 3-methylhistidine, *(b)* free tyrosine, and *(c)* glycogen in red (R.Q.) and white (W.Q.) quadriceps portions of the muscle in rats after 12 h of swimming.

Reprinted from A. Viru 1987.

pressed protein degradation in active muscle fibers (Varrik et al. 1992).

Branched-Chain Amino Acid—Leucine

This group of amino acids is made up of leucine, valine, and isoleucine (essential amino acids that cannot be synthesized in the body). It is peculiar to the branched-chain amino acids that may be oxidized in skeletal muscles. This process begins with an enzymatic transfer of an amino group. As a result of the transamination of branched-chain amino acids, glutamate and α-ketoacid are formed. Glutamate can donate nitrogen to pyruvate or to oxaloacetate and form alanine or glutamine, respectively. The next step in the catabolism of branched-chain amino acids is catalyzed by the enzyme dehydrogenase, which also acts as a decarboxylase. The dehydrogenase step is rate limiting in the catabolism of branched-chain amino acids during exercise and recovery (Kasperek 1989). This step is slightly different for each of three branched-chain amino acids. The 2-ketoisocapronate is the ketoacid formed as a result of leucine transamination. The carbon atoms of leucine, remaining at the end of enzymatic transformation of 2-ketoisocapronate, are converted either to acetyl-coenzyme-A (CoA) or to acetoacetate. They are both introduced into three carbon acids cycle and oxidized. Isoleucine forms acetyl-CoA and succinyl-CoA. The first is oxidized; the second can be converted into glucose. Valine produces only succinyl-CoA and, thereby, gives material only for gluconeogenesis (see Brooks et al. 1996).

Although six amino acids (alanine, aspartate, glutamine, isoleucine, leucine, and valine) are used in skeletal muscles, branched-chain amino acids are the most important, providing oxidation substrates during exercise. The total oxidation of 1 mole of leucine, isoleucine, and valine yields 43, 42, and 32 moles of ATP, respectively (see Graham et al. 1995). This range involves energy released in the oxidation of branched-chain amino acids and amino acids used for gluconeogenesis. Amino acids may contribute from 3% to 18% of the total energy required during prolonged exercise (Poortmans 1984; Dohm 1986; Brooks 1987).

Experiments monitoring the production of $^{13}CO_2$ or $^{14}CO_2$ from an administered labeled branched-chain amino acid showed that the most often oxidized amino acid is leucine. Leucine oxidation increases during exercise in rodents (Lemon et al. 1982; Hood and Terjung 1987a) and humans (Hagg et al. 1982; Wolfe et al. 1982; Knopik et al. 1991). Working skeletal muscle as the site of increased leucine oxidation was convincingly demonstrated on perfused rat hindlimbs during electrical stimulation. The calculations, founded on the rise in muscle leucine oxidation, showed that the leucine oxidation in muscles reasonably determines the rate of whole-body leucine oxidation (Hood and Terjung 1987b).

The extent of the oxidation of branched-chain amino acids depends on the intensity and duration of exercise (Millward et al. 1982; Henderson et al. 1985). A linear relationship between exercise intensity and leucine oxidation was found in the range of 25% to 89% $\dot{V}O_2$max. In this range, 54% of the flux of leucine was oxidized (Millward et al. 1982). At 100% $\dot{V}O_2$max, the leucine oxidation would account for approximately 80% of whole-body leucine flux (Babij et al. 1983).

Leucine oxidation has been shown to be greater in trained rats at the same VO_2 (Dohm et al. 1977; Henderson et al. 1985). The obvious reason is the increased activity of branched-chain oxoacid dehydrogenase in trained muscles (Dohm et al. 1977). However, in experiments on electrically stimulated perfused rat hindquarters, training reduced the relative contribution of leucine oxidation to VO_2 (Hood and Terjung 1987b).

Intramuscular and extramuscular sources provide branched-chain amino acids for oxidation. Intramuscular sources consist of an intramuscular pool of free amino acids, and liberated free amino acids result in protein breakdown in other tissues. The latter extramuscular source is considered significantly larger than the intramuscular one (Graham et al. 1995). The extramuscular source consists of an inflow of branched-chain amino acids from blood plasma into muscle fibers. An intense inflow of branched-chain amino acids from blood plasma into the muscle fibers has been evidenced (see Poortmans 1984; Graham et al. 1995). The plasma pool of free amino acids depends on the protein degradation in various tissues, including resting skeletal muscles, and the absorption of amino acids from the digestive tract. Human liver has a low activity of branched-chain amino acid transferase. Therefore, the rate of liver uptake of ingested branched-chain amino acids is modest, and they quickly affect the circulatory levels of these amino acids (Wahren et al. 1975). In prolonged exercise, a possibility for a release of branched-chain amino acids from the liver exists (Ahlborg et al. 1974). The inflow of branched-chain amino acids into the skeletal muscles may compensate for their oxidation without change in the amount of their intramuscular pool (MacLean et al. 1991). In some studies, even an increase of free branched-chain amino acid contents has been found in skeletal muscles (Dohm et al. 1981; Eller and Viru 1983). These results indicate the possibility that exercise intensity and/or duration may influence the overflow of branched-chain amino acids. Rennie et al. (1980) reported that at least in men, the increased muscle pool of branched-chain amino acids appears at the beginning of exercise, whereas during prolonged exercise, the pool decreases (Rennie et al. 1980, 1981).

Similar changes have also been found in plasma levels of branched-chain amino acids during exercise: an increase at the beginning of exercise that is substituted by a decrease after a certain amount of work is done (Rennie et al. 1980). Increases of branched-chain amino acids in intensive short-term exercises are attributable to altered splanchnic exchange rather than augmented release from peripheral tissues (Felig and Wahren 1971).

At the finish of the Stockholm Marathon, a significant reduction of the content of branched-chain amino acids was found (Parry-Billings et al. 1990). However, the results of the Colmar Ultra Triathlon showed that the level of valine but not of leucine and isoleucine decreased significantly (Lehmann et al. 1995). After the Boston Marathon, no changes were observed in the blood levels of leucine, isoleucine, and valine (Conlay et al. 1989).

Assessment of leucine dynamics is important in understanding amino acid metabolism. In training, the main significance of monitoring the blood plasma leucine level is to establish the critical shortage in providing branched-chain amino acids, which leads to exaggerated protein breakdown in muscles. However, to draw convincing conclusions, it is necessary to know how much the leucine concentration in blood plasma has to decline to enhance protein catabolism.

Alanine

Alanine is mainly synthesized in muscle tissue by combining pyruvate and an amino group. During exercise, pyruvate is formed as a result of glycogenolysis or degradation of blood borne glucose. Perriello et al. (1995) found that 42% of the alanine released by human muscle was formed at the expense of the pyruvate derived from blood glucose. The rest of the alanine released by muscle was synthesized from pyruvate derived from muscle glycogen. Amino groups for alanine synthesis are released in the oxidation of branched-chain amino acids (see figure 3.1). In this way, alanine acquires its metabolic function. Alanine is an essential substrate for gluconeogenesis in the liver. After deamination of alanine, released amino groups are used for urea synthesis. The remaining carbon skeleton provides material for glucose synthesis. These processes form the glucose-alanine cycle (Felig 1973): the liver sends glucose that degrades to pyruvate to muscles, releasing energy for anaerobic ATP resynthesis. A part of pyruvate combines with amino groups, and the formed alanine works as a vehicle, transporting NH_3 to the liver to avoid ammonia accumulation and to provide a substrate for glucose synthesis (figure 3.6). During exercise, the alanine output augments in working muscles with a concomitant increase in their blood level and alanine use in the liver for gluconeogenesis and urea formation increases after urea formation (for a review, see Felig 1977; Viru 1987).

In alanine synthesis, the amino group from precursor amino acids is first transferred to α-ketoglutarate to form glutamate. Glutamate is a common substrate whose metabolism will be determined by the factors controlling the synthesis and release of either alanine or glutamine. The alanine synthesis, as well as its metabolism in liver, is catalyzed by alanine-aminotransferase. Factors inhibiting this enzyme block de novo alanine formation (Garber et al. 1976). Cortisol is an inductor of the synthesis of alanine-aminotransferase. Insulin inhibits the enzyme (Felig 1973). The lack of glucocorticoids in adrenalectomized rats excludes the rises in alanine levels in blood plasma, FOG fibers, and liver induced by 3 h of swimming. The normally found increase of alanine-aminotransferase activity was absent in muscles and was substituted by suppressed activity in the liver in exercising adrenalectomized rats (Viru A. et al. 1994).

Training increases the alanine aminotransferase activity in skeletal muscle. Consequently, the muscles adapting to endurance training have an increased capacity for generating alanine from pyruvate (Mole et al. 1973).

In top athletes, blood alanine concentration increased dramatically after 200- and 400-m running competitions (Weicker et al. 1983). To establish the significance of exercise intensity, male volunteers exercised at intensities 25%, 50%, 75%, and 100% of $\dot{V}O_2$max. During the first two workloads, blood alanine concentration did not increase significantly. At intensities of 75% and 100% $\dot{V}O_2$max, the increase of alanine level was significant (Babij et al. 1983). Berg and Keul

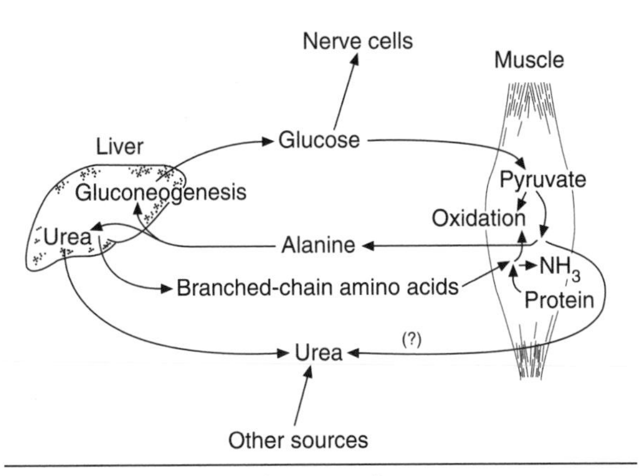

Figure 3.6 Glucose-alanine cycle.
Reprinted from A. Viru 1987.

(1980) compared the effect of prolonged exercise (cross-country skiing, running, swimming). Alanine concentration increased significantly in exercise with a duration of 70 or 80 min. In exercises lasting several hours, alanine concentration decreased significantly.

The study by Felig and Wahren (1971) established that the released alanine and the increased rate of glycolysis are in proportion to energy demands. Biopsy specimen studies showed that within the first 10 min of submaximal exercise, muscle alanine content increased (by 50% to 60%) in association with a decrease (by 50% to 70%) in the precursor glutamate content. Afterwards, up to the end of 90 min of exercise, alanine slowly returned to the initial level, whereas glutamine remained at a low level (Bergström et al. 1985). Similar results were obtained by Sahlin et al. (1990) and Van Hall et al. (1995). The diminished alanine release is associated with a decrease in muscle glycogen content (Van Hall et al. 1995). The low level of glutamate in muscles may be explained by conversion of its carbon skeleton through α-ketoglutarate into other tricarboxylic acid cycle intermediates (Sahlin et al. 1990; Van Hall et al. 1995). In this way, alanine-aminotransferase reaction in skeletal muscle is essential not only for alanine synthesis but also for providing the tricarboxylic acid cycle with α-ketoglutarate.

In vitro, the progressive decrease in alanine and glutamine synthesis was noted during prolonged incubation. The decrease was prevented by the addition of amino acids to the incubation medium (Garber et al. 1976). Therefore, after a prolonged period of muscular activity in vivo, the decreased alanine production may be related to the decreased supply of branched-chain amino acids because of their inadequate outflow from the liver. In turn, the inadequate outflow of branched-chain amino acids may be related to a decrease in the liver protein content during prolonged exercise (Kasperek et al. 1980).

In rats, after 12 h of swimming, a significant decrease of alanine concentration was observed in active muscles and liver in association with increased levels of leucine, valine, isoleucine, and glutamine (Eller and Viru 1983). Thus, an insufficient availability of branched-chain amino acids was not observed. Therefore, some other factors were responsible for the decreased alanine production in this experiment. First, the activity of alanine-aminotransferase might be inhibited, and, therefore, glutamate was mainly used for glutamine synthesis. This possibility was confirmed by the increased glutamine content in skeletal muscle. Other alternative explanations are the suppressed rate of oxidation of branched-chain amino acids and, thereby, the reduced release of the amino group and/or diminished availability of pyruvate in relation to the drop in muscle glycogen store and to hypoglycemia. When glycogen depletion occurs during prolonged exercise, the decrease in muscle pyruvate content (Sahlin et al. 1990) may limit the participation of the alanine-aminotransferase reaction (Wagenmakers 1999).

In humans, the determination of the arteriovenous difference of alanine showed that alanine uptake by splanchnic area (probably by liver) increased during prolonged exercise. During the first 40 min of exercise, the increase was moderate, but during the following exercise lasting 4 h, it became more pronounced (Felig and Wahren 1971; Felig 1977).

Because alanine has an essential function in metabolism, data on its pattern are important for understanding the metabolic situation during exercise. However, it is a task for detailed study, not for routine monitoring of training.

Glutamine

In deamination or transamination of amino acids, the basic idea is to convert amino acids into glutamine. In mitochondria, the oxidative deamination converts glutamine into α-ketoglutarate under the catalytic action of glutamate dehydrogenase. NAD and NH_3 participate in this reaction. Finally, they change to NADH and ammonium ions (NH_4^+), respectively. α-Ketoglutarate is related to the oxidation process; it is an intermediate of the tricarboxylic acid (TCA) cycle. NADH can yield several ATP molecules as a result of mitochondrial oxidative phosphorylation.

Glutamine synthesis consists in combining glutamate with NH_4. As a result, glutamine carries two amino groups from working muscles to the liver or kidneys for disposal. Thus, glutamine is well adapted for preventing ammonia toxicity during exercise (Brooks et al. 1996). Studies with [^{15}N]-leucine have convincingly demonstrated that amino groups of branched-chain amino acids incorporate into the α-amino nitrogen of alanine (Wolfe et al. 1982) and glutamine (Darmaun and Dechelotte 1991) in humans. The carbon skeletons of six amino acids (leucine, isoleucine, valine, glutamine, aspartate, and asparagine) are different from the synthesis of alanine and are used for de novo synthesis of glutamine, but not pyruvate as in the case of alanine synthesis (Chang and Goldberg 1978; Wagenmakers et al. 1985).

A point has been raised that glutamine is more important than alanine as a vehicle for transport of protein-derived carbon and nitrogen from muscles through plasma to the sites of gluconeogenesis or further metabolism (Wagenmakers 1999). In the postabsorptive state, glutamine accounts for 71% of the amino acid release and 82% of the nitrogen release from muscles (Elia et al. 1989). However, from the results of Williams and coworkers (1998) alanine, rather than glutamine, is the predominant N carrier involved in the transfer of N from muscle to the liver, at least during exercise of moderate intensity.

Glucocorticoids (cortisol in humans and corticosterone in rats) are involved in the control of glutamine metabolism. They increase glutamine efflux from skeletal muscles (Darmaun et al. 1988) and increase glutamine synthetase mRNA expression (Max 1990), and thereby the activity of the enzyme (Ardawi and Jamal 1990).

Exercises may result in various changes in the plasma glutamine concentration. Short-term sprint exercise increased the glutamine level, whereas marathon races decreased the level. Between these two extremes, several exercises do not cause changes in plasma glutamine concentration (Parry-Billings et al. 1990). Lack of changes in plasma glutamine concentration in several studies (e.g., Poortmans et al. 1974) is consistent with the understanding that the glutamine level in blood alters only in the extremes of the exercise continuum. In regard to short-term exercise, the intensity is important. In cycle exercises at 25% to 100% $\dot{V}O_2max$, the plasma glutamine level increased exponentially with the working rate (Babij et al. 1983). Katz et al. (1986) found an elevated plasma glutamine level after 4 min of exercise at 100% $\dot{V}O_2max$ in association with an increased rate of glutamine release from skeletal muscle. According to the calculations, the released amount of glutamine constituted only a minor proportion of the total glutamine store. Therefore, muscle glutamine content did not decrease (Katz et al. 1986).

In prolonged exercise, the decrease of glutamine concentration is common (Brodan et al. 1976; Decombaz et al. 1979; Rennie et al. 1981). However, the increase in plasma glutamine concentration was also found in exercises that are neither "prolonged" nor "extremely intensive": in 45-min exercise at 80% $\dot{V}O_2max$ (Eriksson et al. 1985), in 90-min exercise at 70% $\dot{V}O_2max$ (Maughan and Gleeson 1988), or in 70-min exercise at 75% $\dot{V}O_2max$ (Sahlin et al. 1990). Sahlin et al. (1990) established that the elevated plasma glutamine level was associated with a twofold increase in the rate of glutamine release from skeletal muscle and a modest (10%) decrease in muscle glutamine content. Bergström et al. (1985) observed an elevated glutamine level in skeletal muscle after the first 10 min of exercise at 70% $\dot{V}O_2max$, followed by a subsequent decrease. Rennie et al. (1981) found muscle glutamine content equal to 50% of the initial level after 225 min of exercise at 50% $\dot{V}O_2max$. Thus, to suggest the significance of any of the glutamine metabolic functions, it is necessary to take into consideration the possibility of different changes in glutamine levels. Therefore, for such a suggestion, we have to know the actual change in glutamine levels.

Rat experiments confirmed the decrease of glutamine content in skeletal muscles during exercise (Dohm et al. 1981) and also demonstrated the decrease of this amino acid content in myocardium (Eller and Viru 1983) and liver (Dohm et al. 1981; Eller and Viru 1983).

Overload training is associated with a low level of glutamine in the blood plasma (for a review, see Parry-Billings et al. 1992; Rowbottom et al. 1996). Because glutamine produced by skeletal muscles is an important regulator of DNA and RNA synthesis in mucosal cells and immune system cells (see Wagenmakers 1999) and affects the immune function of leukocytes (see Rowbottom et al. 1996) and some other immune responses (see Rowbottom et al. 1996), the decrease in plasma glutamine concentration in overtraining may contribute to immunodeficiency (Parry-Billings et al. 1992).

In the liver, production of antioxidant glutathione requires glutamine as a precursor. Therefore, Rowbottom et al. (1996) suggested that the exercise-induced increased rate of free radical production is related to suppressed levels of glutamine.

Other ideas for use of the plasma glutamine level for training monitoring may arise to get information on protein metabolism and central fatigue. Glutamine contributes to the control of protein synthesis (MacLennan et al. 1987) and degradation (MacLennan et al. 1988). During exercise, the decreased production of glutamine promotes protein degradation. During the recovery period, the increased glutamine production supports the increase in the protein synthesis rate and inhibits protein degradation. Again, glutamine is only one of several controlling factors. However, it is necessary to remember that after the end of strenuous exercises, the glutamine levels in plasma remain suppressed for several hours (Rennie et al. 1981), after the 100-km run for more than 24 h (Decombaz et al. 1979).

Graham et al. (1990) estimated that during an hour of leg extensor work at 80% $\dot{V}O_2max$, the net efflux of NH_3 from working muscle was 4.4 mmol, glutamine was 3.3 mmol, and alanine was 2.5 mmol. These data indicate that glutamine and alanine effluxes together are greater than NH_3. At the same time, these data support the point that glutamine release may be greater compared with alanine release in prolonged exercise. In short-term high-intensity exercises, NH_3 efflux from active muscles may increase up to 300 µmol/min (Graham et al. 1990). When intense exercise resulted in exhaustion in 3 min, the NH_3 efflux reached 500 µmol during exercise, but glutamine and alanine release remained at low levels (Graham et al. 1995). Thus, the "anatoxic ammonia transport" that uses glutamine and alanine may be effective in prolonged but not in short-term, intensive exercise. In this regard, the suggestion of Bannister and Cameron (1990) that intense exhaustive exercises may cause a state of acute NH_3 toxicity with altered CNS function, including motor control, has to be mentioned. Graham et al. (1995) argue against the dramatic effects of NH_3 on CNS functions during exercise, but they consider the contribution of NH_3 in central fatigue an interesting theory.

A question arises whether the glutamine/ammonia ratio in blood plasma provides an opportunity for evaluating central fatigue. As a matter of fact, such opportunity requires special studies. Still, we do not have a convincing background for acceptable conclusions on central fatigue by plasma glutamine/ammonia ratio. We do not know the quantitative characteristics of the ratio, which may be used for diagnostics of central fatigue. However, to understand a general trend, the ratio may be useful. Another question is whether NH_3 accumulation participates in peripheral fatigue. Moreover, NH_3 penetration through the blood-brain barrier is facilitated by high blood pH but not by acidosis appearing in strenuous exercise (see Graham et al. 1995). Last, the brain itself possesses a defense mechanism against ammonia accumulation. The mechanism is the same as in skeletal muscle—synthesis of glutamine.

In brain tissue, glutamine contributes to the neurotransmitter gamma-aminobutyric acid (GABA) synthesis because glutamate is the precursor for glutamine and GABA synthesis. GABA and glutamine act as neurotransmitters involved in the motor command and regulation of locomotion (Cazalets et al. 1994). When rats were submitted to treadmill running until exhaustion, ammonia levels increased in blood, brain cortex, cerebellum, and striatum (approximately 50% more in previously trained rats than in untrained rats). Increased ammonia reasonably stimulated glutamine synthesis in the brain to reduce the ammonia level. Glutamine contents increased in brain structures in trained and untrained rats. Glutamate contents decreased in all structures of trained rats but only in the striatum of untrained rats. In the striatum of trained rats, the GABA level decreased (Guezennec et al. 1998). The authors explained the differences found between trained and untrained rats by the fact that the duration of running was 288 ± 12 min in trained and 62 ± 5 min in untrained rats.

Tryptophan

Tryptophan belongs to the group of essential amino acids. In blood plasma, tryptophan exists in a free unbound state and bound with albumin fractions. The total concentration of tryptophan is low. Tryptophan is a precursor for the synthesis of a neurotransmitter 5-hydroxytryptomine (serotonin). Because serotonin may be involved in the mechanism of central fatigue, the attention to this amino acid is justified regarding training monitoring.

A theory of central fatigue (Newsholme et al. 1992; Krieder 1998) assumes that during prolonged exercise, a reduced blood level of branched-chain amino acids and/or an increase of free tryptophan facilitates the entry of tryptophan into the brain. Tryptophan, even in low concentrations, competes with several amino acids, including branched-chain amino acids, for a transport carrier at the blood-brain barrier. Therefore, the increased ratio of free tryptophan to branched-chain amino acid in plasma facilitates tryptophan entry into the brain (Fernstrom 1983).

The synthesis of serotonin is controlled by tryptophan hydroxylase, the activity of which depends on its substrate, tryptophan. The increased or decreased concentration of tryptophan changes the rate of formation of serotonin (Fernstrom 1983; Chaouloff 1993). The abundance of serotonin may result in altered activity of the serotonergic neurons concerned in the control of motor functions. Accordingly, administration of serotonin antagonists reduces endurance in humans (Wilson and Maughan 1992) and in rats (Bailey et al. 1993b). Serotonin antagonist increased the running time up to exhaustion (Bailey et al. 1993b).

As early as 1964, Haralambie (1964a) determined the tryptophan concentration in the blood serum of athletes in several events after intensive training sessions. Results failed to show the change, except for a decrease in two soccer players. Eleven years later, Cerny (1975) demonstrated an increase of serum tryptophan concentration during the first 40 min of cycle exercise at 60% to 65% $\dot{V}O_2max$. Continuation of the exercise up to 2 h resulted in a dramatic decrease of the tryptophan level. More recently, an increase was found in plasma concentration of free tryptophan after prolonged exercises, including military training or marathon races (Blomstrand et al. 1988), ultraendurance triathlons (Lehmann et al. 1995), cycling for up to 255 min (Davis et al. 1992), and cycling at 40% $\dot{V}O_2max$ to exhaustion (Mittleman et al. 1998). All these studies showed increases in the free tryptophan/branched-chain amino acid ratio.

Rat experiments demonstrated that total tryptophan in plasma did not change during 1 to 2 h of treadmill running, but free tryptophan increased. The increase of free tryptophan was also found in brain tissue and cerebrospinal fluid (Chaouloff et al. 1985, 1986). Blomstrand et al. (1989) and Bailey et al. (1993a) established the increase of serotonin in various brain regions during exercise in association with the increase of free tryptophan in plasma and tryptophan in the brain. In various brain regions, the increase of serotonin level accompanied an elevated content of another neurotransmitter, dopamine. The levels peaked at exhaustion (Bailey et al. 1993a). Thus, exercise-induced penetration of tryptophan into brain tissue and accompanying serotonin production in the brain were confirmed.

The inflow of tryptophan into the brain may be supported by the increased concentration of free tryptophan in the blood plasma (Davis et al. 1992).

Serotonin is metabolized to 5-hydroxyindoleacetic acid (5-HIAA). The increased content of 5-HIAA demonstrated a high turnover rate of serotonin and, thereby, elevated the activity of serotonergic systems. Experiments with rats indicated that the increase of serotonin release is associated with the augmented content of 5-HIAA in the brain region (Blomstrand et al. 1989) and in cerebrospinal fluid (Chaouloff et al. 1986).

It would be incorrect to think that the brain is overflowed with serotonin in the state of fatigue developed during prolonged exercise and that the general abundance of serotonin disturbs the function of all neurons concerned in motor control. Serotonin as a neurotransmitter influences only serotonergic neurons (i.e., neurons possessing specific serotonin receptors). Mostly, serotonin is synthesized in serotonergic neurons. The increased level of serotonin in brain regions indicates an intense synthesis of this neurotransmitter and, thereby, its release into the extracellular compartment. In this situation, the function of serotonergic neurons is favored. The same situation may take place when exogenous agonists of serotonin are bound with the serotonin receptors. Several types of serotonin receptors have been distinguished. Each type possesses its own functional and biochemical characteristics. For example, the study of Bailey and collaborators (1993a) showed that in rats the blockade of serotonin receptors 1C and 2 was effective for delaying fatigue. Human experiments failed to show significant alteration in performance during cycling at 65% of maximal power output after blockade of serotonin 2A/2C receptors (Meeusen et al. 1997). Therefore, even the high level of serotonin in brain tissue does not include the synchronous activity of all serotonergic neurons.

From a methodological aspect, alterations in tissue levels of neurotransmitters are crude measures of activity that do not necessarily reflect a corresponding change in synaptic release (Meeusen and DeMeirleir 1995). More precise information can be obtained with the aid of microdialysis investigation. By use of this method, Meeusen and collaborators (1996) demonstrated that extracellular serotonin and 5-HIAA levels increased in the hippocampus of rats running for 1 h. These responses increased when tryptophan was administrated before running.

Meeusen and DeMeirleir (1995) concluded that interactions between brain neurotransmitters and their specific receptors could play a role in the onset of fatigue during prolonged exercise. The central neurotransmitters can affect motor effector mechanisms at several levels. However, the crucial point is whether the changes in the neurotransmitter level trigger or reflect effects evoked by their release. The interaction of various neurotransmitter effects has to be taken into consideration but not only the "isolated" action of a neurotransmitter.

In conclusion, the tryptophan dynamics, particularly the increased ratio of free unbound tryptophan to branched-chain amino acids, may be used in training monitoring for detection of conditions that promote the development of central fatigue. However, it is not enough for the diagnostics of central fatigue because the latter is related to an integration of several neuronal and neurochemical changes. Serotonin, the synthesis of which is controlled by the availability of tryptophan, is one component of this integral. It may be a predominating component, but it is only a component.

Oxidative Substrates in Blood

Two main oxidative substrates in blood are glucose and free fatty acids. The role of blood is to transport them into tissues and make them available for oxidation in the mitochondria of cells. Evaluation of the dynamics of these substrates makes it possible to know the general conditions of oxidation. The most important is to detect hypoglycemia disturbing oxidation in nerve cells. It is also important to know the availability of free fatty acids during endurance exercises to judge about transfer from carbohydrates to lipids as oxidation substrates.

Additional oxidation substrates in the blood are branched-chain amino acids, the metabolism of which has been already discussed, triglycerols, lactate, and lipoproteins. Lactate can be oxidized during exercise in myocardium and oxidative fibers (only in low-intensity exercise and in modest amounts). The plasma level of triglycerols is low to make a significant contribution. Lipoproteins are rarely used for oxidation in skeletal muscles, but their assessment is useful to determine the antisclerotic effect of training.

Glucose

Blood glucose belongs to the group of rigid homeostatic parameters that have to be maintained at a constant level. Intense or long-lasting deviations of such parameters result in serious metabolic disturbances, including the inability to maintain life activities.

Blood glucose is an essential fuel for various tissues, particularly for the nerve cells. Compared with nerve cells, muscle fibers have a good carbohydrate store (glycogen) that can be used alone or together with blood glucose. In nerve cells, the glycogen store is rather small. Muscle tissue is able to substitute carbohydrates with lipids as substrates for oxidation processes. Lipids cannot be oxidized in nerve cells.

Blood glucose also fulfills an important role in metabolic control. First, the metabolic responses evoked by altered levels of blood glucose contribute to homeostatic regulation. The responses are directed to maintain a constant level of glucose in the blood plasma or to restore the normal level after deviations. However, the so-called glycostatic regulation is not limited by adjusting blood glucose levels. Blood glucose alterations influence hormonal responses that contribute to overall mobilization of energy stores.

A sensitive homeostatic mechanism guarantees correspondence between hepatic glucose output and glucose use by tissues (Newsholme 1979; Jenkins et al. 1985; Hoelzer et al. 1986). The most important regulatory means are the modulation of insulin and glucagon secretion (Vranic et al. 1976; Felig and Wahren 1979; Jenkins et al. 1986; Hoelzer et al. 1986; Wolfe et al. 1986). When the use of glucose by tissues increases, glucose output by the liver increases correspondingly. It is mainly due to the decreased ratio of insulin/glucagon (secretion of insulin is suppressed and release of glucagon increased). This regulatory change also results in decreased glucose use in various tissues but not by nerve cells. Therefore, the reduced glucose use is essential not only for blood glucose homeostasis but also for reserving blood glucose for nerve cells. When a tendency to an increased blood glucose level appears, opposite regulatory means are triggered. Increased insulin secretion inhibits hepatic glucose output and stimulates glucose inflow to tissue, including the return inflow to the liver. In this situation glucagon secretion is held on the basal level.

Experiments on exercising humans confirmed that when production of insulin and glucagon was held artificially constant, no increase in glucose output appeared. The exercise-induced increase in glucose output was the greatest when insulin secretion was simultaneously inhibited and glucagon production stimulated (Marker et al. 1991).

The role of epinephrine in the stimulation of hepatic glucose output during exercise has not been seen (Carlson et al. 1985), although convincing proof exists concerning the essential role of epinephrine in glycogenolysis in skeletal muscles (Richter 1984). However, facts do exist about the contribution of catecholamines in blood glucose homeostasis (Péquignot et al. 1980; Young et al. 1985; Hoelzer et al. 1986).

The gluconeogenesis is under a complex control exerted by insulin, glucagon, catecholamines, and glucocorticoids. Besides the stimulatory influence on gluconeogenesis and the control of the gluconeogenic substrate released from peripheral tissues (Exton et al. 1972), cortisol inhibits glucose transport to adipose and other tissues (Fain 1979). By increasing the rate of gluconeogenesis and inhibiting peripheral glucose use, cortisol elevates the blood glucose level.

In the control of hepatic output of glucose, a certain role belongs to sympathetic and parasympathetic nerves, influencing this process either directly or through changing insulin secretion by β-cells of the pancreas (for a review, see Shimazu 1987). In humans, sympathetic innervation of the liver is more abundant than it is in most animals. Obviously, the effective innervation ensures a rapid increase in glucose output in humans at the beginning of exercise.

The sensitive homeostatic mechanism should ensure a stable euglycemia during the exercise, which has been demonstrated in several articles (Jenkins et al. 1986; Wolfe et al. 1986). However, contrary to that, a number of other articles demonstrated variability and instability in the pattern of glucose response to prolonged exercise (Yakovlev 1955; Flynn et al. 1987; Lavoie et al. 1987). A study of 27 endurance athletes and 34 untrained persons showed a pronounced interindividual variability in the glucose pattern during a 2-h cycle exercise at 60% $\dot{V}O_2max$ (Viru et al. 1995). The results showed that the stable steady glucose level was preceded by an initial adjustment; in 79% of persons studied a transient decrease of blood glucose was found, and in 21% of persons studied an immediate increase was found at the beginning of exercise. In trained persons, the initial decrease was overcome, and a stable glucose level was established more rapidly than in untrained persons (figure 3.7). The initial transient decrease has been previously noticed (Yakovlev 1955; Costill 1984). Newsholme (1979) considered the initial hypoglycemia as a stimulus that homeostatically increases hepatic glucose output. In 25% of persons, the homeostatic stimulus for increased glucose output was so strong that a gradual increase of glucose level continued up to the end of exercise. On the other hand, in 41% of persons, the initial hypoglycemia exerted a weak stimulus; the reduced level of blood glucose (on average, 4.5 mmol/L) persisted to the end of the exercise. The initial decrease was followed by stable glycemia at the initial level in only 13% of the persons.

In a few of the subjects (21%), the blood glucose level increased immediately after the onset of exercise (Viru et al. 1995). The increase should be related to the rapid augmentation of hepatic glucose output initiated by central feed-forward regulation. Accordingly, it has been shown that the hepatic glucose production may exceed the peripheral glucose output in exercising humans (Katz et al. 1986) and that the mismatch between production and use is particularly pronounced in the early period of exercise (Kjaer et al. 1987). In our experiment, the rapid blood glucose rise was followed either by leveling off at levels greater than the initial one (on average, 6.2 mmol/L) or by a gradual decrease of glucose levels to the end of the exercise. These variants appeared in 11% and 10% of persons, respectively. According to the mentioned interindividual variability, five variants of glucose patterns during 2-h exercise were discriminated (figure 3.8) (Viru et al. 1995).

The discussed results are not contradictory to the aforementioned homeostatic mechanism. The results suggest that this mechanism guarantees a correspondence between hepatic glucose output and blood glucose use at diverse levels that are established throughout various

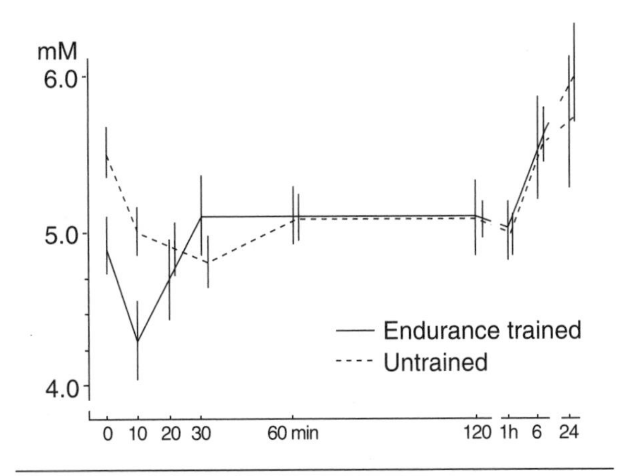

Figure 3.7 Blood glucose dynamics during 2-h exercise at 60% $\dot{V}O_2max$ in endurance-trained (solid line) and untrained (dotted line) men.

Reprinted from A. Viru et al. 1995.

patterns. It should be emphasized that steady-state glucose concentrations in blood may appear not only at the initial level but also at levels either below or above it.

In our study, the pattern of blood levels of insulin, growth hormone, and cortisol were assessed. We failed to establish the link among insulin, cortisol, and glucose patterns during exercise. However, growth hormone might be significant in glucose control. This suggestion is based on the observation that high levels of growth hormone are associated with trends for hyperglycemia, and low levels are associated with hypoglycemic trends (Viru et al. 1995).

During prolonged exercise, hypoglycemia may appear in relation to exhaustion of carbohydrate stores in the body. The main endogenous mechanisms to avoid hypoglycemia are

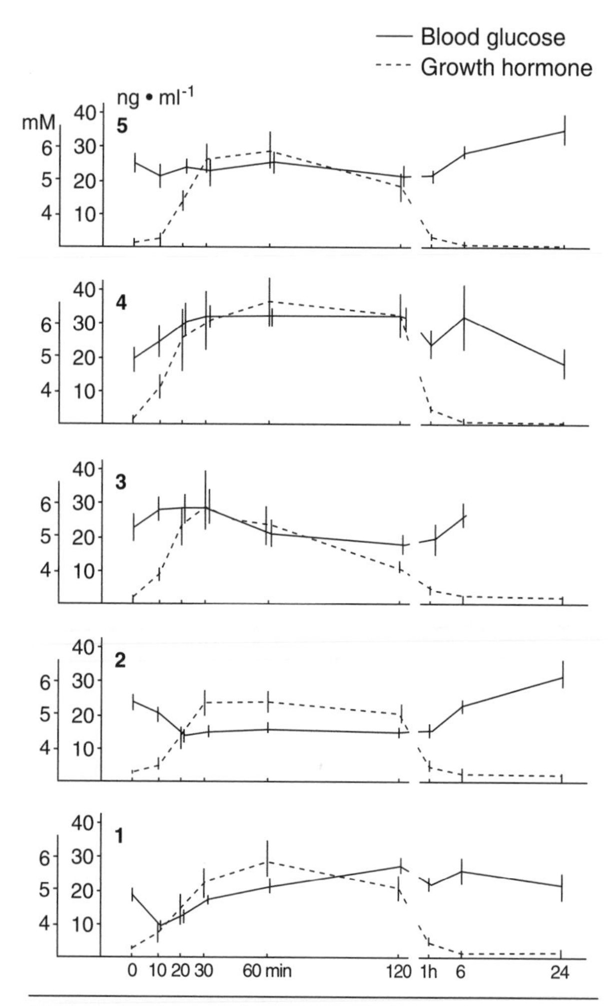

Figure 3.8 Five variants of blood glucose dynamics during 2-h exercise. (Solid lines represent interrupted glucose, and dotted lines represent growth hormone.)

Reprinted from A. Viru et al. 1995

intensive gluconeogenesis and transfer from carbohydrate oxidation to lipid oxidation.

Another mechanism that avoids hypoglycemia in prolonged exercise may consist of transient insulin resistance caused by fat–tissue-deprived tumor necrosis factor-2 (TNF_2) promoting lipolysis. Nevertheless, hypoglycemia may appear (see chapter 1, p. 4) and reach values less than 2.5 mmol/L (Felig et al. 1982). To avoid a decline in performance caused by hypoglycemia, athletes usually consume glucose when they perform exercises lasting several hours. The consumed glucose, on the one hand, replenishes carbohydrate availability and increases possibilities for its oxidation (Van Handel et al. 1980). On the other hand, it alters the hormonal metabolic regulation; blood levels of insulin increase, and blood levels of catecholamines, glucagon, and cortisol decrease. These changes are a typical expression of the glycostatic regulation (Nazar 1981) involving hypothalamic glucosensitive nervous structures (Kozlowski et al. 1981).

After glucose intake, the results of altered hormonal control are reduced hepatic glucose output, increased glucose use by muscle fibers, and suppressed lipolysis. Consequently, carbohydrates were again used instead of lipids as an oxidation substrate. Therefore, there is no more carbohydrate sparing.

Similar consequences are caused by glucose intake before exercise. Although carbohydrate feeding before starting exercise elevated the blood glucose level up to 6.5 to 7.0 mmol/L, a rapid decline during the first 10 to 15 min of exercise led blood glucose concentration to values less than 2.5 mmol/L (Costill et al. 1977). In this situation, muscle glycogen use is exaggerated (Ahlborg and Felig 1977), resulting in an earlier onset of exhaustion. The initial hyperglycemia after preexercise glucose consumption is obviously related to hyperinsulinemia and relative hypoglucagonemia (Ahlborg and Felig 1977). The significance may also have high insulin sensitivity and low sympathoadrenal activity (Kuipers et al. 1999).

Blood glucose determinations are essential for the elaboration of an effectual regime for carbohydrate feeding during competitions. However, this regime is, indeed, effectual if it considers not only the replenishment of carbohydrate store but also its effects on the glucostatic regulation.

Free Fatty Acids and Glycerol

Adipose tissue represents the most voluminous energy store. To use this energy, deposited triglycerols have to be degraded into FFA and glycerol. FFA are released by the action of hormone-sensitive lipase. Remaining glycerol is liberated after additional hydrolysis by monoacylglycerol lipase. A part of released fatty acids is re-esterified to form new triglycerol and constitutes the triglycerol–fatty acid cycle. The other part of FFA is removed from the adipose tissue and is dependent on the plasma albumin concentration (albumin binds FFA in blood plasma) and the blood flow through adipose tissue. Glycerol cannot be reused in adipose tissue. Its small molecules diffuse easily through the membrane of adipose cells into the blood. Therefore, changes in glycerol concentration in the blood provide an opportunity for indirect evaluation of the rate of lipolysis in adipose tissue (for a review, see Bülow 1988; Jeukendrup et al. 1998).

Prolonged exercise causes pronounced increases in the lipolysis rate in adipose tissue (Shaw et al. 1975). This fact has been confirmed with the aid of microdialysis of the extracellular space of subcutaneous adipose tissue (Arner et al. 1990). The main activator of lipolysis during exercise is the sympathoadrenal system. By use of this method in man, Arner et al. (1990) demonstrated that an a-adrenergic inhibitory mechanism modulates lipolysis at rest, whereas a b-adrenergic stimulatory effect is important in exercise. The b-adrenergic effect arises from sympathetic nerve stimulation or epinephrine. This hormone is the main activator of the hormone-sensitive lipase. However, several other hormones also stimulate lipolysis. Experiments on human isolated adipocytes established that at physiological doses, glucocorticoids, thyrotropin, and growth hormone are effective (Coppack et al. 1994). Animal experiments showed the lipolytic action of glucagon and corticotropin. However, disagreement exists as to whether these hormones and thyroid hormones and growth hormone are essential for the lipolytic effect in humans in in vivo conditions (Hales et al. 1978). Glucocorticoids exert a permissive influence, supporting the action of catecholamines and growth hormone. The permissive action of glucocorticoids is related either to the changes in intracellular Ca^{2+} metabolism (Exton et al. 1972) or to the inhibition of phosphodiesterase to favor cAMP accumulation (Lamberts et al. 1975).

The actualization of lipolytic action of catecholamines and other hormones depends on the concentration of insulin. Insulin inhibits the activity of lipase and blocks the action of lipolytic hormones (Fain et al. 1966). Accordingly, the effective mobilization of lipid stores requires a decrease in the blood insulin level.

The decreased insulin concentration is a common manifestation of hormonal metabolic control during prolonged exercise (Galbo 1983; Viru 1985a). At the beginning of prolonged submaximal exercise, a decline in the blood insulin level appears after a lag period of 10 to 15 min. In this regard, it is worth noting that the increase of FFA concentration in blood is also delayed for 10 to 15 min (Hargreaves et al. 1991; Langfort et al. 1996).

A significant metabolic factor in suppressing lipolysis is glucose (Arner et al. 1983). By use of a pancreatic-pituitary clamp to maintain basal insulin level in healthy persons during exercise, Carlson et al. (1991) demonstrated that hyperglycemia (10 mmol/L) suppressed the rates of appearance of FFA and glycerol. The authors concluded that glucose regulates FFA mobilization independent of insulin changes by suppressing lipolysis.

In light to moderate exercise, the inflow of FFA into the bloodstream is supported by the suppressed FFA re-esterification rate. When the FFA oxidation rate exceeded the resting level 10-fold, only 25% of released FFA was re-esterified compared with 70% in resting conditions. During postexercise recovery, the re-esterification process augments rapidly (Wolfe et al. 1990). The significance of increased blood flow and intensified binding of FFA to albumin may be suggested as a cause of decreased FFA re-esterification. Adipose tissue blood flow increases threefold during prolonged exercise (Bülow and Madsen 1978), which directly enhances FFA release (Bülow and Madsen 1981). Experiments on isolated adipocytes demonstrated that the decrease in fatty acid re-esterification appears after administration of glucocorticoids in connection with inhibition of glucose use by adipocytes. However, there was a 1- to 3-h lag before fatty acid release was accelerated by glucocorticoids (Fain et al. 1963). Permeation of

FFA into skeletal muscle is a carrier-mediated process. Carbohydrate availability and muscle contraction are known to regulate the transport of FFA in membrane (Turcotte et al. 1995).

At exercise intensities up to 70% $\dot{V}O_2$max, the increase of FFA level is related to exercise intensity and duration (Pruett 1970a). During exercises of higher intensity (above the anaerobic threshold), lactate accumulation inhibits FFA release. Lactate decreases FFA release by increasing FFA re-esterification without affecting lipolysis (Issekutz and Miller 1962). This is why the FFA level increases during prolonged exercise of moderate intensity but not during short-term high-intensity exercise. It has been shown that during intensive exercise, lactate suppresses FFA release and turnover (Issekutz et al. 1975). A lactate concentration of 2 mmol/L decreases FFA release by 35% to 40% (Shaw et al. 1975).

Glycerol is used for liver glycogenolysis. During a 4-h exercise, the contribution of gluconeogenesis to overall hepatic glucose output increased from 25% in the resting state to 45% during exercise in association with an approximately ninefold increase in glycerol use (Wahren and Björkman 1981).

The relationship between FFA uptake and oxidation in exercising muscles and FFA level in blood plasma has been known since the 1960s (Issekutz et al. 1967; Hagenfeldt and Wahren 1968). In 1974, Ahlborg and coworkers reported that during a 4-h exercise at 30% $\dot{V}O_2$max, the uptake of oleic acid (a fatty acid) by exercising legs was augmented approximately threefold during the first 40 min of exercise and further by 140% during the remaining period of exercise. The uptake of oleic acid accounted for 35% of oleic acid turnover during the entire period of exercise. The blood level of oleic acid and its uptake were parallel throughout the 4-h exercise period.

The fatty acid concentration in plasma is an important factor but not the sole factor determining plasma fatty acid oxidation (Jeukendrup et al. 1998). During exercise, the oxidation of fatty acids declines when availability of FFA is reduced by hyperglycemia and hyperinsulinemia.

An essential result of the transition from carbohydrate use to lipid use and of increased fatty acid oxidation is the sparing effect on carbohydrate stores. During exercise, the elevation of plasma FFA decreases muscle glycogen use (Costill et al. 1977), whereas the inhibition of FFA mobilization by nicotinic acid increases glycogen use (Bergström et al. 1969). However, there are also contradicting results; a twofold increase in plasma FFA had no effect on muscle glycogen use during 60 min of knee-extension exercise (Hargreaves et al. 1991). Nevertheless, the potential for a reduction in glycogen use caused by increased FFA availability and oxidation has to be taken into consideration.

Romijn et al. (1995) showed that at an exercise intensity of 85% $\dot{V}O_2$max, the lipid oxidation rate was less than at 65% during exercising in association with the lack of an increased rate of FFA appearance and a decrease of FFA concentration in plasma. When exercising at 85% $\dot{V}O_2$max FFA concentration was increased by lipid-heparin infusion, the lipid oxidation rate was restored only partially compared with the levels observed during exercise at 65% $\dot{V}O_2$max. The findings confirm that at an exercise intensity greater than the anaerobic threshold (e.g., at 85% $\dot{V}O_2$max), lipid oxidation was impaired because of the failure of FFA mobilization, but this explains only part of the decline in lipid oxidation during intensive exercise.

Animal experiments (see Holloszy 1973; Yakovlev 1977) indicated that as a result of training for improved endurance, blood FFA level and fatty acid oxidations increased in aerobic-anaerobic exercises despite high lactate concentrations. Obviously, the training altered the metabolic control at both levels of adipocytes (re-esterification of FFA by lactate was reduced) and muscle fibers (FFA oxidation was stimulated in high-intensity exercise). The plausible reasons for the latter are enhanced activity of ß-oxidation enzymes (Costill et al. 1979; Holloszy and Coyle 1984) and more favorable conditions for the entry of acetyl units derived from ß-oxidation of fatty acids into the TCA cycle in endurance-trained organisms (Gollnick et al. 1985). In this way, endurance training favors a greater use of lipids. Concomitant sparing of the glycogen stores is considered to be an essential factor for enhanced endurance performance (for a review, see Saltin and Gollnick 1983; Holloszy and Coyle 1984; Coggan and Williams 1995). In endurance-trained organisms, the contribution of fat to the oxidation process is elevated at the same relative and absolute exercise intensities compared with untrained ones (Costill et al. 1977; Henriksson 1977; Jansson and Kaijser 1987). However, as soon as glycogen stores become depleted and carbohydrate oxidation falls below a critical level, the exercise intensity has to be reduced because the rate of ATP resynthesis declines (Newsholme 1989).

Assessment of these favorable metabolic effects of endurance training needs to determine FFA and lactate (it is best to determine glucose and glycerol also) after performance of aerobic-anaerobic competitive exercises. All in all, the patterns of blood levels of FFA and glycerol provide information about the use of lipids during prolonged exercises; glycerol changes inform us about the rate of lipolysis (degradation of triglycerides in adipose tissue), whereas FFA dynamics speak about the availability of the essential substrate for oxidation in working muscles. Additional assessment of lactate and glucose ensures the possibility of understanding why lipid mobilization is not as high as desired.

Lipoproteins

Lipoproteins are vehicles for transporting lipids to the sites of their metabolism in various tissues. Plasma lipoproteins differ by their hydrated density, which determines their floating characteristics in ultracentrifugation. Four main species are usually distinguished: high-density lipoproteins (HDL), low-density lipoproteins (LDL), very low-density lipoproteins (VLDL), and chylomicrons (CM). HDL is further divided into the subclasses HDL_1, HDL_2, and HDL_3.

Lipoproteins consist of protein (apolipoprotein or apoproteins) and lipid constituents. Apolipoproteins include three apo A (apo AI, AII, and AIV), two apo B (apo BI and apo BII), three apo C (apo CI, CII, and CIII), an apo D, and several polymorphic forms of apo E. Apolipoproteins are structure components. Their functions are to ensure the stability of the lipoprotein molecule, to constitute recognition sites for cell membrane receptors, and to act as co-factors for enzymes involved in lipoprotein metabolism. Major lipid constituents are triglycerols, phospholipids, unesterified cholesterol, and cholesterol esters.

The major lipoproteins are spherical particles. Their core makes up triglycerols and cholesterol esters; their surface coat consists of apolipoproteins, phospholipids, and unesterified cholesterol. A continuous exchange of lipids and apoproteins exists between the lipoprotein particles and between lipoproteins and cells. Enzymes such as lipoprotein lipase, hepatic lipase, and lecithin-cholesterol acyltransferase contribute to these exchanges.

The function of VLDL is to transport triglycerols from the liver to other tissues. Under the catalytic influence of lipoprotein lipase, VLDL and chylomicrons degrade, providing glycerol and FFA for peripheral tissues as energy substrates. Surface components of degrading VLDL (cholesterol, phospholipids) are transferred to HDL. Apoproteins of VLDL are retained within the LDL (apo B) and HDL (apo E and apo C) after their breakdown.

LDL interact with a specific receptor on the plasma membrane of various cells by means of apo B. Attachment of apo B to the receptor causes the entire lipoprotein to be transported to the inside of the cell. The uptake of LDL by cells is part of a homeostatic mechanism regulating the intracellular cholesterol metabolism and providing cholesterol for plasma membranes as an essential structural component. However, a high blood plasma concentration of cholesterol in the form of LDL is the most important factor causing arteriosclerosis (for a review, see Stokes and Mancini 1988).

In normal conditions, when the cholesterol concentration inside a cell becomes too great, the production of the cell's LDL receptors decreases. In this way, the further absorption of additional LDL declines. If this mechanism does not work with sufficient effectiveness, the fatty lesions called atheromatous plaques will develop inside the arterial wall. These plaques deposit cholesterol crystals in the intima and underlying smooth muscles in the arterial wall. In the course of time, the crystals grow larger. Simultaneously, the surrounding fibrous and smooth muscle tissue proliferate to form additional layers for longer and larger plaques.

The endogenous mechanism in avoiding increased levels of LDL is related to ingestion of LDL by liver cells and decreased formation of new cholesterol. Excess cholesterol inhibits the liver enzyme system in the formation of cholesterol.

HDL contains apoproteins AI and AII, which have a higher affinity for different receptors of cells than those of the apoprotein B of LDL. HDL can absorb cholesterol crystals that are beginning to deposit in the arterial wall and transfer them back to circulating LDL to be carried to the liver. When a person has a high ratio of HDL to LDL, the likelihood of arteriosclerosis developing is considerably reduced (for a review, see Catapano 1987).

Evidence has been collected that endurance training results in a decrease of total cholesterol and LDL levels and an increase of HDL concentration (for a review, see Dufaux et al. 1982; Tran et al. 1983). The training effect involves increased lipoprotein lipase activity in skeletal muscle and adipose tissue (Nikkilä et al. 1978; Marniemi et al. 1980). Consequently, the determination of total cholesterol and/or HDL and LDL is essential for the evaluation of the antisclerotic effect of training performed for health promotion.

FFA released from VLDL and chylomicrons can be used for energetics of working muscles (Havel et al. 1967). The catalyzing lipoprotein lipase is located on the luminal surface of the vascular wall. The activity of this enzyme changes not only in training but also during acute exercise (Ladu 1991). However, the muscle uptake of FFA released from VLDL-triglyceride appears slowly and accounts for less than 5% of the FFA-derived CO_2 in prolonged exercise (Havel et al. 1967). Therefore, there is no sense in investigating VLDL and chylomicrons as a source of oxidation substrate in training monitoring.

Microdialysis

New methodological possibilities for metabolic studies are enabled by the microdialysis technique. A microdialysis probe can be inserted into the extracellular space of subcutaneous adipose tissue or muscle tissue. The probe consists of a hollow fiber membrane that functions as an artificial vessel. Two kinds of probes are used. One probe consists in a dialysis tube with steel cannulas at each end. The other is a double-lumen cannula in which the microdialysis membrane is glued to the top of the cannula. A neutral dialysis solvent (e.g., Ringer's solution) is pumped through the probe at low speed (1 to 5 μl/min). The outgoing solvent is analyzed for molecules that are collected from the extracellular water space (for a review, see Ungerstedt 1991).

The method has been used in exercise studies in humans. For example, glucose and lactate balance in human skeletal muscle and adipose tissue has been studied with the aid of the microdialysis technique during isometric contraction (Rosdahl et al. 1993). Essential results were obtained about the adrenergic regulation of lipolysis (Arner et al. 1990) during exercise.

Although the microdialysis technique is a promising tool for metabolic studies in man, it is meant for laboratory and clinical studies, not for studies in field conditions and, thereby, not for training monitoring. Moreover, microdialysis also shows some limitations because it only allows analysis of "small" molecules like glucose or lactate but not molecules like TNF-alpha. Development of the microdilution technique may overcome this limitation.

General Remarks

Several metabolic parameters were discussed on that may be used in biochemical monitoring of training. The most usable metabolites for training monitoring are listed in table 3.1. However, each method is useful only in the right place. Therefore, knowledge of a parameter should determine whether it is appropriate for training monitoring.

The value of the information obtained depends on the validity of the analytical method used. Duplicate analyses are the best. However, several conditions (high cost of analyses, limited time to get the sample, or limited amount of the sample, etc.) make this approach limited in its use. Therefore, the quality of analyses is paramount. Field studies should be done only by experienced personnel.

It is important to make analytical procedures as simple as possible thereby minimizing the time lag between sampling and results. In several cases, receiving information quickly is essential. For example, information on lactate changes in athletes received during the training session makes it possibile to change exercise intensity or the duration of rest intervals. The same information received several hours or days after the session may have significance for further planning of a similiar session, but offers nothing in the current situation to monitor training. However, the use of express methods presupposes knowing their methodological

Table 3.1
Metabolites and Substrates Used for Biochemical Monitoring of Training

Metabolite	Origin	Possible area in training monitoring
Lactate	End product of anaerobic breakdown of glycogen or glucose Substrate for oxidation and glycogen synthesis	• Assessment of the anaerobic threshold • Index of intensity of anaerobic-glycolytic or aerobic-anaerobic exercises • Index of use of the anaerobic working capacity • Maximal level is used for assessment of anaerobic glycolytic power
Ammonia	Result of degradation of AMP in FG fibers Possible additional source—oxidation of branched-chain amino acids	• Index of ATP resynthesis by combining 2 ATP and formation AMP • Indirect index of activity of FG fibers
Urea	End product of protein (amino acid) breakdown	• Index of the influence of prolonged aerobic exercises • Index of recovery process
Tyrosine	Protein degradation, mainly in muscle tissue	• Index of intensity of protein catabolism in muscles
3-Methylhistidine	Product of degradation of myofibrillar proteins (myosin, actin)	• Index turnover of contractile proteins • Assessment of trainable effect in power and strength-training sessions
Alanine	Product of combining NH_3 groups (released in oxidation branched-chain amino acids) and pyruvate in muscles	• Assessment of the rate of glucose-alanine cycle, which makes link between carbohydrate and protein metabolism in energy attaining of muscular activity
Leucine	Branched-chain amino acid, oxidizable in muscles	• Index of metabolism of branched-chain amino acids
Tryptophan	Precursor for synthesis of neurotransmitter serotonin	• Diagnostics of central fatigue and of a central mechanism related to overtraining
Glutamine	Amino acid, essential for the optimal functioning of several tissues, as well as for normal immunoactivity	• Diagnostics for fatigue and overtraining, mainly to detect a possible background for changes of immunoactivities
Free fatty acid	Product of lipolysis (breakdown of triglycerols in adipose tissue); used in muscles as an oxidation substrate	• Assessment of the magnitude of use of lipid as oxidative substrate (blood level of free fatty acids is proportional to their use in oxidation)
Glycerol	Product of lipolysis; used in liver for gluconeogenesis	• Assessment of the intensity of lipolysis in adipose tissue
Glucose	Normal constituent of blood, supplied by liver to blood Substrate for oxidation and glycogen synthesis	• Index of carbohydrate use • A factor of metabolic control

errors. The conclusions made based on the results of express methods are valid only if the changes found exceed the quantitative estimation of the methodological error. Accuracy, linearity, and reliability have been demonstrated for the express determination of lactate with the Accusport Lactate Analyzer, at low and high lactate levels (Fell et al. 1998).

Another consideration is whether the substrate/metabolite changes depend on the mass of working muscles. When blood catecholamine and lactate levels were compared in exercises of the same intensity but performed either with one or two legs, the dependence of changes on the amount of working muscles was apparent. A 10-minute two-leg exercise induced twice as much lactate and four times as much norepinephrine and epinephrine concentrations in arterial blood as a one-leg exercise. During a one-leg exercise, when the epinephrine concentration was raised with the aid of infusion of the hormone, the lactate released from working muscles increased significantly but did not reach the level found after a two-leg exercise without infusion. Arterial blood glucose concentrations were similar during one-leg and two-leg exercises (Jensen-Urstad et al. 1994). However, the dependence on body mass may not be the same as the dependence on the amount of working muscles. There may be reasons to suggest relationships between exercise-induced changes in blood constituents and body mass. First, the blood volume (and accordingly the plasma volume) is physiologically adjusted to the body mass. Thus, in resting conditions, the concentration of blood constituents should not be dependent on the body mass. However, during exercise, the situation is a little different. The total metabolite release from muscles into the bloodstream depends on the amount of active muscles, on the one hand, and on plasma volume on the other hand. If we assume that plasma volume depends on body mass, the metabolite change will be less pronounced during cycle exercise (the power output does not depend on body mass) in persons with higher body mass but a similar mass of working muscles. Thus, the change of metabolites in blood should be related to the ratio of total muscle tissue to body mass and/or to the ratio of active muscles to total muscle tissue. However, these subtle differences may appear among individuals of contrasting body build but not in the same person during normal nutrition and usual training. Therefore, in regard to training monitoring, the amount of body or muscle mass has theoretical but not practical meaning for the interpretation of metabolite changes in training monitoring of a person.

Exercise-induced changes in plasma volume require more serious attention. When albumin, transferrin, and sex hormone–binding globulin in swimmers were adjusted for the percent change of plasma volume, their concentration did not alter during the training session and 2-h recovery period. Thus, the changes of concentration of these parameters were, at least partly, due to changes in plasma volume. At the same time urea, uric acid, creatinine, and calcium, when corrected for percent change of plasma volume, still demonstrated significant changes (Kargotich et al. 1997). The results are logical. Proteins and metabolites/substrates bound with plasma proteins cannot pass the vascular membrane, and, therefore, during exercise plasma extravasation directly increases their concentration inside the blood vessels without the actual action of exercise. However, compounds of low molecular mass are freely exchangeable between plasma and interstitial fluid. Thus, the shifts of plasma volume cannot influence their concentration in plasma.

No strict evidence exists that exercise-induced metabolic changes are related to biological rhythms. Galliven et al. (1997) tested this question measuring lactate and glucose responses in exercise at 90% and 70% $\dot{V}O_2$max in the morning and late afternoon in the follicular phase (days 3 to 9), midcycle (days 10 to 16), and luteal phase (days 18 to 26) of the ovarian-menstrual cycle. The results obtained failed to show significant diurnal differences in the magnitude of responses for lactate and glucose. During the ovarian-menstrual cycle, a significant time-by-phase interaction for glucose was found, but integral responses of glucose and lactate were similar across cycle phases. The results of a study by Hackney et al. (1991) found that fat oxidation and use are the greatest at ovulation, whereas carbohydrate use is favored during the midfollicular phase compared with either ovulation or the midluteal phase. The results of Scheen and coauthors (1998) demonstrated that elevations of body temperature induced by 3-h exercise at 40% to 60% $\dot{V}O_2$max were higher in the early morning (at 05:00) than at 14:30 or around midnight. During exercise, the glucose decrease was approximately 50% more pronounced around midnight than in the afternoon or early morning.

Summary

Several metabolites in blood, urine, saliva, and oxidative substrates in blood plasma provide possibilities for evaluation of the metabolic situation in the body and are usable in training monitoring.

Frequently, the desire of a researcher is to assess the degree of contribution of various anaerobic pathways of ATP resynthesis during exercise. The main aim of these assessments is to get an idea about the direction of the training effect when the exercise is systematically repeated. The most clear-cut results can be obtained by use of blood lactate responses to evaluate the contribution of anaerobic glycogenolysis. These results may be used for semiquantitative characteristics. The precise quantitative estimate is impossible because of four cautions: (1) lactate is formed from pyruvate as one of the possible transformations besides others (oxidation, alanine synthesis, and glycogen resynthesis); (2) lactate is formed in muscle fibers, and it diffuses into the blood with a certain time delay; the estimation of the contribution of anaerobic glycogenolysis is exact when blood sampling for lactate determination happens at the moment of actual peak values in blood; (3) during exercise, the lactate level in blood depends on the ratio between lactate inflow into the blood and lactate elimination from the blood. Moreover, one must take into consideration the fact that the magnitude of the lactate response to exercise is not a strict linear relationship with exercise intensity. At the maximal level of power output, the lactate response is not the greatest because of the significant contribution of the PCr mechanism.

Blood lactate is a valuable tool in the assessment of anaerobic capacity, anaerobic threshold, and the intensity of training exercises (again the aforementioned caution has to be considered).

A question has arisen concerning whether the contribution of the PCr mechanism is characterized by the increase of plasma creatine concentration. A strict relationship of the contribution of the PCr mechanism in ATP resynthesis and the creatine level in blood has not been detected. The obvious reason is that the PCr creatine shifts are rapid and mainly caused by the intracellular creatine pool.

It is possible that creatine levels (usually the transformed form of creatinine in urine) can be used for the assessment of muscle mass of the body. However, several more simple and more precise methods exist.

The contribution of the myokinase reaction in ATP resynthesis can be evaluated (again only semiquantitatively) with the aid of the accumulation of the products of AMP degradation (ammonia, hypoxanthine, and uric acid), with the most important being the determination of ammonia. The production of ammonia is most intensive in FG fibers. Therefore, a test for the indirect evaluation of FG fiber composition may be elaborated on the background of ammonia responses to high-intensity exercise. A test has been used for selecting promising sprinters by ammonia responses in sprinting and middle-distance running.

Often, urea responses are used for evaluating the workload of the training session and the recovery dynamics. However, the fact that the urea response is inhibited by high lactate levels and is related to nutritional factors and environmental conditions has to be considered.

The exercise-induced catabolic responses during exercise and in the recovery period can be characterized by blood level and urine excretion of tyrosine and 3-methylhistidine. Because meat consumption results in an inflow of exogenous 3-methylhistidine, the assessment of 3-methylhistidine production has to be made during a meat-free diet or accounting for the inflow of exogenous 3-methylhistidine.

Branched-chain amino acids, and particularly leucine, in blood characterize their availability for additional sources of oxidation substrates. Alanine and glutamine levels in blood plasma are essential for studies of metabolic processes, but their value for usual training monitoring is modest. Special attention should be paid to the ratio of tryptophan/branched-chain amino acids if the task of monitoring is to assess the development of central fatigue.

Oxidation substrates, glucose, and FFA provide information on the availability of carbohydrates and lipids for oxidation. A specific homeostatic mechanism maintains the blood glucose level constant in rest and during exercise. The determination of blood glucose is important to detect the critical duration of exercise after which hypoglycemia develops, and thereby glucose availability for tissues decreases. The most important is the maintenance of euglycemia for nerve cells. Blood glucose determination is also essential when the glucose is taken in during exercise. In these cases, the trend to hyperglycemia points to the exclusion of the hormonal mechanism of mobilization of energy stores and, particularly, lipids.

The detection of FFA provides information on the availability and actual use of lipids for oxidation. When the determination of the blood level of glycerol is added, the rate of lipolysis in adipose tissue will be evaluated more precisely because glycerol is not reused for triglyceride synthesis as are FFA.

CHAPTER

4

Methodology of Hormonal Studies

An essential task of hormones in metabolic control is to interfere with cellular autoregulation and to ensure an extensive mobilization of body resources. Otherwise, the actualization of potential capacities of the body is impossible during the performance of athletes in competition. Hormonal studies provide information on the adaptation to certain levels of exercise intensity and duration, as well as on disorders of adaptation, including the exhaustion of the organism's adaptivity and overtraining phenomena. Hormonal responses can be used for the assessment of the trainable effect of the exercise session and for control of the recovery period. To obtain the requested information and to avoid misunderstandings and incorrect results, several cautions and limitations must be taken into consideration.

General Methodological Considerations

Several questions have to be solved before planning hormonal studies for training monitoring. The first question is "Why is the measurement of hormones necessary?" If the answer confirms the necessity for hormonal studies, the next question is "What is the best choice of body fluid for hormone assessment?" To answer this question, attention needs to be directed at the feasibility of getting the required fluid and the availability of a valid method for hormone determination in the amount of obtainable sample. The main question is whether the possibilities ensure reliable results.

Determining Hormone Levels From Body Fluids

In most cases, hormones are determined in venous blood. In several cases, instead of the venous blood, fingertip or earlobe prick blood, urine, saliva, or sweat has been used. The most valuable information is obtained by assessing hormones in the venous blood. Blood is the medium that hormones are secreted into by the endocrine glands. Hormones are transported to body tissues through the blood.

In blood, most hormone molecules are bound by specific hormone-binding proteins. In most cases, bound hormones are metabolically inactive. They cannot pass through the capillary wall and reach the interstitial compartment. In several cases, the methods used to determine the total hormone concentration combine the bound and the unbound fraction of a hormone. It has been the belief that information on the actual biological effect of hormones can be obtained by free, unbound hormone fraction, but the activity of the gland is reflected by the total hormone level. However, this understanding is not complete because a rapid exchange exists between the bound and the unbound fraction of a hormone in the blood. This fact allows us to suggest that the amount of unbound hormone, which passes into the interstitial fluid, is rapidly substituted by the release of an equivalent portion of hormone from the bound fraction.

Moreover, in stress situations, including exercise, increased hormone secretion usually leads to hormone concentrations in blood that exceed the binding capacity of plasma proteins. Accordingly, the total hormone concentration in blood rises in correlation with the increase of the unbound fraction. Consequently, the total amount of hormone in blood contains sufficient information on the availability of hormones for tissues. In most cases, the assessment of unbound hormone fraction and/or hormone-binding capacity is not necessary in training monitoring, although it may provide essential information on specific aspects of endocrine functions.

Urine

Several methods for hormone determination in urine measure the hormone and its precursors and metabolites together. Therefore, the hormones in urine may be used for the evaluation of a general trend, but they do not provide the same quantitative characteristics as the hormone concentration in blood. Chromatographic procedures have to be used to determine intact hormones and their metabolites in urine separately. The use of highly specific radioimmunoassays or chemiluminescence assays also ensures separate determination of various compounds (intact hormone, its metabolites, and precursors) in blood and urine. At least three conditions need to be accounted for:

- Renal excretion of hormones depends first on the blood level of free unbound fraction.
- Most molecules of hormones or their metabolites are conjugated with glucuronic or sulfuric acids before their renal excretion; therefore, the value of hormone determination depends on whether only unconjugated fraction or both conjugated and unconjugated fractions are determined.
- Renal excretion of hormones depends on kidney blood flow and rate of diuresis, and both of these are reduced during exercise.

The last condition makes it necessary not to use hormone concentrations in urine but rather hormone excretion expressed in excreted hormone per time unit (usually per hour). For the calculation of hormone excretion, the measured concentration must be multiplied by urine volume, excreted during the time studied, and divided into the time of urine collection. Another approach is to use the creatinine level in urine as the reference value for the evaluation of hormone concentration in urine. According to this approach, urine concentrations are expressed as a ratio between hormone and creatinine excretion. However, this approach cannot be appreciated in exercise studies because the exercise can alter the creatinine production and the rate of its excretion (see chapter 3, p. 40).

In clinics, the usual period for measuring hormone excretion is 24 h. In exercise studies, this period is not suitable because it includes the exercise and the recovery period. If there are opposite changes in hormone secretion during and after the exercise, the total 24-h sample may sum up the actual increase and decrease into an unchanged hormone excretion. However, if we try to get a urine sample only for the exercise time, the volume of the sample will be rather small, and a methodological error is caused by a low ratio between excreted urine and volume of urine retained in the bladder. Another problem is related to the possibility that produced hormones may not be excreted during exercise because of renal retention.

To assess the possible time delay in the arrival of a hormone (corticosteroids) into the urine at the beginning of exercise, experiments have been conducted in dogs without urinary bladders (ureters were brought on the body surface for urine collection). A rapid increase of 17-hydroxycorticoid excretion was found at the beginning of running. The highest level of excretion was established at the end of the first hour or at the beginning of the second hour after the start of the running. Hence, the effect of the exercise on the adrenocortical function is reflected in urinary excretion of corticosteroids if at least 2-h samples are used for urine collection (Viru 1975b). Because the 2-h sample collected during and after exercise may be small to avoid the aforementioned methodological error, 3-h samples would be better. When the time for urine collection is more than 3 h, another error is caused by the circadian rhythm of endocrine function or by the possibility of opposite changes during and after exercise.

The study on dogs without bladders indicated an undulating pattern of the arrival of corticosteroids into the urine. Undulating patterns were found before, during, and after the exercise, and it was synchronous to undulating changes in diuresis (Viru 1975b). This fact emphasizes the significance of renal function for the arrival of corticosteroids into urine. Human studies revealed significant correlations between diuresis rate and excretion of corticosteroids, including exercise and postexercise periods (Donath et al. 1969a; Viru 1975b). When the antidiuretic effect of exercise was avoided by voluntary intake of water (Israel 1969) or when diuretics were used (Wegner et al. 1965), the corticosteroid excretion increased in parallel with the rise in urine output.

The hormone assessment in urine has meaning for steroid hormones and catecholamines but not for protein hormones because the large molecule of protein hormones can be filtered into urine only in the case of increased permeability of the renal capillary membrane. Therefore, the appearance of protein hormones in urine should reflect the renal permeability but not the actual levels of hormone output and blood concentrations.

Sweat

At high rates of perspiration, hormones (mainly steroid hormones and catecholamines) may penetrate into the sweat. According to our results, during exercise the cortisol concentration in sweat remained lower than that in blood. The total amount of eliminated cortisol is not enough to explain decreases in urinary excretion of cortisol during prolonged exercise (Viru 1975b). Hormone assessment in sweat does not have any purpose for training monitoring.

Saliva

Several attempts have been made to assess steroid hormones in saliva to study exercise-induced alterations in concerned endocrine systems (Cook et al. 1987; Booth et al. 1989; Port 1991; Stupnicki and Obminski 1992). The possibility of frequent and easy specimen collection is considered an advantage of saliva samples. However, only free, unbound steroids can enter saliva. The reported results indicate that the steroid concentration in saliva is not affected by the usual variation of saliva production rate (Fergusson et al. 1980). However, this conclusion must be reinvestigated regarding situations related to strong inhibition of saliva production, such as vigorous emotional strain or pronounced dehydration. Both situations are possible in athletes. The disadvantages of saliva hormone assessment are that the mucosa in saliva would falsely increase the saliva steroid level and that a steroid-metabolizing enzyme has been found in the salivary gland (see Tremblay et al. 1995).

Skeletal Muscle Tissue

The measurement of hormone concentration in biopsy samples of muscle tissue has meaning when hormone receptors are assessed in the same samples. The last task requires an increase in the amount of muscle tissue specimen. Nevertheless, a source for methodological error remains for hormone receptor studies because of the possible cellular membrane damage in biopsy sampling.

Blood Sampling

Blood for hormone analysis may be obtained from veins, arteries, or capillaries. Venous blood is the specimen of choice. However, venous occlusion with a tourniquet will cause fluid and low–molecular weight compounds to pass through the capillary wall into the interstitial fluid. Small changes in the blood concentration of the substance occur if the tourniquet is used for less than a minute, but notable changes may be seen after 3 min. A 15% increase in protein-bound hormones has been seen after 3-min stasis. Therefore, a uniform procedure for blood collection must be used throughout an investigation, minimizing the effect of stasis (Trembley et al. 1995). In exercise testing, blood sampling through a venous catheter is advisable. In cases of venous puncture, vacuum tubes should be used.

Several possibilities of methodological error appear when capillary blood is collected. First, to get the necessary amount of plasma (usually 50 to 200 μl are required for determination of a hormone), the fingertip or earlobe prick has to be sufficient to get free blood outflow. Otherwise, additional pressure is necessary to get enough blood. When this request is satisfied, a question arises as to whether the pain caused is less in a fingertip prick than in a venous puncture. Furthermore, the additional pressure to obtain the required volume of capillary blood results in two possible effects influencing the hormone concentration in the sample. One is the effect of stasis, similar to the action of a tourniquet in venous sampling. The other is the possibility that interstitial fluid may dilute the blood sample. This influences the concentration of protein hormones and protein-bound hormones. An additional source of methodological error is the influence of hemolysis on hormone concentration. In several cases, free hemoglobin may influence hormone determination.

Lehmann and Keul (1985) showed that epinephrine, norepinephrine, and dopamine concentrations were higher in earlobe blood than in venous blood at rest and during graded exercise. They assumed that the difference was related to significant neuronal release of catecholamines into the earlobe prick blood.

In whole blood, the hormone concentrations are in the same range as in plasma when venous blood is analyzed. Equal hormone concentrations in plasma and in whole blood are explained by the equilibrium between hormone level in plasma and erythrocytes. However, a question remains as to whether the equilibrium persists in exercise. Moreover, it should be taken into consideration that erythrocytes metabolize steroid hormones.

Hormone Concentration in Blood

Conventional guidelines for applied and clinical exercise testing recommend that standardized testing conditions prevail. Accordingly, wherever possible, researchers should ensure that exercise testing is performed at 22° C and at a relative humidity of 60% or less at least 2 h after the individual has eaten, smoked, or ingested caffeine, at least 6 h after an individual has consumed any alcohol, and at least 6 h after any previous exercise.

Other conditions that must be taken into consideration are nutritional status (e.g., high- or low-carbohydrate diet; see Galbo 1983), emotional strain (e.g., Kreutz et al. 1972; Péquignot et al. 1979; Vaernes et al. 1982), sleep deprivation (Vanhelder and Radomski 1989), circadian and seasonal rhythm, and posture. Throughout the day, most blood hormones exhibit cyclic variation. A seasonal (circannual) variation in hormone levels has also been established. For example, plasma testosterone shows peak levels in summer and a nadir in winter (Smalls et al. 1976). Hormone responses to exercise are varied to some extent in various phases of the ovarian menstrual cycle (see chapter 5, p. 84 and pp. 107–109).

With the subject in an upright position, circulating blood volume of an individual is 600 to 700 ml less than with the individual in a recumbent position. When an individual goes from a supine to a standing position, water and filterable substances move from the intravascular space to the interstitial compartment, and a reduction of about 10% occurs in blood volume. Because only protein-free fluid passes through the capillaries to the interstitial compartment, blood levels of nonfilterable substances such as protein and protein-bound hormones will increase (approximately 10%). The normal decrease in blood volume from lying to standing is completed in 10 min, whereas the increase of blood volume from standing to lying is completed in approximately 30 min (Trembley et al. 1995). The effect of posture on blood hormone levels is also related to the necessity for the regulation of vascular tone: going from the supine to the standing position results in pronounced increases in levels of norepinephrine, aldosterone, angiotensin II, renin, and vasopressin.

The action of a previous training regime is an essential problem in athletes. Hormonal studies must not be performed the day after a training session of high volume or intensity or after competition. The best time for hormonal studies is in the morning after 1 or 2 days of rest. However, even in these cases, one should be sure that the recovery from previous strenuous training has been completed.

Specimen Storage

After sampling, a rapid separation of erythrocytes from the specimen is important because erythrocytes at room temperature can alter the plasma concentration of steroid hormones. Erythrocytes degrade estradiol to estrone and cortisol to cortisone and absorb testosterone (see Trembley et al. 1995). Plasma has an advantage over serum because it can be removed from the erythrocytes faster and subsequently be put into cold storage. For plasma, heparin is the preferred anticoagulant over ethylenediamine tetraacetic acid (EDTA) because it causes the least interference with most tests. EDTA may cause decreases in thyrotropin, lutropin, and estradiol levels in the range of 10% to 25%. Samples collected with EDTA yield high free testosterone results (Trembley et al. 1995). The use of a suitable anticoagulant is particularly important when protein hormones or catecholamines will be determined. In several cases, the preservation of these hormones requires EDTA.

Many hormones, particularly protein hormones, are thermally labile. Serum or plasma samples should be stored frozen at –20° or –40° C. Separation of plasma or serum should be done at 4° C. Winder and Yang (1987) found that once a blood sample is obtained and placed on ice, the catecholamines appear to be quite stable for periods up to 1 h after the collection. Nevertheless, they recommended centrifugation of blood samples at 4° C within 5 min of the time of collection. Then plasma may be transferred to small polyethylene tubes and frozen. Repeated freezing and thawing of serum, plasma, or urine samples should be avoided.

Hormone Analysis

The most valid method for hormone determination is radioimmunoassay (RIA). This method possesses high analytical sensitivity and specificity. To avoid the use of radioactive isotopes, fluorescence immunoassay, enzyme immunoassay, and chemiluminescence immunoassay are recommended. The latter is becoming the technique of choice in replacing RIA for most hormone determinations. In regard to determination of monoamines (epinephrine, norepinephrine, serotonin, etc.), the high-performance liquid chromatographic assay has proved to be valid.

Interpreting Results

When reliable results of hormonal responses have been obtained, their benefit for training monitoring depends on knowledge of the function of the determined hormone in the control of metabolism and interrelation of the function of measured hormone with the actions of other hormones. It is necessary to know the relations of the detected hormone level to actual secretion rate of the hormone by the gland. It is also necessary to know time characteristics and conditions for actualization of the metabolic effects of assessed hormone. Furthermore, researchers have always clarified what the determinants for the studied hormonal response were and whether any conditions modulate the action of the main determinants.

Private and Attending Regulation

Hormonal regulation is actualized on two levels: the level of production of signal molecules (synthesis and secretion of hormones by endocrine cells) and the level of reception of signal molecules (cellular receptor proteins, which specifically bind hormones on the cellular membrane, in the cytoplasm, or in the nucleus). Both levels are regulated according to metabolic, homeostatic, and adaptational requirements. Therefore, the hormone level in blood does not contain the entire information on the effects of hormonal regulation. On the other hand, the hormone level in blood is not a measure of hormone secretion because it actually expresses the ratio between hormone inflow into the blood (secretion by endocrine gland) and outflow from the blood into the tissues (depends on dynamic balance between bound and unbound fractions and on the intensity of hormone degradation in tissues).

Hormone concentration in the blood determines the tissue supply of hormones. The quantity of hormones arriving at the tissues is divided among various sites (figure 4.1). Most hormone molecules are bound by cellular proteins. A dynamic equilibrium exists between the free hormone content in the extracellular compartment and the hormone content bound by cellular proteins. A part of the arriving hormone content is bound by sites connecting the hormone with enzymes catalyzing its metabolic degradation. This represents a loss of hormone. Besides metabolic degradation, there are also biotransformations of hormones from more active forms into less active or inactive ones, or vice versa. The remainder of the active hormone content is divided into the fraction bound by specific cellular receptors of this hormone and the hormone content bound unspecifically by other proteins. The amount of hormone bound specifically by corresponding cellular receptors is the determinant of the hormone effect.

When the hormone inflow into tissues increases, more hormones can be specifically bound by their own receptors. On the other hand, when the number of hormone receptors increases, an enhanced portion of hormone can be bound specifically, despite the unchanged hormone content in the tissues. In this connection, it is valid to discriminate between "private regulation" and "attending regulation" (Viru 1991). The private regulation consists in the control of producing signal molecules (hormones). The attending regulation consists of:

- modulating influences on the number of receptors (the binding sites), on the affinity of receptor proteins to the hormone, and on the postreceptor metabolic processes;
- regulation of the metabolic situation in the cell, acting by way of other receptors on cellular metabolism and thus changing the actualization of private regulation;
- regulation of protein synthesis, acting on the synthesis of structure and enzyme proteins that contribute to the actualization of private and attending regulation (figure 4.2).

Other hormones may support or aggravate the actualization of the private action of a hormone. A possibility exists for the blockade of hormone action. It may take place either by way of competition for the binding with the receptor protein between the hormone and other similar compounds or by way of inhibiting the postreceptor metabolic processes. An example of the competition for binding with the receptor protein is the interaction between cortisol and testosterone in muscle tissue. When testosterone molecules occupy specific binding sites for cortisol, they exert an anticatabolic effect, reducing the action of cortisol on protein degradation. However, when cortisol occupies the receptors for testosterone, it exerts antianabolic action, reducing the induction of muscle protein synthesis by testosterone (Mayer and Rosen 1977). Another example is the competition for receptors between cortisol and progesterone. It has been found that glucocorticoids restore the working capacity in adrenalectomized rats, inducing synthesis of regulatory protein(s). However, when a large dose of

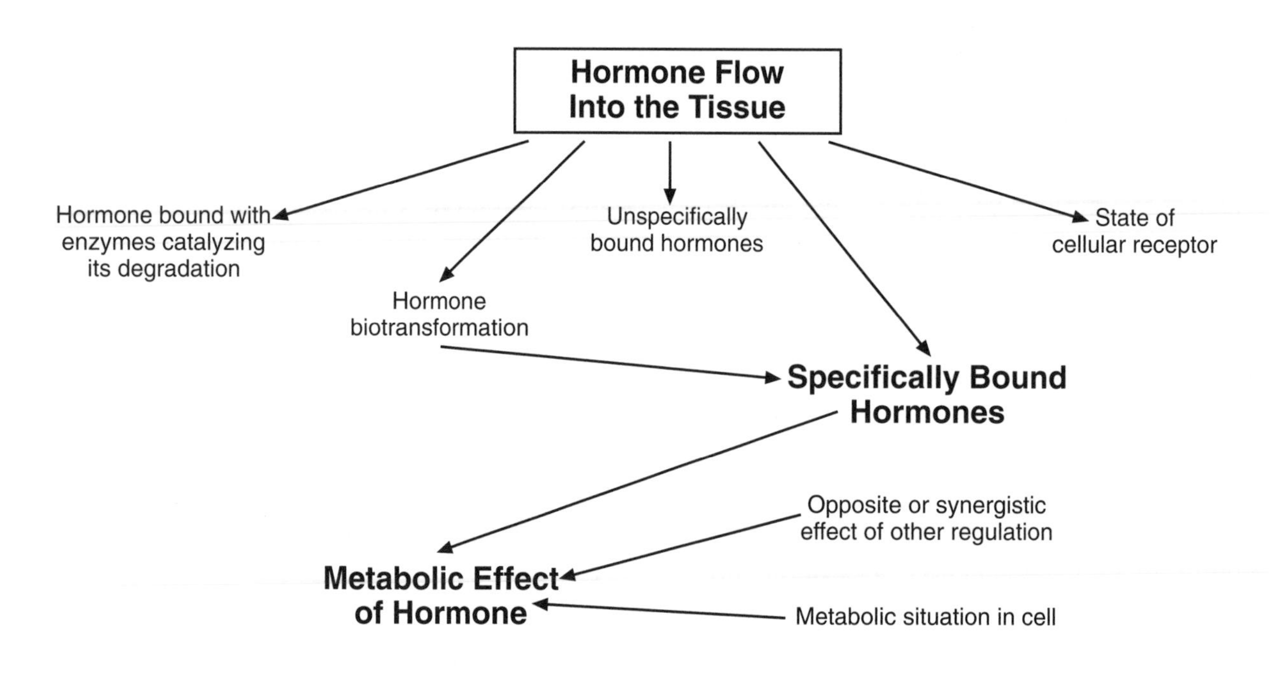

Figure 4.1 Distribution of a hormone in tissues and the formation of its metabolic effect.

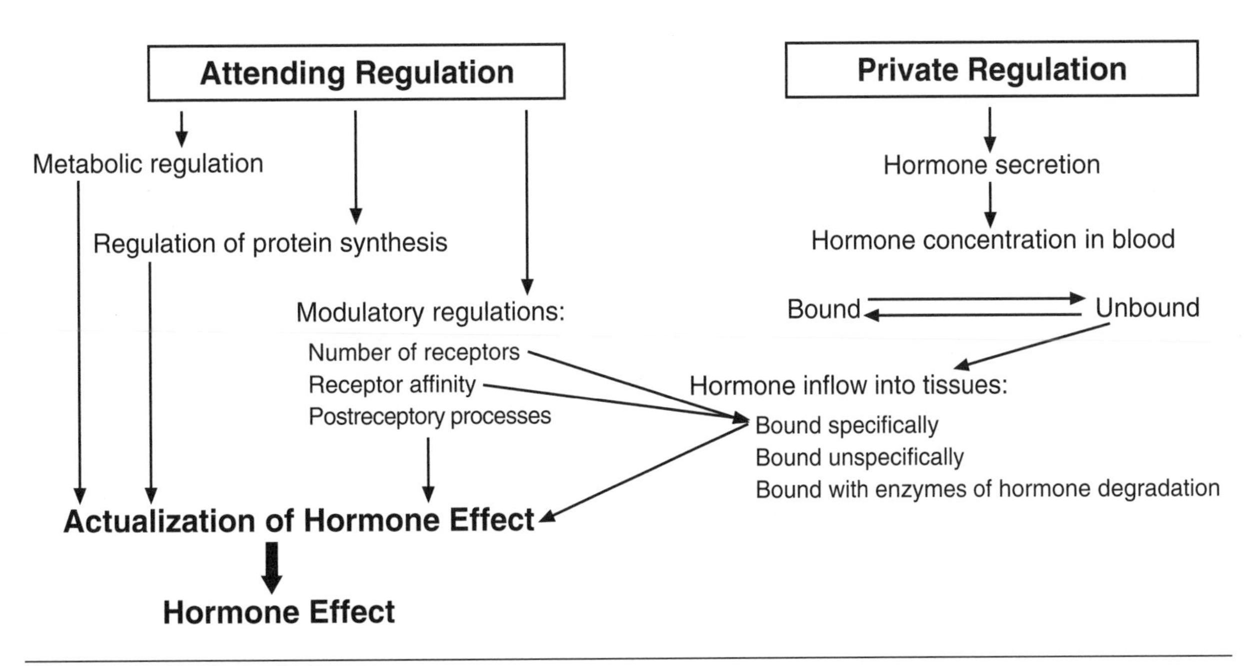

Figure 4.2 Two levels of regulation of hormone effects (private and attending regulation).

progesterone was administrated together with the glucocorticoid dexamethasone, the glucocorticoid effect on working capacity was inhibited, obviously because of the ability of progesterone to be bound to glucocorticoid receptors (Viru and Smirnova 1985).

Insulin blocks the action of lipolytic hormones (epinephrine, growth hormone, glucagon) on triglycerol hydrolysis in adipose cells. The action is related to insulin influence on postreceptor processes. Insulin activates cAMP-phosphodiesterase and thereby stimulates the degradation of cAMP. On the other hand, cAMP accumulation is the essential link in postreceptor processes in the action of lipolytic hormones. In contrast, cortisol potentiates the epinephrine metabolic effects by changes in intracellular Ca^{2+} shifts and by inhibition of cAMP-phosphodiesterase activity (Fain 1979).

Experiments on rats provide evidence that an exercise-induced rise in the alanine levels of blood plasma, slow-twitch fibers, and liver depends on glucocorticoid action on the activity of alanine-aminotransferase. Adrenalectomy excluded the rise in alanine levels and increased activity of the enzyme, but substitution therapy of adrenalectomized rats with glucocorticoids restored the exercise-induced changes (Viru et al. 1994). However, infusion of testosterone reversed the exercise effect on alanine levels (Guezennec et al. 1984). Consequently, when exercise is performed under conditions related to an elevated testosterone level in the blood, cortisol fails to increase alanine production.

In conclusion, in examining an increase in hormone concentration, we cannot conclude the existence of metabolic consequences if we do not know the interrelations with other hormones and the state of cellular hormone receptors. When the actual metabolic influence of a hormone in exercise has not checked out, the conclusions found only on changes at the level of a single hormone are simply speculations.

Determinants and Modulators of Hormonal Responses

Variability in hormonal responses frequently resulted in the combined influence of several factors. Misevaluation of hormonal responses is avoided if the determinants and modulators of hormone responses in exercise are taken into account and used for interpretation (figure 4.3).

The main determinants are exercise intensity and duration, adaptation of a person to the performed exercise, and homeostatic needs (Viru et al. 1996). It is possible to define the threshold intensity of exercise as the minimum intensity needed to evoke hormonal changes in the blood. The dependence of the magnitude of

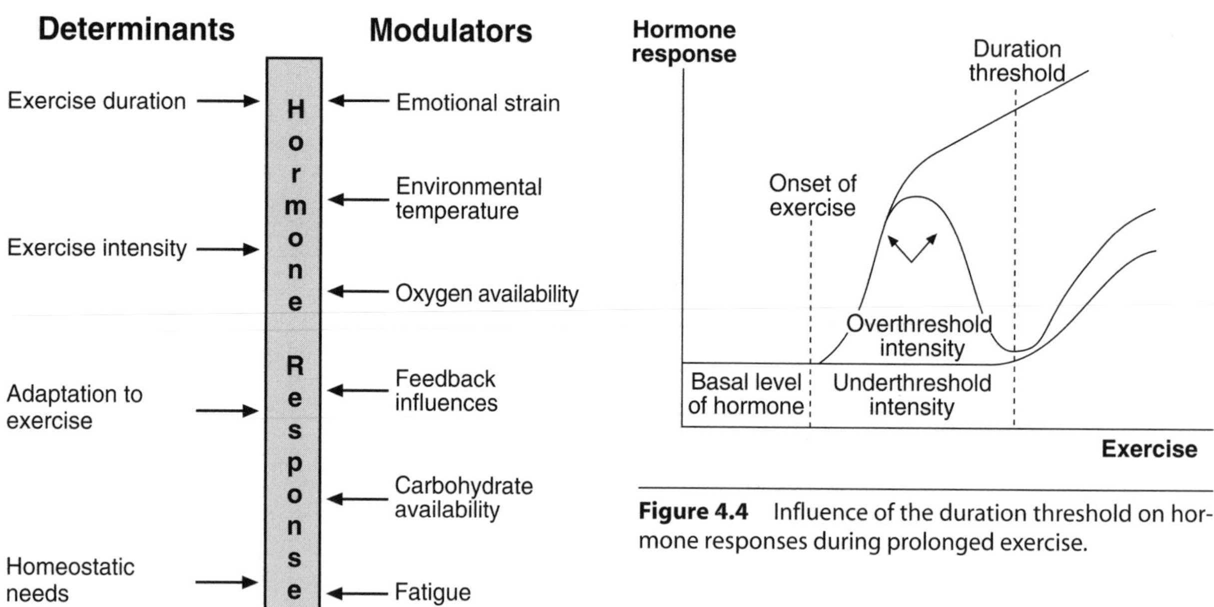

Figure 4.3 Determinants and modulators of hormone responses during prolonged exercise.

Figure 4.4 Influence of the duration threshold on hormone responses during prolonged exercise.

hormonal responses on exercise duration has also been demonstrated. Therefore, the threshold duration of exercise also comes into focus. It follows from the fact that exercise of underthreshold intensity may result in hormonal responses when a given amount of work has been done. At intensities greater than the threshold, the threshold duration is expressed by a further increase of hormonal response or by secondary activation of the endocrine system (figure 4.4). Systematic adaptation to exercise (training) induces an increase in threshold intensity in terms of power output (thereby the training reduces or totally eliminates previously observed hormonal responses in submaximal exercise) and an improved functional capacity of endocrine systems, making possible pronounced hormonal responses in extreme exercises (Viru 1985b, 1995). The significance of exercise intensity and fitness appears first in activities of endocrine systems responsible for rapid mobilization of energy reserves and protein resources. The activities of hormones controlling the water-electrolyte balance depend first on shifts in this balance.

The actualization of the effect of the main determinants is modulated by several conditions (figure 4.3). The most striking among them are the influence of emotional state, environmental conditions, diet (availability of carbohydrates), and biorhythms. The modulation of hormonal responses by fatigue is also possible (see Viru et al. 1996). To avoid circadian differences, hormone measurements should be performed at the same time of day. If this is not possible, it should be taken into consideration that circadian rhythm influences the exercise-induced hormone responses (Scheen et al. 1998).

Triggering and Controlling Hormone Responses

Analysis of the dynamics of hormone levels in the blood during exercise enables one to discriminate fast responses, responses of a modest rate, and responses with a lag period. Fast responses are characterized by a rapid increase in the concentration of hormones in the blood plasma within the first few minutes of exercise. Responses of a modest rate are characterized by a gradual increase in the hormone concentration. The increase may continue to the end of exercise. In other cases, the gradual increase during the first period of exercise is followed by leveling off on a constant value. An initial lag period (about 5 to 30 min) preceding the hormone response has been described for some hormones. It has been suggested that two types of mechanisms activate the endocrine function at the beginning of exercise. One of them is responsible for rapid activation, the other for delayed activation. The mechanism of rapid activation has to be connected with the functions of the nervous centers and a high rate of transfer of nervous influences to endocrine glands either with the aid of autonomic nerves or through the secretion of hypothalamic neurohormones. Not only fast responses but also responses of a modest rate require the contribution of the mechanism of rapid activation at least within the first minutes of exercise. If the duration of exercises is more than a couple of minutes, the magnitude of hormone change depends on the mechanism of delayed activation. This mechanism should be a cumulative one and ensure the correspondence of hormone levels with the actual needs for metabolic alterations. The responses, which occur after the lag period, are probably caused by the lack of activity of the mechanism of rapid activation (Viru 1983, 1995).

The importance of nervous discharge from cerebral motor centers has been shown in experiments in which tubocurarine was used. This compound in the used dose (a bolus of 0.015 mg/kg α-tubocurarine) causes a partial peripheral neuromuscular blockade. As a result, it weakens the skeletal muscles. Therefore, a stronger voluntary effort is necessary to produce a certain work output compared with normal conditions. The increased voluntary effort was confirmed by the higher rate of perceived exertion in this experiment (Galbo et al. 1987). The "stronger" central motor command is associated with exaggerated catecholamine, growth hormone, and corticotropin responses during exercise compared with the exercise performed at a similar level of oxygen uptake without neuromuscular blockade (Galbo et al. 1987; Kjaer et al. 1987; Kjaer 1992).

Galbo (1983) assumes that during continuous exercise, hormone responses are modulated by impulses from receptors sensing temperature, intravascular volume, oxygen tension, and glucose availability. This way the mechanism of delayed activation is stimulated. This opinion has been verified by the results demonstrating that an essential stimulus for the activation of endocrine functions is created by feedback of nervous impulses from receptors located in skeletal muscles. Proprioceptors sensing the muscle tension and metaboreceptors reacting to metabolite accumulation are significant. The use of small doses of epidural anesthesia to block the thin sensory afferent fibers (mostly from metaboreceptors) and leave almost intact the thicker efferent fibers and subsequently motor function, the essential role of nervous feedback from muscles, was shown for corticotropin and β-endorphin responses but not for insulin, glucagon, and catecholamine responses (Kjaer et al. 1989). The lack of influence of a charge from muscle receptors on insulin and glucagon and partly on catecholamine response can be explained by the great significance of glucostatin regulation for exercise-induced responses of these hormones. In regard to catecholamines, feed-forward nervous action may be highly significant because catecholamines are responsible for the mobilization of body resources. Therefore, the influences from higher nervous centers may be decisive without additional aid from the feedback charge.

Examples of Misevaluating Results

In several cases, an increased cortisol level has been considered to be information on overall intensification of protein degradation, a sign of general catabolism. However, several studies provided evidence that muscular activity inhibits the catabolic action of glucocorticoids (Hickson and Davis 1981; Seene and Viru 1982). Therefore, the link between increased blood level of cortisol and activation of catabolic processes is not the same in exercise as in resting conditions. At the same time, we cannot say that cortisol completely loses its catabolic action during exercise. Obviously, we are close to the truth if we assume that during exercise the catabolic influence of cortisol is limited and depends on the interrelation with the action of several other regulators. In fact, during exercise the rate of protein degradation is mainly elevated in less active muscles (Varrik et al. 1992).

Growth hormone is known to stimulate the hydrolysis of triglycerol in adipose cells. However, experiments on isolated adipose cells indicated that the actualization of the growth hormone effect requires at least 2 h (Fain et al. 1965). Therefore, when anyone establishes the exercise-induced increase in growth hormone level, he or she can hope to have a growth hormone effect 2 h later than the onset of increased hormonal level. Accordingly, during 2-h cycling, the administered exogenous human growth hormone caused a more pronounced increase in the blood level of the hormone but no increase in free fatty acid concentration in blood (figure 4.5) (Toode et al. 1993). In conclusion, suggestions about the metabolic effects of a hormone must be founded on the knowledge of time characteristics in actualization of the hormonal influence on the metabolic process. Even if a correlation is found between hormonal response and metabolic change during exercise, the causal relationship can be suggested only if time characteristics make the actualization of the hormone effect real.

In some cases the correlation between hormone level and performance or metabolic characteristics is founded on earlier hormone effects on muscle tissue. It was found that the basal testosterone level in serum significantly correlates with the performance in countermovement jumps and the 30-m dash. This fact was interpreted by the causal relationship between the development of fast-twitch fibers and individual differences in blood testosterone concentration (Bosco et al. 1996). The effect of testosterone

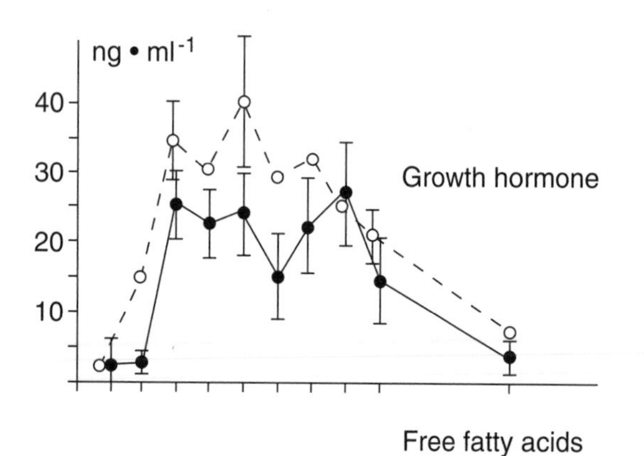

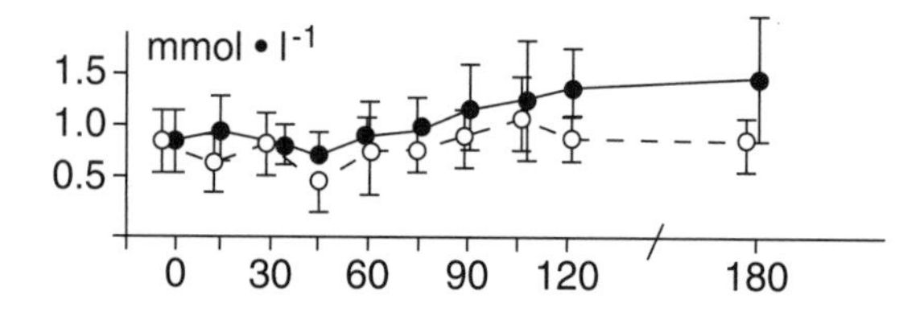

Figure 4.5 Blood levels of growth hormone (hGH) and free fatty acids (FFA) during 2h of cycling. The solid lines represent blood levels after saline injections, and the dotted lines represent levels after hGH injections.

Reprinted from K. Toode et al. 1993.

on fast-twitch fiber development has been evidenced in pubescent animals (Dux et al. 1982). Accordingly, it is possible to believe that a certain genotypic (or phenotypic) peculiarity exists that is characterized by an elevated testosterone level. The latter might promote the development of fast-twitch fibers in puberty.

In another situation, it has been shown that an increase of hormone concentration is not ultimate evidence of the augmented secretion of the hormone. During exercise, the degradation rate of testosterone (Sutton et al. 1978) and estrogens (Keiser et al. 1980) decreases. It obviously contributes to the increase of the blood level of these hormones. Another factor that may increase hormone level without increased secretion is the decrease of plasma volume because of extravasation of blood plasma. A 1-min intensive exercise bout may decrease the plasma volume by 15% to 20% (Sejersted et al. 1986). Therefore, it is desirable to calculate plasma volume changes by hemoglobin concentration and hematocrit value in hormonal studies. Increased hormone concentration also may appear without change in secretion because of the "washout" of hormones from the gland caused by an increased rate of blood flow. In prolonged exercise, dehydration and rehydration may both influence hormone levels. However, when blood concentration of hormones increases, regardless of the mechanism, increased interaction with the receptor is possible because the latter depends on the actual concentration of hormone in the extracellular compartment.

Each hormone response has its own dynamics. If we measure the hormone level only once during or after exercise, we may not obtain the actual picture about the response. Moreover, if the exercise does not require the mobilization of an endocrine function close to its maximal possibilities, individual variability may appear in a pattern of hormone responses, as has been described for cortisol (figure 4.6) (Viru et al. 1992) and β-endorphin (figure 4.7) (Viru et al. 1990).

The activity of endocrine systems is controlled by feedback inhibition. Therefore, the initial level of hormone may suppress or promote the hormone responses. Accordingly, from the augmented preexercise level, the blood cortisol concentration usually drops or does not

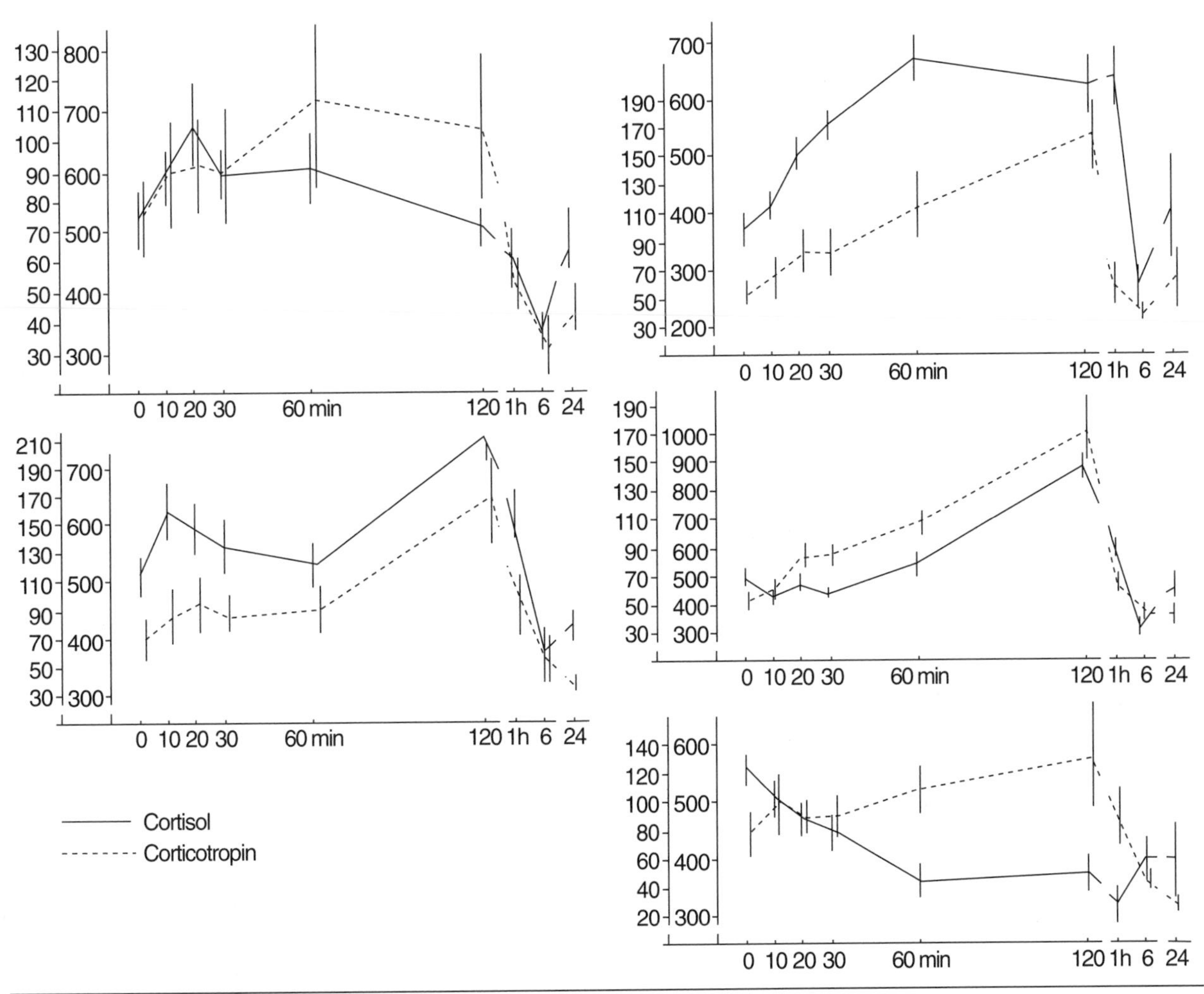

Figure 4.6 Five variants of blood cortisol (solid lines) (mmol/L) and corticotropin (dotted lines) (pg/l) changes during a 2-h cycle ergometer exercise.

Reprinted from A. Viru et al. 1992.

elicit any change during exercise (Few et al. 1975; Brandenberger et al. 1982).

Many hormones are secreted into the blood in an episodic manner. Secretion bursts may be separated by rest periods, with a duration from 5 to 30 min (or even more). This peculiarity of endocrine functions has to be taken into consideration in evaluating basal hormone levels. Frequent blood sampling is the best to characterize the actual dynamics (secretory bursts and decreased hormonal levels between them).

Summary

The most valuable information can be obtained by assessing hormones in venous blood. Several possibilities of methodological error appear in earlobe or fingertip prick samplings. However, most of these sources of error may be minimized by carefully following the methodological recommendations. In blood sampling, the effects of stasis (action of tourniquet in venous sampling or application of additional pressure in sampling capillary blood) and blood hemolysis have to be avoided. Blood sampling in field conditions may be substituted with assessments of hormones in saliva and urine. However, several limitations for these approaches exist. In saliva, there are only unbound fractions of hormones. Therefore, the hormone concentrations in blood and saliva are different. The disadvantages of saliva hormone assessment arise because of hormone binding by mucosa and steroid-metabolizing enzymes in the salivary glands.

When one is determining hormones in urine, it is necessary to separate native hormones and

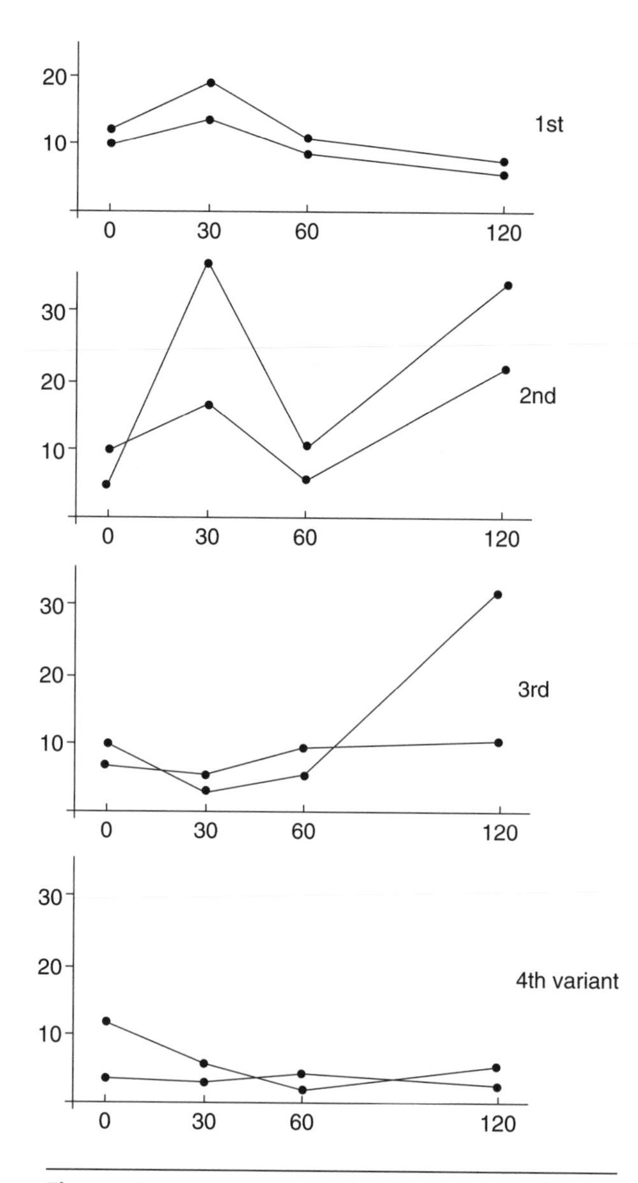

Figure 4.7 Four variants of blood β-endorphin changes (pmol/L) during a 2-h cycle ergometer exercise.

Reprinted from A. Viru et al. 1990

their metabolites. Attention should also be paid to the effect of altered kidney blood flow and rate of diuresis. Because exercise induces pronounced changes in the rate of diuresis, it is meaningless to take the hormone concentration in blood into consideration. Instead, hormone excretion has to be calculated expressed as excreted hormone amount per time unit (usually per hour). In exercise studies, the optimal period of urine collection is 3 h.

Assessment of hormone in muscle tissue with the aid of biopsy sampling or in sweat does not make sense in training monitoring.

The most valid methods for hormone determination are radioimmunoassay, chemiluminescence assay, and, in the case of catecholamines, high-performance liquid chromatographic assay.

When interpreting results obtained, one has to consider that the exercise intensity and duration, adaptation to the performed exercise, and homeostatic needs are the main determinants of hormonal responses. The effect of the main determinants may be modulated by conditions of external environment and nutrition and by emotional strain, fatigue, biorhythms, and initial hormone level. It is also essential to take into consideration that no strict relationship exists between blood hormone level and secretion rate of the hormone by the endocrine gland and the metabolic effect exerted by the hormone.

CHAPTER

5

Hormones As Tools for Training Monitoring

Each hormone has certain functions in metabolic control. Metabolic control doesn't depend solely on hormone level but also on certain conditions for hormone reception in cells and on collaboration with other metabolic regulators, including other hormones. Knowing an effect of a hormone does not guarantee knowledge about the actual role of the hormone in metabolic control. Each hormone has its own pattern of exercise-induced responses. The time of sampling must be synchronized with this pattern to get complete information. One must consider these options to effectively use hormones in training monitoring.

Sympathoadrenal System

The sympathoadrenal system is part of the autonomic nervous system. The higher sympathetic centers are located in the hypothalamus. They govern the activity of the lower sympathetic nervous structures, which send nerve charges to various internal organs, including the adrenal medulla. N. splanchnicus mediates the stimulation of the adrenal medulla. In response, epinephrine and norepinephrine are released from endoplasmic granules of the adrenomedullary cells. Because both catecholamines are stored in granules in the definite form, the secretion response of epinephrine and norepinephrine is rapid. This way, epinephrine and norepinephrine levels in the blood increase rapidly at the beginning of exercise.

Sympathetic nerve endings also produce norepinephrine, which operates as a neurotransmitter (mediator) in the synapses. From the sympathetic nerve endings located in the wall of blood vessels, norepinephrine releases into the bloodstream. This amount of norepinephrine is greater than that from the adrenal medulla. Therefore, the norepinephrine level is greater than that of epinephrine. In the brain, norepinephrine acts as a neurotransmitter of adrenergic neurons. Exercise also causes changes in the brain norepinephrine content.

Epinephrine and norepinephrine are products of the same biosynthetic pathways. They are produced throughout several intermediate compounds. In sympathetic nerve endings, the formation of norepinephrine is the end of the biosynthetic process. In the adrenal medulla, a part of norepinephrine releases into the blood. However, the second part of the norepinephrine molecules is further methylated into the epinephrine.

Catecholamines in Metabolic Control

The metabolic and functional effects of epinephrine and norepinephrine are similar but not the same. According to widespread opinion, the effects of epinephrine are stronger than those of norepinephrine in metabolic control, whereas regarding vasomotor action, norepinephrine dominates. In several cases, their effects are qualitatively different. The reason is that different adrenoreceptors mediate the effects of two catecholamines. Norepinephrine excites mainly α-receptors, but it excites the β-receptors to a slight extent as well. Epinephrine excites both types of receptors about equally. The relative effects of norepinephrine and epinephrine on different effector organs are determined by the distribution of various types of receptors on cellular membrane in tissues. For example, epinephrine because of its greater influence on β-receptors, has a greater effect on cardiac stimulation than norepinephrine. At the same time, epinephrine causes only weak constriction of blood vessels of the muscles compared with the much stronger constriction caused by norepinephrine through α-receptors. In this way, norepinephrine greatly increases the total peripheral resistance and, thereby, greatly elevates arterial pressure, whereas epinephrine raises the arterial pressure to a lesser extent but increases the cardiac output considerably. Most of the metabolic effects of epinephrine are, indeed, 5 to 10 times as great as the effects of norepinephrine. However, in metabolic control, more complicated relationships may exist. Earlier, it was mentioned (chapter 3, p. 53) that in humans the α_2-adrenergic inhibitory mechanism modulates lipolysis at rest, whereas the β_1-adrenergic stimulatory mechanism is predominant during exercise. Stimulation of muscle glycogenolysis appears after the administration of epinephrine or norepinephrine. These effects were prevented by the blockade of β-adrenergic but not α-adrenergic receptors (Nesher et al. 1980). Furthermore, Richter and coauthors (1983) showed that β-receptor blockade abolished epinephrine action on muscle glycogenolysis but not on the increase of oxygen uptake by muscle tissue. α-Receptor blockade eliminated the action on oxygen uptake and only lessened the epinephrine-induced increase in glycogenolysis.

Sympathoadrenal System in Exercise

At the beginning of an exercise, skeletal muscles are involved in working under the influence of the "central motor command." This nerve charge originates from the motor zone of the brain cortex and reaches the spinal motoneurons by the pyramidal tract. The aforementioned experiments on partially curarized persons indicated that the central motor command is also involved in triggering the fast activation of the sympathoadrenal system (chapter 4, p. 69). Obviously, collaterals from the nervous pathway of the central motor command activate the centers of the autonomic nervous system in the hypothalamus. The involvement of hypothalamic centers has been demonstrated in rat experiments. Scheurink et al. (1990) interfered with the activity of hypothalamic adrenoreceptors with the aid of administration of adrenoblockers through permanent bilateral cannulas into either the ventromedial or lateral hypothalamus just before the onset of swimming. Results showed that hypothalamic α- and β-adrenoreceptors are involved in the control of the increase of the epinephrine and norepinephrine concentration in blood.

DiCarlo et al. (1996) demonstrated a significant increase in lumbar sympathetic nerve activity within the first 25 s of treadmill running despite very low intensity of the exercises (rats ran on treadmill with a velocity of only 6 m/min).

In humans, an increased concentration of catecholamines is detectable in blood within the first 30 s after the start of an exercise. Macdonald and coauthors (1983) tested the action of 30-s pedaling at maximal rate against a 14.7 N load. The anaerobic ergometer test was preceded by warm-up exercises, which modestly increased catecholamine concentrations. The 30-s anaerobic test caused a further 4.6-fold increase in norepinephrine level and 6.5-fold increase in epinephrine level (figure 5.1). Although the first postexercise blood sample was obtained between 30 and 90 s after the end of an exercise, the hormone concentrations showed a high rate of catecholamine response. In the experiment of Kraemer and coauthors (1991a), the delay in sampling was avoided by the use of a Teflon cannula.

Immediately after cycle ergometer exercise, a small, but significant, increase of norepinephrine concentration was found when 100% of maximal leg power was used and persons were able to work only for 6 s on average. The comparison of responses with exercises performed

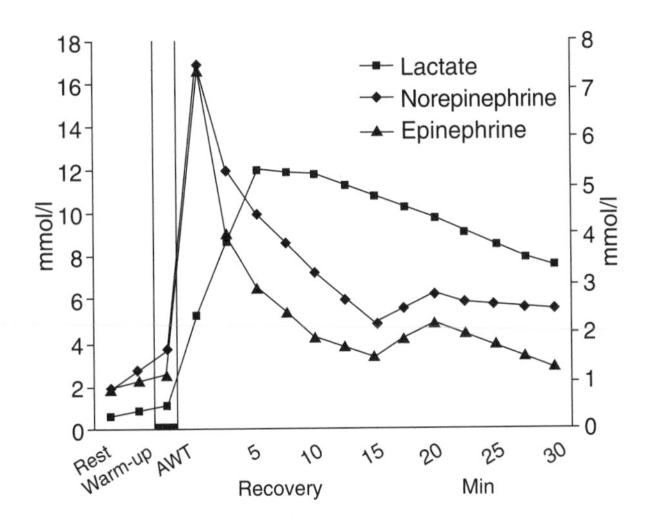

Figure 5.1 Postexercise dynamics of lactate, norepinephrine, and epinephrine levels in blood after 30-s sprint cycling on a cycle ergometer. AWT = anaerobic work test. Modified from Macdonald et al. 1983.

at 100%, 73%, 55%, and 36% of maximal leg power (average durations 6 s, 10 s, 47 s, and 3 min 31 s, respectively) showed that the response increased with prolongation of exercise. Approximately a 2.5-fold increase of epinephrine concentration was found immediately after the exercise of 55% power, which lasted 47 s. A similar significant response of epinephrine was also revealed after the exercise at 36% of maximal leg power. Epinephrine response was triggered also by 6-s exercise at 100% of leg power, but the elevated hormone level appeared only 15 min after the exercise. Even a short-term isometric effort (handgrip exercise) may be enough to evoke blood catecholamine response (Watson et al. 1980).

The results of Brooks et al. (1988) showed a high response rate of catecholamines after 30 s of dash (running distance average 167 m): a sixfold rise in norepinephrine and a sevenfold rise in epinephrine levels. Plasma volume was reduced only by 11%, indicating that the increased catecholamine concentrations were not caused by extravasation of plasma. Several other articles confirm the fast catecholamine responses in intensive exercises (Péquignot et al. 1979; Kindermann et al. 1982; Fentem et al. 1985; Schwarz and Kindermann 1990).

Exercise Intensity

If the exercise duration is at least 2 min, catecholamine responses are clearly dependent on exercise intensity. This dependence is expressed by the modest increase of norepinephrine concentration up to exercise relative intensity of 60% to 70% $\dot{V}O_2$max. The epinephrine increase is usually insignificant or not present at all. When exercise intensity is greater than this threshold value, a sharp rise appears in the concentration of both catecholamines with further increase of exercise intensity. The most pronounced responses were revealed in supramaximal exercises (Vendsalu 1960; Häggendal et al. 1970; Galbo et al. 1975; Lehmann et al. 1981; Jezova et al. 1985; Kraemer et al. 1990a). Lehmann et al. 1981 found that the threshold for catecholamine responses is associated with exercise intensity, causing a pronounced increase of blood lactate level (more than 4 mmol/L). This fact was confirmed by Schwarz and Kindermann (1990). When exercise duration was 45 min (or the highest possible duration) and intensities were 85%, 95%, 100%, and 105% of individual anaerobic threshold, epinephrine and norepinephrine concentrations increased during all exercises.

Responses of norepinephrine were modest, and those of epinephrine were not significant when exercise intensity was less than or equal to the intensity at the individual anaerobic threshold. In these exercises, the steady state of lactate was observed. At exercise intensity greater than the individual anaerobic threshold, lactate, norepinephrine, and epinephrine concentrations increased continuously up to the end of the exercises. Concentrations of all three compounds were significantly greater in all stages of exercise than at lower exercise intensities (Urhausen et al. 1994). In conclusion, the pronounced catecholamine responses are common for aerobic-anaerobic and anaerobic exercises.

When exercises were performed with the arms, the plasma catecholamine response was less than in one-legged exercises at all submaximal and supramaximal exercise intensities. The catecholamine responses were the highest in two-legged exercises. However, in all three variants of exercise, the responses appeared at intensities approximately greater than 60% $\dot{V}O_2$max (Davies et al. 1974).

Exercise Duration

During prolonged aerobic exercises, the significance of a factor related to the amount of work done appears. It is reflected by a gradual increase of norepinephrine and epinephrine levels (Galbo et al. 1975; Schnabel et al. 1982; Friedmann and Kindermann 1989). According to the results of Galbo and coauthors (1977a), during 3 h of exercise, the rise of catecholamine levels is most pronounced within the last hour, particularly for epinephrine, when the exhaustion point is reached.

Strength Exercises

Strength exercises evoked catecholamine responses even when performed with a limited amount of muscles and motor units (e.g., the catecholamine response was already established after the first minute of handgrip at 30% of maximal voluntary strength [Sanches et al. 1980]). Pronounced activation of the sympathoadrenal system has been found after a strength-training session (Guezennec et al. 1986; Kraemer et al. 1987).

Training Effects on Catecholamine Responses in Exercise

The sympathoadrenal response depends on the level of adaptation to exercise. In more fit persons (Vendsalu 1960; Bloom et al. 1976; Lehmann et al. 1981) and in the same individuals after a training period (Hartley et al. 1972a, 1972b; Cousineau et al. 1977; Winder et al. 1978, 1979; Hickson et al. 1979; Péronnet et al. 1981), the increases of norepinephrine and epinephrine are less pronounced or do not appear at all during the same level of submaximal exercise. The reduced responses appear in short-term and prolonged exercises (Hartley et al. 1972a, 1972b). However, the effect of strength training on exercise-induced responses seems to be less pronounced as has been shown by a comparison of catecholamine changes in cyclists and weightlifters during incremental cycle exercise (Lehmann et al. 1984). However, the correspondence of test exercises to training exercises may have significance. The catecholamine response is lower after training when trained muscles are used in the test exercise (Davies et al. 1974).

Winder et al. (1978) established that the training-induced decrement in the plasma catecholamine response to submaximal exercise appears after the first week of endurance training. A major portion of the change was seen after the first 3 weeks of a 7-week training period. Thus, this is an early adaptation in training. In sports training, such an adaptation should take place repeatedly because the workloads of training exercises increase by steps or continuously.

In every step of incremental exercises, the higher level of blood epinephrine and norepinephrine appeared when persons were less trained (Vendsalu 1960) and, therefore, exhibited a higher percentage of oxygen uptake from the maximum (Lehmann et al. 1981). Obviously, the training-induced reduction of the catecholamine response in submaximal exercises is related to the increase of the threshold intensity in terms of mechanical power output. Because the maximal oxygen uptake also increases, the relative intensity of exercise measured as a percentage of $\dot{V}O_2max$ at threshold intensity may remain the same after the training period as it was before the training. The results of Winder et al. (1978) showed that before the training a test exercise (1483 ± 83 kpm/min) corresponded to 95% $\dot{V}O_2max$ and caused a pronounced increase in blood plasma level of epinephrine. During 7 weeks of endurance training, $\dot{V}O_2max$ increased by 22% and therefore the same test exercise corresponded to 77% of $\dot{V}O_2max$ after the training period. The postexercise lactate level was 3.8 instead of 7.1 mmol/L before the training. In posttraining testing, the increases of epinephrine and norepinephrine were significantly lower than before training. However, after training, the intensity of the test exercise was increased (the relative intensity corresponded to 100% $\dot{V}O_2max$), and the catecholamine responses were particularly pronounced; the postexercise level of epinephrine was 2.3-fold and of norepinephrine 1.4-fold greater than after submaximal test exercises before training. Accordingly, Péronnet et al. (1981) found that when the test exercise after the training corresponded with the same relative intensity (% of $\dot{V}O_2max$) as before the training period, the blood norepinephrine response remained approximately the same as before training. However, in another study, Winder et al. (1979) noticed a decrease in the epinephrine response despite the fact that the relative intensity of the test exercise was the same before and after training. Therefore, one must not exclude the alteration of threshold intensity in relation to $\dot{V}O_2max$.

Another important result provided by incremental exercise tests was the more pronounced increase of epinephrine and norepinephrine concentration at supramaximal exercise levels in trained persons compared with untrained ones. This fact was first established by Vendsalu in 1960. Although he used a fluorometric method for catecholamine determination, which is not a highly specific procedure, in essence it showed that training may increase the opportunity for catecholamine release in supramaximal exercises. Accordingly, Häggendal et al. (1970) found that a well-trained person was able to reach a higher level of power output and exhibit the highest norepinephrine response compared with an untrained person during an incremental exercise test.

The training-induced increase of the possibility for mobilizing sympathoadrenal activity in strenuous exercises has been confirmed in several studies. In middle-distance runners, the epinephrine and norepinephrine concentrations became greater after exhaustive running as the training progressed throughout a 5-month period (Banister et al. 1980). A comparison of eight athletes and eight sedentary persons showed that an exhaustive treadmill run (7 min at 60%, 3 min at 100%, and 2 min at 110% $\dot{V}O_2max$) caused approximately twofold greater epinephrine and 43% greater (not significant) norepinephrine responses in athletes (Kjaer et al. 1986). Bullen et al. (1984) tested the catecholamine responses during a 60-min cycle exercise in women (intensity increased gradually from 70% to 85% $\dot{V}O_2max$). The 2-month endurance training period increased postexercise levels of epinephrine and norepinephrine significantly. Näveri et al. (1985a) found that sprinters exhibited a ninefold increase of epinephrine and norepinephrine levels after running 3 × 300 m. The rate of epinephrine clearance did not explain the greater epinephrine response in athletes than in sedentary persons during heavy exercise (Kjaer et al. 1985).

The training-induced increase in the capacity to secrete epinephrine has been confirmed by more pronounced epinephrine responses in trained athletes than in sedentary persons under the influence of glucagon (Kjaer and Galbo 1988), insulin-induced hypoglycemia (Kjaer et al. 1984), hyperbaric hypoxia (Kjaer and Galbo 1988), or exercise in hypoxia (Kjaer et al. 1988). Zouhal et al. (1998) found a more pronounced increase of blood plasma epinephrine concentration in sprinters than in untrained persons, whereas no difference was found in norepinephrine levels. The authors explained the result by a higher responsiveness of the adrenal medulla of sprinters to the same sympathetic input.

An 8-week period of leg-strength training did not increase the catecholamine response in maximal cycle exercise (Péronnet et al. 1986).

The background for increased capacity to secrete catecholamines consists of hypertrophy of the adrenal medulla (Hort 1951; Eränko et al. 1962), augmented storage of epinephrine and norepinephrine (Gorokhov 1969; Östman and Sjöstrand 1971; Matlina et al. 1976; Matlina 1984), and increased activity of enzymes contributing to catecholamine biosynthesis (Bernet and Denimal 1974; Matlina et al. 1976; Parizková and Kvetnansky 1980) found in trained rats. The increase of medulla in the adrenal glands predominates in training with short-term, high-speed dashes. In endurance training with prolonged aerobic exercises, the cortex and medulla increased (Viru and Seene 1985).

Modulating Factors

Emotional strain is a highly effective factor acting on the sympathoadrenal response. Increased blood catecholamine levels appeared before a test exercise (Mason et al. 1973a) and before a marathon race (Maron et al. 1975) as an anticipation phenomenon. Exercise-induced epinephrine and norepinephrine responses are highly pronounced in persons with high trait-anxiety (Péronnet et al. 1982) or increased emotionality (Péquignot et al. 1979).

The concentration of epinephrine and, less conspicuously, of norepinephrine is inversely related to the glucose level in the plasma during exercise (Galbo 1983). The administration of glucose reduces the epinephrine response (Galbo et al. 1977a). Differences in catecholamine responses appear in relation to carbohydrate-rich versus poor diets (Jansson 1980; Galbo et al. 1979a; Nazar 1981; Jansson et al. 1982). When the subjects fasted for 15 or 79 h, the plasma catecholamine responses were enhanced in exercise (Péquignot et al. 1980; Galbo et al. 1981a).

In acute hypoxia (14% O_2 in inhaled air), the norepinephrine concentration increased significantly at a workload that did not elicit any increase in normoxia (Galbo 1983).

Conversely, in hyperbaric conditions (1.3 ATA) (Fagraeus et al. 1973) or breathing a gas mixture containing 100% O_2 (Hesse et al. 1981), catecholamine concentrations were lower during exercise than when the same exercise was performed in normoxia.

During swimming in "hot" (33° C) and in "cold" (21° C) water, epinephrine and norepinephrine concentrations were higher than during swimming in the thermally neutral (27° C) water (Galbo et al. 1979b). A special experiment showed that cold skin or cold core temperatures might increase the norepinephrine response during exercise (Bergh et al. 1979). Catecholamine levels in plasma are higher at exercise after sodium depletion versus a sodium-repleted condition (Fagard et al. 1978).

In an upright position, the increase of blood norepinephrine levels is more pronounced than during the comparable exercise in the supine position (Galbo 1983). When exercise was performed with the hands, one leg, or two legs at the same percent of $\dot{V}O_2max$, the catecholamine response was inversely related to the effective muscle (plus bone) volume used to perform the exercise (Davies et al. 1974).

Pituitary-Adrenocortical System

The pituitary-adrenocortical system consists of corticotropes that are located in the anterior lobe of the pituitary gland (adenohypophysis) and secrete corticotropin (adrenocorticotrophic hormone, ACTH) and of cells of the zona fasciculata of the adrenal cortex, producing glucocorticoids (cortisol and corticosterone). In a lesser amount, glucocorticoids are produced also by the zona reticularis, the deep layer of adrenocortical cells. The system makes a whole; corticotropin controls the development of the adrenal cortex (mainly the zona fasciculata) and stimulates the production of glucocorticoids. Glucocorticoids are not stored in the adrenal glands; their secretion rate is related to their biosynthesis rate. In turn, the stimulation of glucocorticoid production by corticotropin is controlled with the aid of feedback inhibition exerted by increased blood levels of glucocorticoids.

Secretion of corticotropin is stimulated by a hypothalamic neurosecretion compound called either corticoliberin or corticotropin-releasing factor (CRF). Experimental studies indicate that corticotropin secretion may also be stimulated by vasopressin or by direct influence of the sympathetic nerves. The production of corticoliberin is controlled by the collaboration of hypothalamic and other neurons, which may diverge by their neurochemical specificity. Important regulator neurons for corticoliberin production are located in the amygdala and hippocampus. Neurons of the amygdala exert stimulatory influence, and serotonergic neurons of the hippocampus exert inhibitory influence. The production of corticoliberin is also controlled with the aid of feedback inhibitory influences exerted by blood levels of corticotropin and cortisol.

In blood, more than 90% glucocorticoids are bound with plasma proteins, mainly with cortisol-binding globulin (transcortin) and, to a lesser extent, with albumin. Because the total binding capacity of these proteins is not high, the doubling of cortisol concentration in the blood as the result of increased adrenocortical activity causes a 4 to 10 times increase in a free, unbound fraction of cortisol. This is a typical result of intense exercise.

The most active glucocorticoid hormone is cortisol. It accounts for approximately 95% of all glucocorticoid activity. The other glucocorticoid—corticosterone—has much less potential than cortisol and accounts for approximately 4% to 5% of total glucocorticoid activity.

Glucocorticoids

Cortisol has a wide spectrum of metabolic effects. Therefore, it influences the control of several metabolic pathways. Because the cortisol level in blood increases sharply in various stress situations (environmental influences, emotional strain, exercise, trauma, infection, poisons, intoxications, almost any debilitating disease) and contributes to the adaptive metabolic arrangements, it is justifiably called the adaptation or stress hormone.

Most cortisol metabolic effects need the binding of hormone molecules with specific cellular receptors located in the cytoplasm. The formed steroid-receptor complex will be activated and translocated into the cell's nucleus, where it will induce formation of a specific mRNA to transcribe the synthesis of the related enzyme molecules. This is a time-consuming process. Therefore, the metabolic effects of cortisol appear after a lag period, the duration of which is in some cases about half an hour and in other cases more than an hour.

The main cortisol metabolic effects, which are actualized through induction of synthesis of enzyme proteins, are the following.

Stimulation of gluconeogenesis by the liver. Stimulation of gluconeogenesis results from the induction of enzymes required for this process and from the mobilization of amino acids from extrahepatic tissues, mainly from muscles (both striated and smooth muscles) and lymphoid tissue. As a result, more free amino acids are available to enter into the gluconeogenetic process of the liver. The glucocorticoid effect on the adipose tissue is also essential, resulting in the increased release of glycerol. Glucocorticoids are also capable of stimulating hepatic glucose production from lactate. Therefore, the formation of glucose is increased, and a marked increase in hepatic glycogen storage occurs. The latter is favored by direct induction of glycogen synthase and/or by the induction of glycogen synthase phosphatase catalyzing the activation of glycogen synthase. The glycogen synthase phosphatase is normally inhibited by glycogen phosphorylase a. Glucocorticoids cause the appearance of a protein factor in the liver, which cancels this inhibitory effect.

Stimulation of glucose-alanine cycle. Cortisol is the inducer of alanine-aminotransferase, which catalyzes alanine formation in skeletal muscles and alanine deamination in the liver. The latter process makes it possible to use the nitrogen-free residue of alanine for the glucose formation.

Decreased glucose use by the cells. Cortisol moderately reduces the rate of glucose use by cells everywhere in the body. It seems to be related to the influences of the entry of glucose into the cells and glucose degradation. It has been suggested that these influences are related to the depression of oxidation of nicotinamide-adenosine dinucleotide (NADH). At least in adipose and hepatic tissue, glucocorticoid inhibition of glucose transport requires concomitant synthesis of a regulative protein(s).

Reduction of cellular protein stores, except those of the liver. Reduction of cellular protein stores is caused by decreased protein synthesis and increased protein degradation. The antianabolic effect (decreased protein synthesis) is in an obvious relationship with suppression of formation of RNA in many extrahepatic tissues, especially in muscle and lymphoid tissue. Coincidentally, the liver protein amount increases with reduced protein contents elsewhere in the body. Plasma proteins produced by the liver also increase.

Increased free amino acid pool. Experiments on isolated tissues have demonstrated that cortisol depresses amino acid transport into muscle and other extrahepatic cells. At the same time, catabolism of protein continues to release amino acids. In this way, cortisol mobilizes amino acids from the nonhepatic tissues. As a result, plasma concentration of amino acids increases in association with enhanced transport into hepatic cells. The consequences of the enhanced use of amino acids by the liver are an increased rate of amino acid deamination, increased formation of urea, augmented gluconeogenesis, and increased protein synthesis in the liver, including blood plasma proteins. It has been assumed that for many of the metabolic effects of cortisol, the creation of the free amino acid pool is essential.

Stimulations of erythropoiesis. A high rate of secretion of cortisol is associated with an increased count in the red blood cell count, whereas anemia is common in adrenocortical insufficiency.

Influence on behavior. Cortisol effects include alterations in mood, changes in detection and recognition of sensory stimuli, changes in sleep, and changes in the extinction of previously acquired habits. The behavioral alterations are obviously related to glucocorticoid-induced changes in neural activity (for a review, see McEven 1979).

Anti-inflammatory effects. Damaged or infected tissue usually becomes inflamed. In certain cases, the inflammation is more damaging than the trauma or disease itself. The function of cortisol is to inhibit the inflammatory process and, thereby, avoid the harmful results of exaggerated inflammation. Administration of large amount of glucocorticoids is used to suppress the inflammation or to reverse its effects. The glucocorticoid effect is related to the stabilization of lysosomal membranes, decreased permeability of the capillaries (prevents loss of plasma into the tissues), and reduced migration of leukocytes into the inflamed area (inhibition of phagocytosis of the damaged cells). In addition, cortisol suppresses the immune system, lowers the fever, and blocks the inflammatory response to allergic reactions.

Effect on immunity. Cortisol causes the involution of lymphoid tissue, including lymphoid nodes and thymus. In blood, the number of lymphocytes and eosinophils decreases under the influence of glucocorticoids. The involution of lymphoid tissue is related to inhibited glucose transport and use in cells of this tissue. The result of involution of lymphoid tissue is the decreased output of T-cells and antibodies from the lymphoid tissue with a following decline in the level of immunity for almost all foreign invaders of the body. In normal conditions, cortisol contributes to the polyfactoral control of immunity. Taking into account the whole of the control mechanism, it is not wise to make conclusions about the immunity or immunoactivity only knowing the changes in the blood level of cortisol.

Some metabolic effects of cortisol appear rather rapidly. In these cases, neither the formation of steroid-receptor complexes nor the induction of protein synthesis is real. In several cases of rapid effects, cortisol influences postreceptor processes that follow the stimulation of adrenoreceptors. The site of the action is assumed to be either calcium shifts (Exton et al. 1972) or inhibition of cAMP-phosphodiesterase activity (Manganiello and Vaughan 1972). In both cases, the accumulation cAMP, the second messenger of the adrenergic effects, is promoted by calcium shifts essential for cAMP formation and by inhibition of phosphodiesterase activity to avoid rapid degradation of the cAMP. A possibility exists that cortisol increases the activity of protein kinase, essential for the actualization of adrenergic effects (Lamberts et al. 1975). In all cases, the outcome is that cortisol supports and/or amplifies the catecholamine action. In 1952, Ingle established this effect of glucocorticoids and named it the permissive action. Later, this effect was discussed when referring to the role of glucocorticoids as biological amplifiers (Granner 1979).

Glucocorticoids potentiate the action of glucagon and epinephrine on glyconeogenesis and glycogenolysis in liver, skeletal, and heart muscle; the action of epinephrine, growth hormone, and corticotropin on lipolysis in adipose tissue; the effects of epinephrine and glycogen on amino acid transport in the liver; and the effects of catecholamines on cardiovascular function (see Granner 1979).

Pituitary-Adrenocortical System in Exercise

The pituitary-adrenocortical system is rapidly activated at the start of an exercise. It was mentioned earlier that corticotropin response is triggered by collateral nervous charge from the motor command, sent by the pyramidal tract from the motor cortex to the spinal motoneurons. At the level of hypothalamus, the collateral charge obviously activates the neurosecretory cells, producing corticoliberin. The latter, reached by the portal system of the blood vessels directly from the hypothalamus to the pituitary gland, stimulates the production and release of corticotropin. The corticotropin reaches the adrenal cortex by means of the circulation. In response, the rate of biosynthesis of glucocorticoids increases. After an unexpected short interval, the glucocorticoid secretion by the adrenal cortex is augmented. Taking into account the high rate of glucocorticoid response at the onset of exercise, it has been suggested that besides transpituitary activation (with the aid of corticotropin), glucocorticoid output might also be stimulated by the parapituitary pathway (e.g., by the local release of epinephrine or the influence of sympathetic nerves reaching the adrenal glands or simply by increased blood flow through the adrenal glands.) No convincing evidence exists for these possibilities, but there are no results excluding them.

The increased blood level of corticotropin is characterized by a higher rate than that of cortisol (Buono et al. 1986; Karelson et al. 1994). The increase in blood cortisol level follows the onset of corticotropin response after an interval of less than 1 min. Usually, the initial increase of corticotropin concentration lasts only a couple of minutes (in some cases 5 to 15 min), and afterwards the corticotropin level returns to the initial level, whereas this time cortisol continues to increase in the blood. When the intensity of prolonged exercises is close to the anaerobic threshold or a little bit less, the cortisol concentration may decrease to less than the initial values during the second half hour in association with the lack of stimulation by corticotropin (the corticotropin level is close to the initial values at that time). However, during the second hour of an exercise, new rises of corticotropin and cortisol appear (Viru et al. 1992a). The secondary rise is a stable one and is more pronounced in endurance-trained athletes than in untrained persons or in athletes of nonendurance events (figure 5.2). The expression of a secondary rise is the high cortisol levels in well-trained athletes after marathon races (Maron et al. 1975; Dessypris et al. 1976), triathlon competitions (Jürimäe et al. 1990b), and other ultraendurance exercises (Sundsfjord et al. 1975; Keul et al. 1981; Zuliani et al. 1984).

Determinants of Corticotropin and Cortisol Responses

The corticotropin (Farrell et al. 1987; Rahkila et al. 1988; Schwarz and Kindermann 1990) and cortisol (Davies and Few 1973; Port 1991) responses depend on the intensity threshold, which is close to the anaerobic threshold (Rahkila et al. 1988; Port 1991; Gabriel et al. 1992a). However, different from the catecholamine responses, the further increase of exercise intensity above the threshold is not accompanied by a parallel increase of blood cortisol concentration (Port 1991). Some results indicate the possibility that high-intensity anaerobic exercises may suppress the cortisol response (Barwich et al. 1982; Port 1991; Karelson et al. 1994). The reason may be an inhibitory action of a high concentration of hydrogen on the adrenocortical function. In short-term, high-intensity exercises, the blood cortisol response increases with exercise duration (Hartley et al. 1972a, 1972b; Weicker et al. 1981; Kindermann et al. 1982; Kraemer et al. 1989b; Snegovskaya and Viru 1993a).

Short-term strength and power exercises also activate the pituitary-adrenocortical system. Increased corticotropin and cortisol levels were found after holding a 20-kg weight for 5 min in one hand. The corticotropin increase was established immediately after the end of the effort; the cortisol concentration was significantly greater than the initial level 5 min after the effort, and the highest level appeared 15 min after (Few et al. 1975). However, a certain threshold also exists in these exercises; three consecutive bouts of static handgrip at 30% maximal voluntary contraction did not evoke significant increases of corticotropin and cortisol level (Nazar et al. 1989).

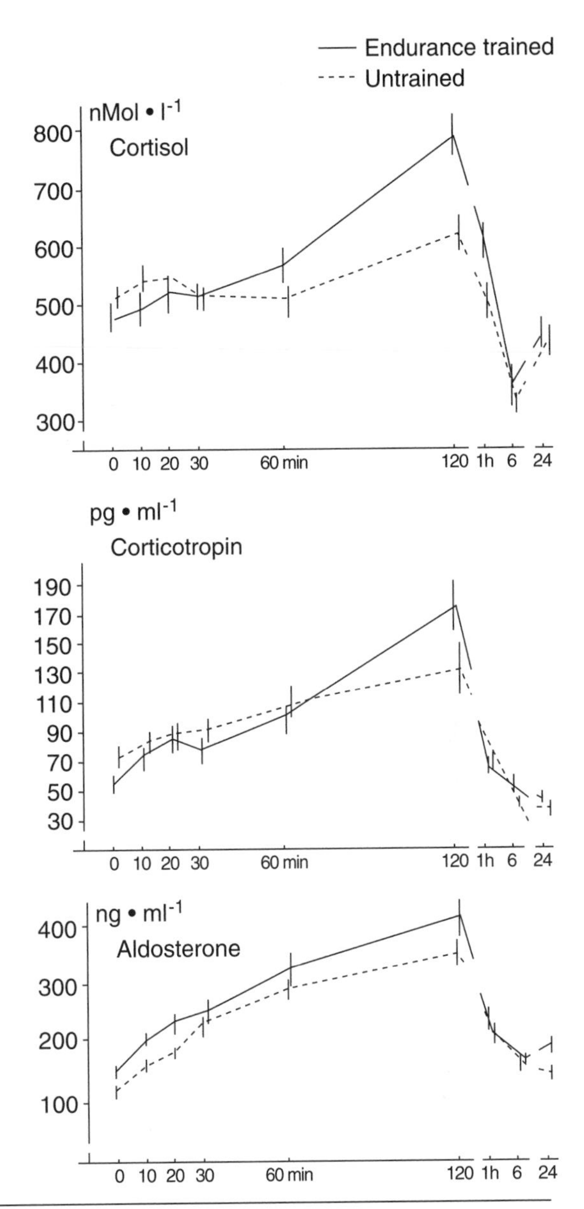

Figure 5.2 Dynamics of cortisol, corticotropin, and aldosterone during 2 h of exercise at 60% of $\dot{V}O_2max$ in trained (solid line) and untrained men (dotted line).
Reprinted from A. Viru et al. 1992.

Kraemer et al. (1989b) studied the effect of a high-rate application of muscle force for high-speed cycle movement. When 100%, 73%, 55%, or 36% of maximal leg power was executed, the possible duration of exercise was 6 s, 16 s, 47 s, and 3 min 19 s, respectively. Immediately after exercise, the corticotropin concentration was increased when exercise duration was 3 min 19 s (36% of maximal power was used). Plasma cortisol seemed to be increased only 15 min after this exercise. No response was detected after exercises that lasted 6 to 47 s.

Immediately after a 60-s period of consecutive vertical jumps (Bosco test), significant increases were found in the concentration of corticotropin (by 39%) and cortisol (by 14%) (Bosco et al. 1996).

In submaximal exercises, the typical training effects are reduced responses of corticotropin and cortisol caused by the increase of the intensity threshold measured in terms of power output (see Galbo 1983; Viru 1985a, 1995). The difference in cortisol responses is obvious when endurance-trained athletes and unfit persons are compared (Bloom et al. 1976; White et al. 1976; Sutton 1978). However, when sedentary persons were studied before and after endurance training for 7 weeks, the training effect on cortisol response did not appear (Hartley et al. 1972a, 1972b). After a 4-month physical fitness program, the cortisol response to a submaximal exercise test disappeared in previously physically inactive persons (White et al. 1976).

Training increases the capacity of the adrenal cortex to produce glucocorticoids through inducing adrenal hypertrophy and increments of cellular structures producing glucocorticoids (Viru and Seene 1985). As a result, the magnitude of the corticotropin (Bullen et al. 1984; Farrell et al. 1987) and cortisol (Bullen et al. 1984; Snegovskaya and Viru 1993a, 1993b) responses increased in endurance-trained athletes during supramaximal exercises.

Experiments on rats showed that the corticosterone level drops to less than the initial values when the duration of swimming in water of 32° to 34° C is more than 6 h (Kõrge et al. 1974a; Seene et al. 1978; Viru M. et al. 1994). In rats previously trained in swimming, a stable high corticosterone level persisted in swimming up to 20 to 22 h (Kõrge et al. 1974b; Seene et al. 1978).

Modulating Factors

Earlier (chapter 4, pp. 62-68) it was pointed out that the effect of the main determinants of hormonal responses (exercise intensity and duration and adaptation to exercise) may be modulated by several conditions, called modulators. Many study results confirm the significance of modulators for pituitary-adrenocortical responses in exercise.

Emotions The effect of emotional strain on adrenocortical activity in exercise is evident from the anticipatory changes in the excretion of 17-hydroxycorticoids before competitions (Thorn et al. 1953; Hill et al. 1956; Viru 1964) and increased blood levels of glucocorticoids, before competitions (Maron et al. 1975; Sutton and Casey 1975) or laboratory testing (Hartley et al. 1972b; Mason et al. 1973a). During a competition, the adrenocortical response to exercise may be exaggerated (Hill et al. 1956) but not in all cases (Vinnichuk et al. 1993). After competitive fighting in judo, the blood cortisol level was higher in winners than in losers (Elias 1981). Under the influence of emotional strain, underthreshold exercises may cause adrenocortical activation (Raymond et al. 1972).

Carbohydrate availability The glucostatic mechanism is also extended to control the activity of the pituitary-adrenocortical mechanism; glucose administration prevents the glucocorticoid response in exercise (Nazar 1981). In accordance with this, Bonen et al. (1977) showed that if glucose was ingested 15 min before exercise (80% of $\dot{V}O_2max$), a mild hyperglycemia before exercise was associated with a decrease in the cortisol concentration at the 15th min of exercise. Thereafter, the cortisol concentration increased. In the control experiment, the cortisol concentration was increased during the entire period of exercise. Another article from this team showed that the previous depletion of glycogen reserves enhanced the blood cortisol response to the endurance exercise at 80% $\dot{V}O_2max$ (Bonen et al. 1981). Fasting for 59 h also enhanced the cortisol level, both before and 5 min after exercise at $\dot{V}O_2max$ (Galbo et al. 1981a). If the previous diet had been lipid rich, the blood cortisol response to exercise was more pronounced than in the case of a carbohydrate-rich diet (Galbo et al. 1979a).

In prolonged exercises, the inhibitory effect of glucose administration may not appear. Vasankari et al. (1991) reported that the consumption of a carbohydrate solution (total carbohydrate amount, 105 g) during running for 36 km at a heart rate of 160 to 175 beats/min increased the postexercise cortisol level in endurance athletes.

Environmental conditions In hypoxic conditions, the underthreshold exercises are capable of causing a pronounced increase in blood cortisol level (Sutton 1977). In cases of overthreshold exercises, the same work output performed in hypoxic conditions greatly increased the glucocorticoid response (Davies and Few 1976). An underthreshold exercise evoked an exaggerated cortisol response when the working leg was made ischemic (blood supply reduced by 15% to 20%) by the application of 50 mmHg external pressure over the exercising leg (Viru M et al. 1998).

During swimming at different water temperatures, cortisol concentration increased only at 33° C in conjunction with increased core temperature. At temperatures of 27° C (core temperature did not change) and 21° C (core temperature decreased), the cortisol response was suppressed (Galbo et al. 1979a). When a submaximal exercise was performed in a thermochamber, the body temperature increased during the first 20 min up to 38.8° C. Simultaneously, the cortisol concentration declined. Further in the course of exercise, the body temperature leveled off, and the cortisol level increased (Few and Worsley 1975). Prolonged moderate exercise in a hot, humid environment is associated with the increase of core temperature and blood cortisol level. When fluid was consumed, the increase of rectal temperature was minimal, and no significant cortisol response was found (Francis 1979).

Muscle mass During 30 min of exercise, increases of plasma cortisol and blood lactate level and heart rate were significantly less when using both legs than for one-leg exercise (Few et al. 1980). In one-leg exercise, the power output per unit volume of the leg muscles is greater than in the two-leg exercise. Therefore, the central motor command also should be stronger. A greater metabolic strain on working muscles was evident by increased lactate accumulation. However, the stronger motor command was also associated with more pronounced lactate accumulation and, thereby, with increased stimulation of local metaboreceptors in working muscles in the case of the one-leg exercise.

Fatigue The significance of fatigue as a factor modulating exercise-induced hormonal responses is problematic. Publications from the 1950s and 1960s point out the changes in excretion and blood levels of hormones at the final stages of prolonged exercises, which were interpreted as fatigue effect. In 1953, Rivoire et al. reported that in prolonged exercises, the initial increase in excretion of corticosteroids is followed by a decrease. This observation was confirmed in several studies (Bugard et al. 1961; Viru 1977; Kassil et al. 1978). Measurement of corticosteroids in blood by use of the color reaction specific for 17-hydroxysteroids (Staehelin et al. 1955) or fluorometric assay (Viru et al. 1973; Keibel 1974) also demonstrated that after an initial rise, the corticosteroid level declined during prolonged exercises. Accordingly, animal experiments showed increases after short-term exercises and decreases after prolonged exercises in the blood level of glucocorticoids (Viru and Äkke 1969; Kõrge et al. 1974a). Similar changes were found in the adrenal content of glucocorticoids (Viru and Äkke 1969). After the administration of exogenous corticotropin, the manifestations of subnormal adrenocortical activity disappeared in humans (Viru 1977) and guinea pigs (Viru and Äkke 1969). Therefore, the reason for these manifestations was, obviously, related to insufficient endogenous stimulation of adrenocortical cells. Experiments with hippocampectomized rats showed that they were able to maintain an increased blood level of corticosterone in prolonged swimming, which caused a subnormal corticosterone level in sham-operated animals (Viru 1975a). Possibly, hippocampal serotonergic neurons were involved in the inhibition of the activity of hypothalamic neurosecretory cells responsible for activation of the pituitary-adrenocortical system. It has been suggested that fatigue triggers central inhibition of functions responsible for the mobilization of body resources. In this way, the fatal exhaustion of body resources is avoided (Viru 1975a).

In the 1970s, radioimmunoassays provided highly specific and valid methods for hormone determination. The previous results were forgotten or considered to be doubtful. By the use of radioimmune methods, high cortisol levels were found at the finish of a marathon race and other prolonged exercises (see p. 81). Nevertheless, the question remained whether fatigue arising during prolonged exercises influences hormonal responses. Some results have been published indicating a possible link between fatigue states and a low level of cortisol in the blood. Dessypris et al. (1976) reported that different from other persons, low cortisol levels were found in a marathon runner who collapsed after a 15-km run. A low cortisol concentration was recorded the day before the competition in a rower who collapsed during a regatta (Urhausen et al. 1987). The results of Feldmann et al. (1992) show that during the first 2 days of a 6-day Nordic ski race (distance per day 44 to 61.5 km), skiing induced increased blood levels of cortisol in association with high levels of corticotropin. However, from the third day, blood samples obtained after skiing did not show any more cortisol responses. Attenuated cortisol responses have also been found in monitoring training with great workloads and in overtraining (see chapter 9, pp. 203, 215-216).

Biological rhythm Galliven et al. (1997) did not find any differences in magnitude of corticotropin and cortisol responses to an intensive exercise (90% $\dot{V}O_2max$) performed in the morning and late afternoon. Differences were found between hormonal responses to exercises at 70% $\dot{V}O_2max$ performed by women in the follicular (days 3 to 9), midcycle (days 10 to 16), and luteal (days 18 to 26) phases of the ovarian-menstrual cycle. Cortisol responses were similar in follicular and luteal phases during 30 min of exercise (Bonen et al. 1983). Lavoie et al. (1987) showed that phase-dependent differences in cortisol response appear at the second half of a 90-min exercise on a bicycle ergometer at 63% $\dot{V}O_2max$; in the midluteal phase (days 20 to 23), the blood cortisol level was significantly greater than in the midfollicular phase (days 6 to 9). Accordingly, a greater postexercise cortisol level was found in middle-distance runners, aged 15 to 17 years old, tested in the luteal phase compared with those tested in the follicular phase (Szczepanowska et al. 1999).

Cortisol and Metabolic Control During Exercise

The preceding discussion (see chapter 4, p. 65) emphasized that the actual metabolic effect of a hormone depends not only on its increased level but also on the cellular reception of the hormone. Various other factors may influence the same metabolic pathways. Time characteristics of the metabolic effect of a hormone also have to be taken into consideration. Therefore, additional experiments are needed for convincing evidence of a hormone in metabolic control during exercise. The most favorable pathway is to record the metabolic disturbances when either hormone response or reception is experimentally altered. Still, this opportunity is used in a limited number of studies. Extirpation of the gland producing the hormone is one possibility for animal experiments. A shortcoming of this approach is that the hormone insufficiency induces metabolic disorders without the exercise. Therefore, the metabolic changes during exercises are altered not only by the lack of exercise-induced hormone response but also by performing the exercise on the level of prior existing metabolic disorders. The situation is improved if the effects of hormone insufficiency can be removed or reduced with the aid of a substitution administration of the missing hormone.

During exhaustive running, glycogen depletion, found in the soleus and tibialis anterior muscles of sham-operated rats, was suppressed in adrenalectomized rats (Struck and Tipton 1974). Gorski et al. (1987) confirmed reduced glycogen use in the soleus muscle but not in the quadriceps femoris muscle of adrenalectomized rats. According to our results, adrenalectomy reduces glycogen depletion in red (FOG fibers) and white (FG fibers) portions of the quadriceps femoris muscle (Viru M. et al. 1994). Interpretation of these results may be the lack of epinephrine after extirpation of the adrenal glands (medulla and cortex). However, Struck and Tipton (1974) showed that the muscle glycogen depletion was less aggravated in adrenal demedullated rats (medulla was eliminated, cortex intact) than in animals without the whole adrenal glands. Obviously, in addition to the lack of epinephrine, the adrenocortical insufficiency was also important. The latter was evidenced in the following results. In warm water (32° to 34° C), adrenalectomized rats were able to swim for approximately 9 h. When dexamethasone (a high-potential synthetic glucocorticoid) was injected after 6 h of swimming, the total duration of swimming increased up to 12 h (figure 5.3). In this case, the glycogen contents in FG and SO fibers were lower at exhaustion than in adrenalectomized rats treated with saline and exhausted after 9 h of swimming, and pronounced hypoglycemia was avoided (Viru M. et al. 1994).

The significance of glucocorticoids for glycogen use in skeletal muscles can be explained by the permissive action of glucocorticoids on the glycogenolytic effect of epinephrine. In adrenocortical insufficiency, the aggravated muscle glycogenolysis was also reflected in the lack of lactate accumulation in the blood. The lactate response was found in running adrenalectomized rats treated with glucocorticoids and intact rats. However, a blunted lactate response was also revealed in adrenal demedullated rats (Malig et al. 1966).

During prolonged swimming, the liver glycogen dropped significantly more in adrenalectomized rats than in sham-operated or adrenal demedullated rats (Struck and Tipton 1974). The possible reason was the low rate of hepatic gluconeogenesis in adrenocortical insufficiency. In addition, the significance might also belong to disorders of lipolysis, which led to increased carbohydrate use in prolonged exercise. In an obvious relationship with intensive exhaustion of hepatic glycogen store, hypoglycemia was more pronounced in adrenalectomized rats than in adrenal demedullated rats (Struck and Tipton 1974). Glucocorticoid therapy diminished the decline of blood glucose in adrenalectomized rats during prolonged running (Malig et al. 1966). When, after 6 h of swimming, dexamethasone was administered to adrenalectomized rats, the continuous decline of blood glucose level stopped. Furthermore, a rather stable glucose level persisted at the expense of a further use of liver glycogen (figure 5.3). Thus, after 12 h of swimming, the glucose level was higher and liver glycogen content lower than in saline-treated adrenalectomized rats at exhaustion, which appeared in this group after 9 h of swimming (Viru M. et al. 1994).

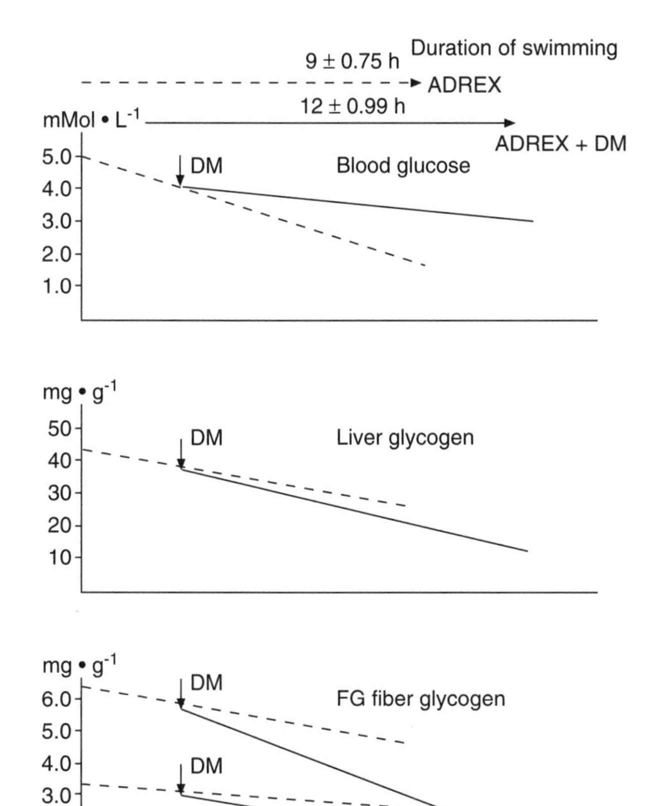

Figure 5.3 Dynamics of blood glucose, liver glycogen, and FG and SO fibers in adrenalectomized rats when swimming to exhaustion. DM = injection of dexamethasone during exercise.

By results of Viru M et al. 1997.

In normal rats, a prompt decrease of liver glycogen was observed during the first 4 h of swimming. From the fifth to twelfth hours of swimming, the hepatic glycogen content did not decrease significantly. This was possibly due to the predominance of lipid oxidation and intensive gluconeogenesis. Both might be related to the high level of blood corticosterone, which persisted until the eighth hour of swimming. By the twelfth hour of swimming, the corticosterone level dropped to the initial level and to values below the initial level. At this stage of exercise, hepatic glycogen content dropped again. As early as 1949, Ingle and Nezamis showed that after the administration of adrenocortical extract to normal rats, the decrease in their liver glycogen content was less pronounced during a prolonged period of muscle contractions. These results confirm the suggestion that during prolonged exercise, the rate of gluconeogenesis is related to the availability for glucocorticoids.

The significance of glucocorticoids for glycogen synthesis has also been demonstrated during the postexercise recovery period. Postexercise supercompensation for cardiac glycogen was enhanced with dexamethasone treatment, eliminated by adrenalectomy, and restored in adrenalectomized rats who had been given daily doses of dexamethasone (Poland and Trauer 1973). In adrenalectomized rats, a slow rate of glycogen postexercise repletion was found not only in the myocardium but also in skeletal muscles and the liver. Dexamethasone treatment restored the glycogen repletion rate in adrenalectomized rats. The blockade of protein synthesis excluded the dexamethasone effect, suggesting that the glucocorticoid effect was mediated by the synthesis of a regulatory protein, most likely of glycogen synthase (Kõrge et al. 1982).

In the control of lipid mobilization, glucocorticoids contribute through their permissive influence on lipolytic action of epinephrine and growth hormone. In this way, adrenalectomy abolishes or decreases significantly (Malig et al. 1966; Struck and Tipton 1974; Gorski et al. 1987) the increase in the plasma level of free fatty acids during exercise. Cortisone treatment reversed this change (Malig et al. 1966).

Cortisol is well known by its catabolic action. In exercise, the catabolic effects of glucocorticoids (suppression of protein synthesis; augmented release of amino acids; elevated protein degradation rate in muscle, lymphoid, and connective tissues; enhanced activity of myofibrillar proteases in muscle; and release of 3-methylhistidine in muscle tissue) are subjected, actually, to a polyfactoral control. First, it is necessary to take into consideration the fact that muscular activity exerts a protective effect against the catabolic action of glucocorticoids on muscle tissue (Hickson and Davis 1981; Seene and Viru 1982; Czerwinski et al. 1987). This effect obviously rules out the possible harmful action of the high level of glucocorticoids. Evidence is obtained that the exercise-induced catabolism is not extended to contractile proteins (Dohm et al. 1987; Varrik et al. 1992) and is mostly related to less active or nonactive muscle fibers (Varrik et al. 1992). Muscular activity inhibits the stimulating effect of glucocorticoids on myofibrillar alkaline proteases (figure 5.4) (Seene and Viru 1982). Here a substantial role may be played by the opposite action of testosterone (see chapter 4, p. 69).

The increased level of glucocorticoids in the blood provides a stimulus for increased alanine production. This effect and cortisol influence on gluconeogenesis makes this hormone important for determining the rate of the glucose-alanine cycle. In adrenalectomized patients with Cushing's disease, an exercise caused a less pronounced increase in the blood plasma alanine concentration than it did in normal persons (Barwich et al. 1981). Accordingly, adrenal insufficiency excluded the exercise-induced rise in alanine levels of blood plasma, oxidative muscles, and liver found in normal rats (Viru et al. 1994).

The action of cortisol on the alanine metabolism is actualized through the induction of alanine aminotransferase. In normal rats, 4 h of swimming increased alanine-aminotransferase activity in SO fibers but not in FG fibers and liver. Adrenalectomy abolished the increase in SO fibers and resulted in a decrease of enzyme activity in the liver. In another experiment, 3 h of swimming caused a parallel increase of liver alanine-aminotransferase activity and blood level of corticosterone. After adrenalectomy, low activity of the enzyme persisted in rest and after exercise. However, the dependence of exercise-induced changes on glucocorticoids was confirmed by the increase of enzyme activity in hepatic tissue of adrenalectomized rats treated with corticosterone (Viru et al. 1994).

The role of glucocorticoids in the use of free amino acids by the liver during exercise was demonstrated in an experiment on rats. Swim-

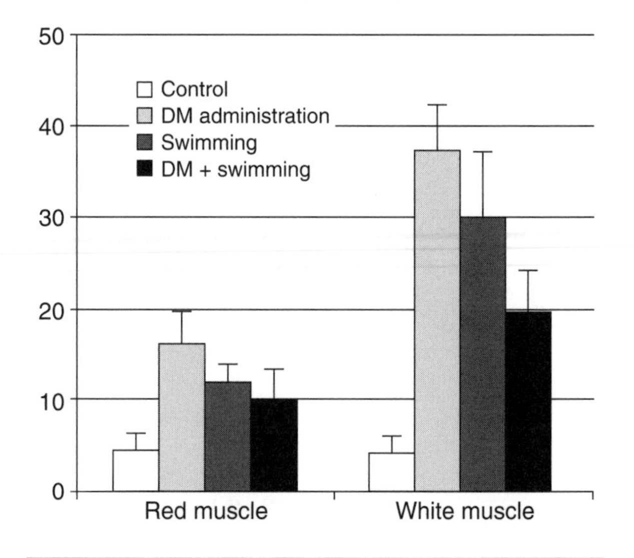

Figure 5.4 Activity of alkaline protease (release of tyrosine moles per 1 mg of DNA) in muscles of various types after chronic administration of dexamethasone (DM), exhaustive swimming, and the combined action of dexamethasone and exhaustive swimming.

By results of Seene and Viru 1982.

ming for 3 h caused an increase in the blood corticosterone level accompanied by a decreased free amino acid level in the liver of sham-operated rats. In adrenalectomized rats, these changes were not observed, but they appeared again when corticosterone was administered to adrenalectomized rats before swimming (Viru and Eller 1976).

In adrenalectomized rats, arginase activity in the liver decreased during exercise in conjunction with the lack of elevation of the urea levels in the blood, liver, and skeletal muscle. In normal rats, exhibiting almost twice the arginase activity, urea concentration increased in response to exercise (Viru et al. 1994). The significance of corticosteroids in urea production during exercise was further tested by assessment of ^{14}C-urea content in liver tissue after administration of $NaH^{14}CO_3$. In intact rats, swimming for 30 min, 3 h, or 10 h caused increased ^{14}C-urea content in the liver. Concentration of urea in the liver and blood elevated simultaneously. In adrenalectomized rats, the basal rate of urea production was insignificantly reduced. Swimming for 3 h resulted in a decrease of liver ^{14}C-urea (Litvinova and Viru 1995a).

Thus, adrenal hormones have an essential role for urea synthesis during exercise. Glucocorticoids are also essential for urea elimination in the postexercise recovery period. The postexercise period is characterized by an increased renal clearance of urea in rats after swimming sessions of various durations. The postexercise increase in urea renal clearance was absent in adrenalectomized rats (Litvinova et al. 1989).

During exercise, glucocorticoids also seem to contribute to the control of Na^+-K^+ pump function on the plasma membrane of heart and skeletal muscle fibers. In rats, swimming for 90 min elevated the activity of Na^+-K^+ ATPase in the microsomal fraction of the myocardial cells of untrained rats. Furthermore, continuation of an exercise from 6 to 10 h was accompanied by return of the enzyme activity to the initial level. After the extreme duration of an exercise (more than 16 h), the Na^+-K^+ ATPase decreased significantly. The decrease accompanied a reciprocal increase of Na^+/K^+ ratio in the intracellular compartment. The dynamics of plasma corticosterone concentrations and the activity of Na^+-K^+ ATPase were approximately parallel during these exercises (Kõrge et al. 1974a). Rats, exhausted by the strenuous training regime, showed a decreased level of corticosterone in the blood and diminished Na^+-K^+ ATPase activity in skeletal muscles (Kõrge et al. 1974b).

The experiment on adrenalectomized rats confirmed that the link between glucocorticoid levels and Na^+-K^+ ATPase activity was a causal one. Adrenalectomy abolished the swimming-induced increase of the Na^+-K^+ ATPase activity in myocardial cells. Instead, low enzyme activity persisted in adrenalectomized rats during exercise (Kõrge and Roosson 1975).

Two possibilities exist for interpreting the glucocorticoid action on the Na^+-K^+ pump. First, glucocorticoids may exert a permissive (supporting) effect on the activation of Na^+-K^+ pump function by epinephrine. Another possibility consists of the induction of Na^+-K^+ ATPase synthesis by glucocorticoids. The latter may be responsible for changes in enzyme activity during prolonged exercises.

Misunderstanding of Cortisol Function

In several articles dealing with cortisol responses in training, cortisol function was understood only in stimulating catabolic processes. In these articles, the wide spectrum of metabolic adjustments induced or supported by cortisol is ignored. Also, the adaptive significance of catabolic changes in exercise is usually ignored. Therefore, the cortisol response evaluated has been incorrectly considered a "bad" one. Moreover, the exercise-induced increase in cortisol concentration is thought to provide information on dysadaptation. The next misunderstanding is that training reduces cortisol responses to avoid cortisol-induced dysadaptation.

These misunderstandings contradict the results of physiological and biochemical studies presented earlier. The wide spectrum of cortisol contributions to metabolic control makes the hormone essential for working capacity and good performance level. In part, a great amount of evidence indicates that normal adrenocortical function is a prerequisite for physical working capacity (for a review, see Viru 1985b). More precisely, the relationship between glucocorticoid availability and the opportunity to perform prolonged muscular exercise was obtained by Ingle in a series of studies (e.g., Ingle and Nezamis 1949; Ingle et al. 1952). Additional evidence shows that when chronic administration of dexamethasone resulted in atrophy of the adrenal glands (their weight reduced by 41%), the maximal duration of swimming decreased by 62% (Smirnova and Viru 1977). The significance of these results consists of the fact that the daily administration of dexamethasone avoided the actual glucocorticoid insufficiency, but the atrophied adrenal glands were not able to increase the glucocorticoid production in response to exercise.

The significance of glucocorticoid response to prolonged exercise was tested in rats by Sellers et al. (1988). Adrenalectomized rats were given a subcutaneously implanted corticosterone pellet at the time of adrenalectomy. This manipulation avoided actual adrenocortical insufficiency, but these rats could not respond to exercise by increased glucocorticoid production. To simulate the exercise-induced response, a group of adrenalectomized-implanted rats were treated with a corticosterone injection immediately before the exercise bout. Other adrenalectomized rats were treated with corn oil (placebo) before the exercise. The duration of treadmill running until exhaustion was in 136 ± 6 min in sham-operated rats compared with 114 ± 9 min in adrenalectomized-implanted rats injected with corticosterone, and 89 ± 8 min for adrenalectomized-implanted rats injected with corn oil.

However, the short-term exercise test did not demonstrate the dependence of the physical working capacity and maximal oxygen uptake on cortisol responses. The pharmacological blockade of adrenocortical response during test exercises did not decrease the levels of maximal oxygen uptake and power output at a heart rate of 170. These indexes of the aerobic working capacity also did not change when the cortisol response was stimulated with the aid of a corticotropin injection before the test exercise (Viru and Smirnova 1982). Hence, the cortisol response is highly significant for the performance level of prolonged exercises but not for aerobic power.

The significance of normal adrenocortical function for the working capacity is emphasized by the fact that adrenalectomy excludes the training effect. When adrenal glands were extirpated in previously trained rats, their working capacity dropped to the level found in sedentary adrenalectomized rats (Kõrge and Roosson 1975; Viru and Seene 1982).

As was noted, muscular activity inhibits the catabolic action of glucocorticoids. Therefore, the link between an increased blood level of cortisol and the activation of the catabolic processes during exercise is not the same as in resting conditions. The results of Fimbel et al. (1991) showed that after administration of moderate doses of glucocorticoids the decrease of FG fiber weight did not exceed 12%. Training did not prevent this glucocorticoid effect but enhanced muscle oxidative capacity and maximal oxygen uptake and the delayed appearance of exhaustion during submaximal exercise bouts in glucocorticoid-treated rats.

The catabolic action of glucocorticoids is essential in posttranslational control for adjusting the number of synthesized protein molecules to actual need. Probably more important is the glucocorticoid catabolic action to create an increased pool of free amino acids. In this way, glucocorticoids provide necessary "building materials" for the adaptive protein synthesis. Therefore, glucocorticoids favor, but do not aggravate, the training effects. The catabolic effect of glucocorticoids may also be essential to increase protein turnover rate in previously active muscles during the recovery period after the training session.

The contribution of glucocorticoids to training effects was confirmed by the results of an experiment. During swimming training of rats, the adrenocortical response to the swimming bout was inhibited with the aid of administration of dexamethasone on the first 2 days of every week. Five weeks of training significantly increased the maximal duration of swimming with an additional load of 3% body weight in control rats. In dexamethasone-treated rats, training failed to increase maximal swimming duration (Viru 1976b). In this experiment, the manipulation used did not cause the actual adrenocortical insufficiency; the blood level of corticosterone remained constant. The obvious reason for the lack of a training effect was the aggravated response to the swimming bouts on the first 2 days of every week. On these days, the response on training influence was the most important because the swimming duration was increased at the onset of every training week. In the untreated rats, a correlation (r = .638; p <

.05) was found between maximal duration of swimming and corticosterone level in the blood after exercise (Viru 1976b).

Experiments with adrenalectomized rats also failed to show the training-induced increase of working capacity in adrenal insufficiency (Tipton et al. 1972; Viru and Seene 1982).

Last, attention should be paid to the results showing that training reduces or abolishes adrenocortical responses only in submaximal exercises, whereas responses to supramaximal exercises even increase (see this chapter, p. 82).

Pancreatic Hormones

The islets of Langerhans in the pancreas secrete two hormones, insulin and glucagon, into the blood. The pancreas is also the source of several gastroenteropancreatic hormones. Although the significance of most gastroenteropancreatic hormones for metabolic adjustments in exercise still needs to be evaluated, insulin and glucagon are important regulators of metabolism. Their role is essential in metabolic control during exercise and postexercise recovery. Therefore, exercise-induced changes in insulin and glucagon levels provide information on control of carbohydrate, lipid, and protein metabolism.

Insulin

The main function of insulin is related to storing excess energy substance. In conditions of increased use of body energy resources (starvation, muscular exercise), insulin secretion is suppressed. The excess of carbohydrates in the diet stimulates insulin release.

Insulin is synthesized in the beta cells of the islets of Langerhans. Insulin circulates almost entirely in an unbound form. Insulin half-life in blood plasma is short—an average of about 6 min. This means that insulin molecules are cleared from the circulation within 10 to 15 min after their secretion into the blood. In target tissues, a portion of insulin is bound with receptor proteins on the cellular membrane. The remaining portion of insulin is degraded mainly in the liver and to a lesser amount in the kidneys.

The membranes of muscle cells, adipose cells, and many other types of cells become highly permeable to glucose within a couple of seconds after insulin binds with its membrane receptors. This allows rapid entrance of glucose into the cells. After insulin binding, the cell membranes also become more permeable for many amino acids and for potassium, magnesium, and phosphate ions.

Insulin causes changes in the activity of several intracellular enzymes within 10 to 15 min. The activity changes of enzymes are mainly related to activation through the phosphorylation of the enzymes. There are also slow effects that occur for hours. These are related to changes in rates of the translation process at the ribosomes to promote the formation of new proteins.

Metabolic Effects

The main effects of insulin in metabolic control are the following (for a review, see Felig et al. 1987).

Increased permeability of muscle membrane. In resting conditions, muscle membranes are only slightly permeable to glucose. Under the influence of elevated levels of insulin in blood plasma, the membranes become highly permeable. As a result, glucose enters muscle fibers, and muscle tissue begins to prefer the use of carbohydrates instead of fatty acids. In exercise, the entry of glucose into muscle fibers may happen even at low levels of blood insulin because of the release of certain specific proteins ($GLUT_1$, $GLUT_2$, $GLUT_3$, $GLUT_4$, $GLUT_5$) during muscle contractions (for a review, see Sato et al. 1996) that promote glucose transport into the muscle fibers.

Storage of glycogen in muscle. After meals, insulin promotes glucose transport into the muscle cells in abundance. The same appears during postexercise recovery even without an exogenous supply of carbohydrates at the expense of further glucose output by the liver. At the same time, insulin also stimulates storing of the glucose in the form of muscle glycogen.

Promoting liver uptake, storage, and use of glucose. After meals, absorbed glucose is stored almost immediately in the liver in the form of glycogen. This change is considered to be one of the most important of all the effects of insulin. When food is not available and the blood glucose level starts to decline, liver glycogen degrades back into glucose. The increased glucose output by the liver and its release into the blood keeps blood glucose from decreasing.

Conversion of carbohydrates to lipids. When the quantity of glucose entering the liver cells is more than can be stored as glycogen, insulin promotes the conversion of the excess glucose into fatty acids in the liver. The same happens in the adipose tissue. Insulin promotes glucose transport through the cell membrane into the adipose cells. A part of this glucose is used for the synthesis of fatty acids. Glucose also forms α-glycerophosphate, which supplies the glycerol. Glycerol combines with fatty acids to form the triglycerols that make the storage of fat in adipose cells.

Inhibition of lipolysis. Insulin inhibits the hormone-sensitive lipase activity even at normal levels of the hormone. This way the basal rate of lipolysis and the effect of lipolytic hormones are suppressed. In low levels of insulin, the enzyme releases from inhibition how the hormone-sensitive lipase in fat cells may become strongly activated. The results are hydrolysis of stored triglycerols and the release of large quantities of fatty acid and glycerol into the circulation. In exercise, the decrease of the blood level of insulin is the main event that activates lipid mobilization and use (see chapter 3, p. 53).

Promoting protein synthesis and inhibiting its breakdown. Insulin stimulates active transport of many amino acids (most of all valine, leucine, isoleucine, tyrosine, and phenylalanine) into the cells. Insulin has a direct effect on the ribosomes, increasing the translation rate of mRNA. Insulin also increases the rate of transcription of selected DNA genetic sequences in the cell nuclei, forming increased quantities of RNA. Insulin inhibits protein catabolism and in this way decreases the rate of amino acid release from the cells, especially from muscle cells. In exercise, the low level of insulin favors stimulation of the glucose-alanine cycle, increasing the rate of alanine release from muscles and glucose output by the liver. Insulin also increases plasma amino acid concentrations and enhances urea production.

Insulin is essential for growth. First, insulin is essential for growth because of stimulating and supporting the transport of amino acids into cells and promoting of the translation process of protein synthesis. Second, it is important because increased carbohydrate availability and use related to insulin actions exert a protein-sparing effect.

Blood glucose level is the most important factor controlling insulin secretion. Plasma insulin concentration increases about 10-fold within 3 to 5 min after an acute elevation in blood glucose level. The rapid insulin response is related to the immediate dumping of preformed insulin from the beta cells of the islets of Langerhans. The high rate of insulin secretion cannot be maintained. Furthermore, within the following 5 to 10 min, insulin concentration decreases about halfway toward basal values. The secondary rise in insulin secretion appears after 15 min. Now the secretion rate reaches a new plateau for 2 to 3 h. The new secretion rate may be even greater than found immediately after the rise in blood glucose. The secondary increase results from the additional release of preformed insulin and from activation of the enzyme system responsible for insulin synthesis.

As a result of the high insulin level, blood glucose concentration decreases. The turnoff of insulin secretion is almost equally as rapid as its initial rise. Within a few minutes after the reduction in blood glucose level back to basal, insulin concentration drops. Thus, blood glucose concentration provides an important feedback mechanism for the control of the insulin secretion rate.

Several amino acids, mainly arginine and lysine, exert a stimulating action on insulin secretion. Amino acids have only a modest and short-term effect on insulin secretion if the blood glucose level remains constant. However, amino acids strongly potentiate the rise of insulin secretion caused by blood glucose elevation.

Several hormones (glucagon; growth hormone; cortisol; and, to a lesser degree, progesterone and estrogen) are also capable of stimulating insulin secretions directly or potentiating a glucose-induced rise of insulin secretion.

A possibility exists that insulin secretion is controlled by antagonistic influences of the autonomic nervous system; parasympathetic nerves stimulate and sympathetic nerves inhibit the secretion. Experiments with various adrenoblockers showed the significance of sympathetic influences reaching the pancreas by means of α-adrenoreceptors for the decreased insulin release in exercise (Galbo et al. 1977b; Järhult and Holst 1979). Accordingly, several results show that the rise in blood levels of norepinephrine seems to be the factor causing the drop in insulin secretion with exercise (Galbo 1983).

Insulin in Exercise

Unlike other hormones, insulin secretion decreases in response to exercise (for a review, see Galbo 1983; Viru 1985a 1992). An exception appears only in short-term supramaximal exercises, causing an increase that is accompanied by a transient hyperglycemia (Hermansen et al. 1970; Adlercreutz et al. 1976). A lag period of 10 to 15 min is typical before the insulin level declines (Hunter and Sukkar 1968; Pruett 1970a). Insulin decrease in prolonged exercises is intensity-dependent. Exercise intensities of 40% $\dot{V}O_2max$ (Hartley et al. 1972a) or 47% $\dot{V}O_2max$ (Galbo et al. 1975) were enough to cause a decline in insulin concentration. When the load was 50% or 70% $\dot{V}O_2max$, a significant reduction of the insulin concentration was detected without any further dependence on the exercise intensity (Pruett 1970a). Galbo and collaborators (1975) found the insulin decline in exercises at 47% or 77% $\dot{V}O_2max$. At an intensity of 100% $\dot{V}O_2max$, the insulin level tended to increase. After exercise with near maximal oxygen uptake, the insulin concentration decreased during the exercise but increased immediately and dramatically on cessation of the exercise (Pruett 1970b).

In ergometer exercise, the addition of arm work to the leg work did not alter the dynamics of blood insulin (Green et al. 1979).

During a 2-h exercise at 60% $\dot{V}O_2max$ after the first 30 min characterized by a prompt decrease, insulin concentration stabilized (figure 5.5). Furthermore, only a minimal decline appeared (Viru et al. 1992a). The insulin pattern was similar during a 4-h exercise at 50% $\dot{V}O_2max$ (Luyckx et al. 1978). However, Wahren et al. (1975) reported that the insulin concentration decreased steadily during a 4-h exercise at 30% $\dot{V}O_2max$.

Training decreases or eliminates the insulin decline in submaximal exercises (Hartley et al. 1972a 1972b; Rennie and Johnson 1974; Sutton 1978) and in exercise at 100% $\dot{V}O_2max$ (Hartley et al. 1972a). In rats, swimming training reduced the insulin secretion rate of isolated islets of Langerhans (Galbo et al. 1981b).

Exercises inhibit the augmented secretion of insulin after the administration of glucose (Pruett and Oseid 1970; Luyckx et al. 1978). However, glucose administration during exercise avoids lowering the insulin concentration in the blood (Bonen et al. 1977) or even substituting it with elevated insulin levels (Ahlborg and Felig 1976; Koivisto et al. 1981). A fat-enriched diet (Galbo et al. 1979a) or fasting (Stock et al. 1978; Galbo et al. 1981a) increases the reduction of the insulin level during exercise.

Hypoxia enhances the blood insulin response to exercise. A simulated altitude of 4550 m caused a more prompt insulin drop in 20-min exercise than the same exercise at normal barometric pressure (Sutton 1977).

Convincing evidence demonstrates that insulin plays an essential role in metabolic control during exercise. In exercise, the most important metabolic effects of insulin are related to maintaining euglycemia and regulating liver glucose output and lipolysis in adipose tissue. Maintaining euglycemia is done in balance with glucagon and some other hormones. Stimulation of

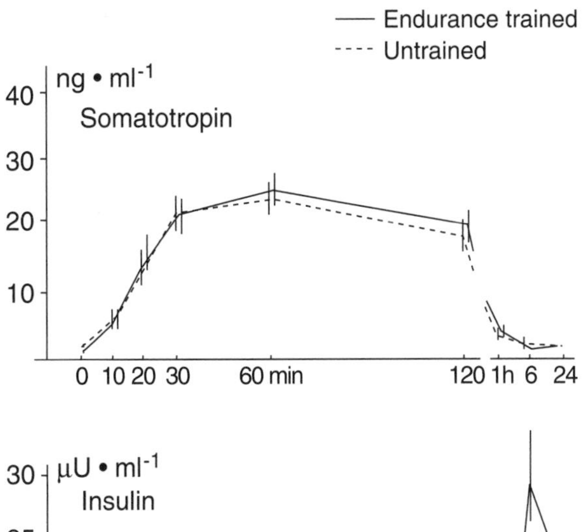

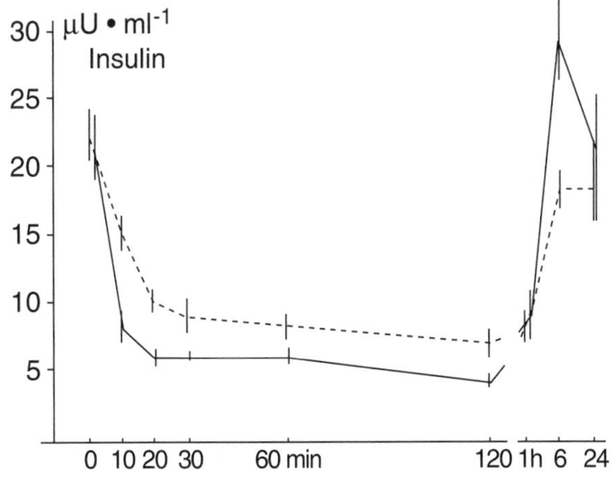

Figure 5.5 Growth hormone (somatotropin) and insulin dynamics (average values) in endurance-trained (solid line) and untrained men (dotted line).

Reprinted from A. Viru et al. 1992.

glucose output by the liver is founded on the effect of hypoinsulinemia. This effect is potentiated by the increased secretion of glucagon. The decreased blood level of insulin is an important determinant for increased lipolysis in exercise. It has been demonstrated that a decrease in insulin level is associated with an increase in free fatty acids in plasma during prolonged exercise (Wahren et al. 1975) (figure 5.6). Hypoinsulinemia is considered to be an important event necessary to reserve blood glucose for the fueling of nerve cells.

In a limited number of studies, acute effects of exercise on insulin receptors were investigated in skeletal muscles and adipose tissue. Results showed that binding of insulin may not change or be reduced in skeletal muscles (Bonen et al. 1985). A decrease in the number of binding sites in skeletal muscles was observed after exhaustive exercise (Pedersen and Bak 1986), but no changes of insulin binding were found in adipose tissue (Koivisto and Yki-Järvinen 1987).

A common effect of endurance training is an enhanced sensitivity to insulin (Johansen and Munck 1979; Sato et al. 1986; Mikenes et al. 1989). According to the results of rat experiments, this change is in conjunction with the increased number of insulin receptors in skeletal muscles (Dohm et al. 1987).

Glucagon

α-Cells of the islets of Langerhans secrete a hormone called glucagon. Most of its functions are diametrically opposed to those of insulin. The effects of glucagon are mediated through the same cascade of postreceptor events as epinephrine. The activation of adenyl cyclase in the hepatic cell membrane, which causes the formation of cAMP, is the main effect. Amplification of the cAMP effects by glucagon is also possible. In muscle tissue, glucagon fails to activate adenyl cyclase. Therefore, it does not contribute to the control of glycogenolysis in muscle cells. In adipose tissue, glucagon activates hormone-sensitive lipase.

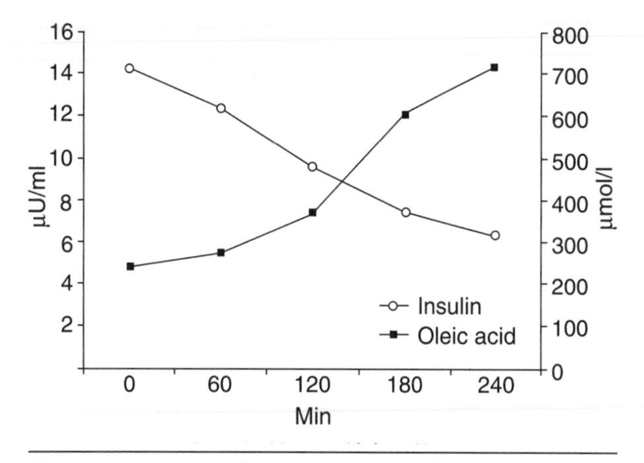

Figure 5.6 Arterial concentration of insulin and oleic acid during prolonged exercise at 30% $\dot{V}O_2$max in men. Adapted from J. Wahren et al. 1975.

Metabolic Effects

The main metabolic effects of glucagon are as follows (for a review, see Felig et al. 1987).

- Stimulation of glycogenolysis in the liver, which results in increased glucose output and elevated blood glucose concentration. The latter appears within minutes after the increase of glucagon in the blood. Infusion of glucagon for about 4 h causes such intensive liver glycogenolysis that the liver stores of glycogen become totally depleted. The augmented glycogenolysis under the influence of glucagon contributes to the continued hyperglycemia found during infusion of the hormone.
- Stimulation of gluconeogenesis in the liver. Glucagon increases the extraction of amino acids from the blood by the liver. In this way, glucagon provides increased availability of amino acids to be converted into glucose.
- Stimulation of lipolysis in adipose tissue. Simultaneously, glucagon inhibits the storage of triglycerides in the liver, preventing extraction of fatty acids from the blood by the liver.

Some evidence has been obtained that glucagon promotes amino acid transport into cells and other tissues in addition to hepatic tissue. In this way, glucagon contributes to protein synthesis.

Blood glucose level is by far the most potent factor controlling glucagon secretion. When blood glucose concentrations drop to hypoglycemic levels, a several-fold increase has been found in the blood glucagon concentration. Hyperglycemia decreases the plasma level of glucagon. After a high-protein meal, concentrations of amino acids stimulate the release of glucagon from the pancreas into the circulation. The most effective amino acids are alanine and arginine. Because alanine outflow from muscle fibers is increased during exercise, in this situation alanine may contribute to increased glucagon secretion in muscular activity. The alanine effect on glucagon secretion may contribute to feed-forward stimulation of the rate of the glucose-alanine cycle; an increased glucagon level augments the conversion of alanine to glucose in the liver.

Sympathetic nervous stimulation of the function of α-cells of islets of Langerhans may also contribute to the stimulation of glucagon secretion (for a review, see Steffens and Strubbe 1983).

Glucagon in Exercise

In prolonged exercise, glucagon concentration in blood may increase up to fourfold to fivefold. Because it may appear without a necessary fall in blood glucose concentration, the glucagon response is not triggered by hypoglycemia. Therefore, an increased amino acid concentration in blood and sympathetic nerve stimulation of the islets of Langerhans may have significance. The significance of sympathetic nerves in increasing glucagon secretion during exercise has been evidenced in rats (Luyckx et al. 1975) and dogs (Harvey et al. 1974) but not in humans (Galbo et al. 1976; Galbo 1983). Experiments on demedullated rats pointed to the significance of epinephrine in the glucagon response to prolonged swimming (Richter et al. 1980).

In prolonged exercises, the glucagon response appears after a long lag period. It has been reported in several articles that the increase of blood glucagon was detected in the second hour of exercises at intensities of 30% to 50% of $\dot{V}O_2max$ (Ahlborg et al. 1974; Luyckx et al. 1978). At an exercise intensity of 60% $\dot{V}O_2max$, the lag period was approximately 45 min (Winder et al. 1979). In short-term exercises, the glucagon response appears only at high intensities (Galbo et al. 1975; Näveri et al. 1985a). Blood sampling at the end of various running events showed a moderate increase of glucagon after 100 m, 1500 m, and 10,000 m but a prompt increase after a 25-km race (Weicker et al. 1981). The glucagon response was found after swimming for 1000 yd but not for 200 yd (Hickson et al. 1979). Thus, exercise duration seems to be more important than intensity.

In more-fit persons, the glucagon response is less pronounced (Bloom et al. 1976; Hickson et al. 1979). Longitudinal experiments confirmed the results of cross-sectional studies (Gyntelberg et al. 1977; Winder et al. 1979).

Although glycogen responses appeared in exercises during which euglycemia was maintained, the significance of the glucostatic mechanism is in question. When the blood glucose decline was avoided or euglycemia was restored by glucose infusion, the glucagon response to exercise was reduced (Galbo et al. 1977a, 1979a; Luyckx et al. 1978). It was also possible to prevent or inhibit the rise in blood glucagon level during prolonged exercise by the administration of glucose, despite the maintenance of euglycemia (Ahlborg and Felig 1976).

A fat-rich diet (Galbo et al. 1976) or fasting (Galbo et al. 1981a) increases the blood glucagon response to exercise compared with a carbohydrate-rich diet.

The glucagon response may be modulated by environmental temperature or atmospheric pressure. No rise of glucagon was found during swimming for 60 min in water of 21° C, although a pronounced increase was found in the glucagon level during swimming in water of 27° or 33° C (Galbo et al. 1979b).

Actualization of the three main metabolic effects of glucagon (augmented liver glucose output, increased rate of glycogenolysis, and increased rate of lipolysis) and maintained euglycemia in exercise has been discussed earlier (see chapter 3, p. 50). All these effects do not depend solely on the glucagon response but mainly on the ratio between the opposite effects of glucagon and insulin (Vranic et al. 1975; Wahren 1979). The actual contribution of glucagon is shown in an experiment that used inhibitors of glucagon secretion (Issekutz and Vranic 1980; Richter et al. 1981).

Somatostatin

This regulative peptide is produced by delta cells of islets of Langerhans. A peptide of identical molecular structure is also produced by the hypothalamic neurosecretory cells. These cells reach to the pituitary gland by the portal blood vessels and inhibit the release of growth hormone. Secretion of pancreatic somatostatin is stimulated by increases of blood glucose, amino acids, and fatty acids in blood. Increased concentrations of several gastrointestinal hormones released from the upper gastrointestinal tract in response to food intake increase the release of somatostatin. Hence, food ingestion results in factors that stimulate secretion of pancreatic somatostatin.

By its local action within the islets of Langerhans, somatostatin inhibits secretion of insulin and glucagon. Somatostatin released into the circulation decreases the motility of the stomach, duodenum, and gallbladder, as well as the secretion and absorption processes in the gastrointestinal tract.

According to the listed effects, the main function of pancreatic somatostatin is to extend the time over which food nutrients are assimilated into the blood. Through the suppression of insulin and glucagon secretion, somatostatin also decreases the use of absorbed nutrients by the tissues. This way food is available over a longer time.

During prolonged moderate exercises, somatostatin concentration increases gradually (Hilsted et al. 1980). It has been suggested that somatostatin may have significance in the suppression of insulin secretion during exercise. During exercise, exogenous administration of somatostatin reduces blood levels of insulin and glucagon in man (Chalmers et al. 1979; Björkman et al. 1981).

Growth Hormone and Growth Factors

A little less than a quarter century ago, growth hormone started to attract attention among athletes for the purpose of enhancing training effects. Although there was no striking experimental evidence to prove that growth hormone could enhance training, athletes believed in the effectiveness of this manipulation. Administration of growth hormone was considered doping.

Actually, growth hormone produced in the body is an important factor ensuring normal growth of children and adolescents. Endogenous growth hormone is also an essential regulator of metabolism in the bodies of adolescents and adults. Adjustments in acute exercise performance and development of training effects require the contribution of endogenous growth hormone.

Body tissues also produce several growth factors that are only in part related to growth hormone action. Growth factors are also essential for adaptation to muscular activity.

To decide whether growth hormone and growth factors may be used as tools for training monitoring, we have to discuss their significance in metabolism and particularly in adaptations during exercise.

Growth Hormone

Growth hormone (also called somatotropin or somatotropic hormone) releases into the circulation from the anterior lobe of the pituitary gland. The stimulus for growth hormone release is somatoliberin (growth–hormone releasing factor) produced by neurosecretory cells of the hypothalamus. Hypothalamic somatostatin (growth–hormone inhibiting factor) suppresses growth hormone secretion by the pituitary gland.

The main function of growth hormone is to cause the growth of almost all tissues of the body that are capable of growing. It promotes increased cell size, augments mitosis with an increase in the number of cells, and specifically differentiates bone growth cells and early muscle cells.

Metabolic Control

Growth hormone contributes a wide spectrum to metabolic control (for a review, see Felig et al. 1987).

- Enhanced amino acid transport through the cell membranes.
- Enhanced mRNA translation to promote protein synthesis by the ribosomes.
- Increased nuclear transcription of DNA to form RNA. It is a delayed effect and appears over more prolonged periods (24 to 48 h) than other metabolic effects of growth hormone.
- Reduced degradation of proteins and amino acids. This way growth hormone acts as a potent "protein sparer."

These effects are essential for promoting growth. If sufficient energy, amino acids, vitamins, and other necessities for growth are available, growth hormone actualizes its main function: it causes the growth of tissues.

The growth-promoting effect of growth hormone requires the contribution of insulin. Without insulin growth hormone fails to cause growth. The growth-promoting effect of exogenously administered growth hormone is potentiated by a simultaneous injection of insulin. Insulin, as well as carbohydrates, provides the energy needed for growth.

However, the contribution of growth hormone in the control of protein metabolism is only a part of the metabolic effects of growth hormone. These include the following.

- Stimulation of lipolysis.
- Decreased use of glucose for oxidation by blocking the glycolytic breakdown of glucose and glycogen.
- Enhanced glycogen deposition in the cells (under the influence of growth hormone, cells become maximally saturated with glycogen).
- Increased blood glucose concentration. During the first 30 min after growth hormone administration, cellular uptake of glucose is enhanced. Subsequently, decreased transport of glucose into the cells appears. This is the result of an excess of glucose taken up by cells and of the difficulty to use it. A pronounced increase of blood glucose follows. In clinics, this condition is called "pituitary diabetes."
- Increased secretion of insulin as a consequence of increased blood glucose. However, growth hormone is also capable of a direct stimulatory effect on the β-cells of the islets of Langerhans. A possibility for overstimulation of insulin secretion arises, and β-cells can become exhausted. Because of that, growth hormone is considered to exert a diabetogenic effect.

Much of the growth hormone effects are actualized through somatomedins. Those regulatory proteins have a strong effect on all aspects of bone growth. The liver is the main site of production of somatomedins. Often, they are called tissue growth factors. The most important growth factor is somatomedin-C. The pattern of its concentration in blood plasma is usually, but not always, parallel to the dynamics of growth hormone. Most growth effects of growth hormone on bone and other tissues are plausibly actualized by somatomedin-C and other growth factors. However, results have been obtained that also show a direct effect of growth hormone on increasing growth in some tissues. Therefore, it is thought that the somatomedin mechanism is an alternative means of increasing growth but not always a necessary one.

Somatomedin-C also promotes glucose transport through cell membranes. Therefore, it is called "insulin-like growth factor" (IGF-I).

Growth hormone releases rapidly from the blood into the interstitial fluid. The half-life of the hormone in blood is less than 20 min. By contrast, somatomedin-C is strongly bound to a carrier protein. Therefore, somatomedin-C is released slowly from plasma to tissues. Its half-life in plasma is about 20 h. This way, somatomedin-C greatly prolongs the growth-promoting effects of growth hormone. The significance of that prolongation becomes understandable when we take into consideration the episodic character of growth hormone secretion into the blood. Growth hormone is secreted by short-term bursts of intensive release of the hormone into the circulation. These secretory bursts last only several minutes. Between secretory bursts, growth hormone concentration decreases to low levels.

Secretion of growth hormone is evoked by starvation, hypoglycemia, low levels of free fatty acids in the blood, exercise, emotional strain, and trauma. Growth hormone secretion increases during the first 2 h of deep sleep.

Growth hormone secretion is controlled almost entirely by hypothalamic somatoliberin and somatostatin. The activity of corresponding hypothalamic neurosecretory cells is controlled by a number of neuronal systems having different neurochemical specificity. Adrenergic, dopaminergic, serotoninergic, and other neurons participate in the control. Growth hormone secretion is subjected to the feedback control exerted by the actual growth hormone level in the blood.

Growth Hormone in Exercise

Exercise studies show that growth hormone response is common and stable for prolonged exercise (see Viru 1992). Compared with responses of other anterior pituitary hormone responses, growth hormone response to a 2-h bicycle ergometric exercise or to 60 km of skiing is the most stable (Viru et al. 1981). Despite the episodic character of growth hormone secretion, growth hormone response during exercise is characterized by a continuous increase of blood level during the exercise or by a leveling off after the initial increase (figure 5.5). In a limited number of cases, a decline in growth hormone concentration has been found in the second hour of exercise (Viru et al. 1992a).

In submaximal prolonged exercise, a lag period precedes the initial increase (Lassarre et al. 1974; Sutton and Lazarus 1976; Karagiorgos et al. 1979). From our results, the lag period appeared in 70% of untrained and in 65% of trained persons (Viru et al. 1992a). In short-term intensive exercises, the lag period may not appear (Buckler 1973; Näveri et al. 1985a; Nevill et al. 1996).

In submaximal exercises, the magnitude of growth hormone response depends on exercise intensity (Hartley et al. 1972a; Sutton and Lazarus 1976; Vanhelder et al. 1984a, 1985; Näveri 1985; Näveri et al. 1985b). Threshold intensity for growth hormone response is approximately 60% to 80% of $\dot{V}O_2max$ (Hartley et al. 1972a; Sutton and Lazarus 1976; Näveri 1985). Chwalbinska-Moneta et al. (1996) found a good agreement among intensity thresholds for growth hormone, epinephrine, and norepinephrine. All thresholds were close to the anaerobic threshold determined by the lactate increase in the blood. Schnabel et al. (1982) reported that 50 min of exercise at the individual anaerobic pressure increased growth hormone level.

Comparison of supramaximal anaerobic exercise and prolonged submaximal aerobic exercise showed that exercise duration is a stronger determinant of the magnitude of growth hormone response than intensity (Hartley et al. 1972a, 1972b; Kindermann et al. 1982; Snegovskaya and Viru 1993a). However, intensive anaerobic intermittent exercises (interval training session or repeated supramaximal exercises) have a strong stimulatory effect on growth hormone release (Adlercreutz et al. 1976; Karagiorgos et al. 1979; Vanhelder et al. 1984a, 1987).

Short-term strength and power exercises have to be reported to evoke growth hormone response. Immediately after 1 min of continuous jumping (the Bosco test), growth hormone remained at the basal level in the blood (Bosco et al. 1996). During three consecutive bouts of static handgrip sustained for 3 min, growth hormone level did not change significantly; an increase was detected 10 min after the end of the exercise (Nazar et al. 1989). A training session, consisting of repeated high-resistance exercises, elicited increased levels of growth hormone (Vanhelder et al. 1984b, 1985; Jürimäe et al. 1990a; Kraemer et al. 1991b), depending on the total work done (Kraemer et al. 1990b; Häkkinen and Pakarinen 1993; Cotshalk et al. 1997). When exercises were performed either with arms or legs but at the equivalent oxygen uptake, growth hormone response was greater in arm exercise (Kozlowski et al. 1983).

As a result of training, growth hormone response to submaximal exercises decreases or disappears (Sutton et al. 1969; Hartley et al. 1972a; Buckler 1973; Rennie and Johnson 1974; Koivisto et al. 1982). However, in exercises with great demands on the body because of either intensity or duration, growth hormone increases to higher levels in endurance athletes than in untrained or less-trained persons (Hartley et al. 1972b; Kraemer et al. 1993). Growth hormone response to 7-min all-out exercise on a rowing apparatus increased even in well-trained rowers with a year of training together with an increase of performance level (Snegovskaya and Viru 1993a).

Glucose intake (Bonen et al. 1977) or high-carbohydrate diet (Galbo et al. 1979a) suppresses growth hormone response. A low-carbohydrate, high-fat diet (Galbo et al. 1979a) or fasting (Galbo et al. 1981a) increases the response. Prior depletion of the body's glycogen deposit enhanced growth hormone response to endurance exercise (Bonen et al. 1981).

Great increases in plasma growth hormone level have been found while exercising under hypoxic conditions (Sutton 1977; Raynaud et al. 1981). Conversely, Vasankari et al. (1993) found lower growth hormone levels after skiing races at middle altitude than at sea level. Low environmental temperature suppresses and high temperature exaggerates growth hormone response in exercise (Buckler 1973; Few and Worsley 1975; Frewin et al. 1976; Galbo et al. 1979b).

The possibility exists that fatigue may modulate growth hormone response. After a certain amount of muscle work is done, growth hormone concentration begins to decline (Hunter et al. 1965; Hartog et al. 1967; Buckler 1972). Zuliani et al. (1984) found similar dynamics of growth hormone during 24 h of skiing: a gradual increase during 18 h and afterward and a drop to the initial values during the last 6 h. Dynamics such as these may explain why Sutton et al. (1969) failed to find increased growth hormone concentrations after a marathon race or why Dufaux et al. (1981a) observed a decreased somatotropin level after running 34 km. However, other results showed a high growth hormone level at the finish of 60 km of skiing (Viru et al. 1981) or 100 km of running (Keul et al. 1981).

During an incremental exercise test, growth hormone response was virtually absent in persons with high scores on the Beck Depression Inventory and the negative attitude subcomponent (Harro et al. 1999).

Several metabolic effects of growth hormone were listed earlier. However, serious experimental evidence confirming their actualization during exercise is not available. Of course, metabolic effects related to the stimulation of growth are not for acute adjustments during exercises. A number of growth hormone effects may be essential during the recovery period, first, in the promoting effect of growth hormone on the translation process of protein synthesis and enhancement of glycogen repletion. Augmented transport of amino acids into the cells is essential during recovery and exercise. However, we can only suggest the contribution of growth hormone in the complex of other hormones and metabolic regulators. During exercise, growth hormone may contribute to the control of glucose use and lipolysis.

The lipolytic effect of growth hormone requires a lag period of at least 1 to 2 h (Fain et al. 1965). Therefore, the rise in growth hormone blood level after the onset of exercise will actualize this lipolytic action only at the end of the second or during the third hour of exercise (see chapter 4, pp. 69-70).

Somatomedins in Exercise

It has been assumed that growth hormone causes the formation of several somatomedins (growth factors) in the liver. The most important is somatomedin-C, also called insulin-like growth factor I. However, questions still exist as to whether somatomedins are formed only under the influence of growth hormone and whether growth hormone metabolic effects depend only on formation of somatomedins. The most plausible fact is that growth hormone itself is directly responsible for increased growth in several tissues and that the influence through somatomedins is an alterative means of increasing growth but not always a necessary one. The possibility also exists that somatomedins prolong the growth-promoting effects of a burst of growth hormone secretion.

Somatomedin-C promotes glucose transport through cellular membranes. However, in regard to adaptation to muscular activity, the importance of somatomedin-C is its contribution to the development of muscle hypertrophy (for a review, see Adams 1998). During increased muscle loading, myofibers up-regulate the expression and secretion of somatomedin-C, which stimulates anabolic processes in muscle fibers (Adams 1998). However, growth hormone administration to elderly persons doubled the concentration of circulating somatomedin-C but had no effect on protein synthesis rate or on strength gain in training (Yaresheski et al. 1995).

Heavy-resistance training sessions are capable of stimulating the formation of somatomedin-C in sufficient amounts to be expressed by its elevated blood level in men and women (Kraemer et al. 1990b 1991b). Increases in somatomedin-C concentration varied among heavy-resistance exercise protocols and did not constantly follow growth hormone changes (Kraemer et al. 1990b). An eight-station, heavy-resistance exercise protocol (three sets of 10 RM with a 1-min rest) resulted in a pronounced increase of growth hormone concentration without significant changes in somatomedin-C level. Despite the pronounced growth hormone rise, somatomedin-C changes were not found during a 24-h period after the exercise session (Kraemer et al. 1995a). Twelve weeks of high-volume resistance training did not change the resting level of somatomedin-C in the blood (McCall et al. 1999).

A short-term but intensive exercise (1-min consecutive jumps) did not change the somatomedin-C level in blood plasma (Bosco et al. 1996).

Leptin

Adipose tissue produces leptin, which is considered a hormone. It can be argued that it is not a hormone, because it is not produced by a collection of secretory cells (endocrine gland). However, its action, not located in the tissue where it is produced but in the brain, makes it possible to use the term "hormone."

The plasma level of leptin is proportional to the amount of body fat (Lonnqvist et al. 1995; Maffei et al. 1995; Caro et al. 1996). Leptin represents a signal to the brain, the consequences of which are hormonal and neuronal responses that reduce food intake. A decrease in leptin level stimulates food intake and increases body stores (Friedman 2000). A possibility for postexercise decrease of leptin has been found (Landt et al. 1997), although this change was not confirmed by results of another study (Torjman et al. 1999). Training effects on leptin concentration are not established (Hickey et al. 1996, 1997).

Leptin measurement has a meaning in training monitoring when exercising is associated with pronounced changes in food intake. The results would indicate nutritional disorders. Also, in an overtraining state, leptin may provide useful information for handling this state.

Thyroid Hormones

The function of the thyroid gland is controlled by pituitary thyrotropin. The pituitary-thyroid system is formed this way. The activity of this system is stimulated by thyroliberin (thyrotropin-releasing hormone) secreted by hypothalamic neurosecretory cells. At the same time, the circulating level of thyroid hormones has a strong feedback effect on the secretion of thyrotropin.

Thyroid glands secrete three hormones: thyroxin, triiodothyronine, and calcitonin. Biosynthesis of thyroid hormones except calcitonin consists of the iodination of tyrosine. Therefore, in iodine insufficiency in food or drinking water, formation of thyroid hormones is aggravated up to the lack of hormone synthesis. Thyroxin accounts for the greatest portion of thyroid secretion. The amount of secretion of triiodothyronine is almost 10 times less than that of thyroxin. However, most thyroxin converts eventually to triiodothyronine in the tissues. Because triiodothyronine is about four times as potent as thyroxin, the main metabolic effects of thyroid hormones are induced by triiodothyronine. Calcitonin does not contain iodine, and its metabolic effects are completely different from those of iodine-containing thyroid hormones. Calcitonin is an important hormone controlling calcium metabolism.

Because of the high affinity of plasma-binding proteins for thyroid hormones, triiodothyronine and, particularly, thyroxin reach the cells slowly. The half-life of thyroxin in the blood is approximately 6 days, and that of triiodothyronine is 1 day. A long latent period is characteristic for thyroid hormones before their metabolic effects appear. The latent period for thyroxin is 10 to 12 days and for triiodothyronine is 6 to 12 h. Accordingly, it is impossible to believe the contribution of thyroid hormones in metabolic adjustments during exercise. However, the metabolic effects of thyroid hormones are, plausibly, essential during delayed recovery after exercise. Most of all, the essential effect of thyroid hormones consists of long-term stimulation of the biogenesis of mitochondria and synthesis of oxidative enzymes. Thyroid hormones are also essential for long-term adaptation of ionic pumps in cellular membranes. By means of control of the adaptive synthesis of various enzymes, thyroid hormones contribute to long-term changes of carbohydrate and fat metabolism. Attention should also be paid to heart adaptation in an overload situation controlled by thyroid hormones.

Exercise-induced changes of thyroid hormones and thyrotropin in blood are modest if observable at all (for a review, see Galbo 1983; Viru 1985a). Therefore, a question arises as to where the changes reflect the actual increase of thyroid activity. It is plausible that modest increases are related to extravasation of blood plasma or to increased blood supply of the gland (the "washout" phenomenon). However, the possibility has to be considered that the exercise triggers an increase in the activity of the pituitary-thyroid system, which develops at low speed and reaches a level of elevated thyroid hormone production in the recovery period. This possibility has been supported by results of a rat experiment. After 30 min of swimming, thyrotropin level was greater than that in the sedentary control from 1.5 h to 12 h and thyroxin and triiodothyronine levels from 1.5 h to 48 h (Konovalova et al. 1997). The fact that chronic hypoxia increases exercise-induced changes is in agreement with this possibility. Mild exercise caused elevated levels of thyroxin and triiodothyronine at an altitude of 3650 m but not in conditions of normoxia (Stock et al. 1978). An increase of thyrotropin level has been observed in anticipation of muscular exercise (Mason et al. 1973b).

Hormones Regulating Water and Electrolyte Balances

Several hormones contribute to the balance of water and electrolytes in the body. Vasopressin is responsible for maintaining the normal amount of water in the body and the necessary blood volume. Aldosterone and atrial natriuretic factor control the normal concentrations of sodium and potassium in the extracellular fluid. Blood plasma calcium level is maintained by the opposite influences of parathormone and calcitonin.

The significance of the Na^+-K^+ pump for the restoration of the intracellular ionic balance after excitation was mentioned previously. Epinephrine activates the Na^+-K^+ pump function, whereas thyroid hormones contribute to the long-term adaptation of the Na^+-K^+ pump (figure 5.7). Glucocorticoid insufficiency is associated with an increase of sodium and water in the intracellular compartment as a result of the contribution of glucocorticoid in the control of Na^+-K^+ pump function, at least by their permissive action on the epinephrine effect. Glucocorticoids also exert a stimulating action on diuresis in conditions of hyperhydration.

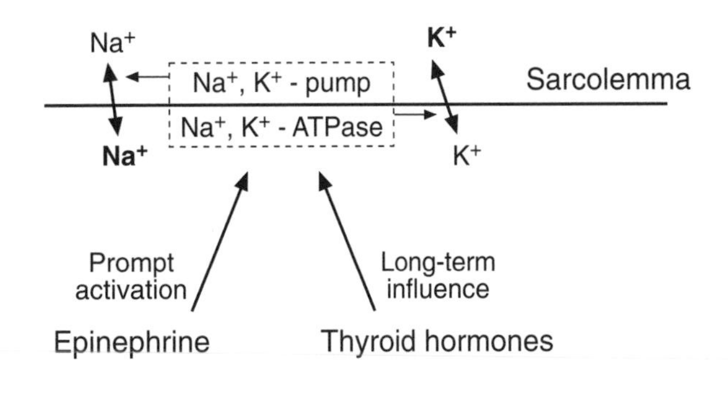

Figure 5.7 Activation of the Na^+-K^+ pump by epinephrine (adrenaline) and thyroid hormones.

Vasopressin

According to the site of formation, vasopressin belongs to the group of hyperthalamic neurosecretory hormones. Cell bodies of the supraoptic nucleus are the site of vasopressin synthesis. The hormone is transported in combination with "carrier" proteins (neurophysins) to the nerve endings in the posterior lobe of the pituitary gland. When nerve impulses are transmitted along the fibers from the supraoptic nucleus, the hormone releases from the secretory granules in the nerve endings into the adjacent capillaries. At the same time, the hormone immediately releases from its binding with neurophysin.

Vasopressin exerts antidiuretic and vasoconstrictor effects. The first effect is stimulation of reabsorption of water in renal tubules, as a result of which the loss of water into urine decreases (antidiuresis). Such water conservation is essential in the loss of water from perspiration and in the loss of a great amount of blood. The simultaneous vasoconstrictor effect is an additional tool for maintaining blood pressure despite the decrease of blood volume.

Vasopressin secretion is controlled by osmoreceptors sensing the osmotic pressure of blood plasma. The amount of osmotic pressure is proportional to the concentration of solute in the number of molecules or ions (osmolality). Osmolality increases either because of reduction of the amount of solvent (e.g., water) or because of increases in the number of ions or substrate molecules. This way hypohydration and increase of sodium concentration cause increased vasopressin secretion.

In exercise, particularly in prolonged exercise, increased blood level of vasopressin is a common response (Melin et al. 1980; Wade and Claybaugh 1980; Convertino et al. 1981, 1983). The response is in relationship to exercise intensity (Wade and Claybaugh 1980). However, exercises lasting several hours exert a strong influence on vasopressin release despite their moderate intensity. Running a marathon race caused a 1.6-fold rise in blood vasopressin level (Dessypris et al. 1980). During 4 h of exercise, plasma volume loss and subsequent increase in vasopressin concentration were prevented by progressive rehydration (Brandenberger et al. 1986). Therefore, vasopressin response is related to dehydration and changes in osmolality. Thus, exercise intensity might influence the vasopressin secretion not by direct neural influence on the supraoptic nucleus but through osmolality changes, which were related to the increase in perspiration rate with exercise intensity. This possibility has been confirmed by results showing similar relationships of plasma vasopressin and lactate concentrations and blood volume reduction with exercise intensity (El-Sayed et al. 1990). However, vasopressin increase in blood plasma was also found after three consecutive bouts of static handgrip (Nazar et al. 1989). At least in this case, it was unlikely that an increase in plasma osmolality and a reduction of blood volume were the main factors stimulating the vasopressin release. The prompt response in static handgrip points to the significance of the neural mechanisms.

The training effect on vasopressin response in exercise has not been convincingly demonstrated (Melin et al. 1980; Geyssant et al. 1981; Convertino et al. 1983).

Aldosterone and the Renin-Angiotensin System

The outer layer of the adrenal cortex—the zona glomerulosa—produces mineralocorticoids. The most important among them is aldosterone. The main function of aldosterone is to control the tubular reabsorption of sodium and the tubular secretion of potassium in the kidney. Aldosterone stimulates the reabsorption of sodium in renal tubules. Because of this, sodium excretion decreases. However, sodium level in blood plasma increases only very little. This is because sodium retention causes osmotic absorption of almost equivalent amounts of water from renal tubules into the plasma. Consequently, an effect of aldosterone is to increase the extracellular fluid volume. In this way, aldosterone may increase arterial blood pressure.

Simultaneously, with sodium retention, aldosterone increases potassium excretion with urine and, thereby, avoids possible increases in the extracellular concentration of potassium. In pathological conditions that cause high aldosterone levels for prolonged periods, an excessive loss of potassium ions from the extracellular compartment into the urine leaves an abnormally low level of potassium in blood plasma (hypokalemia). When the potassium ion concentration falls below one-half of normal, severe muscle weakness develops because of disturbances of transmission of action potentials in muscle fiber membranes.

To summarize, aldosterone causes sodium to be conserved in the extracellular fluid and potassium to be eliminated from the body to maintain the normal Na^+/K^+ ratio in the blood plasma and interstitial fluid (figure 5.7). These homeostatic effects are essential during exercise because intensive perspiration causes extrarenal loss of sodium from the extracellular fluid and protein catabolism and glycogen degradation release additional portions of potassium into the extracellular fluid.

Aldosterone causes a secretion of hydrogen ions in exchange for sodium reabsorption in renal tubulus. However, tubular secretion of hydrogen is much smaller than potassium secretion. Therefore, the homeostatic effect of aldosterone to prevent acidosis in anaerobic exercises cannot be considerable. During exercise, the blood supply of the kidneys is decreased. Therefore, the amount of excreted hydrogen ions should be low.

Aldosterone has almost the same effect on sweat glands as it has on the renal tubules. Therefore, sweat is hypotonic compared to blood plasma.

A highly specific receptor protein for aldosterone is located in the cytoplasm of the tubular cells in the kidneys. Aldosterone-receptor complex diffuses into the nucleus, where it induces a specific portion of DNA to form mRNA for synthesis of specific proteins in ribosomes. The proteins synthesized constitute a mixture of enzymes (one of them is Na^+-K^+ ATPase) that channels proteins into the membrane of tubular cells (allows rapid diffusion of Na^+ from the tubular lumen into the cell) and receptor proteins. Forty-five minutes is required before the rate of Na^+ transport begins to increase under the influence of aldosterone. Aldosterone effect reaches maximum after several hours (for a review, see Felig et al. 1987).

Secretion of aldosterone is controlled by concentrations of potassium and sodium ions in the extracellular fluid, the renin-angiotensin system, and corticotropin. The most potent are influences of potassium concentration and the renin-angiotensin system. A slight percent change in potassium concentration causes a severalfold change in aldosterone secretion.

The renin-angiotensin system consists of

- release of renin from the juxtaglomerular cells of the kidney under the influence of diminished blood flow to the kidney or sympathetic nervous impulses;
- renin, which acts enzymatically on angiotensinogens (a blood plasma globulin) to release angiotensin I; and
- conversion of angiotensin I into angiotensin II under the catalyzing action of the converting enzyme.

Renin persists in blood plasma for 30 to 60 min. During this period, it causes the formation of angiotensin I. Conversion of angiotensin I to angiotensin II needs a few seconds. Secretory response of aldosterone also occurs rapidly. Then, the renin-angiotensin system increases aldosterone release without any substantial delay. Vasoconstrictor action of angiotensin II results in increased arterial pressure within 10 min after severe hemorrhage.

Corticotropin enhances overall steroid synthesis in the adrenal cortex. Although corticotropin action is specifically related to production of glucocorticoid to a modest degree, the action is also extended to the biosynthesis of aldosterone.

Exercise results in increases in blood level of aldosterone. Several studies demonstrated aldosterone response in prolonged aerobic exercise (Sundsfjord et al. 1975; Costill et al. 1976b; Kosunen and Pakarinen 1976; Melin et al. 1980; Geyssant et al. 1981; Keul et al. 1981). Our results showed that in 2-h cycling at 60% $\dot{V}O_2$max, aldosterone response belongs to the group of stable responses of moderate rate (figure 5.2, Viru et al. 1992a). However, short-term anaerobic exercises (e.g., running 3 × 300 m) can evoke a high increase of aldosterone concentration, with a maximum 30 min after exercise (Adlercreutz et al. 1976).

By use of an incremental exercise test, Buono and Yeager (1991) showed that threshold intensity for aldosterone response is lower than that for cortisol. Aldosterone response correlated significantly with potassium, corticotropin, and angiotensin II increases.

Exercise-induced aldosterone increase in blood follows rises in plasma renin activity and level of angiotensin II (Sundsfjord et al. 1975; Costill et al. 1976b; Kosunen and Pakarinen 1976; Melin et al. 1980). The same could be observed after running 3 × 100 m (Adlercreutz et al. 1976). When exercises were performed in hypoxic conditions, plasma renin activity increased with workloads but less than in normoxic conditions. Aldosterone concentration increased at normal atmospheric pressure, but no change was found in a simulated altitude of 3000 m (Bouissou et al. 1987). Lower aldosterone level and plasma renin activity during exercise performed at altitude conditions rather than at normal conditions had also been previously reported (Maher et al. 1975).

The relationship between renin and aldosterone responses was confirmed by experiments in which renin release was blocked by administration of methyldopa. After 6 days of drug administration, reduced response of plasma renin activity to 30 min of exercise was accompanied by a similar reduction of aldosterone concentration (Sundsfjord et al. 1975).

Plasma renin activity increases simultaneously with concentrations of norepinephrine and epinephrine and lactate (Kotchen et al. 1971; Fagard et al. 1977). All four responses are similarly related to exercise intensity (Kotchen et al. 1971; Wade and Claybaugh 1980). Blockade of β-adrenoreceptors greatly diminished renin and also angiotensin and aldosterone responses in exercise (Lijnen et al. 1979). These results confirm the significance of the sympathoadrenal system in activation of the renin angiotensin–aldosterone system.

Several other influences contribute to the control of the renin-angiotensin system in exercise. After salt loading, virtually no increases in plasma renin activity (Aurell and Vikgren 1971) and aldosterone level (Cuneo et al. 1988) were observed. After salt depletion, plasma renin activity roughly doubled during exercise (Aurell and Virken 1971; Fagard et al. 1977). In the heat, renin and angiotensin increases during prolonged exercise were significantly diminished when sweat losses of sodium and potassium were compensated by the intake of a sodium-potassium solution in equivalent amounts (Francis and MacGregor 1978). However, sodium concentration, sodium/potassium ratio, and plasma osmolality may be the only modulating factors. It has been suggested that a more important determinant of renin production in exercise may be volume and pressure changes in the blood vessels, including changes of intrathoracic blood volume (Kirsch et al. 1975).

After inhibition of the angiotensin-converting enzyme, the elevation of renin activity was significantly more pronounced than during exercise performed in normal conditions. Obviously, the release of renin is controlled by the negative feedback of angiotensin II (Khouhar et al. 1979).

In exercise-induced plasma renin activity, increases in circadian variations have been observed: at 4:00 A.M. the response was less pronounced than at 4:00 P.M. However, the results failed to show a concomitant difference in aldosterone response (Stephenson et al. 1989). In the early follicular phase, aldosterone response to exercise is more pronounced than in the middle of the luteal phase. Differences in vasopressin and renin responses were not significant (De Souza et al. 1989).

Atrial Natriuretic Peptide

Atrial natriuretic peptide (ANP, also called atrial natriuretic factor) is a regulatory peptide that is stored in the cytoplasmic granules in heart atria. When atria become stretched, the peptide releases from the atrial walls into the circulation. Heart failure is associated with an excessive increase in atrial pressure accompanied by as much as a 5-fold to 10-fold increase of circulating levels of ANP. Independently, on atrial stretch, release of natriuretic peptide appears when the rate of atrial contraction increases.

ANP increases natriuresis and, thereby, renal excretion of water. It contributes to blood pressure regulation and relaxation of vascular smooth muscles. The peptide inhibits aldosterone biosynthesis and the release of renin. Consequently, this peptide is opposite the renin-angiotensin–aldosterone system. However, the effects of ANP are considered to be short term (e.g., under the influence of ANP, the kidney increases sodium and water excretion threefold to 10-fold, but this is only a temporary effect).

During exercise, ANP level in the blood increases (Cuneo et al. 1988; Follenius and Brandenberger 1988). The increase may be more than three times the basal level (Cuneo et al. 1988). After a marathon race, high levels of ANP persisted for 12 h (Lijnen et al. 1987). High ANP levels also persisted in military cadets during a 6-day military training course (Opstad et al. 1994).

Sodium loading increased ANP response but suppressed plasma renin and aldosterone responses to exercise (Cuneo et al. 1988).

A marathon race that reduced plasma volume by 7.4% caused a 2.5-fold increase in ANP concentration. Simultaneously, concentrations of aldosterone and vasopressin and renin activity increased (Altenkirch et al. 1990). At the end of 18 km of swimming in the sea, plasma renin activity and aldosterone concentration were unchanged, but ANP and vasopressin levels decreased (Bonifazi et al. 1994).

In heat, ANP responses to exercise were not significant, although blood levels of aldosterone, cortisol, and renin increased (Kraemer et al. 1988). Rehydration, restoring the plasma volume, decreased ANP response to exercise, as well as vasopressin, corticotropin, and cortisol concentrations during exercise in heat (Follenius et al. 1989). When a 10-min exercise was performed in an anti-gravity suit, the responses of aldosterone, corticotropin, and norepinephrine were suppressed, but the responses of ANP and epinephrine increased (Guezennec et al. 1989).

An incremental exercise (4-min stages at 30%, 60%, 80%, and 100% of $\dot{V}O_2max$) caused parallel increases of ANP and aldosterone levels and plasma renin activity. Simultaneously, plasma K^+ concentration increased and blood volume decreased (Mannix et al. 1990). The authors explained the parallel increases of ANP and aldosterone levels by the responses of ANP and the renin-aldosterone system to independent stimuli. ANP response might be related to atrial distention caused by increased venous return and an increased heart rate. Decreased plasma and blood volumes decreased renal perfusion pressure and activated the renin-aldosterone system.

Reduced pO_2 in inhaled air suppressed the exercise-induced response of aldosterone but did not change ANP response (Schmidt et al. 1990).

The results presented show that in normal conditions ANP and aldosterone levels increase during exercise, although their action on sodium and water excretion is the opposite. The final effect of these hormones on water-electrolyte metabolism obviously depends on their ratio, which may be changed diversely depending on the conditions of the exercise.

Calcitonin and Parathormone

Calcium level in extracellular fluid is controlled by the opposite actions of calcitonin and parathormone. When calcium concentration falls in the extracellular fluid, secretion of parathyroid hormone (parathormone) increases and secretion of calcitonin decreases by the direct effects of the calcium level on the parathyroid gland and on the thyroid cells producing calcitonin (parafollicular cells in the interstitial tissue between follicles of the human thyroid gland). An increased calcium concentration in blood plasma directly inhibits parathormone secretion and stimulates release of calcitonin. In turn, parathormone increases release of calcium from bones into the extracellular fluid. When parathyroid activity is suppressed (e.g., by a high calcium level), bones fail to show calcium outflow. Yet, calcium deposit continues; calcium is removed from the extracellular fluid, and decrease of calcium concentration is promoted until the parathormone level increases in the plasma.

Calcitonin promotes deposition of calcium in the bones and, thereby, decreases calcium concentration in the extracellular fluid.

Control of the extracellular calcium level is extended with the aid of the influence of these hormones on calcium reabsorption from renal tubules and calcium absorption in the gut. Both processes are stimulated by parathormone and inhibited by calcitonin.

The calcitonin feedback mechanism operates more rapidly, reaching peak activity in less than an hour. In contrast, 3 to 4 h are required for peak activity of the parathormone feedback mechanism. Thus, calcitonin acts mainly as a short-term regulator of calcium ion concentration. A little later on, calcitonin effects will be overridden by a more powerful parathormone control mechanism.

Exercise Effects

In 1978, Cornet et al. reported that a treadmill exercise at 50% $\dot{V}O_2max$ caused an elevated calcium content in plasma and increased renal reabsorption of calcium. However, the concentration of parathormone in blood plasma did not change. The results of another study showed that at the end of a 20-min exercise, an increased level of calcium ions was observed in association with an increase in calcitonin and a decrease in parathormone levels (Aloia et al. 1985). Obviously, observed changes in the concentration of hormones expressed homeostatic responses to the increased calcium level.

The effect of various exercises on the parathormone level has been studied in several other articles, with controversial results; some articles indicate an increase (Ljunghall et al. 1986, 1988), while others indicate a decrease (Vora et al. 1983; O'Neil et al. 1990). In one article, both possibilities were observed (Brandenburger et al. 1995). The most recent article shows a 50% increase of parathormone plasma concentration after 60 min of bicycling (Tsai et al. 1997).

Rat experiments point out the significance of these hormones for endurance capacity. In thyroid-parathyroidectomized rats treated with triiodothyronine (absence of calcitonin and parathormone), maximal duration of running was only 30% of the duration in the thyroid-parathyroidectomized rats treated with triiodothyronine, calcitonin, and parathormone. When only calcitonin was absent (i.e., operated rats were treated with triiodothyronine and parathormone), duration of running was 50%; when only parathormone was absent (i.e., operated rats were treated with triiodothyronine and calcitonin), running time was 75% of the control rats (Tsõbizov 1978).

Sex Hormones

Attention to sex hormones was raised several decades ago, when synthetic anabolic steroids became available for athletes. They believed they could increase the effectiveness of strength training (mainly muscular hypertrophy) by using these drugs. The use of these drugs was justifiably considered doping. Because the main substance of anabolic preparations is derived from testosterone, trainers became interested in the function of endogenous testosterone.

However, the use of testosterone for training monitoring is not enough to believe the positive significance of testosterone for muscle strength. It is necessary to know the overall function of testosterone in metabolic control, time parameters of the testosterone action, and the testosterone pattern during exercise.

Less attention has been paid to female sex hormones, although their level alters significantly during the ovarian-menstrual cycle. Therefore, relationships of effects of other steroid hormones may be influenced by female sex hormones.

Pituitary-Testicular System

The pituitary-testicular system consists of gonadotropic hormones (gonadotropins), lutropin (luteinizing hormone), and follitropin (follicle-stimulating hormone) secreted by the anterior lobe of the pituitary gland and testosterone secreted by the testes.

The functions of the testes are the production of spermatocytes and the male sex hormone testosterone. Production of spermatocytes (spermatogenesis) takes place in Sertoli cells under the stimulating influence of follitropin. Without this stimulation, the process of spermatogenesis will not occur. Leydig cells located in the interstitium of the testes secrete testosterone. Lutropin stimulates the secretion of testosterone. A minimal amount of estrogens is formed from testosterone by the Sertoli cells under the influence of follitropin. The estrogens formed are thought to be essential for spermatogenesis. An essential function has also been found for testosterone in spermatogenesis.

The system is regulated by feedback mechanisms (influence of blood level of testosterone on gonadotropin secretion) and by central nervous system action reaching the pituitary gland with the aid of the hypothalamic neurosecretory hormone gonadoliberin (gonadotropin-releasing hormone). Its function is to stimulate secretion of lutropin and follitropin. The testosterone feedback inhibitory action is addressed to the hypothalamic neurosecretory cells to suppress release of gonadoliberin. At the same time, testosterone also acts directly on the anterior lobe of the pituitary gland, inhibiting release of lutropin.

Metabolic Action of Testosterone

Besides the function of testosterone in sexual behavior and the development of male sex characteristics, it also contributes to hormonal metabolic control. Almost all metabolic effects of testosterone result from the increased rate of protein synthesis in the target cells. Testosterone enters the cell and may be converted to dihydrotestosterone; this happens in the prostate gland. Conversion of testosterone to dihydrotestosterone is not found in muscle fibers and bone cells. Testosterone in muscle and bone and dihydrotestosterone in several other cells are bound with the specific cytoplasmic receptor. The steroid-receptor complex translocates into the nucleus, where it induces DNA-RNA transcription. Within 30 min, RNA polymerase is activated, and the concentration of RNA begins to increase in the cells. After that, there follows a progressive increase and the synthesis of certain cellular proteins.

Anabolic action Testosterone is considered the main anabolic hormone because it stimulates protein synthesis mainly in muscle and skeletal tissues, accounting for more than half of the body's mass. Ontogenetic development of skeletal muscles is deeply related to testosterone metabolic effect. Testosterone effect ensures the male type of body build developing from the final stages of puberty and characterized by the increased musculature over that of the female. Under the influence of testosterone, fast-twitch fibers develop most of all. Therefore, male adolescents exhibit an intensive improvement of muscle strength and power output. Increased testosterone level after sexual maturation also warrants good faculties for strength, power, and speed training.

Effect on bone growth and calcium retention Testosterone increases the total quantity of the bone matrix and causes calcium retention. The increase in bone matrix is believed to result from the general protein anabolic function of testosterone in combination with deposition of calcium salts. Therefore, after puberty, the bones grow considerably in thickness and also deposit additional amounts of calcium salts. Simultaneously, testosterone has a specific effect on the pelvis: it narrows the pelvic outlet, lengthens it, causes a funnel-like shape, and greatly increases the strength of the entire pelvis for load bearing.

Other effects During adolescence and early adulthood, testosterone increases basal metabolism by approximately 5% to 10%. Testosterone is capable of increasing the rate of erythropoiesis. It also exerts a modest influence on sodium reabsorption in renal tubules.

It is thought that testosterone influences nervous function. The expression of that is aggressive behavior. Little is known about the mechanism of testosterone influence on nerve tissue. In birds, testosterone regulates the number of acetylcholine receptors in muscles. It has been suggested that testosterone also contributes to control over the calcium-handling mechanism in muscle.

Exercise Effects of Male Sex Hormones

Data on exercise-induced changes of testosterone are variable (see Viru 1992a; Hackney 1996). Galbo et al. (1977c) found a significant increase of testosterone concentration after the last stage of incremental exercise when exercise intensity was 100% $\dot{V}O_2$max. The mean increase was only 13% (individuals vary between 1% and 24%) and no change of lutropin was found. The authors suggested that the increased testosterone level was caused by a reduction of plasma volume. When a 20-min trial of exercise at 75% of $\dot{V}O_2$max was repeated over 10-min rest intervals, after the first two trials, testosterone level increased 31% (individual variations, 14% to 50%), which was followed by a decrease after subsequent repetition. Again, lutropin failed to show any change. Similar, but not exactly the same, dynamics were found after treadmill exercise (Bruce protocol, total duration 12 to 15 min). Immediately after exercise, testosterone concentration was modestly increased, but lutropin level remained unchanged. In the postexercise period, testosterone and lutropin levels declined, with nadir values between 60 and 180 min after exercise. Because exercise induced an increase of plasma corticoliberin concentration, the authors suggested that corticoliberin depressed lutropin secretion. No change was found in follitropin concentration (Elias et al. 1991).

According to the results of Jezova et al. (1985), testosterone concentration did not increase either during 6-min incremental exercise (heart rate, 126 to 156 bpm) or during three repetitions of 2-min moderate exercise (heart rate, 133 to 150 bpm), but rose after two repetitions of vigorous exercise (heart rate, 165 to 177 bpm) for 4.5 min each.

Wilkerson et al. (1980) confirmed the possibility that the increase of testosterone concentration is due to reduced plasma volume. They determined testosterone concentration in blood plasma after 20 min of exercises at 30%, 45%, 60%, 75%, and 90% $\dot{V}O_2$max. The concentration of testosterone increased in correlation with exercise intensities. However, the reason was hemoconcentration. When the testosterone amount in total blood plasma was calculated, no significant change appeared. Another possibility for the increases in testosterone concentration without stimulation of secretion by lutropin is related to testosterone degradation. It has been shown that during exercise, the elimination rate of testosterone from blood decreased (Sutton et al. 1978).

A vigorous anaerobic exercise, running 3 × 300 m, resulted in an immediate increase of testosterone and lutropin concentration. Postexercise levels of testosterone at 0.5 to 6 h were less than the initial levels, and the lutropin level corresponded to the initial values (Adlercreutz et al. 1976). A short-term strenuous jumping exercise (the Bosco test consisting of 1-min consecutive vertical jumping) increased testosterone concentration by 12% (Bosco et al. 1996a).

Kindermann and Schmitt (1985) reported that during short-term exercise testosterone level increased, but during prolonged exercise it declined. Increased levels of testosterone, lutropin, and follitropin were found in athletes after running 800 m. After 36 km of skiing, testosterone concentration was increased in the three best skiers but decreased in others (Schmid et al. 1982). Kuoppasalmi (1980) showed that testosterone concentration is increased immediately after a 2-min anaerobic exercise but not after a 45-min aerobic exercise. Webb et al. (1984) studied testosterone dynamics during a 2-h exercise at 70% $\dot{V}O_2$max. Testosterone concentration increased in men during the first 30 min. Afterward, a decline followed. At the end of exercise, the testosterone level was less than the initial values. In women, testosterone concentration increased until the end of the exercise. The reason is that in women testosterone production reflects the intensity of adrenal steroidogenesis but not gonadal function.

A fall in testosterone level was found after a marathon race in combination with variable changes of the lutropin level. However, the best marathon runner did not show any decrease in testosterone, but the increase in lutropin concentration was more than twice the basal level. A runner who collapsed after running 15 km had very low testosterone and lutropin values (Dessypris et al. 1976).

Guiglielmini et al. (1984) found an increased testosterone level in competitive walkers after a 20-km race (by 52%), in middle-distance runners after a 1-h training session (by 38%), and in marathon runners after the race (by 45%), but a decreased level was found in ultramarathon runners (by 32%).

Hackney et al. (1995) compared testosterone and lutropin dynamics in a 1-h continuous aerobic exercise at 65% $\dot{V}O_2max$ and a 1-h intermittent exercise (2-min anaerobic bouts at 110% $\dot{V}O_2max$ followed by 2-min aerobic bouts at 40% $\dot{V}O_2max$). Both exercises evoked a significant increase of testosterone concentration (the magnitude of responses was almost equal). No change was found in lutropin concentration. Our data did not allow us to establish any common variants in the testosterone dynamics during a 2-h exercise (Viru et al. 1992a).

Heavy-resistance exercise evokes a testosterone response. An increase of testosterone concentration has been found after four series of six squats at 90% to 95% of 6 RM and after four series of 9 or 10 squats at 60% to 65% of 6 RM (Schwab et al. 1993). High testosterone levels are common after heavy-resistance training sessions in strength-trained persons (Häkkinen et al. 1988b; Kraemer et al. 1990b, 1992; Volek et al. 1997), untrained university students (Jürimäe et al. 1990a), and junior weightlifters (Kraemer et al. 1992). Testosterone response depends on the workout loads in the training session (Häkkinen and Pakarinen 1993; Cotshalk et al. 1997). In women, as opposed to men, weightlifting (Weiss et al. 1983) or other heavy-resistance training sessions (Kraemer et al. 1991b, 1993a) did not increase the blood level of testosterone.

Testosterone response depends on psychological factors. In cycling at 80% $\dot{V}O_2max$, the testosterone increase was reduced in persons with high trait-anxiety rather than low-anxiety, although lutropin level was higher in high trait-anxiety men (Diamond et al. 1989). According to an interesting observation, testosterone concentration was increased in winners but decreased in losers after a tennis match (Booth et al. 1989) or judo fighting (Elias 1981).

Endurance training results in a decreased basal testosterone level (Hackney 1989, 1996). However, the endurance-trained athletes responded to an incremental treadmill exercise test from the lowered initial level by a pronounced testosterone increase (Hackney et al. 1997).

The metabolic effects of testosterone, mediated by their cytoplasmic receptor, appear after a lag period of 1 or more hours. Therefore, it is impossible to believe that testosterone has an effect on performance during short-term intensive dynamic or acyclic power or strength exercises, although a correlation has been established between the initial level of testosterone and performance in the vertical jump and sprinting (Bosco et al. 1996b). It has been suggested that the possible effect of testosterone on intracellular calcium shifts may not need the induction of regulatory protein throughout the formation of the steroid-receptor complex and its influence on the genome. According to this suggestion, it is believed that testosterone may act on calcium movement similar to glucocorticoid permissive action. However, no strict evidence exists for this mechanism of testosterone action and for the time characteristics of this mechanism. Therefore, it is better to suggest the preconditioning action of the initial level of testosterone on performance of the neuromuscular apparatus. It may be similar to the action of testosterone on aggressiveness (Olweus et al. 1980). Long-term preconditioning may be related to testosterone influence on the development of fast-twitch fibers established in animal experiments (Dux et al. 1982). Thus, those persons who have high levels of testosterone from the pubertal period are preconditioned for muscular performance, which is based on the activity of fast-twitch fibers. This possibility is confirmed by the high basal levels of testosterone in qualified sprinters (Bosco and Viru 1998).

The long-term action of testosterone on muscular adaptation is a well-established fact (see chapter 2, p. 16). Therefore, testosterone responses during a training session and testosterone's pattern in the postexercise recovery period are essential means for heavy-resistance training effects on skeletal muscles.

Pituitary-Ovarian System

The pituitary-ovarian system consists of two pituitary (lutropin and follitropin) and two ovarian (estrogens and progestins) hormones. In blood plasma, estrogens are presented in three fractions: β-estradiol, estrone, and estriol. The estrogenic potency of β-estradiol is 12 times that of estrone and 80 times that of estriol; the major estrogen is β-estradiol.

By far, the most essential progestin is progesterone. Another progestin, 17α-hydroxyprogesterone is secreted in small amounts.

The female reproductive function is also controlled by prolactin (produced in the anterior lobe of the pituitary gland) and by oxytocin (formed in hypothalamic neurosecretory cells and released into the circulation from the posterior lobe of the pituitary gland).

The major quantities of estrogens and progesterone are secreted by the ovaries. Minute amounts of these hormones are released from the adrenal cortex into the circulation as by-products of the biosynthesis of adrenocortical steroids. Therefore, estrogens and progesterone are in men's blood but in low concentrations. Because progesterone is secreted mainly during the luteal phase of the ovarian-menstrual cycle, in the follicular phase, the amount of progesterone in blood originates from the adrenal cortex. During pregnancy, large amounts of progesterone are secreted by the placenta.

The sites of production of estrogen are cells of the follicle walls. A co-function of follitropin and lutropin stimulates the growth and development of follicles and secretion of estrogen into the follicles. In women with a normal 28-day cycle, 14 days after the onset of menstruation ovulation occurs (a rapid swelling of the outgrowing follicle followed by its rupture with a discharge of the ovum). Before ovulation, the lutropin level increases promptly, reaching peak values approximately 16 h earlier. The follitropin level also increases.

Shortly before ovulation, the secretion of progesterone begins. The following increase in progesterone production is related to a rapid change of secretory cells of the follicle into lutein cells. These cells together constitute the corpus luteum. Lutein cells also secrete estrogens but in lesser amounts than progesterone. Lutropin stimulates the formation of lutein cells and the secretory activity of the corpus luteum.

Estrogen and Progesterone

The main function of estrogen is to cause cellular proliferation and growth of the tissues of the sexual organs and other tissues related to reproduction. Estrogens are responsible for the initiation of the development of breast glands and the deposition of fat in the breasts.

Estrogens result in increased osteoblastic activity. Estrogen deficiency after menopause diminishes osteoblastic activity in the bones, decreases bone matrix, and reduces deposition of bone calcium and phosphate. Estrogens increase the basal metabolic rate three times less than testosterone. Their metabolic effect is also deposition of fat in subcutaneous tissues.

The most important function of progesterone is to promote the secretory changes in the uterine endometrium after ovulation to prepare the uterus for implantation of the fertilized ovum. At the same time, progesterone is the gravidity hormone. It ensures the normal progression of pregnancy. Progesterone and prolactin also cause determinate growth and regulate the function of the breast glands.

Female Sex Hormones in Exercise

Exercise-induced changes of β-estradiol and progesterone, as well as of lutropin and follitropin, are variable (see Cumming and Rebar 1985; Viru 1985a, 1992). Because the initial level of these hormones depends on the phases of the ovarian-menstrual cycle, exercise-induced changes cannot be unanimous if the cycle phases are not taken into account.

Several years ago it was reported that in 20-min running at 60% to 65% or 80% to 95% $\dot{V}O_2$max, blood progesterone and β-estradiol concentrations increased in the luteal phase. In the follicular phase, no significant change of progesterone level was detected. β-Estradiol response was found only at the highest exercise intensity in this phase (Jurkowski et al. 1978). The results of Bonen et al. (1979) indicated that estradiol, progesterone, and lutropin responses increase with exercise duration and are most pronounced when exercise duration is greater than 40 min. Estradiol and progesterone responses were more pronounced in the luteal phase, whereas lutropin response was more pronounced in the follicular phase. However, a pronounced increase in lutropin, estradiol, cortisol, androstenedione, and dehydroepiandrosterone levels was found in trained and untrained women when a 15-min exercise was performed in the early follicular phase (Cumming and Rebar 1985).

More recent studies showed that plasma estradiol, progesterone, testosterone, androstenedione, prolactin, and corticotropin levels increase in 15-min exercises with intensities of 60%, 70%, or 80% $\dot{V}O_2max$, irrespective of the phases of the ovarian-menstrual cycle. However, higher exercise intensity was necessary for evoking estradiol response in the luteal than in the follicular phase in female marathon runners and for progesterone and lutropin responses in the follicular phase in untrained women (Keizer et al. 1987a). Free testosterone response was similar in both phases in untrained and trained women, but no response of free estradiol was found (Keizer et al. 1987b).

To interpret these results, two points must be considered. First, the significance of different intensities and duration of exercise may vary in the follicular and luteal phases. Second, in the follicular phase, estrogen response depends to a degree on the growth and development of follicles; progesterone response is possible only at the expense of the release of progesterone from the adrenal cortex.

Moreover, the combination of regulating actions undoubtedly includes the feedback influence exerted by the level of β-estradiol. Thus, in 15- to 17-year-old female runners, lutropin and β-estradiol levels were elevated after an incremental exercise up to an intensity of 100% $\dot{V}O_2max$ in the luteal but not in the follicular phase. Thus, in the follicular phase, the higher initial level probably inhibited the response. However, progesterone response appeared only in the follicular phase because of progesterone, which originates from the adrenal cortex (Szczepanowska et al. 1999). Lack of progesterone response in the luteal phase suggests two possibilities: either the corpus luteum is not capable of responding by increased progesterone production in exercise or the high progesterone level participated in the feedback inhibition of progesterone secretion.

During exercise, elevated prolactin level has been frequently observed in males (Viru et al. 1981) and females (Keizer et al. 1987a), although a pronounced variability exists in responses. In girls aged 15 to 17 years, prolactin response appeared only in the luteal phase in correlation with an increased cortisol level (Szczepanowska et al. 1999). The correlation with cortisol suggested that a certain relationship exists between adrenocortical activity and secretion of prolactin in exercise.

A study of girls aged 11 to 14 years showed that exercise-induced β-estradiol response appears at the beginning of puberty (stage 2 of sexual maturation by Tanner). The magnitude of β-estradiol response was lowest at the final period of sexual maturation (stage 4 by Tanner), when the basal level of estrogens was significantly increased compared with less maturated girls. The greatest magnitude of response was observed when sexual maturation was completed (stage 5 by Tanner) despite the high basal level. A question arises whether the pronounced response of β-estradiol from the high initial level reflects the functional maturation of the pituitary-ovarian system in girls (Viru et al. 1998).

Significant responses of progesterone appeared in stage 4 and of testosterone in stage 5 by Tanner (Viru et al. 1998). Obviously, the adrenarche (production of androgenic steroids by the adrenals) was essential for significant responses of both hormones, whereas the cyclic appearance of the corpus luteum might also support progesterone response.

It can be suggested that during exercise the changes of female sex hormones are related to the necessity to ensure a certain homeostasis in reproductive activities of women in a condition of high-energy expenditure caused by the muscular activity.

Almost nothing is known about the contribution of female sex hormones in adaptation to muscular activity. It seems justified to ask whether such a contribution exists at all. Yet, some facts allow us to suggest that female sex hormones may interfere with metabolic control exerted by steroid hormones in exercise.

One such possibility arises from the fact that a high level of progesterone and to a lesser degree of estrogens competes with other steroid hormones for their metabolic receptors (Bell and Jones 1979). This way, progesterone may reduce the metabolic effects of cortisol and, to some extent, even of testosterone. It has been shown that glucocorticoid administration does not increase the working capacity of adrenalectomized rats when progesterone is simultaneously administered in high doses (Viru and Smirnova 1985). Regarding these results, it is necessary to ask whether the increase of the progesterone blood level after ovulation is sufficient to produce competition between progesterone and cortisol for the glucocorticoid receptor. The possibility of competition between cortisol and progesterone for the glucocorticoid receptor on hormonal responses in exercise is indirectly supported by the following fact. During the luteal phase, a high progesterone level was compensated by increased levels of cortisol before and after exercise compared with the follicular phase characterized by a low progesterone level (Szczepanowska et al. 1999). Similarly, one may ask whether the increase of the estrogen level just before ovulation is sufficient for the competition between β-estradiol and cortisol for the glucocorticoid receptor. We still do not have experimental confirmations.

Another possibility is related to the results based on experiments. They showed that estrogens increase the sensitivity of tissue for anabolic action of testosterone (Danhaive and Rousseau 1988). Again, without experimental confirmation, it is only possible to speculate that before ovulation, the training effects on muscle tissue increased with the elevation of estradiol level. Such experiments would provide evidence about the interrelations between female sex and other steroid hormones, which in turn would provide knowledge that would help to guide training in women.

In female athletes, the significance of testosterone for muscle adaptation exists. Dehydroepiandrosterone produced by the adrenal cortex can be converted into testosterone and, therefore, a certain amount of testosterone is also in the blood of women. The low level of this hormone is obviously compensated by the estrogen-dependent increase of muscle-tissue sensitivity to the action of testosterone.

Endogenous Opioid Peptides

Endogenous opioid peptides (EOPs) are compounds similar to morphine but produced in the body. They are breakdown products of three large protein molecules: pro-opiomelanocortin, proenkephalin, and prodynorphin.

Pro-opiomelanocortin is the precursor for corticotropin, melanocyte-stimulating hormone, β-lipotropin, and β-endorphin. The latter is one of the most important EOPs found in the central nervous system and peripheral blood. In the central nervous system, the main sites of β-endorphin production are several hypothalamic neurons (mainly in the median eminence and areas of nucleus arcuatus and nucleus paraventricularis) and neurons and synapses located in other brain structures. The contribution of β-endorphin in the regulation of blood pressure, pain perception, and thermoregulation is actualized through neurons in the medial eminence of the hypothalamus, midbrain, and rostral parts of the medulla oblongata.

β-Endorphin of the peripheral blood originates from the anterior lobe of the pituitary gland, which is rich in pro-opiomelanocortin. Under the influence of corticoliberin, a breakdown of pro-opiomelanocortin occurs, resulting in the release of corticotropin and β-endorphin. Vasopressin has a similar influence. Corticoliberin and vasopressin potentiate the action of each other. The production of endorphin has also been found in the adrenal glands and pancreas.

The most important products of the breakdown of proenkephalins are leu-enkephalin and met-enkephalin. These peptides have been found in the hypothalamus; structures of the brain limbic system; basal ganglions; sensory areas of the brain cortex; and structures of the brain analgesia system, including dorsal horns of the spinal cord (a pain inhibitory complex). Encephalins are present in nerve terminals surrounding the nucleus tractus solitarius, dorsal vagal nucleus, and nucleus ambiguous. These areas are involved in autonomic circulation control. Peripheral enkephalins originated mainly from the adrenal medulla.

The products of prodynorphin breakdown are dynorphins. Dynorphin is important, although found only in minute quantity, because it is an extremely powerful opiate having a painkilling effect 200 times greater than that of morphine. Dynorphins are found in the hypothalamus (in the area of nucleus paraventricularis) and several other structures related to the control of the autonomic nervous system and in the spinal cord, where they modulate the pain sensitivity together with enkephalins. From the area of the nucleus paraventricularis, dynorphins release together with vasopressin. Production of dynorphins has also been established in the adrenal cortex.

Although there are certain exceptions, EOPs as a rule are not able to pass through the blood-brain barrier. Therefore, opioids produced in the brain and those originating from the pituitary gland, the adrenal glands, or other peripheral sites consist of two separate pools. These two pools have different locations of action and therefore different functions.

Opioids of the brain pool modulate the activity of several central regulatory mechanisms. Therefore, brain opioids are involved in reduction of pain sensitivity, emotional behavior, food intake, glycostatic regulation, thermoregulation, control of cardiovascular and respiratory functions, and regulation of immunoactivities. Mostly EOPs actualize their functions acting as neurotransmitters in the nerve endings of opioidergic neurons. The specific opioid receptors are located either on the presynaptic terminals of axons or on the neuron's soma and dendrites.

The release of peripheral opioids from the pituitary gland and adrenal medulla is concomitant to the secretion of corticotropin and catecholamine, respectively. This fact suggests that peripheral opioids should contribute to the adaptation processes. β-Endorphin and enkephalins modulate the various regulatory influences of sympathetic nerves. Opioid receptors, present on peripheral sympathetic and cardiovascular structures, confirm the possibility of their peripheral effects.

In stress situations, brain and peripheral opioid subsystems are highly active. Obviously, the modulating effects of brain and circulating opioids are essential for effective adaptation. Moreover, the central effects of opioids on the brain structure help a person to endure distress.

Exercise Effects on Peripheral EOP Subsystem

Data about the activation of this subsystem in severe exercise began to accumulate during the early 1980s. In 1980, Fraioli et al. reported that running on the treadmill with an increased speed up to the level of $\dot{V}O_2$max caused an increase of plasma β-endorphin level (almost fivefold), with simultaneous elevation of plasma corticotropin content. The next year, Gambert et al. (1981) confirmed a simultaneous rise in blood levels of corticotropin and β-endorphin during a 20-min run. Several publications followed, confirming the exercise-induced increase of β-endorphin concentration in plasma (Carr et al. 1981; Colt et al. 1981; Farrell et al. 1982). Rahkila et al. (1988) convincingly demonstrated that β-endorphin response appears when exercise intensity is over a certain threshold. Intensity thresholds for β-endorphin and corticotropin are the same. In principle, this result was confirmed by Schwarz and Kindermann (1990).

Comparison of the effects of six heavy-resistance exercise protocols showed that the plasma β-endorphin level increased in a repetition of 10 RM exercise series with a 1-min interval between series but not in protocols of lower workload (Kraemer et al. 1993b).

Brooks et al. (1988) showed that sprint running for 30 s was enough to evoke β-endorphin response. According to the results of Schwarz and Kindermann (1990), β-endorphin and corticotropin concentration in blood was significantly increased after 5 min but not immediately after a 1-min anaerobic exercise test. In this case, the magnitude of responses was less than in more prolonged incremental exercise.

During a 2-h bicycling exercise, several variants of the dynamics of β-endorphin responses appear. The most frequent variant was the initial increase during the first 20 min, then a reduction that was followed by a secondary increase up to the end of exercise (Viru et al. 1990). A pronounced increase of β-endorphin concentration has been found after marathon races (Heitkamp et al. 1993).

Carr et al. (1981), and Farrell et al. (1987) found that exercise-trained women and men showed more pronounced β-endorphin response than untrained persons. Kraemer et al. (1989a) reported that after 10 weeks of sprint training, β-endorphin response increased in the maximal exercise test. After endurance training, the response was the same as before the training, whereas combined endurance and sprint training resulted in reduced β-endorphin response to maximal exercise.

β-Endorphin is metabolized to γ- and α-endorphins. In trained persons, a 2-h bicycling exercise caused a β-endorphin level increase together with increases in blood levels of γ- and α-endorphin. These changes were not found in untrained persons. Their levels of γ- and α-endorphin were significantly higher than in trained persons before and during exercise. Obviously, training influences the metabolism of endorphins (Viru and Tendzegolskis 1995).

The exercise effect of blood levels of enkephalins is less studied in humans. An incremental treadmill test (20 min running at 60%, 70%, and 80% $\dot{V}O_2$max) significantly increased met-enkephalin and β-endorphin levels. After 10 weeks of running training, met-enkephalin response, unlike β-endorphin response, diminished (Howlett et al. 1984). Exercises also induce an increase of blood concentration of proenkephalin peptide F if the exercise intensity is 75% or 100% of $\dot{V}O_2$max (Kraemer et al. 1985, 1990a).

Exercise Effects on Brain EOP Subsystems

Rat experiments confirmed that exercises induce significant changes of β-endorphin, leu-, and met-enkephalin contents in several brain structures, as well as the pituitary and adrenal glands (Blake et al. 1984; Orlova et al. 1988; Tendzegolskis et al. 1991). Exercise training increases β-endorphin content in the brain cortex and striatum and in the pituitary and adrenal glands (Orlova et al. 1988). Changes are also found in opioid receptor occupation in the brain after acute and chronic exercises (Christie 1982; Christie and Chester 1983; Sforzo et al. 1986).

In rats, running increases the β-endorphin concentration in the cerebrospinal fluid (Hoffmann et al. 1990b). In rats, hypothalamic β-endorphin release increased in association with augmentation in the pain threshold during prolonged exercise (Shyu et al. 1982). Blockade of the opioid receptor with naloxone also removed the jogging-induced increase in the pain threshold in men (Haier et al. 1981). Psychological testing before and after a noncompetitive marathon run failed to confirm that naloxone reversed mood changes associated with running (Markoff et al. 1982). However, exercise-induced joy and euphoria were reversed, in part, by naloxone (Janal et al. 1984).

In conclusion, EOPs modulate the transmission of pain impulses, influence behavioral and mood response, and contribute to the regulation of various functions. In this connection, it has been speculated that not only decreased pain sensitivity but also positive feelings (runners high) are related to increased activity of the EOP system.

Summary

Exercises intense and/or prolonged enough (exceeding the thresholds of intensity or duration) evoke a generalized alteration in the functions of the endocrine system. A wide spectrum of changes of hormones in blood may be recorded. Most of them express the activation of the mechanism of general adaptation, with the aim to mobilize the body's energy reserves and protein resources and to adjust immune activities and other defense faculties to the necessary level. Increased activity of the sympatho-adrenal system is the main tool for mobilizing energy reserves. Mobilization of energy reserves is supported, controlled, and balanced by actions of cortisol (permissive action) insulin, glucagon, and growth hormone. Mobilization of protein resources consists of the creation of a free amino acid pool, which may be used as an additional energy source (through oxidation of branched-chain amino acids and their use for gluconeogenesis and for supporting the tricarboxylic acid cycle). The main role belongs to the cortisol action that is balanced by testosterone and insulin. Activation of the endogenous opioid system synchronously with increased activity of the pituitary-adrenocortical system provides the possibility for modulating neuronal and hormonal responses but mainly for mood changes promoting the performance of serious efforts and helping to endure the accompanying possible unpleasantness.

Another function of hormonal responses in exercise is to contribute to homeostatic regulation. The well-balanced actions of insulin and glucagon are crucial for maintaining a constant level of blood glucose. This interplay is supported by other hormones. The main functions of vasopressin, aldosterone, and atrial natriuretic peptide are to controls water and electrolyte balances. The ratio between parathormone and calcitonin ensures constancy of the calcium level in the blood.

Several hormones fulfill their main function during the recovery period. Besides hormones controlling the replenishment of energy resources, testosterone, thyroid hormones, growth hormones, insulin, and cortisol are essential to control adaptive protein synthesis.

Exercise-induced hormonal responses provide comprehensive opportunities for use in training monitoring. Mostly, however, the information that is obtainable is essential for analysis of metabolic changes in the bodies of athletes. Hormonal responses may be used for evaluation of the intensity of training sessions and for other similar applied tasks. In these cases, the researcher has to decide whether the information available from the less-complicated and less-expensive testing procedures is enough for effective guidance of training or whether hormonal studies are necessary. Nevertheless, several situations and tasks for training monitoring necessarily need hormonal studies. Then, the researcher has to know what the function of the contributing hormones is and what is known about the metabolic role of the hormones overall as well as in the particular situation. It is wise to never overvalue the role of a single hormone or ignore the interrelationships between several hormones in metabolic control.

CHAPTER

6

Hematological and Immunological Indexes and Water-Electrolyte Balance

In addition to metabolites, substrates, and hormones, the means for monitoring training have been extended to blood constitution, indexes of immune activities, and water-electrolyte balance.

Blood is an important part of the internal environment of the body. Blood studies make it feasible to obtain information about outflow of metabolites from tissues and about exhaustion of essential substrates. The oxygen transport function of the blood is important in the adaptation to muscular activity. Blood studies provide an opportunity to characterize the effectiveness of the homeostatic regulation necessary for maintaining constancy of the rigid parameters of the internal environment of the body (e.g., temperature, content of ions and water, osmotic pressure, pH, partial pressure of O_2, and glucose level). Exercise performance depends to a great extent on the effectiveness of homeostatic regulation. Therefore, in this chapter, in addition to discussing opportunities to use blood parameters in training monitoring, special attention is paid to regulation of water and electrolyte balance in exercise.

A special part of blood studies is aimed at providing information on immune activities. Immune activities constitute an essential part of adaptation processes, including adaptation to muscular activity. Immunological studies establish an essential link between performance improvement and health conditions in athletes. These studies inform trainers of the "health cost" of high performance and allow them to direct attention to possible risks.

Hematological Indexes

Adaptation to muscular activity is related to changes in total blood and blood plasma volumes. In several cases, these changes are essential for improved performance. At the same time, changes of plasma volume influence concentrations of blood constituents and thereby modify the results of the determination of metabolites, substrates, and hormones in blood. Muscular activity also induces alterations in blood cell counts and in the specific distribution of various cell types. Related studies of white blood cells (leukocytes) provide information on immune activities, whereas studies of red cells (erythrocytes) characterize improvement (or impairment) of the oxygen-transport function of the blood. Erythrocytes contain hemoglobin, protein-binding oxygen. The composition of this protein includes iron. Therefore, adaptation to endurance exercises is related to iron metabolism.

In training studies, analysis of hematological indexes has to include plasma proteins, buffer systems, blood antioxidants, and the entrance of enzyme protein released from muscle tissue in the blood plasma. Muscle enzymes in blood plasma are considered to be related to disorders of the plasma membranes of muscle fibers.

Blood Volume

In a normal adult, the total amount of blood is close to 5 L (about 6% of body mass). On average, 3 L of this are plasma (25 to 45 ml/L · kg body weight) and 2 L are blood cells (mostly erythrocytes). These values are individually variable, depending on sex, weight, training effects, and other conditions. The ratio between plasma and erythrocytes is usually expressed by the hematocrit value. This value is obtained by centrifuging blood in a hematocrit scale. The graduated tube allows investigators to estimate the percent of the erythrocyte mass (the average value for men is 48% and for women, 42%, with a range of 39% to 55% and 36% to 48%, respectively). The hematocrit value is correct if red cells become packed tightly in the lower part of the scale. Approximately 3% to 8% of plasma remains trapped among the cells. Therefore, true and measured hematocrit values have to be distinguished. The true value constitutes on average 96% of the measured hematocrit. Thus, to get the true value, one has to multiply the measured hematocrit by 0.96.

In the capillaries, arterioles, and other small vessels, the hematocrit value of the blood is less than in the large arteries and veins. This is because erythrocytes cannot flow near the walls of thin vessels as easily as plasma can. As a result, the ratio of plasma to cells is greater in small vessels compared to the ratio in large vessels. To get the average value of hematocrit, the measured hematocrit value obtained in blood samples from large vessels is multiplied by 0.87. The correcting factor is 0.91 if the true hematocrit value is used for the calculation of body hematocrit.

Blood volume measurement is determined by diluting radioactively labeled red cells in the circulation. ^{51}Cr binds tightly with erythrocytes. Therefore, this isotope is most frequently used for labeling red cells. Radioactivity of a blood sample allows us to calculate the total blood volume using the dilution principle after waiting until the injected labeled cells are equally distributed in the circulation.

Measurement of plasma volume is found by labeling plasma proteins (e.g., ^{131}I-protein) and measuring their dilution after injection into the blood. Another possibility is an injection of a vital dye (usually T-1824, called Evan's blue) that strongly attaches to plasma proteins and diffuses throughout the plasma with only small amounts leaking into the interstitium. When the plasma volume is known, the total blood volume may be calculated by the formula:

$$\text{Blood volume} = 1 - \frac{\text{Plasma volume} \times 100}{\text{Hematocrit}}$$

In this formula, the body hematocrit value has to be used.

An alternative technique has been proposed for the determination of hemoglobin mass and blood volume (Burge and Skinnar 1995). The method involves rebreathing the air containing a certain amount of carbon monoxide (CO). The dilution and distribution of CO in blood provide the possibility of calculating the hemoglobin mass and blood volume.

Blood Cells

The main site for the formation of blood cells is bone marrow. In circulating blood, all the cells are derived from the pluripotential hemopoietic stem cells (PHSC) of the bone marrow. Reproduction of these cells continues throughout a person's life. A portion of these cells remains like the original PHSC and is retained in the bone marrow. Larger portions of the reproduced stem cells differentiate further and form, after several stages of conversion, definite blood cells. Different pathways exist for producing erythrocytes, leukocyte-granulocytes (neutrophils, eosinophils, basophils) and monocytes, megakaryocytes (the latter divide into the blood platelets), and lymphocytes.

Erythrocytes

The main function of erythrocytes is to transport the oxygen carrier, hemoglobin. In blood capillaries surrounding lung alveoli, hemoglobin combines with oxygen. The resulting oxyhemoglobin is the oxygen transport form in arterial blood. In the condition of lowered pO_2 as it exists in peripheral tissues, oxyhemoglobin dissociates, releasing oxygen to be consumed by cells. 2,3-Diphosphoglycerate (present in high concentrations in erythrocytes) regulates the release of O_2 from oxyhemoglobin by decreasing the affinity of hemoglobin for oxygen.

Erythrocytes contain a large quantity of carbonic anhydrase. This enzyme catalyzes the reactions between carbon dioxide and water. These reactions are rapid thanks to the high catalyzing activity of carbonic anhydrase. Therefore, large quantities of CO_2 may be transported from tissues to lungs in the form of bicarbonate ions.

Hemoglobin is also an acid-base buffer. In this way, erythrocytes are responsible for most of the buffering capacity of blood.

The concentration of erythrocytes in blood for a normal average man is 5,200,000 cells per μl of blood and for a normal average women, 4,700,000 cells (individual variations range from 4.3 to 5.9×10^6 for men and 3.5 to 5.5×10^6 for women). Red blood cells may not contain more than 34 g hemoglobin per all cells because a metabolic limit exists for the cell's hemoglobin-forming mechanism. In normal persons, the percentage of hemoglobin is almost near the maximum in each cell. A lower level of hemoglobin in red cells is related to a certain deficiency of hemoglobin formation in the bone marrow. Whole blood of a normal healthy man contains 15.8 g of hemoglobin per all blood on average (range, 13.9 to 16.3 g). In normal healthy women, this value averages 13.5 g per all blood (range, 12.0 to 15.0 g).

Each gram of hemoglobin combines with approximately 1.39 ml of oxygen. Thus, in healthy sedentary men, about 21 ml of oxygen can be transported in each milliliter of blood, and in normal healthy women, 19 ml of O_2.

The total mass of erythrocytes is carefully controlled and kept at a constant level within narrow limits of changes in each person. An adequate number of erythrocytes have to be available. Erythrocyte concentrations do not increase so much as to impede blood flow. Tissue oxygenation is the main determinant for the related homeostatic mechanism. Erythropoietin (EPO) is the main tool for actualization of the corresponding homeostatic responses. Any decrease of tissue oxygenation stimulates the production of EPO. The main site for EPO synthesis is the kidney (produces 80% to 90% of the total amount of EPO, the remainder is produced by the liver). In a hypoxic atmosphere, EPO level in blood begins to increase within minutes or hours. The maximum will be reached within 24 h. The intensive production of EPO continues as long as the hypoxic stimulus exists. The process will end either by removing low pO_2 or by increasing the amount of circulating erythrocytes enough to eliminate tissue hypoxia. At the same time, the rate of EPO production decreases. However, actualization of EPO action in forming new erythrocytes depends on nutritional status, particularly vitamin B_{12} and folic acid.

Synthesis of hemoglobin is related to iron metabolism. In blood, iron combines with apotransferrin (a β-globulin) to form transferrin. The latter is the transport form of iron. Release of iron from transferrin may happen in any tissue because it is linked loosely with globulin. In the cytoplasm of hepatocytes, iron combines with apoferritin and forms ferritin. Ferritin constitutes storage for iron and is also found in blood plasma (figure 6.1).

Transferrin, reaching the bone marrow, binds with receptors on the cell membrane of erythroblasts. This step is followed by the release of iron. In mitochondria, iron is used for the synthesis of the heme molecule. Finally, each heme molecule combines with a long polypeptide chain (a globulin synthesized by ribosomes), forming a subunit of hemoglobin. Four subunits together form a definite molecule of hemoglobin.

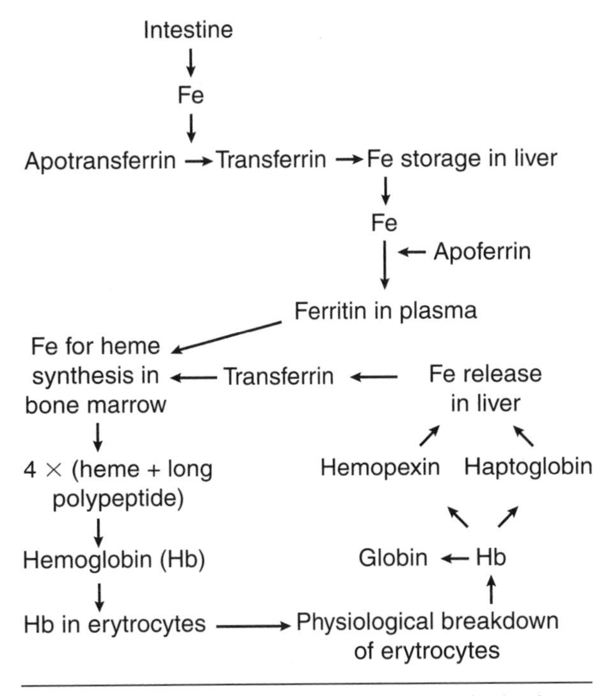

Figure 6.1 Schema of iron metabolism in the body.

In circulation, life of an erythrocyte is approximately 120 days. When senescent erythrocytes break up, the liberated hemoglobin is bound to a specific carrier protein, haptoglobin. Some of the plasma hemoglobin is cleared intravascularly into globin and heme, and the latter binds to another carrier protein, hemopexin. Hemoglobin-haptoglobin and heme-hemopexin complexes are cleared from the circulation by the liver and catabolized by hepatic parenchymal cells. Iron is transported into the plasma, where it is bound to transferrin. Transferrin carries iron to the cells of bone marrow for heme synthesis. Subsequently, iron is either incorporated into heme or stored as ferritin.

Leukocytes

Leukocytes constitute a mobile unit of the body's defense system. They may be transported to areas of serious inflammation and ensure a rapid defense against infectious agents.

The types of leukocytes are polymorphonuclear neutrophils (50% to 70% of the total number of leukocytes), polymorphonuclear eosinophils (0% to 3%), polymorphonuclear basophils (0% to 1%), monocytes (1% to 10%), and lymphocytes (20% to 40%). Neutrophils, eosinophils, and basophils have a granular appearance and are therefore called granulocytes. Granulocytes and monocytes ingest invading organisms (phagocytosis). Lymphocytes (as well as occasionally appearing plasma cells) are related to the immune system (see pp. 130-131 of this chapter).

The number of leukocytes in the blood of an adult human is, on average, 7000 cells/µl of blood (range, 4000 to 9000 cells/µl).

Megakaryocytes also belong to the group of white blood cells. In blood, fragments of these cells, called platelets or thrombocytes, are in abundance (150,000 to 300,000 platelets/µl). Thrombocytes are significant in the formation of clots within blood vessels.

Exercise Effects on Blood Volume and Cells

In training monitoring, attention should be paid first to acute exercise and training effects on blood volume, erythrocytosis, leukocytosis, hemoglobin, and iron metabolism. Together with changes related to essential adaptive changes, a possibility exists for a harmful consequence called sports anemia.

Plasma Volume

Since the 1930s, it has been known that plasma volume decreases during exercise (see Kaltreider and Meneely 1940). Later, the phenomenon was convincingly confirmed (Åstrand and Saltin 1964; Saltin 1964a, 1964b; Pugh 1969; Van Beaumont et al. 1973; Lundswall et al. 1972; Altenkirch et al. 1990). Exercises as short as 30 to 60 s may be effective. When a short-term anaerobic exercise was performed at a rate close to the highest possible, plasma volume decreased by 15% to 20% (Sejersted et al. 1986). The 30-s Wingate test caused a reduction of plasma volume by 17.4% ± 2.6%. When the test was repeated after 10 min, plasma volume reduction continued. The magnitude of the total response of two tests was 20.1% ± 3.1% (Whittlesey et al. 1996). In women, the Wingate test resulted in a 10% to 15% reduction of plasma volume (Rotstein et al. 1982). Five repetitions of high-speed running for 35 to 60 s with rest periods of 4 to 4.5 min reduced plasma volume by 20% to 25% (Hermansen et al. 1984). Simultaneously, a corresponding increase appeared in muscle water content (Hermansen and Vaage 1977).

Hemoconcentration develops from the onset of exercise as a result of an extravasation of plasma (figure 6.2). Most likely, the corresponding mechanism is related to the fast rise of mean blood pressure. This way, the hydrostatic pressure in capillaries increases, resulting in an elevated fluid filtration rate. If the pressure at the venous end of capillaries exceeds the osmotic pressure, the fluid reabsorption rate remains lower than the filtration rate. The comparatively low fluid reabsorption may be supported by the increased osmolarity of the extravascular fluid because of the accumulation of the metabolites of anaerobic metabolism (Lundswall et al. 1972). On the other hand, increased intramuscular fluid pressure that builds up during the muscle contraction acts in the opposite direction. In prolonged exercise, fluid loss by perspiration is also a factor acting on the plasma volume. However, when total body water declines during exercise, the level of plasma water is relatively well maintained (Kozlowski and Saltin 1964).

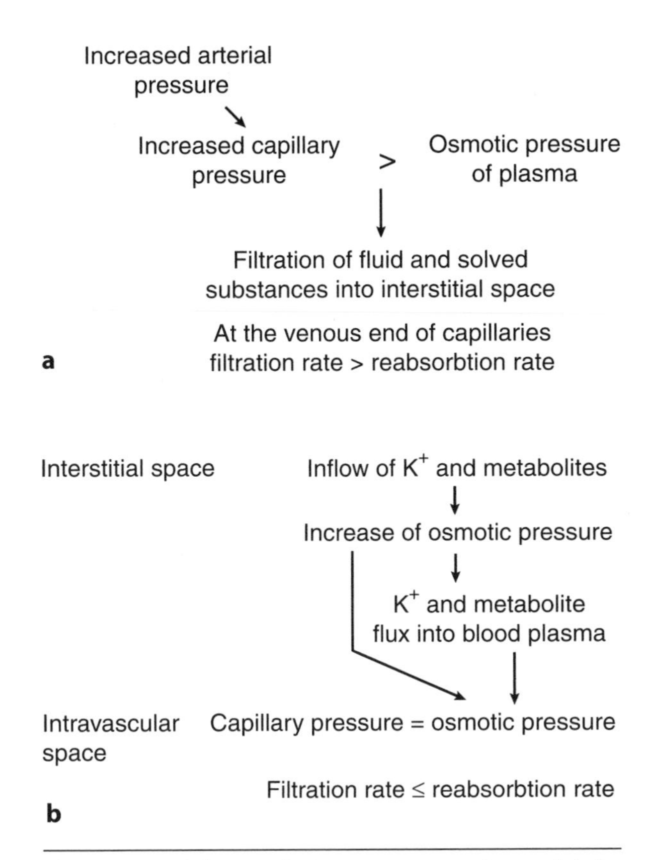

Figure 6.2 Schema of plasma extravasation at *(a)* the beginning of exercise and *(b)* the termination of this process when exercise is continued.

During prolonged exercises, plasma loss continues up to a certain amount of plasma. Possibly, the extravasated fluid fills the available extracellular space and, therefore, the interstitial pressure begins to resist further fluid shift out of the blood vessels (Sjögaard and Saltin 1982). Obviously, equilibrium was established between interstitial fluid pressure and intracapillary pressure (Greenleaf et al. 1977).

The changes of hematocrit do not precisely reflect the changes in plasma volume. Hematocrit changes are always less than changes in blood volume (Kaltreider and Meneely 1940; Åstrand and Saltin 1964; Saltin 1964a; Pugh 1969; Van Beaumont et al. 1973). Conflicting data have been obtained when exercise-induced changes in hematocrit and plasma protein concentrations were compared (Joye and Poortmans 1970; Poortmans 1970; Senay 1970). Therefore, neither hematocrit values nor plasma protein concentration provides valid information on exercise-induced changes in plasma volume.

Accordingly, Dill and Costill (1974) proposed a formula for estimating changes in blood plasma (table 6.1). In these calculations, both hematocrit value and hemoglobin concentrations were used. The authors used venous blood and corrected for 4% plasma trapped with the packed red cells. The hematocrit value has to be expressed in hundredths (the total blood is equal to 1.00). The initial value of blood volume, that is, blood volume before exercise (BV_B), was considered to be equal to 100.

Knowlton et al. (1990) compared these calculations, made on the bases of fingertip and venous blood after 20-min exercises at 100 and 200 Watts. Although a significant correlation was found between the results obtained with the aid of two different sources of blood collection, the plasma volume shift was greater in venous blood (–8.04%) than in fingertip blood (–6.25%). The authors proposed a formula for correcting values obtained by analysis of fingertip blood: $PV\% = 0.8662 \times \text{Fingertip}$. Using this formula, the standard error of plasma volume change was ±2.6%.

For an advanced method of the measurement of plasma volume in exercise that uses Evans blue spectra, the reader is referred to the article by Farjanel et al. (1997).

Erythrocytes

According to a popular opinion in the 1920s and 1930s, exercise causes erythrocytosis. In vigorous exercise, the concentration of circulating erythrocytes may be raised by 25% (DeVries 1974). At first, this change was explained as the mobilization of the blood depot because the deposited blood is rich in cells and poor in plasma compared to circulating blood. In dogs, spleen volume decreased 70% to 87% during exercise (Barcroft and Stephens 1927). This meant that approximately an additional 20% of the erythrocytes were released into the circulation. However, Dill et al. (1930) argued against the spleen's contribution to erythrocytosis during exercise because the spleen does not function as a blood depot in humans. However, the exercise-induced increase in the red blood cells might be related to the release of stored cells from other sites. The vessels in organs and tissues with low blood flow rate at rest may have significance in storing blood cells. A certain amount of red cells may be "washed out" from bone marrow

Table 6.1
Formula for Indirect Assessment of Changes in Blood Plasma Volumes

Index	Formula
Blood volume after exercise	$BV_A = (Hb_B/Hb_A)$
Volume of erythrocytes after exercise	$CA_A = BV_A(Hct_A)$
Plasma volume after exercise	$PA_A = BV_A - CA_A$
Change in blood volume in %	$\Delta BV\% = 100\ (BV_A - BV_B)/\ BV_B$
Change of volume of erythrocytes in %	$\Delta CV\% = 100\ (CV_A - CV_B)/\ CV_B$
Change of plasma volume in %	$\Delta PV\% = 100\ (PV_A - PV_B)/\ OV_B$

BV = blood volume; CV = volume of erythrocytes; PV = plasma volume; Hb = hemoglobin; HcT = hematocrit; delta (Δ) = shift
Subscript B and A refer to prior and subsequent exercise.
From Dill and Costill 1974.

because of the exercise-induced increased rate of circulation. In any case, the main reason for the increased erythrocyte count, as well as the elevated concentration of plasma proteins, is hemoconcentration during exercise.

Nylin (1947) showed that the intravascular erythrocyte mass might not change during exercise. This result confirms the viewpoint that the main factor causing increased erythrocyte concentration is decreased plasma volume. Lack of an increase in the total volume of red cells was confirmed by Oscai et al. (1968).

When one takes into account the time characteristics of erythropoiesis, it is difficult to believe that an intensive production of new cells could happen during exercise of less than several hours.

For the stimulation of erythropoiesis, the level of EPO has to increase. Elevated EPO activity was found in skiers (De Paoli Vitali et al. 1988) and runners (Vedovato et al. 1988) after a long-lasting event. Radioimmunological determination of EPO showed that a 60-min exercise at 60% $\dot{V}O_2$max and an incremental exercise test until exhaustion did not have an immediate effect on plasma EPO level either in normoxia or in normobaric hypoxia. Yet, 3 h after a 60-min exercise in hypoxia, an increase of erythropoietin concentration was observed. Twenty-four hours after exercises in normoxia, erythropoietin level tended to increase, but the response was not statistically significant (Schmidt et al. 1991). The results of Engfred et al. (1994) confirmed the lack of EPO responses in acute exercise. During exercise at 85% $\dot{V}O_2$max to exhaustion, EPO changes were observed neither before nor 5 weeks after training in normoxia or hypoxia. Immediately after a marathon race, erythrocyte concentration was increased because of hemoconcentration. The level of EPO started to increase 3 h after the race. Thirty-one hours after the run, this change was more pronounced (figure 6.3) (Schwandt et al. 1991). If an exercise had a direct effect on EPO production in the kidney, one would expect an increased serum EPO level not before 90 min after the beginning of the stimulus (Eckardt et al. 1989). Accordingly, the results referred to EPO response during the recovery period. Therefore, exercise-induced erythropoietic response is delayed and appears several hours after exercise. Late erythropoietic response has been confirmed by the release of immature reticulocytes from bone marrow into the peripheral blood within a few days after different types of exercises (Schmidt et al. 1988, 1989, 1991). Metabolic acidosis inhibits EPO response to hypoxia (Eckardt et al. 1990).

Hemoglobin and Iron Metabolism

Because erythrocyte mass does not increase in acute exercise, there is no reason to expect an elevation of hemoglobin mass. Immediately after ultramarathon races for 56 or 160 km, red

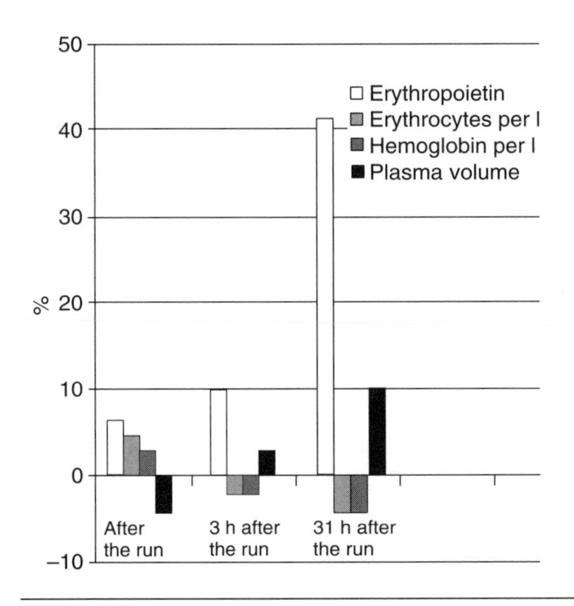

Figure 6.3 Postrace dynamics of erythropoietin, erytrocyte, and hemoglobin concentrations and plasma volume in marathon runners.

Adapted from J. Schwandt et al. 1991.

cell count and hemoglobin level increased as a result of hemoconcentration. Later, during the postrace recovery period, hemodilution followed, which was the greatest 48 h after the 160-km race (Dickson et al. 1982).

It has been mentioned (p. 115 of this chapter) that there is a possibility that the oxygen transport function of hemoglobin is supported by 2,3-diphosphoglycerine, which promotes oxyhemoglobin dissociation. However, recent results do not confirm the exercise-induced changes in 2,3-diphosphoglycerine concentrations. Changes were not found during exercises performed at 85% $\dot{V}O_2$max to exhaustion either before or after 5 weeks of training in normoxia or hypoxia (Engfred et al. 1994).

After 56- and 160-km ultramarathon races, serum ferritin levels were markedly elevated. Return to prerace levels was observed 6 days after the 56-km race. During the eighth day, ferritin level decreased further. A possibility exists that these changes reflect hemoconcentration during and immediately after the race, which was later substituted by hemodilution. However, because ferritin concentration increased after the 160-km race by 1.5-fold, one must not exclude the possibility of actual changes in iron metabolism during prolonged exercises (Dickson et al. 1982). Davidson et al. (1987) found an elevation of serum ferritin level immediately after a marathon race. Haptoglobin level increased insignificantly after the 160-km race. Poortmans (1970) observed significant changes either in transferrin or haptoglobin concentration immediately after 60 min of bicycling at 67% of $\dot{V}O_2$max. Transferrin, but not haptoglobin, concentrations increased 30 min after exercise.

Several studies established haptoglobin reduction after a single bout of prolonged running (Casoni et al. 1985; Davidson et al. 1987) or swimming (Selby and Eichner 1986).

Sports Anemia

Several studies showed that running might result in an appearance of free hemoglobin in plasma in augmented quantities. This phenomenon has been considered as a sign typical of sports anemia. Naturally, sports anemia is also expressed by low hemoglobin level in total blood (Casoni et al. 1985).

Increased free hemoglobin concentration in plasma has been found after swimming for 10 km together with a modest increase of haptoglobin level (Bichler et al. 1972). The task of haptoglobin is to bind the released hemoglobin to prevent its urinary excretion. This study did not show evidence that a failure at the haptoglobin level was responsible for the increase of free hemoglobin concentration. The authors suggested that the increased plasma level of free hemoglobin was related to mechanical influence on erythrocytes or to structural deficiencies of red blood cells. In agreement, authors of several studies suggested that exercise might lead to an intravascular hemolysis (Dufaux et al. 1981b; Eichner 1985; Selby and Eichner 1986). The reason for the intravascular hemolysis may be "footstrike trauma" and/or consequences of iron depletion (Hunding et al. 1981; Magnusson et al. 1984; Miller 1990; Clement and Sanichuk 1984).

The significance of the "footstrike hypotheses" was supported by the results of Miller et al. (1988). They found a more pronounced increase in plasma-free hemoglobin and decrease of haptoglobin levels in downhill running than in uphill running. In downhill running, mean foot impact was 11% higher than in uphill running. They concluded that mechanical trauma to red cells at footstrike is a major cause of hemolysis during running. On the other hand, disorders in iron metabolism may appear in intensive endurance training, leading to iron loss. It has been estimated that iron loss reaches 1.75 mg/day during a competitive training program (Haymes and Lamanca 1989).

A general assumption proposes that a low ferritin level indicates lowered iron stores (Newshouse and Clement 1988). According to this assumption, ferritin level in blood plasma is in equilibrium with the ferritin content in tissues, where iron bound to ferritin constitutes the main iron store. Thus, iron loss in runners (e.g., caused by "footstrike trauma") is associated with a lowered ferritin level (Dufaux et al. 1981b; Magnusson et al. 1984; Casoni et al. 1985). Dufaux et al. (1981b) did not find the ferritin change in rowers and cyclists or in untrained control persons. Dickson et al. (1982) confirmed that swimmers had higher serum ferritin levels than ultramarathon runners, but there was no significant difference between runners and untrained persons. Pellicia and DiNucci (1987) observed higher levels of erythrocytes, hemoglobin concentration, iron-binding capacity, serum iron, and ferritin in male swimmers than in control persons. Female swimmers also had a greater iron pool.

Pizza et al. (1997) performed a longitudinal study with runners. During the period of intensive training, runners had lower hematoglobin, hematocrit, and red cell numbers. Serum ferritin did not change during this period but seemed to be lowered after 3 weeks of taper and 4 days after the championship. In swimmers, no changes in red cell indexes were found during the period of intensive training, but serum haptoglobin tended to be reduced after the initial stage of training. Another longitudinal study confirmed that ferritin level declined only during the first 6 months in male runners, and it did not change any more during the following 12 to 14 months. In female runners during the first stage of training, ferritin level increased and then leveled off (Kaiser et al. 1989). In advanced stages of training, reduction of ferritin does not appear in all runners, and sports anemia is not common for all runners.

A study with top-level soccer players did not show differences in serum iron, serum total iron-binding capacity, percent of transferrin saturation, and serum ferritin compared with untrained control persons (Resina et al. 1991). These authors pointed out a lower level of haptoglobin in soccer players than in control persons. They suggested that this difference indicated a disorder in iron metabolism in soccer players.

A study of 39 athletes of the Polish national team showed that the concentration of hemoglobin, erythrocytes, ferritin, and transferrin was lower in endurance athletes, and reticulocyte counts were higher than in control persons. No significant changes were found compared with strength athletes and control persons (Spodark 1993).

Exercise Action on Leukocytes

In the 1920s, a pronounced increase in leukocyte count was observed during exercise. This phenomenon was called myogenic leukocytosis. In the first stage, myogenic leukocytosis was found in short-term intensive or prolonged low-intensity exercises; a moderate increase in the total leukocyte count appeared mainly due to an increased number of lymphocytes. The next stage (appears mainly in severe prolonged exercises after the first stage) is characterized by an increase of total leukocytes up to 20,000 or more per milliliter of blood. The second stage is characterized by an increased number of neutrophils in combination with a decreased number of lymphocytes and eosinophils (Egoroff 1924).

In the late 1930s and 1940s, interest in exercise-induced lymphopenia and eosinopenia increased. This was because these phenomena were evoked by the administration of adrenocortical extract or synthetic glucocorticoids. Therefore, lymphopenia and eosinopenia were used as indirect indicators of enhanced adrenocortical activity during exercise.

More recent studies confirmed both phases of myogenic leukocytosis. The results obtained after exercises of short duration were in accordance with the first stage (Bieger et al. 1980). After a 32-km race (Moorthy and Zimmerman 1978), 56-km ultramarathon race (Dickson et al. 1982), and 24-h relay race (Williams and Ward 1977), appearance of the second phase (very pronounced leukocytosis because of neutrophilia, in combination with lymphopenia and eosinopenia) has been confirmed. Davidson et al. (1987) observed an increase in leukocyte counts up to 20,000, with a pronounced increase of neutrophils and monocytes, in combination with lymphopenia in runners after a 3-h marathon race.

During exercise, the concentration of platelets increases (Dawson and Ogston 1969; Davidson et al. 1987). This change (myogenic thrombocytosis) is not due to hemoconcentration because the concentration of platelets may double. Increases in fibrolytic activity of blood (Bennett et al. 1968) were seen in relation to exercise intensity and duration (Rosing et al., 1970). Despite high fibrinolytic activity, hypercoagulability follows strenuous exercises (Poortmans et al. 1971). Bärtsch et al. (1995) showed a balance between coagulation and fibrinolysis in long-term exercise.

Training Effects

In 1949, Kjellberg et al. convincingly demonstrated that training increases the plasma volume (figure 6.4). A short-term endurance training experiment showed an increase in plasma volume of 10% to 19%. In a cross-sectional comparison, athletes had 41% to 44% more plasma than untrained persons. In principle, these results were confirmed by many researchers. After a few days of endurance training, a pronounced increase in plasma volume was found (see Brooks et al. 1996). Changes in blood cell and hemoglobin mass are less rapid (Sawka et al. 2000) (figure 6.5). Therefore, during the initial period of training (and at the onset of training after a relief period), hemoglobin concentration decreases. In these cases, hemodilution caused by increased plasma volume conceals the actual increase in the mass of blood, cells, and hemoglobin. Therefore, it is incorrect to conclude that an athlete's functional state deteriorates from decreased hemoglobin concentration. In related training stages, concentrations of other constituents of plasma also decrease as a result of normal adaptation, which begins by an increase of plasma volume.

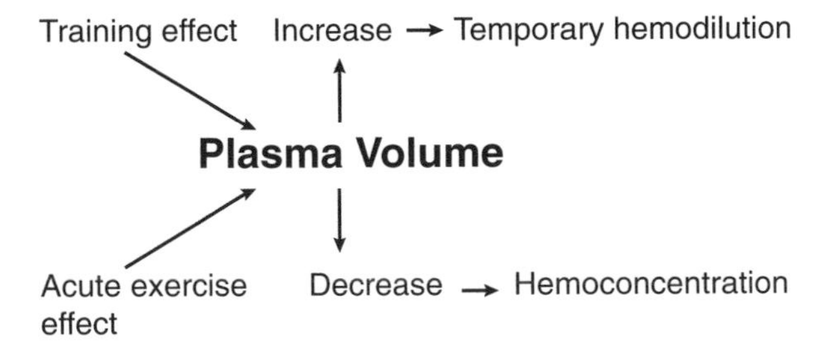

Figure 6.4 Opposite effects of acute exercise and training on plasma volume.

During more advanced endurance training, red cell mass may increase. In some athletes, an increase of erythrocyte mass has been found, in others no changes were detected (see Brooks et al. 1996). Oscai et al. (1968) reported a modest increase of blood volume as a result of an augmentation of plasma without a change in red cell volume. The results of Green et al. (1991) were similar. A pronounced increase in red cell volume has been found by Brothershood et al. (1975) and Remes (1979). Schmidt et al. (1988) found that the blood volume increase is more pronounced than the change in erythrocyte volume. The increase in erythrocyte volume is in accordance with increases in the total amount of hemoglobin (Kjellberg et al. 1949). Core et al. (1997) pointed out the limited capability to further increase either total red cell volume or hemoglobin mass in qualified athletes in the course of further training.

In a limited number of studies, training effects on plasma EPO have been investigated. The results vary. Berglund et al. (1988) reported normal levels of EPO in cross-country skiers in the

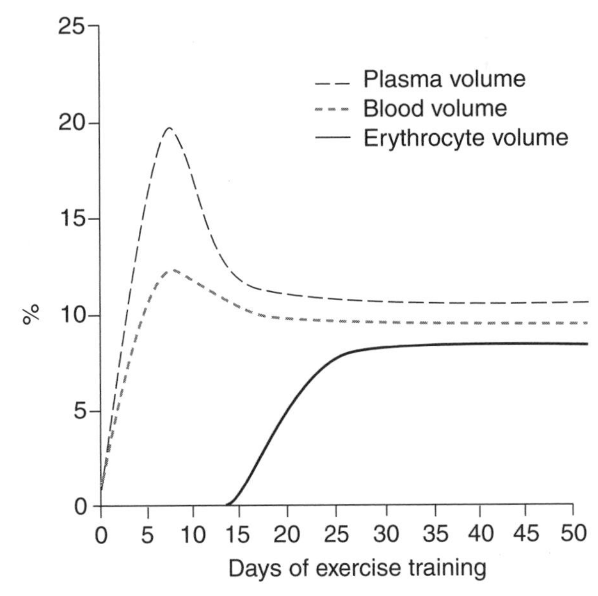

Figure 6.5 Dynamics of blood volume, plasma volume, and erythrocyte volume during the initial period of training. The changes are indicated as percentages.

From Sawka et al. 2000.

resting state. Engfred et al. (1994) did not find an increase in EPO levels either in training in normobaric or in hypoxic conditions. Increased EPO activity in serum was found in a group of skiers tested before the competition but not in another group tested during an intensive and voluminous training stage (De Paoli Vitali et al. 1988). The training effect on the plasma EPO level is not sufficiently studied to draw general conclusions. Further studies are necessary to establish conditions, which make training effective to increase the EPO concentration in blood.

Training-induced erythropoietic effect is reflected by the release of immature reticulocytes from bone marrow into circulating blood within a few days after various vigorous exercises (Schmidt et al. 1988, 1989, 1991).

Plasma Proteins

Plasma proteins are responsible for several tasks. One group performs a transport function. They bind hormones, lipids, and mineral compounds, ensuring their circulation without metabolic changes. Mostly, specific interrelationships exist among species of plasma proteins and the compounds transported. Each protein species has a certain binding capacity. Plasma proteins are also used for defense. These proteins, called immunoglobulins, act as circulating antibodies. Plasma protein fibrinogens (and some others) are essential for blood clotting. Plasma proteins and hemoglobin in erythrocytes contribute to blood buffer systems. Some peptides (glutathione) are related to antioxidant systems. Although most proteins cannot pass the capillary membrane, some enzyme proteins delivered from liver, skeletal- muscle, or myocardium cells can be found in the blood plasma.

The total amount of plasma proteins constitutes the colloid-osmotic pressure. Colloid-osmotic pressure is the part of plasma osmotic pressure that depends on dissolved protein molecules, which do not diffuse through the capillary membrane (or do not readily pass through the capillary pores). The obstacle for diffusion of proteins at the capillary pores is the reason that protein concentration in plasma is about three times as much as in interstitial fluid (7.3 g/dl vs. 2 to 3 g/dl). The difference in osmotic pressures (in plasma by 28 mmHg higher) causes fluid to return to the intravascular compartment at the venous end of the capillaries. Therefore, the necessary amount of water is maintained in the intravascular compartment.

During exercise, hemoconcentration results in an increase in the concentration of plasma proteins. However, various proteins change diversely. Poortmans's article (1970) can be used as an example. He found an increase of some proteins in subjects during 1 h of cycling, whereas the protein levels of others remained constant. Albumin concentration increased up to 12% above the resting level during the first 30 min. During the second 30 min, a slight decrease followed. The reason for diverse changes of individual proteins is that in addition to the general influence of hemoconcentration, individual proteins are controlled according to their function.

It is necessary to consider not only the total concentration of proteins in plasma but also the intravascular protein mass in the maintenance of colloid osmotic pressure. The latter mostly determines the water-binding capacity of the intravascular compartment. A special study showed that:

- trained persons have significantly higher intravascular protein mass than untrained persons;
- the intravascular protein mass highly correlates with plasma volume;
- immediately after a 32-km run, concentration of plasma proteins increased by 11.9%, colloid osmotic pressure by 25.1%, and intravascular protein mass was elevated only slightly (Röcker et al. 1975).

The authors suggested that the relative constancy of the intravascular protein mass was essential for the increased colloid osmotic pressure and to regain intravascular volume after exercise without fluid intake.

Enzymes and Other Intracellular Proteins in Blood Plasma

In various clinical conditions, several intracellular proteins have been found in blood plasma. These proteins mostly originate from myocardial, hepatic, skeletomuscular, and probably also brain tissue. It has been believed that release of enzymes and other proteins from various tissues into blood might be associated with disruption of the cellular membrane as a result of pathological conditions. The released proteins reach the lymph from the intrastitial fluid. The proteins flow into the blood circulation by means of the lymphatic system (Lindena and Trautschol 1983; Hortobagyi and Denahan 1989). However, diffusion of some proteins in minute amounts through the capillary pores is not excluded.

Intracellular proteins from muscles and other tissues also appear in blood serum during various exercises (Halonen and Konttinen 1962). Among those proteins are creatine kinase and its isozymes MM and MB (for a review, see Hortobagyi and Denahan 1989); lactate dehydrogenase (Munjal et al. 1983; Okhuwa et al. 1984; Pills et al. 1988); hexose phosphate isomerases (Berg and Haralambie 1978); aldolase, alanine-aminotransferase, and aspartate-aminotransferase (Fowler et al. 1962, 1968; Critz and Merrick 1962; Wegmann et al. 1968; Lijnen et al. 1988); alkaline phosphatase (Lijnen et al. 1988); pyruvate kinase (Stansbie et al. 1983); and myoglobin (Demos et al. 1974; Melamed et al. 1982; Munjal et al. 1983; Lijnen et al. 1988).

Changes in various enzyme activities in serum are not parallel. Five minutes after a marathon race, significant elevation was detected in the activities of creatine kinase isozymes MM and MB, aldolase, aspartate-aminotransaminase, lactate dehydrogenase, and alkaline phosphatase but not alanine-aminotransferase. Within 12 h, alkaline phosphatase returned to initial values, but a significant increase was found in the activity of alanine-aminotransferase. Activities of all other enzymes increased continuously and peaked either 12 h (both isozymes of creatine kinase) or 36 h after the race. Furthermore, the activities declined, but within 7 postrace days, the return to initial values was observed only in activities of alkaline phosphatase and lactate dehydrogenase. Myoglobin concentration in plasma increased sharply during the race, with peak values 5 min after. Afterward, myoglobin concentration decreased, but it did not return to the initial level for 7 days (Lijnen et al. 1988).

Most attention has been paid to creatine kinase in serum during exercises. One of the reasons is the relationship between creatine kinase response and the time course of muscle soreness, pain, stiffness, and damage (Armstrong 1986; Byrnes and Clarkson 1986). Immediately after unaccustomed exercise, particularly after one performed with eccentric contraction, injury of the skeletal muscle fibers is evidenced by disruption of the normal myofilament structures in some sarcomeres. This is accompanied by loss of intramuscular protein (e.g., creatine kinase) into the plasma, indicating damage to the sarcolemma, delayed-onset soreness, and temporary reduction of muscle force (see Armstrong 1990).

Increased activities of creatine kinase and lactate dehydrogenase have been found after 15 s or 60 s of cycling on an ergometer at maximal rate, with peak values 3 min after the end of exercise (Pills et al. 1988). Similar response was found in activities of both enzymes after running 400 m (Okhuwa et al. 1984). Increased creatine phosphokinase activity after intensive exercises was confirmed and found in combination with elevated activities of pyruvate kinase (Stansbie et al. 1983). Other articles showed increased creatine kinase activity after prolonged exercises (Fowler et al. 1968; Berg and Haralambie 1978; Munjal et al. 1983), including marathon races (Haibach and Hosler 1985; Lijnen et al. 1988) and 56-km ultramarathon races (Noakes and Carter 1982). Exercise with eccentric contractions induced two-peak postexercise increases of creatine kinase and lactate dehydrogenase activities, whereas exercise with concentric contractions resulted in a short-term one-peak response (Armstrong et al. 1983).

Tiidus and Ianuzzo (1983) analyzed the significance of exercise intensity and duration with a dynamic leg-extension apparatus. Concentric contractions were performed by extending the knee to 90° of extension. With the aid of this movement, they raised the weights vertically. Weights were lowered slowly to allow eccentric contraction. The authors measured the activity of creatine phosphokinase, lactate dehydrogenase, and aspartate-aminotransferase in serum and the perception of delayed muscular soreness (with the aid of a 10-ball scale). The activity of creatine phosphokinase increased more than the activities of the other two enzymes. Creatine phosphokinase activity increased during 8 h after 30 min of contraction at 90% 10 RM. Peak level persisted up to 24 h after the exercise. A high level was also found 48 h after exercise. Pain sensations increased continuously within 48 h after the exercise. Increased intensity in diapason

35% to 90% 10 RM and an increased number of contractions in diapason 100 to 300, respectively, resulted in increases in serum enzyme activities and muscular soreness. When the total work was held constant, the higher intensity (170 contractions at 80% 10 RM) caused greater serum enzyme intensities and muscular soreness than the greater number of low-intensity contractions (545 contractions at 30% 10 RM). Accordingly, Clarkson et al. (1985) suggested that the quantity of creatine kinase release was related to the overall tension output of the involved muscles. After an extended analysis of the literature, Hortobagyi and Denehan (1989) concluded that intensity and duration thresholds exist for serum enzyme responses.

Responses of serum creatine kinase have been observed after isometric exercises (Mayer and Clarkson 1984; Clarkson et al. 1987) and also after a 24-h period of speed driving.

Several studies confirmed the relationship between serum enzyme responses and delayed muscular soreness after exercises, particularly in eccentric muscular contraction (figure 6.6) (Byrnes et al. 1985; Evans et al. 1986; Hyatt and Clarkson 1998). Twenty-four hours after eccentric, concentric, or isometric exercises, serum creatine kinase response was almost the same (Clarkson et al. 1986). However, eccentric exercise released more creatine kinase with a longer time delay than isometric or concentric exercise (Byrnes and Clarkson 1986). After resistance exercises, serum creatine kinase increase was less pronounced in weightlifters, but soreness was more pronounced than in untrained persons, although the workload in each exercise was individually adjusted at 12 RM. Peak values of enzyme activity appeared in untrained persons four days and in weightlifters five days after the exercise session (Vincent and Vincent 1997).

To explain the difference in the dynamics of serum enzyme activities and muscular soreness, it is worth recalling the sequence of events after an exercise:

1. Exercise-induced muscular damage
2. Leakage of enzymes into the bloodstream
3. Physiological reaction to the damage resulting in muscular soreness sensations (Tiidus and Ianuzzo 1983)

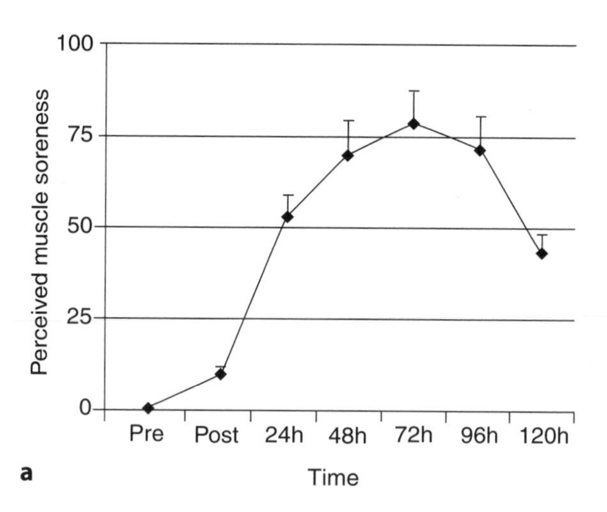

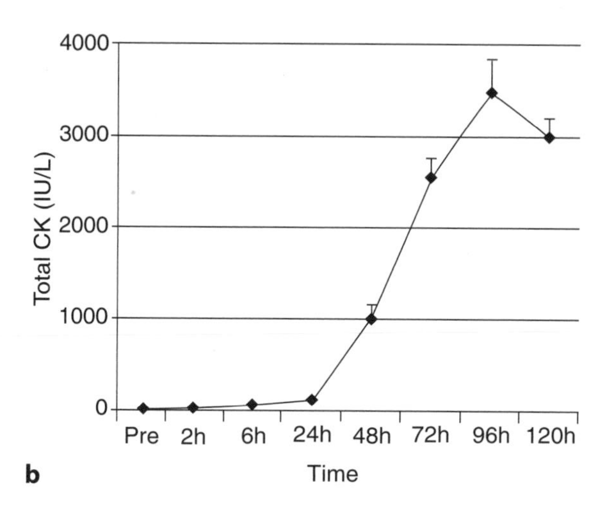

Figure 6.6 Effect of 50 voluntary eccentric contractions of the elbow flexor muscles (performed on a preacher-curl bench with each contraction for 3 to 5 s and 10- to 12- s rest periods) on *(a)* perceived muscle soreness and *(b)* total creatine kinase.

Reprinted from M.N. Sawka et al. 2000.

Exercise-induced muscle damage is usually reduced to the ultrastructural integrity of muscle fibers (Ebbeling and Clarkson 1989; Nosaka and Clarkson 1995). After intense eccentric exercise, disruption in the sarcoplasmic reticulum, t-tubule system, or mitochondria may be implied in man (Hortobagyi and Denahan 1989). Clear changes have been found in the form of a Z-line, streaming immediately after such exercises (Friden et al. 1983). Seven days after eccentric forearm flexion exercise, degenerative fibers and fibers infiltrated by mononuclear cells were detected (Jones et al. 1986).

When the same exercise was repeated 2 weeks or several months later, less soreness was experienced, and no creatine kinase response appeared (Clarkson et al. 1992). A special study showed that accelerated clearance of creatine kinase seems to be a factor contributing to the blunted response of this enzyme after a repeated bout of exercise (Hyatt and Clarkson 1998).

A recent study of a Finnish team showed that bed rest for 17 h after 8-km cross-country running reduced serum creatine kinase response. The authors explained the bed rest effect by reduced lymphatic transport of creatine kinase and release of enzyme from muscle fibers (Havas et al. 1997).

Acid-Base Buffer Systems in Blood

The activity of enzymes depends on the pH of the compartment. Therefore, one of the body's homeostatic tasks is to keep the pH in body fluids at a constant level. Three control systems exist to prevent acidosis or alkalosis.

- All body fluids are supplied with acid-base buffer systems that immediately combine with any acid or base and, thereby, prevent excessive changes in hydrogen ion concentration.
- A change in hydrogen ion concentration of the blood influences the respiratory center; as a result of an altered rate of breathing, the CO_2 removal from body fluids increases or decreases, and the hydrogen ion concentration starts to return to normal.
- The kidneys excrete either an acid or alkaline into urine, thereby helping to normalize hydrogen ion concentration in body fluids.

The most plentiful buffers in the body are the proteins of the cells and plasma because of their high concentrations. Some amino acids in proteins have free acid radicals that can dissociate to form a base plus H^+. Histidine is especially essential for the buffering capacity of proteins. It is also important that the amino acid buffering systems have a pK (log of the constant of dissociation of a buffer) close to 7.4, which is the normal pH of blood. The buffering power of the buffer system is the greatest when the target pH is equal to the pK. Buffering power is directly proportional to the concentration of the buffer substances.

A rapid buffer system is made up of a mixture of carbonic acid (H_2CO_3) and sodium bicarbonate ($NaHCO_3$). When a strong acid (e.g., lactic acid) flows into plasma, it combines with $NaHCO_3$, forming H_2CO_3 and Na-lactate. An inflow of a strong base will be combined with H_2CO_3, forming bicarbonate and water. In blood, H_2CO_3 forms from CO_2 and H_2O. In erythrocytes, H_2CO_3 is converted to HCO_3^- and H^+. Hemoglobin buffers the released H^+. A mechanism is set up when HCO_3^+ is shifted into plasma. Lactic acid released into the plasma dissociates to H^+ and positive lactate ions. Hydrogen ions combine with HCO_3^- and form H_2CO_3. Lactate ions combine with sodium from $NaHCO_3$ and form sodium lactate. H_2CO_3 decomposes to H_2O and CO_2 to be eliminated by the kidneys and lungs, respectively. In this way nonmetabolic CO_2 is formed.

The bicarbonate buffer system is not a powerful buffer because the concentrations of CO_2 and HCO_3 are not great. Moreover, the pK of the bicarbonate buffer is only 6.1. Therefore, approximately 20 times as much of the bicarbonate buffer is in the form of dissolved CO_2. For this reason, the system operates on a portion of its buffering curve, where the buffering power is weak. Despite the limited buffering power, the bicarbonate buffer is exclusively important because each of the two elements of the bicarbonate system can be regulated (CO_2 by breathing and HCO_3^- by kidney function).

The third buffer system consists of $H_2PO_4^-$ and HPO_4^-. $H_2PO_4^-$ reacts with a strong acid, forming NaH_2PO_4 and salt. HPO_4^- reacts with a base, forming Na_2HPO_4 and H_2O. The phosphate buffer has a pK of 6.8, which allows the system to operate near its maximum buffering power. However, its concentration in the extracellular fluid is only one twelfth that of the bicarbonate buffer. Correspondingly, the total buffering power of the phosphate buffer is lower than that of the bicarbonate buffer.

Assessing Training Effects on Acid-Base Buffer Capacity

Since 1922, the Van Slyke apparatus was used for assessment of blood buffer capacity. The apparatus measures the amount of CO_2 released from a blood sample after adding a strong acid (lactic acid, hydrochloric acid, or sulfuric acid). In a normal situation, the amount of released CO_2 is 50 to 70 ml/100 ml of plasma or serum.

This amount of CO_2 has been used as the quantitative measure of the so-called alkali reserve. Acute intensive exercise reduces the amount of released CO_2 irrespective of the diminished quantity of alkali reserve (Dill et al. 1930; Schenk 1930; Robinson and Harmon 1941), obviously as a result of the accumulation of endogenous lactic acid. The results concerning training effects are variable. Steinhaus (1933), Robinson et al. (1937), and Robinson and Harmon (1941) did not find a training-induced increase of alkali reserve. However, several other authors obtained results showing increases of alkali reserve by 10% to 22% (Bock et al. 1928; Schenk 1930; Herxheimer 1933). The training effect of reserve alkali depends on the nature of the exercises used. In female students, a 6-week follow-up study showed that running training three times a week increased alkali reserve when anaerobic interval exercises, sprint exercises, or uphill running were used but not when aerobic training means were used (Viru et al. 1972).

In 1960, a method for the determination of pH, pCO_2, base excess, and standard bicarbonate in capillary blood was published (Siggaard-Andersen et al. 1960). The apparatus constructed was called Micro-Astrup (Radiometer, Copenhagen). Siggaard-Anderson (1963) also published a blood acid-base alignment nomogram, which made it possible to estimate several blood parameters related to pH control. The foundations for this approach were described by Stegmann (1981, pp. 156-157). It is difficult to say why this approach rarely is used for training monitoring now.

Using the Micro-Astrup principle, the relationship between exercise intensity and decrease of standard bicarbonate (corresponding to 40 mmHg of pCO_2) has been demonstrated. Simultaneously, a parabolic increase of lactate concentration appeared. Less pronounced were the increases of pCO_2 in venous blood (Tibes et al. 1974). These results were in accordance with the previously established relationship between exercise-induced changes in blood pH, lactate, standard bicarbonate, and pCO_2 (Bouhuys et al. 1966; Wasserman 1967). Bouhuys et al. (1966) showed that 2 months of training reduced the changes of pH, lactate, standard bicarbonate, and base excess during acute exercise.

An extended study has been performed by Kindermann and Keul (1977). This material included the results obtained from high-level athletes in various events, untrained healthy persons, and patients (those with diabetes or cardiovascular diseases)—480 males and 66 females. This material confirmed the relationship among pH, lactate, standard bicarbonate, and base excess changes in intensive exercises. After running 500 m, speed skating 1000 m, rowing 2000 m, and swimming 100 m, extremely high values of lactate and low values of pH and base excess were recorded. A question remains whether the values of base excess within –20 and –30 mval/L were possible as a result of the rise in buffer capacity because of previous training or whether these extreme values were caused by the increase of blood lactate level up to 15 to 22 mmol/L in athletes with high anaerobic capacity.

Sharp et al. (1983) used the results obtained with the aid of Micro-Astrup (Radiometer) and blood lactate values for regression analysis to calculate the amount of lactate required to cause a change in pH of 1.0 unit. They did not find differences between endurance-trained male cyclists and untrained men when assessing blood buffer capacity in this manner.

For study of muscle buffer capacity, the biopsy sample was homogenized, deproteinized, and titered to pH 6.0 with 0.01 N HCl. Buffer capacity was expressed in µmoles of HCl per pH shift per gram of wet weight of sample. The results showed that 800-m runners had significantly higher muscle buffer capacity than untrained persons or marathon runners. No difference was found between marathon runners and untrained men (McKenzie et al. 1983). A comparison of sprinters, rowers, marathon runners, and untrained persons confirmed that the buffer capacity of the vastus lateralis muscle is higher in anaerobically trained athletes (sprinters and rowers) than in aerobically trained athletes and untrained persons (Parkhouse et al. 1983). In untrained men, an 8-week sprint training (30-s sprints over 4-min rest periods) increased muscle buffer capacity calculated from the changes in lactate concentrations and pH of the vastus lateralis muscle during an incremental exercise (this method was used before by Sahlin 1978). Posttraining buffer capacity was greater than in endurance-training cyclists possessing $\dot{V}O_2max$ of 70.5 ± 2.6 ml/kg · min (Sharp et al. 1986). Bell and Wenger (1988) also reported a positive sprint training effect on muscle buffer capacity in previously untrained or recreationally active persons.

The effect of high-intensity interval training (six to eight repetitions of 5-min cycling at 80% of peak sustained power output separated by 1 min of recovery) was studied in well-trained competitive cyclists (mean $\dot{V}O_2max$, 62 ml/kg · min) by use of the pH titration method. After 4 weeks of training, buffer capacity increased by 16% in the vastus lateralis muscle (Westin et al. 1997). Accordingly, Sahlin and Hendriksson (1984) observed that team athletes have higher muscle buffer capacity than sedentary persons.

The results obtained demonstrate that intensive anaerobic training increases buffer capacity of involved muscles (figure 6.7). In regard to blood buffer capacity, training effect is not strictly evidenced. There may be several reasons for this. First, the methods used might not be adequate for assessment of actual buffering capacity of blood. Another possible reason is that training specificity was poorly accounted for in related studies. However, most of all, the reason might be that the bicarbonate buffer was assessed in blood but not the protein and phosphate buffers. Calculation showed that proteins are responsible for approximately 50% of the buffering in muscles during exhaustive exercise (Hultman and Sahlin 1980), whereas the contribution of the bicarbonate buffer is within 15% to 18% of total buffer capacity (Sahlin 1978). Theoretically, there should be only modest possibilities to increase the capacity of the bicarbonate buffer because the main components (CO_2 and bicarbonate ion) are homeostatically controlled parameters. Increase of protein buffer capacity is possible by the induction of synthesis of proteins that are rich in histidine. Indisol groups of histidine have the main role as intracellular buffers (Hochacka and Somero 1984). In mammalian muscles, half of the total histidine is in proteins, and the remainder is bound by dipeptides or carnosine, anserine, and optidine. Protein-bound histidyl residues, histidine-containing dipeptides, and free histidine are important buffers in skeletal muscle (Burton 1978). Parkhouse et al. (1983) showed that within human skeletal muscle, carnosine level is likely related to the glycolytic capacity of muscle, substantially contributing to buffer capacity, particularly in fast-twitch fibers.

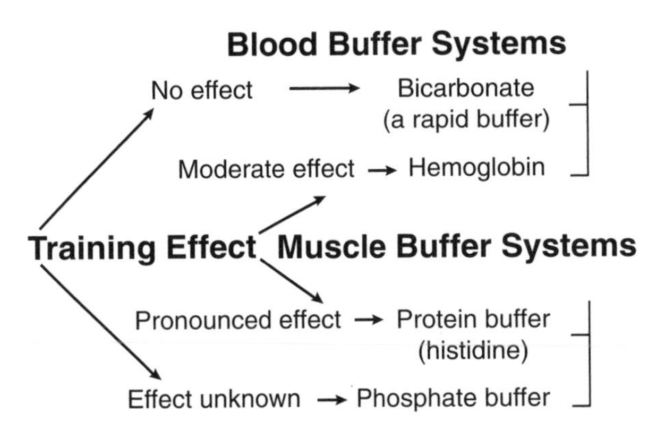

Figure 6.7 Training effects on various buffer systems.

Blood Antioxidant System

In certain conditions, molecules of O_2 convert into an active form of oxygen. This way, so-called free radicals are produced. Free radicals have an unpaired electron in their outer orbital (see Chance et al. 1979). The most important among them are superoxide (O_2^-), hydrogen peroxide (H_2O_2), nitric oxide (NO^-), and hydroxyl radicals (HO^-). Muscular contraction has been shown to generate them (for a review, see Powers et al. 1999). Most free radical oxidants produced by contracting muscle fibers are due to the elevated rate of mitochondrial oxidation (Jackson 1998). During exercise, potential sources of oxidants are also the xanthine oxidase pathway, prostanoid metabolism, and calcium-mediated radical production (Powers et al. 1999).

Increased amounts of free radicals may have serious destructive and even lethal effects on cells. They oxidize polyunsaturated fatty acids, which are essential components of cellular membranes. They also oxidize some of the cellular enzymes, thereby damaging cellular metabolic systems. Free radicals can modify macromolecules in the cell, including nucleic acids, proteins, and lipids (Yu 1994). Nervous tissue is especially susceptible because of its high content of structure lipids.

The harmful effects of free radicals are opposed by the antioxidant system. Tissues contain multiple enzymes that can remove free radicals. These antioxidants are superoxide dismutase, glutathione peroxidase, catalase, and glutathione. The primary function of superoxide dismutase is to convert superoxide radicals into hydrogen peroxide and O_2. Glutathione peroxidase reduces H_2O_2 to form oxidized glutathione and water, with glutathione as the electron donor. Catalase converts H_2O_2 to H_2O and O_2 (for details, see Halliwell and Gutteridge 1989). The activity of all three enzymes is greater in oxidative muscle fibers than in fast-twitch glycolytic fibers (Powers et al. 1994).

The most important antioxidant function of glutathione is to remove H_2O_2 and lipid peroxidase. Glutathione is also involved in reducing a variety of antioxidants to their native structure. Glutathione reduces vitamins E and C radicals (α-tocopherol and semidehydroascorbate, respectively). At the expense of glutathione, vitamins E and C regenerate effectively. Glutathione concentration is high in the liver. Its concentration in lungs, kidney, and heart is half that in the liver (Halliwell and Gutteridge 1989). Slow oxidative (SO) fibers of skeletal muscles contain six times more glutathione than fast glycolytic (FG) fibers (Ji 1995). Erythrocytes contain a comparatively high level of glutathione. In plasma, glutathione level is lower than in erythrocytes (Kretzschmar et al. 1991; Ji et al. 1992).

Intracellular glutathione levels depend on glutathione uptake from blood, intracellular synthesis, use, and regeneration. Much of the synthesis of glutathione occurs in the liver. The liver contains a major reserve of glutathione and supplies a large amount of circulating glutathione. In the liver, available amino acids and hormones control glutathione synthesis. Insulin and glucocorticoids stimulate hepatic glutathione synthesis by induction of glutamyl cysteine synthetase. Epinephrine and glucagons enhance the liver efflux of glutathione (Lu et al. 1990).

Vitamin E serves as an intramembrane antioxidant and membrane stabilizer (Burton and Ingold 1989; Van Acker et al. 1993). Vitamin E directly scavenges most species of free radicals, including superoxide, hydroxyl radical, and lipid peroxides by use of the hydroxyl group to either donate a proton or accept an electron. The resultant vitamin E radical will ultimately react with itself or with another peroxyl radical to form nonreactive degeneration by-products. As a result of these reactions, vitamin E may be lost from tissues. At the same time, vitamin E radical may be regenerated to vitamin E by means of its interaction with ascorbic acid or glutathione (see Tiidus and Houston 1995).

A harmful consequence of the action of free radicals occurs when an imbalance exists between the production of oxidants (free radicals) and the activity of the antioxidant system. In this situation, called oxidative stress, oxidation of cellular components takes place. Oxidation stress is possible when local antioxidant defenses are depleted because of the high level of oxidants or when the rate constants of the radical reactions are greater than the rate constants of the antioxidant defense mechanisms (Buetiner 1993). During strenuous exercises, the possibility of oxidative stress (Alessio 1993; Ji 1995) is evidenced by the accumulation of free radicals in skeletal muscles and oxidative injury to lipids, proteins, and DNA within skeletal muscle fibers (for a review, see Powers et al. 1999). An article by Davies et al. (1982) was among the first to provide arguments on the possibility for oxidative injuries produced by exercise. Reid et al. (1992a, 1992b) established the kinetics of intracellular oxidants and showed extracellular release of radicals in skeletal muscle.

Most, but not all, studies show that endurance training increases the activity of antioxidant enzymes in rat or dog muscles (for a review, see Powers et al. 1999). The training effect is pronounced in oxidative fibers. Fast-twitch glycolytic fibers show a moderate decline in activity of superoxide dismutase (Powers et al. 1994). Human biopsy studies confirmed the training-induced elevated activity of antioxidant enzymes in the vastus lateralis muscle. Jenkins (1988) found that the activity of superoxide dismutase in trained persons is greater than in untrained persons. The linear relationship is revealed between the activity of this enzyme and $\dot{V}O_2max$. Hellestein et al. (1996) found increased activity of glutathione peroxidase after 7-week sprint cycle training. Conversely, Tiidus et al. (1996) did not find any changes in muscle superoxide dismutase, catalase, and glutathione peroxidase activity during 8 weeks of 35 min of aerobic cycle training (three times weekly), despite significant increases of $\dot{V}O_2max$ and muscle citrate synthase activity. The question remains whether the duration of training or exercise workload was great enough. Rat experiments showed that the training effect on glutathione peroxidase activity was insignificant when the duration of daily running sets was 30 min. The effect appeared in daily running for 60 min and was the highest when daily running duration was 90 min (Powers et al. 1994).

Although the training effect on glutathione content varies among animal species and tissues, the dominating result is that training increases glutathione content in skeletal muscles (for a review, see Powers et al. 1999). The studies of Sen et al. (1992) on dogs and rats showed that training increases glutathione content but not glutathione peroxidase in the liver. Tiidus et al. (1996) failed to show an increase of glutathione content in the vastus lateralis muscle after 8 weeks of endurance training.

A human study showed that long-distance runners have a higher glutathione concentration in blood plasma than untrained persons. The difference in the activity of plasma lipid peroxidase was not found. An incremental ergometer test did not change these parameters in untrained persons but lowered both indexes in runners (Kretzschmar et al. 1991).

In another human study, plasma total antioxidant status was measured using 2,2'-azino-di-[3-ethylbenzothiazoline sulfonate] (ABTS) incubated with a peroxidase (e.g., metmyoglobin) and H_2O_2 to produce the radical cation $ABTS^+$. Antioxidants in the added sample cause suppression of this colored product to a degree, which is proportional to their concentration (Miller et al. 1993). Twenty nine women (aged 66 to 82) were studied using this method. However, no significant correlations were established between the plasma total antioxidant status with $\dot{V}O_2max$ or with mean habitual daily energy expenditure. In this study, superoxide dismutase and glutathione peroxidase activities in erythrocytes and glutathione peroxidase activity in plasma were also assessed. A significant negative correlation of $\dot{V}O_2max$ and mean habitual daily energy expenditure was found for red blood cell superoxide dismutase and glutathione peroxidase activities but not for the plasma activity of glutathione peroxidase. Lipid peroxidation was assessed by spectrofluorometric measurement of thiobarbituric acid reactive substances. Results showed significant correlations of this index with $\dot{V}O_2max$.

Several other studies provided results that according to formation of malondialdehyde, a lipid peroxidation by-product, or to thiobarbituric acid reactive substances, lipid peroxidation increases in skeletal and heart muscles and plasma after exhausting exercise (Sen 1995). A 14-min incremental exercise test did not increase the production of malondialdehyde, but exercise at the level of anaerobic threshold for 30 min did (Sen et al. 1994). In trained persons, increase of malondialdehyde is less pronounced after exercise than in sedentary persons (Alessio and Goldfarb 1988; Alessio 1993). A set of resistance exercises increased plasma malondialdehyde and creatine kinase activity, with peak values 6 and 24 h after exercising. Vitamin E intake significantly reduced peak values, obviously by the membrane protection effect of vitamin E (McBride et al. 1998).

Several articles confirm the training effect on erythrocyte activity of antioxidant enzymes (Mena et al. 1991; Roberson et al. 1991). Other studies failed to detect the training effect on erythrocyte enzyme activities (Alessio and Goldfarb 1988). A bout of cycling for 22 km did not modify the antioxidant enzyme activity in erythrocytes, but the activity of superoxide dismutase increased and that of catalase decreased in professional cyclists after 2800 km for 20 days (Mena et al. 1991). Ohno et al. (1986) did not find any change in superoxide dismutase activity of erythrocytes after brief exercise in sedentary young men.

A marker of free radical oxidation is pentane in expired air. An increased amount of pentane was found expired during an exercise performed at 75% $\dot{V}O_2max$ (Dillard et al. 1978). Application of this method requires inhalation of air completely free from pentane. Usually, in cities, air pollution causes inhalation of pentane. Although the inhaled pentane amount is modest, it induces an error in pentane determination in exhaled air. In conclusion, vigorous acute exercises may result in free radical accumulation, whereas exercise training results in possibilities for inhibition of oxidative stress by increasing the capacity of antioxidant systems (figure 6.8).

Immunological Indexes

Defense against invaders penetrating into the body and against endogenous defective proteins is arranged with the aid of the combined actions of several immune processes. The purpose of the brief overview following this section is to supply elementary knowledge of what various immune processes are for and delineating what

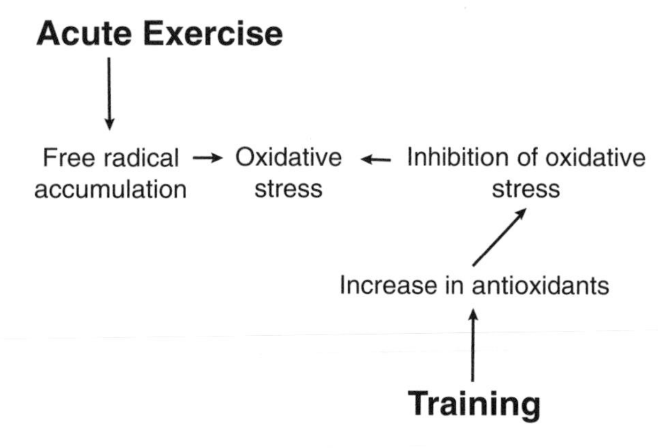

Figure 6.8 Acute exercise and training effects on oxidative stress. Intensive acute exercise favors free radical accumulation and thereby oxidative stress. Endurance training increases levels of antioxidants and thereby inhibits oxidative stress.

information can be obtained from diverse indexes of immune activities for training monitoring.

General Traits of the Immune Response

Immunity is the body's ability to resist (micro) organisms and toxins that tend to damage the tissues and organs. The tools for innate immunity are

- phagocytosis,
- digestive enzymes that destroy the microorganism swallowed into the stomach,
- resistance of the skin,
- compounds in the blood (lysozymes, basic polypeptides, complement complexes) that attack foreign organisms and toxins and destroy them, and
- natural killer lymphocytes that recognize and destroy foreign cells, tumor cells, and some infected cells.

Acquired immunity develops specifically against individual invading agents (lethal bacteria, viruses, toxins, foreign proteins). Acquired immunity consists of humoral immunity (B-cell immunity) and cell-mediated immunity (T-cell immunity).

Humoral immunity is found on circulating antibodies (globulin molecules) that are capable of attacking the invading agent. The mechanism for recognizing the initial invasion and its specific nature is related to antigens. Antigens are certain specific proteins or large polysaccharides that contain each toxin or each microorganism. Antigens perform their role by contact with lymphocytes.

Lymphocytes are not solely in the blood cells. They are most extensively located in the lymph nodes and also in specific lymphoid tissues (spleen, submucosal areas of the gastrointestinal tract, tonsils, and adenoids in the throat and the pharynx). In bone marrow, lymphocytes constitute the reserves for those in circulating blood. Some lymphocytes migrate to and are preprocessed in the thymus. These are T-lymphocytes, which are responsible for cell-mediated immunity. Other lymphocytes are destined to form antibodies. These cells are called B-lymphocytes. They are responsible for humoral immunity. When a specific antigen comes in contact with T- and B-lymphocytes, certain T-lymphocytes become activated to form T-cells and certain B-lymphocytes form antibodies. Furthermore, T-cells and antibodies react specifically against the invading agent, which was recognized by the particular type of antigen that initiated the development of T-cells and antibodies.

The produced antibodies circulate throughout the body. Antibodies (immunoglobulins) are divided into five classes: IgM, IgG, IgA, IgD, and IgE. IgG antibodies make up approximately 75% of the antibodies in a normal person. Antibodies either attack the invader directly or activate a complementary system, which destroys the invader.

T-cells are classified into three groups: helper T-cells, cytotoxic T-cells, and suppressor T-cells. The majority of T-cells are helper T-cells. They form a series of protein mediators (lymphokines) that act on other cells of the immune system and on bone marrow. The most important lymphokines are interleukin-2, interleukin-3, interleukin-4, interleukin-5, interleukin-6, granulocyte-monocyte colony-stimulating factor, and interferon-γ Without the lymphokines from the helper T-cells, the remainder of the immune system is almost paralyzed. Interleukin-1 is secreted by macrophages. It promotes the growth and reproduction of specific lymphocytes.

Cytotoxic T-cells attack cells directly. This type of T-cell is capable of killing microorganisms and even some of the body's own cells. Receptor proteins on the surface of cytotoxic cells enable them to bind tightly to the microorganisms or cells containing the specific binding antigen. Then the cytotoxic cell secretes hole-forming proteins (perforins). After that, the cytotoxic cell releases cytotoxic substances directly into the attacked cell. The latter becomes greatly swollen and shortly thereafter usually dissolves entirely.

Exercise and Training Effects on Immune Activity

Several studies and reviews have shown evidence that exercise training alters susceptibility to illness. Moderate training, such as jogging and other recreational activities, does not increase the risk of infectious illnesses; on the contrary, it enhances the body's resistance to pathogenic agents. Hard training regimes in top athletes together with physical and emotional strain during competitions elevate the rates of infectious illnesses, especially upper respiratory illnesses (see Roberts 1986; Shephard 1986; Mackinnon 1992; Nieman 1994a, 1994b). At the same time, the health-promoting effect of systematic exercising is comprehensively discussed (Bouchard et al. 1993). The expression of the health-promoting effect is a pronounced decrease of mortality in middle age (Blair et al. 1989), including the decreased risk of several cancers (Kohl et al. 1988). The health-promoting effect is possibly related to training-induced changes in the body that are simultaneously essential for improved sport performance capacity and enhanced capacity of the general adaptation mechanism (Viru and Smirnova 1995). However, the hard training necessary to achieve top performance may exhaust the body's adaptivity. Therefore, a cascade of changes follows that makes the body susceptible to several pathogens, particularly those against which the body fights through immune activities.

Innate Immunity

Several studies provided evidence that acute exercise enhanced phagocytic activity. This result has been obtained regarding connective tissue macrophages after a 15-km exhaustive run in endurance athletes, in association with an elevated macrophage lysosomal enzyme content (Fehr et al. 1989). After 1 h of cycling at 60% $\dot{V}O_2$max, neutrophil oxidative activity increased for 6 h in trained and untrained persons (Smith et al. 1990). Increased activation of neutrophils was also found after running 2 km or 10 km (Schaefer et al. 1987). In contrast, the phagocytic activity of blood monocytes decreased after short-term maximal running (Bieger et al. 1980). An investigation by Gabriel et al. (1992b) showed that after an ultradistance run, the phagocytic capacity of the blood is increased, but the phagocytic activity per a circulating neutrophil is decreased. In mice, running to exhaustion enhanced the growth-inhibitory (cytostatic) but not cytotoxic (killing) activity of murine macrophages against tumor cells (Lotzerich et al. 1990).

Exercise effects on total complement titer and on serum C3 and C4 were found to be diverse if they existed at all (see Mackinnon 1992). C-reactive protein level increased for 4 days after 2 h or 3 h of running (Liesen et al. 1977). Nine weeks of training decreased C-reactive protein level in rest and eliminated the response to 2-h running (Liesen et al. 1977).

B- and T-Lymphocytes

A pronounced increase appears with exercise in the population of B-cells, with rapid return to basal levels (Steel et al. 1974; Bieger et al. 1980). This response is greater in untrained persons compared with athletes (Ferry et al. 1990). During endurance running, changes in B-cells were slight in experienced marathon runners (Nieman et al. 1989b). An increased number of B-cells was found after 40 min of exercise at 85% $\dot{V}O_2$max but not at 50% $\dot{V}O_2$max (Nieman et al. 1994).

Brief maximal exercises increase the population of T-cells in the circulation, mainly in untrained persons. During prolonged exercise, the response is diverse, with a dominating decrease of the T-cell population (Steel et al. 1974; Moorthy and Zimmerman 1978; Bieger et al. 1980; Oshida et al. 1988; Nieman et al. 1989b). A 19-min incremental exercise test caused a parallel increase in the number of total T-cells and subsets of helper T-cells and suppressor T-cells (Lewicki et al. 1988). However, Gabriel et al. (1991a, 1992a) did not find changes in the quantity of T-helper/inducer cells after high-intensity exhausting exercises. According to the results of Nieman et al. (1994), a 40-min exercise of high intensity (80% $\dot{V}O_2$max) caused increases in the circulating number of T-lymphocytes, T-helpers/inducers, and T-suppressors. At moderate intensity (50% $\dot{V}O_2$max), exercise of the same duration did not change the number of cells of these subsets. Gabriel et al. (1992b) confirmed that exercises at 85% and 100% of the individual anaerobic threshold increased the number of T-cells, helper/inducer cells, and suppressor cells.

Several lymphokines originate from helper T-cells. Interleukin-2 production is reduced during exercise (Lewicki et al. 1988). Interferon-α activity has been found to increase during 1 h of cycling at 70% $\dot{V}O_2$max (Viti et al. 1985). Interleukin-6 concentration in blood plasma increases during 1 h of cycling at 75% $\dot{V}O_2$max. Tumor necrosis factor-α remained unchanged immediately after the 5-min race but increased 2 h later (Espersen et al. 1990).

Production of interleukin-1 by macrophages increased during and after prolonged exercises (Cannon and Kluger 1983; Lewicki et al. 1988). The resting level of interleukin-1 is greater in endurance runners than in untrained persons (Evans et al. 1986).

Haahr et al. (1991) investigated the action of a 60-min bicycle exercise at 75% $\dot{V}O_2$max on in vitro production of interleukin blood mononuclear cells. Production of interleukin-6 increased. Less pronounced was the increase in production of interleukins-1α and β. The exercise effect on production of tumor necrosis factor-α, interleukin-2, and interferon-γ was not significant. From the results of Tvede et al. (1993), interleukin-2 production by blood mononuclear cells decreased at 75% $\dot{V}O_2$max. Significant changes were not detected 1 h after exercises at 25% or 50% $\dot{V}O_2$max.

Increased interleukin-1β persisted in muscle tissue for up to 5 days after eccentric exercise (Cannon et al. 1989).

The third subset of T-cells constitutes natural killer (NK) cells. Dramatic changes have been found in the number of NK cells. After various exercises, increase of the NK cell number was within 50% to 300% (Brahmi et al. 1985; Lewicki et al. 1987; Pedersen et al. 1989; Tvede et al. 1989). The response depends on exercise intensity (Gabriel et al. 1991b, 1992b; Nieman et al. 1994). However, results were published that during prolonged exercises, the number of NK cells remained unchanged (Mackinnon et al. 1988; Berk et al. 1990). By analysis of the results of various studies, Pedersen et al. (1994) concluded that exercise intensity is responsible for the degree of increment in the number of NK cells. Suppressive action takes place in exercises with a duration of 1 h or more. During postexercise recovery, the NK cell number may remain elevated (Pedersen et al. 1990) or decrease and remain low for up to 21 to 24 h (Mackinnon et al. 1988; Berk et al. 1990).

The activity of NK cells may increase most in brief exercises, but after intense exhausting or prolonged severe exercises, NK cell activity is suppressed (Targan et al. 1981; Brahmi et al. 1985; MacKinnon et al. 1988; Pedersen et al. 1988, 1989; see also reviews of Keast et al. 1988; Mackinnon et al. 1989; Pedersen and Brunsgaard 1995).

Tvede et al. (1993) showed that NK cell and lymphokine activated killer cell activities increased during 2 h of exercise at 25% and 50% $\dot{V}O_2$max. Suppression was evident at an intensity of 75% $\dot{V}O_2$max.

After cycle exercise for 60 min at 60% $\dot{V}O_2$max, NK activity rose markedly during exercise but decreased to almost half of the initial value at 30 and 60 min of recovery (Shinkai et al. 1992).

Immunoglobulins

Severe exercises, neither brief maximal (Nieman et al. 1989a) nor prolonged (Hanson and Flaherty 1981; Mackinnon et al. 1989), changed the blood level of IgG, IgA, IgM, and IgE. However, in vitro production of IgG, IgA, and IgM by lymphocytes was decreased when a blood sample was obtained after 15 min of cycling (Hedfors et al. 1983).

After a marathon race, the titer of serum immunoglobulin specific to injected tetanus toxin was slightly increased (Eskola et al. 1978).

In mucosal fluid, the predominating immunoglobulin is IgA, which is essential for effector host defense against microorganisms causing upper respiratory infectious illnesses. Exercises decreased IgA levels of salivary and nasal wash in endurance trained athletes (Tomasi et al. 1982; Mackinnon et al. 1989; Muns et al. 1989; Tharp and Barnes 1990) and athletes in sport games (Mackinnon et al. 1991).

Presented results show that the immune system responds to acute exercises (figure 6.9). More important for training monitoring are alterations in immune activities during training and in overtraining. When the immunological indexes are intended to be used for training monitoring, attention should be paid to flow cytometry. This method enables us to investigate a broad spectrum of immunological parameters (see Gabriel and Kindermann 1995).

Water and Electrolyte Equilibrium

Except for proteins, the other constituents of blood plasma pass freely through the capillary walls. Therefore, their concentrations in blood plasma and interstitial fluid are almost the same. However, significant differences appear if we compare concentrations of Na^+, K^+, Ca^{2+}, Mg^{2+}, Cl^-, HCO_3^-, and sulfate ions in interstitial fluid and the intracellular compartment. This is because the plasma membrane of cells is selectively permeable. As a result, concentrations of sodium, calcium, chloride, and bicarbonate ions are greater outside the plasma membrane, and concentrations of potassium, magnesium, phosphate, and sulfate ions are greater inside the plasma membrane. Thus, intracellular ionic concentrations differ from those in extracellular compartments. In other words, ionic gradients exist between these two compartments.

Normal life activities are possible if ionic concentrations in both compartments are constant, and, therefore, the ionic gradients of Na^+ and K^+ are sufficient to ensure the rapid ionic shifts necessary to trigger major specific functional manifestations of cellular activity. The initial event is the change of plasma membrane permeability. The increased permeability enables an inflow of sodium ions into the intracellular compartment, which is followed by the outflow of potassium ions from the cells into the extracellular compartment. The depolarization initiated by ionic fluxes through sarcolemma also influences Ca^{2+} release from the endoplasmic reticulum (in the case of muscle fibers called sarcoplasmic reticulum). When the depolarizing current in the T-tubules is conducted to the terminal cisternae, stored Ca^{2+} is released from the lumens of the reticulum to endoplasm (sarcoplasm) by means of opening special calcium release channels.

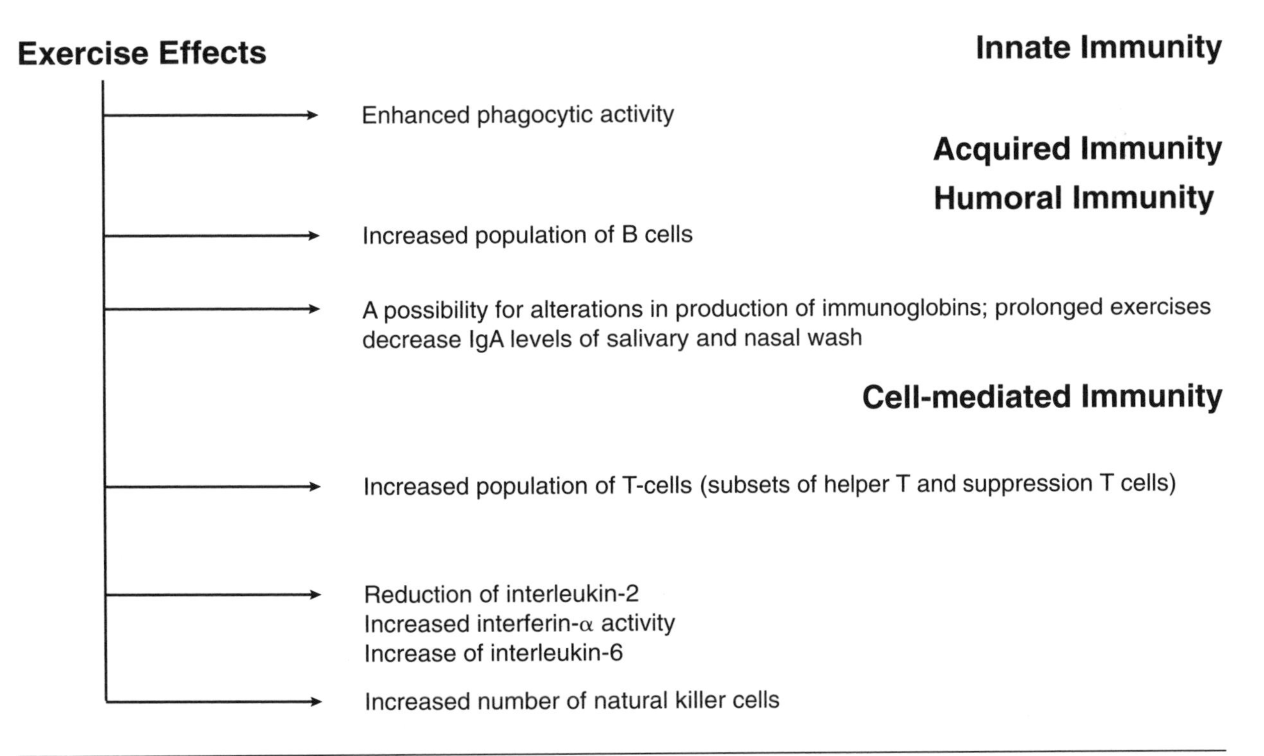

Figure 6.9 Training effects of various indexes of immune activities. These changes are most pronounced after intensive exercise, particularly in untrained persons. The possibility of suppression of immune activities exists in prolonged exercises.

Evidence is accumulating that in various types of fatigue the failure of calcium release is an important factor (for a review, see Allen et al. 1999). Calcium ions are required for actualization of specific cellular activities and for energy processes ensuring energy release.

To repeat the cyclic acts of cellular functional manifestations, the initiating shifts have to be stopped and resting conditions restored. Therefore, opposite ionic shifts (outflow of Na^+ from the cell, inflow of K^+ into the cell, and resorption of Ca^{2+} by the sarcoplasmic reticulum) are necessary. These ionic movements have to go against the ionic gradients and, therefore, require energy expenditure. Thus, the ionic pumps have to work. These ionic movements are related to the degradation of adenosine triphosphate (ATP) and, thereby, to the energy release. Specific ATPases in the membranes catalyze ATP breakdown. Their activators determine the specific nature of these ATPases. In the plasma membrane, the activators are elevated Na^+ concentration inside the cell and increased K^+ concentration outside the cell. The concerned enzyme is called Na^+-K^+–activated ATPase. In the membrane of endoplasmic (sarcoplasmic) reticulum, Ca^{2+}-ATPase is activated by increased concentrations of Ca^{2+} in the endoplasm.

Normally, cessation of the excitation signal causes a rapid relaxation of muscle fibers because of the closing of the Ca^{2+} release channels and the fast uptake of Ca^{2+} by the reticulum. A type of heart failure called "diastole defect" is caused by insufficient relaxation of myocardial fibers as a result of the remaining calcium ion concentration above the normal resting level in the sarcoplasm of myocardial cells after each systole (Meerson 1983). Reduced calcium uptake by the sarcoplasmic reticulum is associated with the increased time required for relaxation of skeletal muscles in man during exercise (Gollnick et al. 1991). However, the Ca^{2+}-ATPase activity (Kõrge and Campbell 1994) and the activity of Na^+-K^+–ATPase (Han et al. 1992) depend on local conditions (microenvironment around the enzyme). Changes in the relaxation rate and sarcoplasmic Ca^{2+} concentration during high-frequency fatigue and recovery can be explained by depression of the Ca^{2+} pump function as a result of changes in substrate/product concentration in the ATPase microenvironment (Kõrge and Campbell 1995). The most important is membrane-bound creatine phosphokinase. The ability to maintain a microenvironment favorable for ATPase function could be depressed in fatigued muscle (for a review, see Kõrge and Campbell 1995; Kõrge 1999).

A decrease in working capacity of the skeletal muscles is also caused by insufficient function of the Na^+-K^+ pump on the sarcolemma of skeletal muscle fibers (Clausen and Everts 1991; Clausen et al. 1993). At rest, only 5% of skeletal muscle Na^+-K^+ pumps are active. During repeated contractions at a high rate, virtually all Na^+-K^+ pumps are called into action (Clausen et al. 1998). In muscles in which the capacity of the Na^+-K^+ pump is reduced, the decline in force development during continuous electrical stimulation is accelerated (figure 6.10), and the subsequent recovery is considerably delayed. The activation of Na^+-K^+ pump function with epinephrine gave opposite effects (for a review, see Clausen and Everts 1991; Clausen et al. 1998). These results make the relationship between Na^+-K^+ pump function and muscle fatigue convincing (see Sejersted 1992; Fitts 1994; Green 1998; Lunde et al. 1998; McKenna 1992, 1999).

Intracellular accumulation of Na^+, caused by the insufficient rate of Na^+-K^+ pump function, increases intracellular osmolality and causes a shift of water into the fibers of skeletal muscles (Kõrge and Viru 1971a) and myocardium (Kõrge and Viru 1971b; Kõrge et al. 1974a 1974b) up to cellular edema during prolonged exercise. Human experiments confirmed that exercise-induced hyperkalemia is associated with a decrease of intracellular K^+ and a rise of intracellular Na^+ in working muscles (Sjøgaard et al. 1985; Medbø and Sejersted 1990). Biopsy samples taken shortly after the end of exercise also confirmed intracellular water accumulation in humans (Sjøgaard and Saltin 1982). The reason may be that besides Na^+ remaining in the cells, metabolites also accumulate within the cells (Lundswall et al. 1972).

Because most metabolites move freely through the sarcolemma, they should also accumulate in interstitial fluid in muscles. Accordingly, fluid accumulation has been found in the interstitial compartment of working muscles (Sjøgaard and Saltin 1982). Sejersted and coauthors (1986) postulated that during short-term high-intensity exercises, muscle cells swell because of the uptake of solute-free fluid. During the early minutes of recovery, fluid is redistrib-

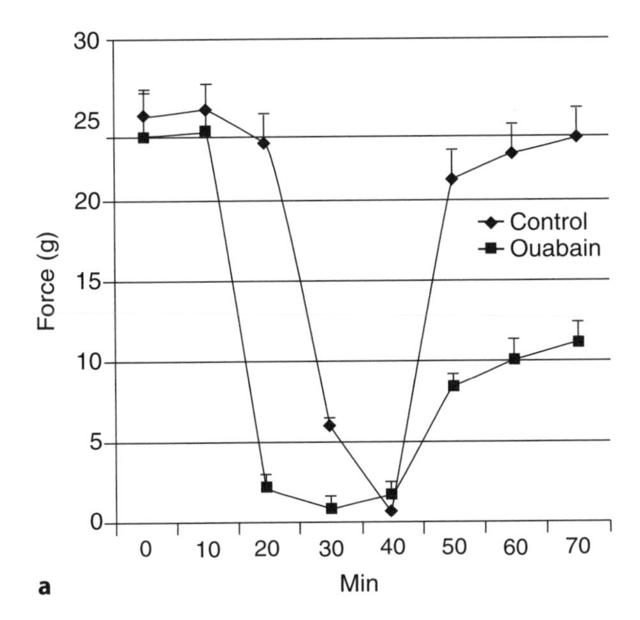

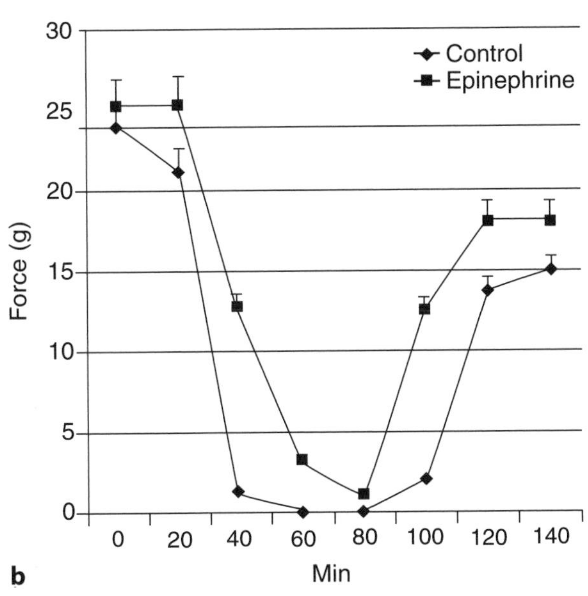

Figure 6.10 Effects of blockade of Na^+-K^+ pump with *(a)* oubain and of stimulation of Na^+-K^+ pump with *(b)* epinephrine on fatigue development in isolated soleus muscle working in buffer solution of high K^+ (12.5 mmol). Note that muscle force decreased at a higher rate when Na^+-K^+ pump function was partially blocked by oubain and at a lower rate when Na^+-K^+ pump function was stimulated by epinephrine.

Adapted from T. Clausen and M.E. Everts 1991.

uted to the interstitium, causing plasma sodium concentration to normalize in the face of reduced plasma volume.

Insufficient functioning of the Na^+-K^+ pump also reduces the ionic gradients of K^+ through increased potassium concentration in extracellular fluid. An elevated plasma potassium level has been found in athletes and untrained persons after severe exercise (for a review, see Saltin and Costill 1988; Lindinger and Sjøgaard 1991). Plasma K^+ rises in relationship to exercise intensity (Green et al. 1993). K^+ increase in plasma is considered a fatigue phenomenon (Sjøgaard 1991). Endurance and sprint training influence these fluxes. After endurance training is reduced, the rate of K^+ loss from contracting skeletal muscles has been found (Kiens and Saltin 1989). Sprint training decreased plasma K^+ response to maximal exercise (McKenna et al. 1993, 1996). These training outcomes apparently have a relationship to the training effects on the Na^+-K^+ pump sites and Na^+-K^+–ATPase activity in skeletal muscles (for a review, see Green 1998; McKenna 1999). Whereas intracellular accumulation of sodium and water is not measurable in training monitoring, blood plasma potassium increase may be used for evaluation of the Na^+-K^+ pump function in athletes during intense exercise.

Several factors influence the extracellular level of sodium and potassium during exercise besides the function of the Na^+-K^+ pump. Increase of potassium concentration in plasma is promoted by release of K^+ during the breakdown of glycogen because of potassium stored together with glycogen (Fenn 1939; Bergström et al. 1973). Bergström et al. (1973) calculated that about 18 mg of potassium has accumulated for each gram of glycogen deposit. Proteins may also bind potassium (Fenn 1936).

Sodium changes in plasma during exercises are modest if noticeable at all. When plasma sodium and potassium changes were corrected to hemoconcentration, significant changes in the sodium level disappeared just after 1-min of exhausting exercise. During postexercise recovery, within 2 to 10 min, actual sodium concentration was close to initial values, but the corrected level of sodium was significantly less than the basal values (Medbø and Sejersted 1985). During five repetitions of short-term (45- to 60-s) exhaustive exercises, the uncorrected concentration of plasma Na^+ rose by 4% to 8%, whereas potassium concentration increased more than 100%. However, during rest intervals lasting 4 to 4.5 min, levels of Na^+ and K^+ decreased rapidly; potassium concentrations dropped to values less than basal (Hermansen et al. 1984).

The ionic gradient of Na^+ may be influenced by the loss of sodium in sweat. This influence is reduced by a salt conservation mechanism at the level of the sweat glands. As a result, sodium concentration in sweat is severalfold less than in plasma. Total osmolality of sweat is 80 to 185 mOsm/L, whereas plasma and muscle osmolality is 302 mOsm/L. The potassium level in the extracellular compartment is controlled by way of a "washout." The main regulatory manifestation is increased renal excretion of potassium stimulated by aldosterone to avoid potassium accumulation in the extracellular compartment and, thereby, to reduce the potassium gradient. At the level of the sweat glands, the conservation mechanism does not work in regard to potassium. Apart from other ions, the concentration of K^+ and Mg^+ is in the same range as in plasma and sweat (Costill et al. 1976c). Consequently, the sweat gland conservation mechanism reduces sodium loss but not potassium loss.

A question arises whether the final result of prolonged "washout" is a potassium insufficiency in body fluid, including the loss of intracellular potassium. Schamadan and Snilvely (1967) suggested that potassium depletion is a possible cause of heatstroke. Knochel et al. (1972) estimated potassium loss of 917 moles in recruits during 4 days of military training in the heat. Critical analysis of these results was made by Costill (1986). He found several shortcomings in the calculations made by Knochel et al. (1972). Costill considered the large potassium loss doubtful. According to his results, exercise dehydration causing body water loss of 4.35 L (5.8% body mass) reduced total body K^+ content less than 1% (44.6 mmol). Although plasma volume declined by 13.7% and muscle water content by 8.0%, muscle K^+ remained unaffected by the large sweat losses (Costill et al. 1976b). In another study, after 2 h of cycling in an environmental chamber (30° C, 46% relative humidity), body water loss by perspiration and urine was an average of 2.4 L and K^+ loss was 22.6 mmol. In this case, the water content decline was 10% in inactive deltoid muscle and 4% in active vastus lateralis muscle. K^+ contents remained unchanged in both muscles, whereas plasma K^+ content increased significantly (Costill et al. 1981).

Intensive perspiration during exercise results in losses of body water. Part of these losses is compensated by the water formed as a result of intensive oxidation rates and another part by the water liberated in glycogen degradation (Olsson and Saltin 1970). However, these amounts of water produced are modest compared with water losses by sweat. When the total loss of body water constitutes 2.2% of body weight, the total loss is distributed among intracellular, interstitial, and plasma fluid in portions of 30%, 60%, and 10%, respectively. When the total loss increased up to 4.1% or 5.8% of body mass, the intracellular fluid loss amounted to half of the total, but interstitial fluid accounted for 38% to 39% of the total loss. The plasma loss remained constant at 10% to 11% of the total (Kozlowski and Saltin 1964).

In 1962, Finnish researchers Ahlman and Karvonen reported that dehydration of only 2% of body mass causes decreased exercise performance. Later, this fact was repeatedly confirmed, and an almost linear relationship between the reduction of body mass and the decline of physical performance was found (see Saltin 1964a; Nielsen et al. 1981; Saltin and Costill 1988). If the loss of body water amounts to 4% to 5% of body mass, the capacity for hard muscular work would decline 20% to 30%.

Several harmful effects are caused by dehydration during exercise. First, the possibilities for thermoregulation are reduced, and body temperature rises to higher levels (Sawka et al. 1998). Cardiovascular effects are caused by the combined action of dehydration and hyperthermia; during 120 min of exercise, declines in mean arterial pressure, cardiac output, and stroke volume and increases in heart rate and systemic vascular resistance appeared after initial adjustments. These changes did not appear or were less pronounced when either hyperthermia was influenced without dehydration (water intake allowed), dehydration without hyperthermia, or dehydration without plasma volume reduction (Conzalez-Alonso 1998; Coyle 1998). Maximal oxygen uptake is unaffected in up to 4% of dehydration (Saltin 1964a; Armstrong et al. 1985). However, after a subject lost 1.2 to 1.6 kg of body water and plasma volume was reduced 9.9% to 12.3%, runners' performance was reduced by 3.1% in the 1500-m, 6.7% in the 5000-m, and 6.3% in the 10,000-m races (Armstrong et al. 1985). Increasing the environmental temperature from 20° C to 40° C reduced exercise time to fatigue at 70% $\dot{V}O_2max$ from 67 ± 1 min (mean ± SEM) to 30 ± 3 min.

Avoiding harmful effects, which lead to decreased performance, requires sufficient hydration (water intake) during exercise (Sawka et al. 1998; Conzalez-Alonso 1998) and rapid rehydration after exercise (Maughan 1998). Hydration/rehydration procedures may influence the electrolyte balance; intake of plain water may reduce electrolyte concentrations in plasma and extracellular fluid (see Maughan 1994).

Despite the strong effects of dehydration exercise performance, thermoregulation, and cardiovascular adjustments, exercise-induced dehydration may not lead to the manifestation of heat illness and heatstroke except in rare cases. Noakes (1998) concluded that he "failed to uncover any firm scientific evidence that dehydration and electrolyte imbalances are neither exclusive nor even important causes of these illnesses currently described in the medical literature."

In conclusion, the water-electrolyte balance is controlled by the Na^+-K^+ pumps on the sarcolemma, Ca^{2+} pump on membranes of the sarcoplasmic reticulum, and by several factors influencing extracellular volume and composition. In the biochemical monitoring of training, information about the functions of these pumps would be highly important. In field conditions, it is impossible to obtain the needed information directly if biopsy samples are not used. However, exercise-induced changes in the plasma potassium level are rather valuable, although certain limitations exist for interpretation of plasma K^+ changes only by outflow of K^+ from working muscles (see McKelvie et al. 1992; Vøllestad et al. 1995). Changes in the plasma sodium level are less informative because they are modest and related to hemoconcentration.

Working capacity depends significantly on the hydration status, which is influenced by perspiration rate and voluntary rehydration. General information on water loss can be obtained by accounting for reduction of body weight and amount of water intake. The decline of plasma volume is even less informative because it depends on several other factors and dehydration effect on plasma volume is limited. A method exists for whole-body sweat collection in man to obtain data on electrolyte loss (Shirreffs and Maughan 1997). This procedure is complicated enough not to be used in the daily monitoring of training. However, it is suitable for episodic studies.

To use plasma electrolyte concentration for water-electrolyte balance, one must keep in mind that determination of both are homeostatically controlled parameters. Their constant level is the aim of a specific homeostatic mechanism. Therefore, a lack of changes in plasma electrolyte levels may not be due to a lack of changes in water-electrolyte metabolism. A lack of changes may be the result of an effective homeostatic regulation. To judge homeostatic regulation, it would be better to assess the responses of related hormones (aldosterone, renin, atrial natriuretic factor, and vasopressin).

Summary

The main hematological parameters that have to be considered in training monitoring are plasma and total blood volume and erythrocyte and hemoglobin mass. When accounting for changes in plasma volume, it is essential to take into consideration extravasation of the plasma. This results in altered concentrations of several constituents of the plasma (mainly proteins and compounds bound by proteins) without other metabolic influences besides plasma extravasation. Increased blood volume and erythrocyte and hemoglobin mass promote oxygen transport and are, therefore, important parameters of the endurance training effect. It is important to know the conditions that favor hemoglobin synthesis and the conditions that may impair the oxygen transport function because of disorders in iron metabolism or immature perishing of erythrocytes (so-called sports anemia). Attention should also be paid to the appearance of muscle enzyme protein and myoglobin in the blood plasma because this indicates increased permeability or disorders of the plasma membrane of muscle fibers.

For deep analyses of muscular adaptation, determination of oxidative stress (ratio between free radicals and antioxidants) must be part of the monitoring tasks.

The capacity of the acid-base buffer system is essential for performance of anabolic exercises. However, the capacity of the bicarbonate buffer system does not increase under the influence of training. Most important is the protein buffer system in muscles.

Exercise induces several changes in the parameters of immune activities (enhanced phagocytotic activity, augmented populations of B- and T-cells, dramatic changes in the number of NK cells, alterations in the level of various immunoglobulins). It is essential to monitor changes in immune activities resulting from muscular activity to know the state of the body's defense mechanisms.

The concentration of ions and the amount of water in the body are homeostatic parameters (rigid constants of the internal environment) that are maintained without significant deviations. Even during strenuous exercises, their changes are modest, if at all, except in a trend to dehydration and accumulation of K^+ in blood plasma. The increased potassium level in the extracellular fluid, including blood plasma, is related to outflow of K^+ from working muscles. It is an indication of disturbed balance between K^+ outflow from muscle fibers at the onset of contraction and inflow of the ion back into the cells. This balance is controlled by the Na^+-K^+ pump in the plasma membrane. This imbalance is related to fatigue.

General Conclusion to Part II

In part II (chapters 3 to 6), several indexes that may be useful as tools in biochemical monitoring of training were considered. It must be remembered that each separate index is only a parameter of a complex process that is controlled by several factors. Therefore, the value of each index depends on the understanding of the essence of the total process and the contribution of the index used to the total process. Moreover, each index has specific limitations in the biochemical monitoring of training. Therefore, it is not wise to seek the general "good" indexes. It is better to know what the index is in its essence and what kind of information it can provide.

Actualization of Biochemical Monitoring of Training

The aim of part III of this volume is to analyze the practical value of tools for monitoring training. Several tests are compared, and their value and limitations are evaluated based on their reliability and practical application. Several results obtained in training monitoring are presented as examples of suitable schemas for biochemical monitoring.

Chapter 7 deals with the assessment of training-induced effects on the body with the aid of biochemical methods. Attention is focused on assessment of the power and capacity of various mechanisms of energy production (adenosine triphosphate resynthesis). Lactate diagnostics are also considered. Several possibilities for analysis of the effectiveness of metabolic control and endocrine function are also presented.

Chapter 8 analyzes how to assess and evaluate the workload of training sessions and microcycles. Its focus is on assessment of the trainable effect of the workloads used. Analysis of microcycle design leads to the biochemical diagnosis of fatigue and monitoring of the recovery processes.

Chapter 9 discusses adaptive changes and exhaustion of the body during the training of high-level athletes and presents various methods for monitoring adaptivity. Such monitoring of sport training has as its goal the achievement of peak performance and the avoidance of overtraining.

CHAPTER

7

Feedback From Training-Induced Effects

Today, many changes that appear in the body during exercise training have been discovered. The following general outcomes of whole-body training may be distinguished (see Viru and Viru 2000):

- Hypertrophy of organs: skeletal muscles, myocardium, adrenal glands, and bones
- Increased functional capacities of organs and functional systems
- Functional stability (a capacity to maintain a necessary functional or metabolic activity during prolonged exercise despite fatigue)
- Functional reserves (the difference between the basal level and the highest possible level of function)
- Energy reserves
- Efficiency of functioning and energy processes
- Perfected coordination, regulation and control of body function, and metabolic processes
- Motor abilities
- Power, capacity, and efficiency of the energy production (adenosine triphosphate [ATP] resynthesis) mechanism

In several cases, assessment of these training outcomes requires biochemical methods. The goal is to acquire information on the metabolic adaptation and its control mechanism to understand the changes at the level of cellular metabolism. First, the need for biochemical methods arises in the assessment of training effects on energy systems. Then when the effectiveness of training disappears, further biochemical studies are necessary to determine why.

Muscle Energetics and Exercise Classification

Preparation for successful performance is accomplished when all aspects that influence the competition are considered. In most cases, the necessary motor potential is founded on well-developed muscle energetics. Therefore, it is essential to know what the energy demands of the physical competition are. To do this, it is useful to have an exercise classification from the point of view of muscle energetics. Furthermore, the physical demands of the competition have to be analyzed. When the physical demands of the competition are known, it is possible to decide with the aid of the classification what is essential to develop in the body and what tests measure the effectiveness of that training.

Competitive exercises may be cyclic (the same movement cycle is repeated continuously) or acyclic (figure 7.1). In sports training, four types of cyclic exercises are frequently delineated: aerobic, aerobic-anaerobic, anaerobic-glycolytic, and anaerobic-alactatic (table 7.1). Later (see p. 157 of this chapter), we will show

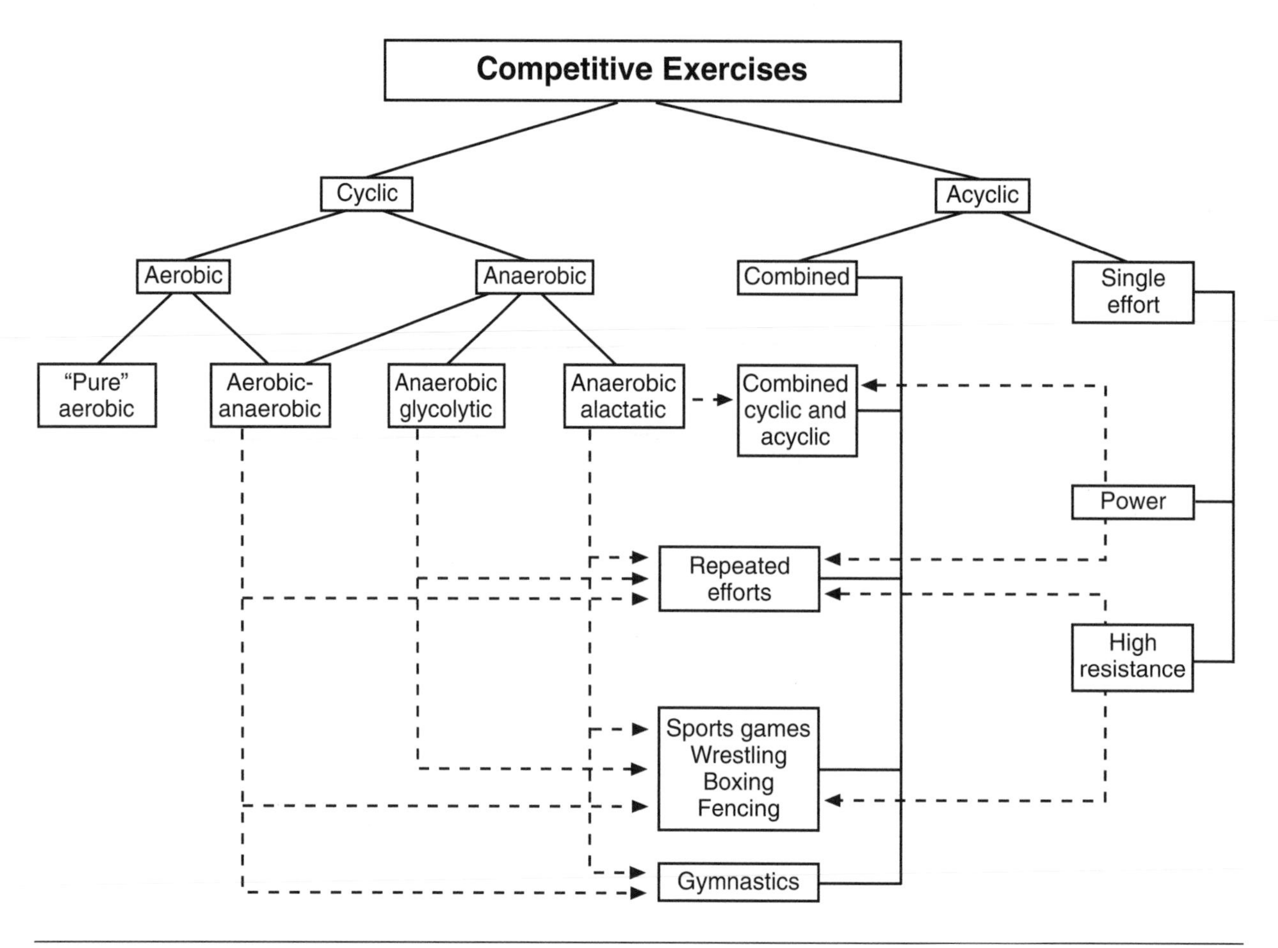

Figure 7.1 Classification of competitive exercises in sports.

that aerobic exercises that use carbohydrates or lipids as the main fuel have to be distinguished from each other.

Acyclic exercises are divided into single effort and combined. Single-effort exercises are founded on the rapid resynthesis of degraded ATP with the aid of the phosphocreatine (alactatic) mechanism. Depending on the ratio between the force used and the duration of its application, the subgroups of high resistive ("pure" strength) and power exercises appear. The combination of cyclic and acyclic activity constitutes combined exercises (e.g., cyclic runup and acyclic jump in long and high jumps, various activities in sports games, wrestling, boxing, dancing or similar other events, performances in gymnastics, and figure skating). In several cases, efforts have to be repeated. Depending on the characteristics of combined exercises, the energetic requirements are supplied by the phosphocreatine mechanism, anaerobic glycogenolysis, and/or oxidative phosphorylation. Therefore, their effectiveness is also related to a certain extent to similar factors as in cyclic anaerobic-alactatic, anaerobic glycogenolytic, or aerobic-anaerobic exercises. In several acyclic activities, static efforts and dynamic movements are combined. For instance, prolonged isometric contractions are performed preceded or followed by movements of auxotonic contraction. In shooting events, isometric contractions are needed to maintain a stable position and so on.

The exercise classification is necessary to clarify the main mechanism(s) of energy production (ATP resynthesis). The outcome of this analysis will show which exercises are necessary in training for increased performance. Once the energy mechanisms are identified, then appropriate methods for monitoring training can be determined. Energy mechanisms have to be evaluated regarding their power (rate of energy production) and capacity (total amount of energy production).

Table 7.1
Characteristics of Cyclic Exercises of Various Intensities

Characteristics	Groups of exercises			
	Aerobic	**Aerobic-anaerobic**	**Anaerobic-glycolytic**	**Anaerobic-alactatic**
Main pathway of ATP resynthesis	Oxidative phosphorylation	Oxidative phosphorylation	Anaerobic glycolysis	Phosphocreatine breakdown
Additional pathways of ATP	—	Anaerobic glycolysis	Oxidative phosphorylation Phosphocreatine breakdown	Anaerobic glycolysis
$\dot{V}O_2$ (L·min)	2.0–2.5	2.5–3.0	>3.0	
$\dot{V}O_2$ (ml·min·kg)	28–35	35–42	>42	
Energy expenditure (kcal·min)	10.0–12.5	12.5–15.0	15.0–62.0	~300
Energy expenditure (METS)	8–10	10–12	12–48	~240
Heart rate (bpm)	140–160	160–180	180–200	150–170
Increase in blood lactate	1.5–2x	2–6x	>6x	2–3x
Maximal possible duration	>40 min	5–40 min	0.5–5 min	10–20 s

Estimated values of power and capacity of ATP resynthesis pathways expressed in rate and total amount of energy release are presented in table 7.2 (Hultman and Harris 1988).

The most important derivation from these results is the qualitative aspect. According to data of Margaria and coworkers (1963), the most rapid pathway for ATP resynthesis is the phosphocreatine mechanism, and the slowest pathway is oxidative phosphorylation. Therefore, it is understandable that by increasing power output, the exercise becomes more anaerobic. The phosphocreatine mechanism is the first pathway for intense ATP resynthesis.

In 1983, Hultman and Sjöholm reported that as little as 1.3 s of intense electrical stimulation was sufficient to degrade phosphocreatine by 11 mmol/kg dm and increase lactate accumulation by 2 mmol/kg dm. Results published by Hultman et al. in 1990 demonstrated that phosphocreatine hydrolysis was at its highest rate within 2 s and anaerobic glycolysis 3 to 5 s from initiation of the electrically evoked contraction rate in men. Thus, single effort for rapid power output or high-resistance performance is found only in energy production by the phosphocreatine mechanism. In short-term cyclic exercises for maximal movement speed or other activities requesting the highest rate of power output, the phosphocreatine mechanism combines with anaerobic glycogenolysis. With further prolongation of exercises, the contribution of oxidative

Table 7.2
Capacity and Power of Mechanisms of ATP Resynthesis in Human Muscles

ATP resynthesis pathway	Power (kcal·kg·min)	Capacity (kcal·kg)	Speed of repayment (half time)
Phosphocreatine splitting	750	100	22 s
Anaerobic glycolysis	350	250	15 min
Oxidative phosphorylation	220		—

Adapted from E. Hultman and R.C. Harris 1988; adapted from E. Hultman et al. 1990.

phosphorylation gradually increases, and the rate of power output decreases proportionally.

Anaerobic Energetics

The three pathways of anaerobic ATP resynthesis are the phosphocreatine mechanism, the myokinase mechanism, and anaerobic glycogenolysis. To evaluate training effects on these pathways of energy production, each pathway has to be characterized by the power and capacity of ATP resynthesis with the aid of the mechanism. In this section, the possibilities for assessing training effects on the power and capacity of anaerobic mechanism are discussed.

Phosphocreatine Mechanism

Because biopsy sampling is not appropriate for testing in field conditions, the phosphocreatine mechanism has to be evaluated by power output tests. Maximal power output in a single effort is frequently assessed with vertical jump tests (see Fox and Mathews 1974). Power is derived from the formula:

$$\text{Power (W)} = 21.67 \times \text{Mass (kg)} \times \text{Vertical displacement (m)}$$

This is an informative index for assessing progress in power athletes. However, it would be wrong to use power output in a single effort for estimation of the power of the phosphocreatine mechanism in ensuring rapid ATP resynthesis. According to data of Greenhaff and Timmons (1998), the rate of ATP production from phosphocreatine increased immediately and reached its highest rate during the first 2 s of the exercises performed at the highest possible rate of energy expenditure. From the third to the fifth second, the creatine phosphate breakdown rate decreases, but the rate remains greater than the rate of anaerobic glycogenolysis. Therefore, estimation of the power of the phosphocreatine mechanism is correct if the measurement has been performed during the first 5 s of the exercise. For this purpose, the Margaria staircase test (Margaria et al. 1966) or the measurement of power output during the first 5 s of consecutive maximal jumping at the beginning of Bosco's test (Bosco et al. 1983; Bosco 1999) is suitable . Similarly, the first 5 s of the Wingate anaerobic ergometer test (Bar-Or 1987) and of treadmill sprinting (Thomson and Garvie 1981; Cheetham et al. 1985) have been used.

If the duration of the test exercise is from 5 to 15 s, the test may be used for assessment of the capacity of the phosphocreatine mechanism. For this purpose, Fox (1975) used 10-s exercise bouts repeated with rest intervals of 30s between repetitions. The anaerobic-alactatic capacity was evaluated by either the number of repetitions up to the decrease of exercise intensity or the total amount of work at the maximum power output. However, it is better to calculate the actual level of power output during a brief initial period of test exercises. In the Wingate test and other similar bicycle ergometer tests, power is computed by the formula:

$$\text{Power (W)} = \text{Wheel revolutions} \times 0.98 \times [60/\text{test duration(s)}] \times \text{Resistance (N)}$$

Bosco et al. (1983) proposed the following formula for calculation of average power during the jumping test:

$$\text{Power} = \frac{9.8 \times \text{Total flight time(s)} \times \text{Duration(s)}}{4 \times \text{Number of jumps} \times [\text{Test duration} - \text{Total flight time(s)}]}$$

Previous attempts to use free creatine and inorganic phosphate for evaluation of the functioning of the phosphocreatine mechanism were mentioned (chapter 3, p. 39). These possibilities are not evaluated systematically. The criticism of this approach is related to the lack of results confirming the quantitative relationship between the use of the phosphocreatine mechanism and increases in blood levels of creatine or inorganic phosphate. Moreover, creatine released in phosphocreatine breakdown may be rapidly reused for phosphocreatine resynthesis and for energy transport from mitochondria to myofibrils. Without answering these questions, this approach for the evaluation of the functioning of the phosphocreatine mechanism cannot be recommended.

Myokinase Mechanism

Another questionable approach is the use of plasma ammonia concentration for evaluation of the energetics in short-term exercises at the highest possible rate. During short-term intensive exercises, accumulation of ammonia in the plasma is indicative of the degradation of AMP. Thus, ammonia accumulation is related to the myokinase reaction (2ADP → ATP + AMP), which is followed by AMP degradation (see chapter 3, p. 37). Although special tests measuring ammonia have not been developed, this method seems to provide a promising approach. We advocate for the use of blood ammonia for the following reasons:

- Ammonia production increases in sprint exercises
- Ammonia production in exercise depends on the amount of fast-twitch fibers
- Sprint training increases the plasma ammonia response to short-term, high-intensity exercises (see chapter 3, p. 37)

The question remains whether the increased ammonia concentration in the blood plasma provides the information either about the power of the myokinase reaction or about the capacity of this mechanism. Theoretically, it is possible to suggest that the power of this mechanism of ATP resynthesis may be evaluated by the rate of ammonia accumulation in the blood, whereas the highest ammonia concentrations may be a measure of the capacity of the myokinase mechanism. Ammonia accumulation in sprint exercises has been suggested as a test for the talent search for sprinting (Hageloch et al. 1990).

This should also be considered in prolonged exercises. Urhausen and Kindermann (1992b) found that in endurance athletes, blood ammonia concentration increased with exercise duration. However, in endurance exercises, the lactate response was related more sensitively to the intensity of the exercise than the ammonia response.

In prolonged exercises, there are two sources of ammonia production by working muscles. In addition to adenosine monophosphate (AMP) degradation, maybe even more ammonia is derived from the oxidation of branched-chain amino acids. In turn, accumulation of the latter portion is inversely proportional to ammonia detoxication with the aid of alanine and glutamine synthesis. Therefore, blood ammonia may be used for suggestions about the myokinase reaction and AMP degradation only in sprint exercises (oxidation rate is comparatively low).

A question arises whether the plasma levels of uric acid or hypoxanthine are more specifically related to AMP degradation than ammonia accumulation. Uric acid and hypoxanthine, as well as ammonia, are formed in AMP degradation. Their levels also increase during exercise (see chapter 3, p. 37). Further studies have to establish the quantitative relationships of all three products with the actual breakdown of AMP. These results provide a possibility to decide what the most informative index for the evaluation of AMP degradation is.

The significance of a valid parameter for assessment of AMP breakdown is not so much necessary for evaluation of the contribution of the myokinase reaction per se. It is important to have information about whether ATP resynthesis lagged behind ATP degradation to adenosine diphosphate (ADP). Quantitative parameters of AMP degradation should provide necessary information because the myokinase reaction is the only one among the ATP resynthesis pathways that results in a reduction of total ATP stores. In the myokinase reaction, ATP is resynthezised as a result of combining two ATP molecules instead of phosphorylation of each ADP molecule in other pathways.

Anaerobic Glycogenolysis

Anaerobic glycogenolysis consists of the degradation of the glycogen molecule or the bloodborne glucose molecule (glycolysis) up to the formation of pyruvic acid (pyruvate). The latter may be oxidized, transformed to lactate, or used for alanine synthesis or glycogen resynthesis. In the degradation of glycogen, energy releases three molecules of ATP for resynthesis. Degradation of a glucose molecule supplies for resynthesis energy of two molecules of ATP. Anaerobic glycogenolysis is highly significant in almost all intensive exercises. In this section, a discussion of the estimation of the power and capacity of this mechanism is presented.

Power

To gain the most information from tests for the power of anaerobic glycogenolysis, one must know the dynamics of this mechanism in ATP resynthesis. Most important is to know the dynamics of the rate of this process. Hultman and Sjöholm (1983) showed that the rate of glycogenolysis reaches maximum between 40 and 50 s, when high-intensity muscle activity is induced by electrically evoked frequent contractions. According to Greenhaff and Timmons (1998), the rate of glycogenolysis-derived ATP regeneration is the highest from 2 s to 20 s of contractions at a high rate of energy expenditure. ATP production from phosphocreatine exceeds that from glycogenolysis during the first 10 s of an electrically induced contraction. Later, most belong to glycogenolysis. However, from the second half of the first minute, the rate of glycogenolysis drops significantly. Consequently, a test of 30-s duration is advisable for the assessment of the power of anaerobic glycogenolysis.

Accordingly, the test for evaluation of the power of anaerobic glycolysis has to last 20 to 50 s. A 30-s Wingate test has been found to be an informative mean (Bar-Or 1981, 1987; Dotan and Bar-Or 1983; Inbar et al. 1996). Verification of the Wingate test needs to establish that the test exercise is actually performed at the expense of anaerobic glycogenolysis. Accordingly, muscle lactate was measured after all-out exercises on an ergometer for 10 s or 30 s. More pronounced lactate accumulation after a 30-s exercise proved that increase of test exercise duration from 10 to 30 s enhances the contribution of anaerobic glycogenolysis (Jacobs et al. 1983). Determination of energy substrates and lactate in working muscles of women showed that during the 30-s test, ATP decreased by 33.5%, phosphocreatine by 60.0%, and glycogen by 23.2%. Muscle lactate increased 6.7-fold. The authors concluded that the Wingate test is a satisfactory test of maximal muscle power that can be generated anaerobically but that the 30-s duration of the test exercise, obviously, does not tax the maximal capacity of anaerobic energy production (Jacobs et al. 1982). Hence, the test may be used more for assessment of the power of the anaerobic mechanism of energy production than for judgments on the capacity of those mechanisms.

Serresse et al. (1988) estimated that in a 10-s maximal ergometer test, the ATP-phosphocreatine mechanism accounted for 53% of total energy expenditure, 44% of anaerobic glycogenolysis, and 3% of oxidative phosphorylation. The values for a 30-s test were 23%, 49%, and 28%, respectively, and for a 90-s test they were 12%, 42%, and 46%, respectively. Thus, the results showed that in a 30-s all-out test exercise, the dominating pathway of energy production is anaerobic glycogenolysis. However, its contribution was only about 50%. Other studies showed that the contribution of aerobic energy during a 30-s test was either 25% (Thomson and Garvie 1981) or 18.5% (Kavanagh and Jacobs 1988) of the total energy expenditure during exercise.

A question arises as to whether the prolongation of a test exercise helps to differentiate the energy production by phosphocreatine degradation and by anaerobic glycolysis. This possibility is revealed from the previously referenced results of Serresse et al. (1988). Withers et al. (1991) demonstrated that muscle lactate concentration peaked after 60 s of exercise. These data argue that a 60-s version of either the Wingate anaerobic test or the Quebec test (originally, a 90-s ergometer cycling test; Simoneau et al. 1983; Boulay et al. 1985) may be the best solution; this was previously affirmed by Szogy and Cherebetiu (1974). However, one has to take into consideration the increased contribution of aerobically produced energy when the test exercise is prolonged. It has been questioned whether the quantification of anaerobic work for calculating anaerobic capacity can be done satisfactorily from performance data (Foster et al. 1995). In several cases blood lactate determination was used after the Wingate and other similar tests (Szogy and Cherebetiu 1974; Jacobs 1980). However, general opinion holds that lactate determination does not significantly enhance the informative values of related tests.

Foster et al. (1995) pointed out the necessity of accounting for the significance of lactate turnover in the accumulation of muscle and blood lactate during exercise. They did not consider the blood lactate response in all-out exercises to be a particularly good index of the contribution of anaerobic glycogenolysis in muscle energetics. In supramaximal exercises, the lactate production rate surpasses several times the lactate elimination rate (see Brooks 1985). Therefore, the greater the exercise intensity, the lower should be the methodological error from blood lactate elimination during exercise (see also chapter 3, p. 34).

Capacity

For assessment of anaerobic capacity, Foster et al. (1995) recommended combining the Quebeck 90-s test or air-braked ergometer test proposed by Withers et al. (1991) with the measurement of accumulated O_2 deficit according to Medbø et al. (1988). When O_2 deficit is used instead of blood lactate levels, the methodological error from lactate turnover (elimination) is avoided. However, a question remains whether 90 s is optimal for testing the capacity of anaerobic glycogenolysis.

Medbø et al. (1988) and Gastin (1994) advocated the use of accumulated oxygen deficit in the assessment of anaerobic capacity. Accumulated O_2 deficit can be calculated by subtracting accumulated O_2 uptake (measured throughout the exercise) from accumulated O_2 demand (measured throughout the exercise and recovery period, see Medbø and Burgers 1990). Gastin (1994) affirms that measurements of blood lactate do not provide a quantitative measure of the anaerobic energy yield. He emphasized that oxygen debt (recovery oxygen uptake above resting metabolic rate) is a valid and reliable measure of anaerobic capacity. For instance, significant differences were found in maximal oxygen deficit and anaerobic performance-based parameters but not in postexercise blood lactate values between sprinters, middle- and long-distance runners, and untrained persons (Scott et al. 1991). According to maximal accumulated O_2 deficit, sprint-trained subjects had a 30% greater anaerobic capacity than untrained or endurance-trained subjects. During 6 weeks of proper interval training, accumulated O_2 deficit after test exercises showed an increase of anaerobic capacity by 10%. However, postexercise peak lactate concentrations remained the same (Medbø and Burgers 1990).

Measurement of oxygen deficit (debt) depends on accurate assessment of the energy cost of work completed. The value of the oxygen deficit as a measure of anaerobic capacity is questionable when energy cost is estimated and not measured (Gastin 1994).

Although blood lactate is an indication of glycogenolysis, it cannot give a precise measure of the anaerobic energy yield (Saltin 1990a). Previously, the significance of elimination of lactate from blood was noted. Table 7.3 summarizes limited factors related to lactate production and further metabolism in skeletal muscles during exhaustive exercise. Ignoring the significance of these factors may reduce the precision of using lactate accumulations as a quantitative measure of anaerobic capacity.

Use of blood lactate responses and oxygen deficit requires that the values measured have to be close to the maximum. Otherwise, conclusions on the capacity of anaerobic glycogenolysis will be doubtful. Actually, the blood lactate responses found during test exercises of 30 to 90 s are lower than those found after competition exercises in high-level athletes. Kindermann and Keul (1977) recorded blood lactate levels after running 400 m (22.0 ± 1.6), 800 m (19.4 ± 1.5), and 1500 m (19.6 ± 1.9 mmol/L), after swimming 100 m (15.3 ± 1.8) and 200 m (16.4 ± 2.6 mmol/L), after speedskating (16.1 ± 1.1 mmol/L), and after rowing 2000 m (15.4 ± 1.9 mmol/L). Svedenhag and Sjödin (1984) found blood lactate concentrations 3 min after races of 800 m (18.4 ± 1.2), 1500 m (18.4 ± 0.7), and 5000 m (15.0 ± 1.4 mmol/L). Nummela et al. (1996a) measured blood lactate immediately and 2.5, 5, and 10 min after the Wingate test. Peak value was 13.2 ± 2.4 mmol/L (mean ± SD).

Lactate responses depended on exercise intensity and duration. Prolongation of intensive test exercise increases the contributions of

Table 7.3
Schematic Summary of Factors Related to Lactate Production and the Fate of Produced Lactate in Skeletal Muscle During Exhaustive Exercise

Event	Limiting factor	Adaptation
Rate of glycolysis lactate production	Glycogen, key activators, LDH_{4-5}	Elevation of glycogen storage, enhanced glycolytic enzyme content including LDH_{4-5}, enhanced activation/less inhibition
Accumulation in muscle fiber	Buffer capacity, pH tolerance	Increased breakdown of CP, elevation of specific amino acids
Transport (if facilitated)	Number of lactate transporters	Increased number of lactate transporters
Uptake by adjacent muscle fibers (FT→ST fibers)	"Own" pyruvate and lactate formation, LDL_{1-2}, mitochondrial capacity, NAD: NADH ratio	Enhanced oxygenation of less active adjacent fibers, enhanced oxidative potential, and LDH^{1-2}
Disappearance by way of interstitial space and blood	Capillary density, muscle perfusion, uptake by other tissues	Capillary proliferation, improved central circulatory capacity, enhanced oxidative potential of nonexercising tissues

Reprinted from B. Saltin 1990a.

anaerobic glycogenolysis and decreases the contribution of the phosphocreatine mechanism. However, at the same time, the contribution of oxidative phosphorylation increases. Thus, the test exercise is not an anaerobic glycolytic exercise anymore but an aerobic-anaerobic exercise. If the task is to evaluate the power of an energy production mechanism, maximal response is not as important as the rate of response. However, in testing the capacity of the mechanism, the value of information depends on recording the lactate values close to their individual maximum. A way to obtain these high values is the use of aerobic-anaerobic exercise with optimal ratio between exercise intensity and duration. This means that the combination of exercise intensity and duration results in the highest lactate response. According to the results of Volkov (1990), the highest value was found at power output 2700 kpm/min, when the duration was 10 min. A high lactate level also appeared after a 2-min exercise at 3900 kpm/min (figure 7.2).

A way to assess the capacity of anaerobic glycogenolysis is detection of maximal lactate values or oxygen deficit (debt) by use of an intense set of anaerobic interval exercises. Volkov (1963) found that the actual maximum of blood lactate concentration and oxygen debt appears after running 400 m four times at the highest possible speed (rest intervals were 4 to 6 min, 3 to 4 min, and 2 min between the first and second, second and third, and third and fourth

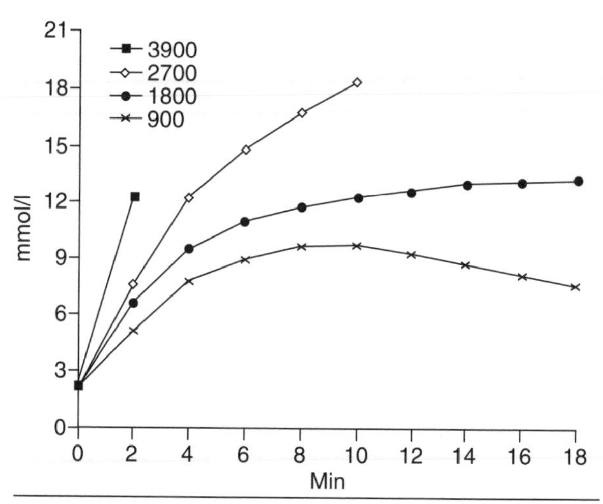

Figure 7.2 Lactate dynamics after cycle ergometer at various power outputs.

By results of Volkov 1990.

bouts, respectively). Volkov (1990) tested 930 qualified sportsmen. Lactate concentration up to 26 mmol/L was found in middle-distance runners and ice-hockey players, up to 24 mmol/L in basketball players and wrestlers. Extremely high lactate concentrations were confirmed after a similar interval exercise set in association with pH values of 6.8 to 7.0 (Gollnick and Hermansen 1973; Fox and Mathews 1974; Fox 1975; Katch et al. 1977; Kindermann and Keul 1977). Hermansen et al. (1984) found a rise of blood lactate on average up to 24 mmol/L (pH 6.9) during 60-s running five times.

A Finnish team of researchers developed a maximal anaerobic running test (MART) using an interval sprint test exercise. The test consisted of a 20-s run with rest intervals of 100 s. Running speed was 14.6 km/h during the first run. Speed increased by 1.37 km/h after each run. Treadmill inclination was 4°. Blood samples for lactate determination were obtained after each run. Power output was calculated for lactate levels 5 and 10 mmol/L, as well as maximal power output. Peak blood lactate concentration was 13.2 ± 2.4 mmol/L after running on a horizontal treadmill (Nummela et al. 1996a). When the treadmill inclination was 7°, the lactate level increased up to 15.4 ± 2.8 mmol/L (Nummela et al. 1996b). Ten weeks of intensive interval training significantly increased maximal power output in sprinters studied with the aid of MART. Peak lactate value did not change (Nummela et al. 1996c).

When incremental exercise tests were used for assessment of the 4 mmol/L lactate threshold (an approximation of the anaerobic threshold), the power output level corresponding to 8 mmol/L was frequently estimated. This value has been used for evaluation of anaerobic capacity without distinguishing the power and capacity of anaerobic glycogenolysis. The shortcomings of this approach are that

- pronounced interindividual differences exist in the ratio of blood lactate level to power output during exercise (see the discussion on individual anaerobic threshold, this chapter, p. 157),
- lactate concentration of 8 mmol/L is only one-half or even one-third of the actual lactate maximal concentration in athletes (particularly in those who specialize in middle distances), and
- power output at 8 mmol/L depends not only on total anaerobic capacity but also on the level of power output at the individual anaerobic threshold.

Green (1994), after a detailed analysis, concluded that "the anaerobic capacity is the maximal amount of ATP resynthesised via anaerobic metabolism (by the whole organism) during a specific type of short-duration, maximal exercise." Although we accept this definition, we do not see any possibility of using the term "anaerobic capacity" in relation to the power output at 8 mmol/L lactate. Green (1994) discriminated between the terms "anaerobic capacity" and "anaerobic work capacity." He defined the latter as "the total amount of external (mechanical) work performed during a specific type of exhausting exercise which is of a sufficient duration to incur a near-maximal anaerobic ATP yield, given that this ATP yield exceeds that from oxidative metabolism." The use of power output at 8 mmol/L also does not fit with this definition.

Green (1994) argues that the terms "maximal anaerobic capacity" and "anaerobic capacity" are not synonymous because the former term relates to a maximal value, which is independent of the type and duration of the exercise. According to Green (1994), maximal anaerobic capacity is only of theoretical interest because athletes need opportunities to use their anaerobic capacity only in their chosen sport event.

In a comprehensive review, Di Prampero (1981) concluded that the H^+ ions and their regulation are primary determinants of anaerobic capacity. The pH shifts induced by intensive exercise may be determined with the aid of blood analysis. In this case, one should recognize that information provided by pH and lactate changes is not the same in essence, although the dynamics of these two parameters may be synchronous. During highly intense anaerobic and aerobic-anaerobic exercise, the increase in H^+ concentration is mainly caused by the production of lactic acid. However, the further fate of H^+ concentration in intracellular and extracellular compartments depends on specific regulatory means, including the significance of buffer systems. Training effects on buffer systems are pronounced in muscle tissue, but they may not appear in blood buffer systems (see chapter 6, p. 125). A limited possibility for evaluation of training effects on buffer systems may appear if the changes of blood pH and lactate are compared during incremental exercise tests up to supramaximal intensities.

The significance of epinephrine in anaerobic glycolysis was discussed (see chapter 3, pp. 19, 25). Because training (particularly sprint training) increases exercise-induced epinephrine responses (see chapter 5, p. 76), it is possible to see the relationship between increased epinephrine production from the very onset of intensive exercise and the mobilization of the anaerobic working capacity. Special study is required to approve this link. It is still only possible to suggest that pronounced catecholamine responses in exercise are essential for high anaerobic working capacity.

Further information on the test of anaerobic work capacity and of power and capacity of the anaerobic energy mechanism and their validity is available in several reviews (Simoneau et al. 1983; Jacobs 1986; Vandewalle et al. 1987; Bouchard et al. 1991; Gastin 1994; Foster et al. 1995).

Aerobic Energetics

Energy release in oxidation provides an opportunity to resynthesize 38 molecules of ATP. It is the most capacious pathway for ATP resynthesis. The use of these possibilities is limited by oxygen transport to working muscles and the oxidative potential (quantity of molecules of oxidation enzymes and their activity). In the whole body, oxygen transport possibilities may be characterized with the aid of maximal oxygen uptake. The indirect measure of the oxidation potential of skeletal muscles is the anaerobic threshold expressed in units of power output. Maximal oxygen uptake ($\dot{V}O_2max$) is the index of aerobic power. Problems related to these considerations are the topic of this section together with a discussion how on assess aerobic capacity.

Oxidative Phosphorylation: Aerobic Power

The rate of oxidative phosphorylation depends on the concentration and activity of mitochondrial enzymes. At the same time, the oxidative potential of skeletal muscle fibers depends on the oxygen supply (figure 7.3). Therefore, in training, effects on the aerobic metabolism have to be distinguished from events in muscle fiber mitochondria and those in oxygen transport from lungs to muscles.

Biopsy samples provide the opportunity to establish an increase in the mass of mitochondria and in the activity of oxidation enzymes in mitochondria. An indirect way to get related information is assessment of the anaerobic threshold. Another important and widely used integral parameter is maximal oxygen uptake ($\dot{V}O_2max$).

Maximal Oxygen Uptake

$\dot{V}O_2max$ measures how much the body can uptake and use oxygen during vigorous exercise. This index seems to quantify integral events at the level of mitochondria and the oxygen transport system. Actually $\dot{V}O_2max$ depends mainly on the oxygen transport rate. Special studies indicated that the reserves for an increase in oxidation rate are several times higher than for increasing the blood flow in working muscles (Saltin 1990b; Saltin and Strange 1992). Consequently, the increase of oxygen uptake is limited, first, by the blood flow through working muscles. Analysis of this question in several articles and reviews led the authors to the conclusion that circulatory factors determine the actual level of $\dot{V}O_2max$ (Clausen 1977; Saltin and Rowell 1980; Åstrand and Rodahl 1986; Cerretelli et al. 1986; Rowell 1986, 1988; Saltin 1990b; Saltin and Strange 1992).

The discrepancy between training-induced increases in $\dot{V}O_2max$ and activity of oxidative enzymes in muscle fibers has been confirmed by the fact that the gain in $\dot{V}O_2max$ does not reflect the pattern of oxidative potential of muscles during training. The results of Henriksson and Reitman (1977) showed a close relationship between improvements of $\dot{V}O_2max$ and the activities of succinate dehydrogenase and cytochrome oxidase in only the vastus lateralis muscle over the first 3 to 4 weeks of endurance training in men. Thereafter, $\dot{V}O_2max$ leveled off, but the activities of the mitochondrial enzymes continued to rise to the end of the 8-week training period. During the following 6 weeks of detraining, enzyme activities dropped, but $\dot{V}O_2max$ remained on the level found at the end of training. Cross-sectional comparison showed that the $\dot{V}O_2max$ of endurance athletes is twice that of untrained persons,

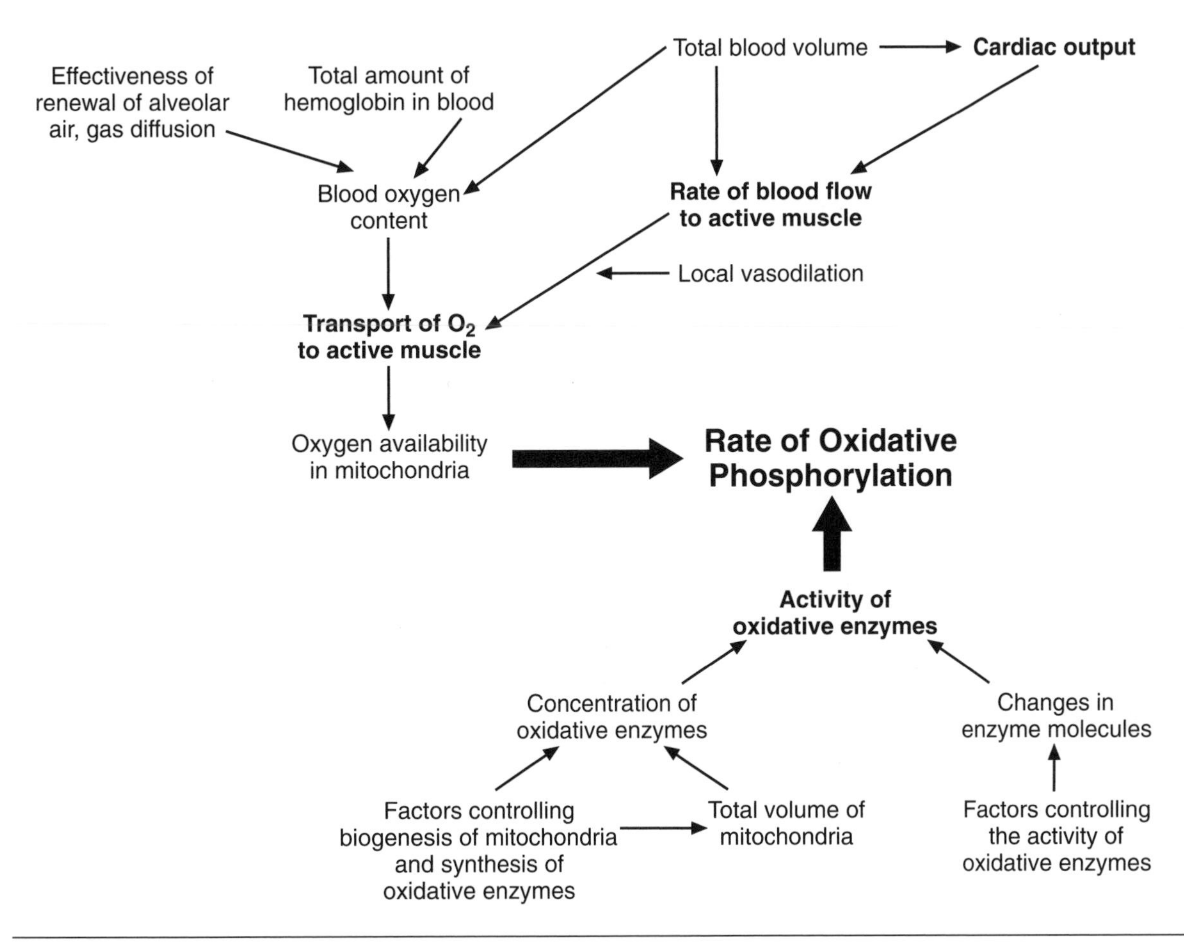

Figure 7.3 Factors influencing the rate of oxidative phosphorylation.

whereas the activity of the mitochondrial enzymes of the muscle in endurance athletes is threefold to fourfold greater than in nonathletes (Saltin and Rowell 1980).

The blood supply of working muscles depends on local vasodilation (Bevegård and Shepherd 1967; Clausen 1977; Åstrand and Rodahl 1986; Ozolin 1986; Rowell 1986, 1988). Local vasoregulation contributes to blood flow enhancement after endurance training (Clausen 1977). Obviously, in this way $\dot{V}O_2$max is inversely related to systemic vascular resistance (Clausen 1977). Accordingly, Snell et al. (1987) found a greater correlation between $\dot{V}O_2$max and vascular conductance during exercise in endurance athletes than in untrained persons.

The increased blood flow rate in working muscles depends not only on local vasodilation but even more on the total rate of circulation and, therefore, on the functional capacity of the heart. Measurements of peak local muscle flow suggest that the vascular conductance per unit of tissue is such that if all the muscles were vasodilated, the available peripheral conductance would far exceed the pumping activity of the heart (Andersen and Saltin 1985). Shephard (1984) provided evidence on the significance of the rate of venous return to the heart.

Complete actualization of the heart's functional capacities requires conditions for effective renewal of the alveolar air and oxygen diffusion in the lungs. Another important presumption for effective O_2 transport is the high total amount of hemoglobin in the circulating blood. Because the hemoglobin concentration in erythrocytes is more or less constant and because limitations exist for the total mass of erythrocytes in blood (see chapter 6, p. 114), the total blood volume is important. However, an increase of total blood volume has positive meaning if the concentration of hemoglobin does not reduce. In this regard, a term "erythrocythemic hypervolemia" has been introduced (Core et al. 1997). Later several investigations proved the significance of blood volume and total amount of hemoglobin for $\dot{V}O_2max$ (see Åstrand and Rodahl 1986). An extended mathematical analysis allowed Shephard (1971) to estimate that at sea level oxygen transport is a limited by-product of hemoglobin concentration and cardiac output.

The main value of measuring $\dot{V}O_2max$ is that it provides information not about a theoretical maximum but an actual peak of oxygen uptake (and use) during performance of an intensive exercise of a certain type. For training monitoring, it is essential that the test exercise for determination of $\dot{V}O_2max$ corresponds to characteristics of the competition exercise and is performed by the same muscles.

As a rule, values of peak oxygen uptake differ not only interindividually but also intraindividually, depending on the mass of the working muscles, the condition of their blood supply, pO_2 in the inspired air, body position, and on the accordance between the test exercise and the main competition exercise (see Shephard 1984; Åstrand and Rodahl 1986). It is also essential to have a sufficiently developed neuromuscular apparatus to perform the test exercise at the level of power output, which enables an increase in oxygen uptake to the highest possible level in the type of exercise used for testing (Jones and McCartney 1986).

The exercise intensity that is enough to evoke the highest possible oxygen uptake is called "maximal" exercise. Exercises with intensities less than the level of $\dot{V}O_2max$ are called "submaximal" and greater than it are called "supramaximal" exercises. However, the meaning of "maximal" has mostly historical, not physiological, value. The actual intensity of the "maximal" exercise is not the greatest possible. Athletes perform a lot of supramaximal exercises during which the power output and the rate of energy production (resynthesis of ATP) is several times greater than during the "maximal" exercise. The many anaerobic pathways for ATP resynthesis (phosphocreatine splitting, anaerobic glycogenolysis) enables one to increase exercise intensity up to the greatest level of oxygen uptake. Blood lactate levels from 8 to 12 mmol/L (Åstrand and Rodahl 1986; Shephard 1992) at $\dot{V}O_2max$ confirm the contribution of anaerobic energy mechanisms. Thus, actual $\dot{V}O_2max$ is revealed in aerobic-anaerobic exercises. Consequently, the term "maximal" does not even mean the highest aerobic exercise.

Determining Maximal Oxygen Uptake

The method of choice for direct measurement of oxygen uptake during incremental exercises is either on a bicycle ergometer or treadmill. Several event-specific procedures are also advisable. Tests for measurement of $\dot{V}O_2max$ should increase exercise intensity up to the leveling off values of oxygen uptake. The "plateau" of oxygen uptake has been defined arbitrarily as an increase in oxygen consumption less than 2 ml/kg · min despite further increase when the intensity of the test exercise increases (Shephard 1992). However, in several cases complications exist in establishing constant values of oxygen uptake (lacking of increase in $\dot{V}O_2$ despite a further increase in exercise intensity). Therefore, several criteria are used to determine that the measured peak $\dot{V}O_2$ is actually the maximum.

The most important criterion used is the lack of further increase in cardiac output or muscle blood flow. However, to determine that is a methodologically complicated task. It is not possible to determine this with the usual tests. Therefore, secondary criteria of $\dot{V}O_2max$ are used such as peak heart rate (estimated by using the formula 220 – age in years), peak respiratory gas exchange ratio (respiratory quotient) higher than 1.10 (some researchers use a criteria of respiratory ratio greater than 1.00), and peak blood lactate of 10 to 12 mmol/L. In older individuals a lactate concentration around 8 mmol/L is satisfactory (Shephard 1984, 1992). Davis (1995) considers blood lactate greater than 8 mmol/L, respiratory quotient greater than 1.0, and heart rate 85% of the age-predicted maximum to be sufficient criteria. One way to to determine that an actual maximum has been reached is to terminate the incremental test with a 1-min spurt at the highest possible rate of pedaling (Pärnat et al. 1975a).

In several cases, the test subject is not able to continue the exercise, and several signs (pallor, bluish-gray facial skin, breathlessness, etc.) confirm acute weakness and the inability to exercise any more. Of course, the test has to be terminated, although the criteria for maximal $\dot{V}O_2max$ failed to appear. In these cases, the researcher cannot consider that the recorded level of oxygen uptake is the maximum. A possibility also exists that a weak neuromuscular system does not allow for the increase of the cardiovascular activity to the actual maximum. Although the power output obtained might be the maximum for such a person, the physiological meaning of the test results is substantially different from that in persons who satisfied the criteria for $\dot{V}O_2max$. Ethically, any kind of testing has to terminate if the test subject refuses further participation. In these cases, secondary criteria for the determination of maximum $\dot{V}O_2max$ must be used.

For critical analyses of various procedures, details of testing, safety assurance, and validity of the result, the reader is directed to the reviews of Åstrand and Rodahl (1986) and Shephard (1984, 1992).

Several methods of indirect assessment of maximal aerobic power are proposed. These tests are designed to avoid maximal effort by using submaximal testing exercises and other measures (e.g., heart rate, endurance performance indexes) instead of recording oxygen uptake (for a critical analysis of several indirect tests, see Åstrand and Rodahl 1986; Shephard 1984, 1992). The basic assumptions that underlie most submaximal predictions of $\dot{V}O_2max$ are

- a relationship between heart rate and oxygen uptake during exercise,
- a known maximum heart rate, and
- a known mechanical efficiency of the performance of the test exercise.

Systematic error of up to 10% and random error of 10% limit the value of the information that can be obtained through such procedures.

Anaerobic Threshold

Threshold intensity of exercise for lactate accumulation in the blood was found in 1930. Owles (1930) was the first who mentioned the "critical metabolic level." Since the article of Wasserman and McIlroy (1964), the concept of anaerobic threshold has been widely used and mostly accepted.

However, some confusion arose in phenomenological point of view. In 1963, Margaria et al. published results showing the appearance of lactate in increasing amounts in the blood only at intensities greater than $\dot{V}O_2max$. The exercise intensities corresponding to the anaerobic threshold in the study of Wasserman and McIlroy (1964) were significantly lower than that at $\dot{V}O_2max$. Jones and Ehrsam (1982) explained the discrepancy in results by the fact that Margaria and coworkers used very short exercise bouts and that the blood lactate was measured a few minutes after cessation of the exercise. Thus, the actual magnitude of lactate responses corresponding to exercise intensity were not detected.

In incremental exercise, the blood lactate response is characterized by a curvilinear increase (Kindermann 1986) (figure 7.4). A measurable increase in the blood lactate curve appears at an exercise intensity of approximately 50% $\dot{V}O_2max$. A sharp increase of lactate concentration begins at intensities 70% to 80% $\dot{V}O_2max$ (Urhausen et al. 2000). The onset of a sharp increase in the lactate curve has been denoted by the terms "lactate break point," "lactate threshold," "onset of blood lactate accumulation," and "aerobic-anaerobic threshold." At this point, the average lactate concentration is about 4 mmol/L. Therefore, several researchers did not analyze the whole lactate curve and interpolated the exercise intensity corresponding to a lactate concentration of 4 mmol/L (Mader et al. 1976; Keul et al. 1979; Skinner and McLellan 1980; Karlsson and Jacobs 1982). The anaerobic threshold, detected by this way, was called the

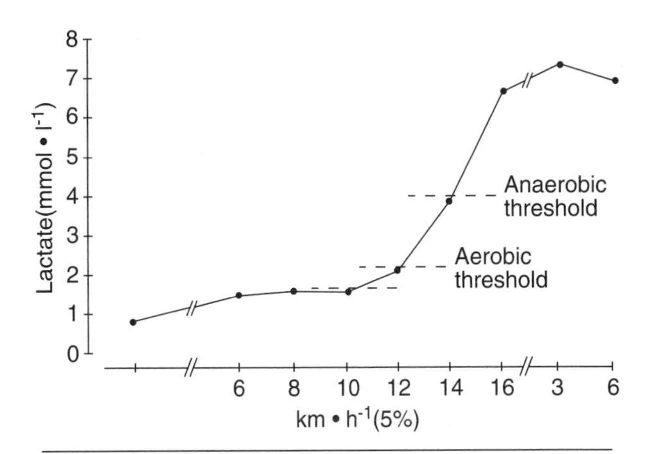

Figure 7.4 Lactate concentration during incremental treadmill running.

Reprinted from W. Vindermann 1986.

lactate 4 mmol/L threshold. On the other hand, detection of anaerobic power by use of the analysis of the whole lactate curve has been indicated by the term "individual anaerobic threshold" (Stegmann et al. 1981, 1982).

Physiological Essence of the Anaerobic Threshold

The term "anaerobic threshold" indicates that its meaning is to denote the transfer from aerobic processes to anaerobic processes in energy production (ATP resynthesis). However, two questions remain. First, why are small amounts of lactate produced during exercises performed at moderate intensities, which obviously require only aerobic ATP resynthesis? Second, what is the reason for the breaking point in the linear elevation of lactate concentration with increasing exercise intensity?

To find an answer to the first question, we have to appeal to the elementary understandings of metabolism. Biochemical reactions are controlled by the activities of various enzymes. If a substrate can transfer into several products, the differences in enzyme activities determine the quantitative characteristic in the substrate conversions. The regulation principle is "more versus less" but not "yes versus no." Accordingly, proportions of pyruvate used as oxidation substrate or for conversion into lactate or alanine (see figure 3.1, chapter 3, p. 35) are different, although all related biochemical reactions are going on simultaneously. Such quantitative regulation ensures smooth transfers and avoids maladjustments caused by less firm switching in and switching off. All in all, the transfer from aerobic to anaerobic energy production is a characteristic example of the smooth quantitative regulation. An increase in exercise intensity augments the rate of glycogenolysis and glycolysis. From a critical level of exercise intensity, the rate of formation of pyruvate begins to surpass the rate of oxidative phosphorylation. Although the rate of oxidative phosphorylation is close to the maximum, lactate formation augments at an increasing rate.

The reason for intensive lactate accumulation in the blood becomes understandable according to the findings of Brooks and collaborators (see Brooks 1985). Even the small amount of lactate produced by working muscles during low-intensity exercises is reflected in the rate of lactate inflow into the blood. However, lactate elimination from the blood equilibrates lactate inflow. Elimination of lactate from the blood means lactate is transported into the sites of lactate oxidation. Those may be the myocardium and resting muscles. A certain amount of lactate may be oxidized in the same muscle but in other fibers possessing a high oxidative potential. The liver consumes lactate for gluconeogenesis as well.

At low exercise intensities, lactate elimination corresponds to the lactate inflow, and lactate concentrations remain at the resting level. An increase of exercise intensity augments the lactate inflow. Now the lactate elimination rate is adjusted to new levels. As an expression of this adjustment, blood lactate concentration stabilizes at an increased concentration, and the lactate level does not increase further despite continuous lactate inflow during continuation of the exercise at the same intensity. Such balancing breaks up when the lactate inflow reaches a certain rate. The main determinant of that breaking point is, obviously, not the rate of lactate inflow but the possibilities for lactate oxidation in working muscles. In this way, it is possible to affirm that the anaerobic threshold expresses the oxidation potential of skeletal muscles of a person.

Using mathematical modeling, Mader confirmed the conclusion on the metabolic origin of the anaerobic threshold and its relationship to the oxidative potential (Mader and Heck 1986; Mader 1991). This point of view has been confirmed by the fact that

- endurance athletes have the lactate threshold at a higher percentage of $\dot{V}O_2max$ than sedentary persons in association with greater activity of oxidative enzymes in skeletal muscles (Sjödin et al. 1982),
- in previously sedentary persons, endurance training shifts the anaerobic threshold to a higher percentage of $\dot{V}O_2max$ (Davis et al. 1979),
- a relationship exists between lactate threshold and in vitro oxygen uptake by incubated muscle samples (Ivy et al. 1980) and between lactate threshold and the activity of oxidative enzymes in the muscles (Sjödin et al. 1981), and

- the running speed at the anaerobic threshold correlates significantly with the percentage of slow-twitch fibers and with their total area in the vastus lateralis muscle (Ivy et al. 1980; Sjödin et al. 1981).

Aunola et al. (1988) found that muscle fiber composition and the activities of citrate synthase succinate dehydrogenase and lactate dehydrogenase explained 74.5% of the variation in lactate threshold and 67.5% of the variation at the 4 mmol/L blood lactate threshold.

A comparison of sprinters and long-distance runners showed that the anaerobic threshold in sprint-trained athletes was 64% $\dot{V}O_2max$ (power output level 154 W) versus 73% $\dot{V}O_2max$ (193 W) in endurance-trained athletes (Hardman et al. 1987). Differences were revealed also when sprint and endurance swimmers were compared (Smith et al. 1984). These data also provide a confirmation of muscle oxidation potential for the anaerobic threshold because the effect of sprint training on the activity of muscle oxidative enzymes is modest if at all (see Viru 1995).

Consequently, the anaerobic threshold assessed by the breakpoint of the lactate curve during incremental exercise reflects a critical event in metabolic adjustment to an increased level of exercise. The physiological meaning of this critical event is that the rate of the oxidative metabolism cannot increase any more. Involvement of the anaerobic mechanism of energy production is unavoidable to perform exercises of higher power output.

This understanding of the anaerobic threshold provides opportunities to use the anaerobic threshold for an evaluation of the oxidative potential of skeletal muscles. Although the anaerobic threshold does not enable a quantitative measure of the possibilities for the oxidative metabolism in skeletal muscle, it may be used as a semiquantitative characteristic.

The level of exercise intensity corresponding to the anaerobic threshold has been defined as either a percentage of $\dot{V}O_2max$ or power output. In training experiments and in training monitoring, the definition of exercise intensity at the anaerobic threshold as a percentage of $\dot{V}O_2max$ is not meaningful. Training may increase $\dot{V}O_2max$; therefore, this expression of the anaerobic threshold makes the values dependent on $\dot{V}O_2max$. However, the idea is to use the anaerobic threshold as an independent index to inform us about the oxidative capacity of skeletal muscles without a relationship to the effectiveness of oxygen transport. Thus, it is more advisable to use power output values or running speed for the definition of exercise intensity at the anaerobic threshold.

Power output or running speed at the anaerobic threshold is in high correlation with performance in long-distance events, particularly in a marathon race. The first articles indicating this relationship were published by Farrell et al. (1979), Sjödin and Jacobs (1981), LaFontaine et al. (1981), Hagberg and Coyle (1983), Lehmann et al. (1983); for a review, see Sjödin and Svedenhag (1985) and Jacobs (1986). A comprehensive analysis of the significance of the anaerobic threshold in the performance of marathon runners has been provided by Mader (1991).

The anaerobic threshold represents a phenomenon that is associated with a change in several other functions.

Lactate threshold is associated with breaking points in dynamics of various parameters, such as the intensity threshold for catecholamine response (Lehmann et al. 1985; Chwalbinska-Moneta et al. 1998), abrupt rise in electromyograph activity of working muscles (Chwalbinska-Moneta et al. 1998), rise in saliva sodium and chloride concentration (Chichardo et al. 1994), reduction in O_2 saturation of hemoglobin (Grassi et al. 1999), and increase in saliva amylase concentration (Calvo 1997). Zarzeczny et al. (1999) reported a significant correlation among lactate, potassium, sodium, calcium, and ammonia thresholds.

Methodological Aspects

Three approaches can be discriminated in assessment of the anaerobic threshold.

- Determination of the individual anaerobic threshold based on analysis of the whole lactate/exercise intensity curve
- Interpolation of the exercise intensity for the blood lactate concentration of 4 mmol/L
- Determination of maximal lactate steady state (MLSS)

The method of choice should be the determination of MLSS. In this way, the highest possible exercise intensity during which lactate persists is on a stable level (Urhausen et al. 1993) (figure 7.5). This method assesses the anaerobic threshold from the point of view of its actual essence. At the same time, this method accounts for the individual peculiarities in the oxidative potential of muscles and in lactate metabolism. Methodological complications arise because of the necessity to use a number of continuous exercises performed at different intensities but lasting at least 20 min each.

An acceptable variant of the determination of MLSS has been proposed by Billat et al. (1994a). Accordingly, only two 20-min exercises have to be performed. They recommend the use of exercise intensities corresponding to 60% and 80% $\dot{V}O_2$max and to rest for 40 min between two test exercises. From their results, the highest steady lactate level was between 4.0 and 6.0 mmol/L. Aunola and Rusko (1992) confirmed the main premise of assessment of MLSS: when a continuous exercise is performed below the anaerobic threshold, blood lactate declines after having reached its peak. At the level of the anaerobic threshold, blood lactate reaches maximal steady state. Above the anaerobic threshold, blood lactate increases continuously. Significant differences have been found in lactate concentrations at MLSS among speedskating, cycling, and rowing (Beneke and van Duvillard 1996).

Stegmann and Kindermann (1982) used 50-min steady exercises for testing the value of the individual anaerobic threshold. They found steady lactate levels within 3.1 to 4.5 mmol/L. In elite triathletes and cyclists, a great variability was found in lactate concentrations during MLSS. In elite triathletes and cyclists, the values varied from 3.2 to 12.2 mmol/L (Hoogeveen et al. 1997). Less variable results were obtained in rowers, cyclists, and speedskaters. Mean lactate values at MLSS were in groups within 3.1 to 6.9 mmol/L (Beneke and van Duvillard 1996).

Using the whole lactate/exercise intensity curve presents a problem in determining whether the breaking point in determined by visual inspection or a computer program is correct because several individual variants appear make the decision doubtful. At least two apparent points of discontinuity in blood lactate response to incremental exercise usually appear (see figure 7.4). The first of these is associated with the sustained increase of blood lactate concentration above resting values. The second

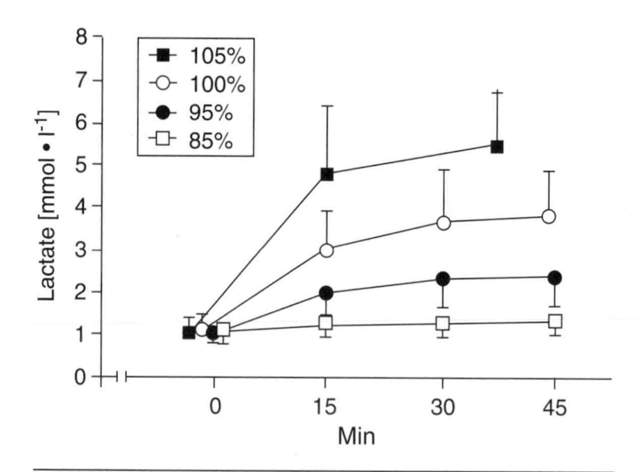

Figure 7.5 Lactate steady states during 45-min exercises at various intensities.

Reprinted from Urhausen et al. 1993.

discontinuity is represented by the onset of an intensive increase in lactate concentration. At the first point, the average blood lactate concentration is 2.0 to 2.5 mmol/L; at the second point, it is 4 mmol/L (Kindermann et al. 1979; Skinner and McLellan 1980). Both groups of authors denoted the second point as the anaerobic threshold. Skinner and McLellan (1980) use the term "aerobic threshold" to refer to the first point.

To avoid observer error in detecting the lactate breaking point, Stegmann et al. (1981) proposed a special trigonometric analysis of the lactate/exercise intensity curve. Stegmann et al. (1982) found that the blood lactate level at the individual anaerobic threshold varies within broad limits among study persons. On this background, they emphasized the need for assessment of the individual anaerobic threshold.

Stegmann and Kindermann (1982) showed that choosing the exercise intensity by the individual anaerobic threshold, the differences from the actual MLSS are less than adjusting the intensity according to the 4 mmol/L lactate threshold. When the intensity for MLSS was set by the exercise level corresponding to the 4 mmol/L threshold, the blood lactate steady state did not appear, the lactate level increased continuously, and most persons were not able to exercise for 50 min.

The 4 mmol/L lactate threshold has been the most widely assessed. Several studies provide evidence for the justification of this approach (Heck et al. 1985, Heck 1990). Despite the wide use of this method, a serious criticism arose from the fact that the individual anaerobic threshold is not at a fixed 4 mmol/L of lactate. Actually, a pronounced variation exists in lactate values at the individual anaerobic threshold. Another source of criticism is revealed from assessment of the maximal blood lactate steady state. Again, this lactate level is variable and not fixed at 4 mmol/L.

Aunola and Rusko (1992) found that MLSS correlated with the individual anaerobic threshold ($r = .83$) but not with the 4 mmol/L blood lactate threshold. The same researcher found poor reproducibility of blood lactate concentration at the power output level predicted by the 4 mmol/L lactate threshold. They concluded that the fixed blood lactate levels of 2 and 4 mmol/L are poor indicators of aerobic and anaerobic thresholds (Aunola and Rusko 1984). Several other articles contain results showing good reproducibility of treadmill running velocity at lactate concentrations of 2.0 mmol/L, 2.5 mmol/L, and 4.0 mmol/L (Weltman et al. 1990; Pfitzinger and Freedson 1998). Test/retest correlation coefficients of running velocity at fixed lactate concentrations and at lactate threshold were all within .89 to .95 (Weltman et al. 1990). Heitcamp et al. (1991) found a high test/retest correlation of running speed at 4 mmol/L. Lactate determinations for anaerobic threshold are discussed by Jones and Ehrsam (1982), Weltman et al. (1990), Foster et al. (1995), and Pfitzinger and Freedson (1998). A general methodological problem is related to the delay of increases in plasma lactate concentration during incremental exercise compared with the increase of muscle lactate concentration (Péronnet and Morton 1994). For further discussion of methodological problems and reliability of determination of anaerobic threshold, see Urhausen et al. (2000).

Aerobic Capacity

Maximal capacity of oxidative phosphorylation depends on

- the amount and availability of substrates for oxidation,
- maintenance of a sufficiently high level of activity of oxidation enzymes for a long time,
- stability in functioning of the oxygen transport system responsible for the supply of oxidation substrates, and
- efficiency of energy processes.

Availability of Oxidation Substrates

According to classical knowledge, lipid oxidation produces more energy per 1 g of substrate than carbohydrate oxidation (9.4 kcal and 4.2 kcal, respectively). At the same time, lipid oxidation requires more oxygen per 1 g of substrate compared with carbohydrate oxidation. Thus, 4.7 kcal is produced per 1 L of O_2 used when the substrate is a fat and 5.0 kcal when the substrate is a carbohydrate. Consequently, the carbohydrate-lipid transfer is beneficial when the total energy expenditure is large and possibilities for complete satisfaction of the oxygen demand exist (the exercise intensity is moderate). The benefit of the transfer to lipid oxidation is problematic when the oxygen demand is greater than the possibility for actual O_2 uptake and total energy expenditure is far from the maximal possibilities.

Taking these considerations into account, the significance of the glycogen store should be understood differently in various exercises. When the contribution of anaerobic energy processes is high, abundant glycogen stores are essential to provide substrate for anaerobic energy production. Produced lactate, either directly or by means of release of hydrogen ions, inhibits the activities of phosphorylase, phosphofructokinase, and ATPase. As a result, a feedback inhibition of glycogenolysis appears together with a decrease of ATP degradation. In this way, muscle contractions stop before the glycogen store is used (figure 7.6).

Another situation exists in aerobic-anaerobic exercises when the rate of anaerobic energy production is less than aerobic energy production and therefore the contribution of oxidative phosphorylation is greater than in anaerobic exercises. Because the lactate concentration increases over a level sufficient to inhibit lipolysis, glycogen and blood-borne glucose have to satisfy the demands of anaerobic glycolysis and oxidative phosphorylation. Several years ago, the biopsy studies of Hultman and coworkers demonstrated that in athletes the endurance capacity for aerobic-anaerobic exercises was related to the glycogen store of skeletal muscles (Bergström et al. 1967; Hultman 1967, 1971; Hultman and Bergström 1973).

The use of muscle glycogen depends on exercise intensity (figure 7.7). The rate of glycogen

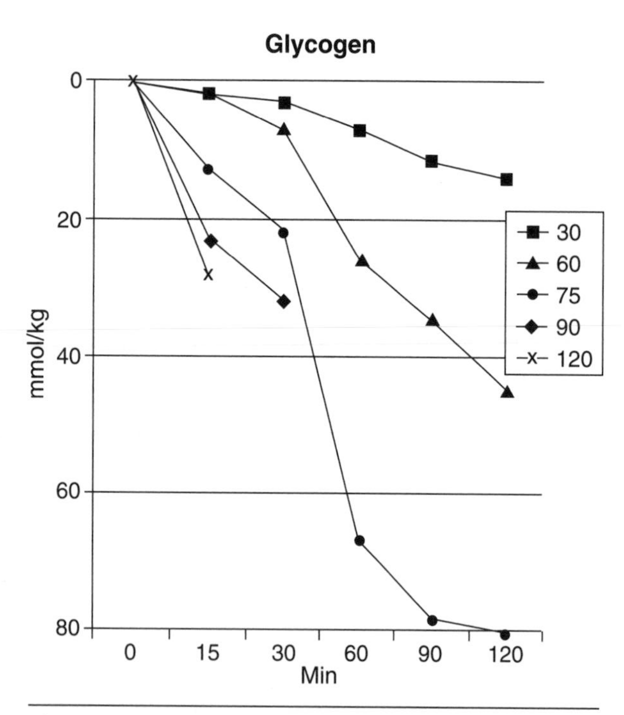

Figure 7.6 Dynamics of muscle glycogen level during prolonged exercise at various intensities.
From Kantola and Rusko 1985.

breakdown increases rapidly above the intensity of 75% $\dot{V}O_2max$ (see Hultman and Spriet 1988). The obvious reason is the inhibition of lipolysis and lipid use at these exercise intensities. By analysis of the patterns of carbohydrate and lipid oxidation, Brooks and Mercier (1994) introduced the crossover concept. Accordingly, the curves of carbohydrate and lipid uses cross at exercise intensities about 70% $\dot{V}O_2max$ (figure 7.8). Obviously, the crossing point is close to the anaerobic threshold.

From the point of view of the use of energy stores, three types of cyclic exercises have to be distinguished. The first type is high-intensity exercises capable of producing power output close to the maximal (anaerobic exercises). These exercises are too short to exhaust glycogen stores. The second type of exercises (aerobic-anaerobic exercises) are performed at comparatively high levels of power output and therefore under conditions of high-oxygen uptake. In these exercises, the glycogen store is a factor limiting performance. The third type of exercises ("pure" aerobic exercises) are found on oxidative phosphorylation; the moderate level of power output (running speed) has to be maintained for a long time. In these exercises, from an energetics viewpoint, performance depends on the effective transfer from carbohydrate to lipid use to spare glycogen use to avoid the hypoglycemia caused by glycogen exhaustion. Table 7.4 presents several characteristics of these types of exercises. This approach is in accordance with the idea of Hawley and Hopkins (1995) to distinguish between the aerobic- glycolytic and aerobic-lipolytic systems.

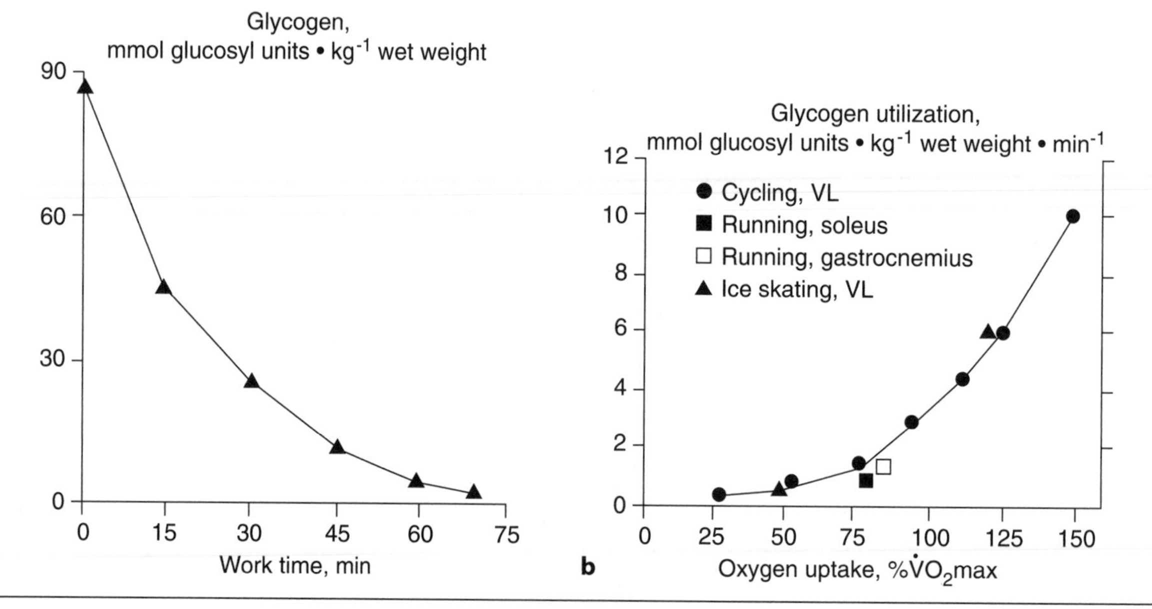

Figure 7.7 *(a)* Decrease of glycogen level during prolonged exercise and *(b)* dependence of glycogen use on exercise intensity in cycling, running, and speedskating in various muscles.
Reprinted from E. Horton and R.L. Terjung 1988.

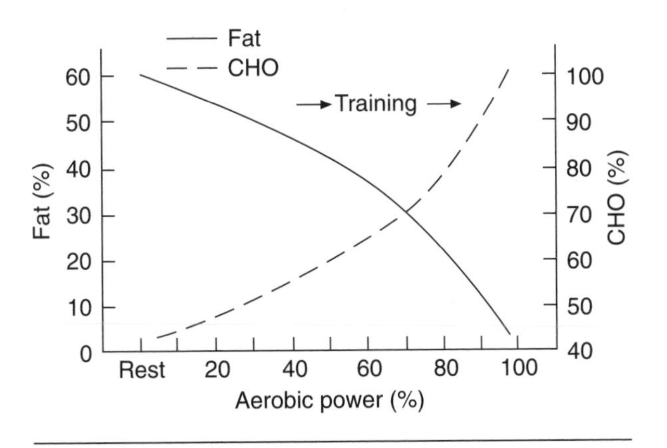

Figure 7.8 The crossover concept in the use of carbohydrates and lipids during continuous exercises of various intensities indicated by % $\dot{V}O_2max$.

From Brooks et al. 1996.

The usual opinion is that the energy store does not limit aerobic capacity. This opinion is not correct in regard to aerobic capacity for aerobic-anaerobic exercises that depend on muscle glycogen stores. In prolonged exercises of moderate intensity ("pure" aerobic exercises), adipose tissue energy reserve of 337,000 kJ is available compared with 7820 kJ of total energy from muscle and liver glycogen and blood glucose (Newsholme and Leich 1983). Therefore, the body's energy reserves seem to be inexhaustible. However, when the liver glycogen reserve is exhausted and hepatic gluconeogenesis cannot compensate with the necessary supply of glucose into the blood, the development of hypoglycemia becomes a factor, sharply decreasing the working capacity (Lavoie et al. 1983). Of course, hypoglycemia can be avoided with the aid of glucose feeding during exercise. However, this external aid may change hormonal regulation and thereby lipid use will decrease (see chapter 5, p. 91-92).

Other Factors

Endurance training increases the activity of mitochondrial enzymes. The question regarding aerobic capacity is whether the high activity of oxidative enzymes persists during prolonged exercises. Rat experiments have shown that after swimming for 10 h, the activity of succinate dehydrogenase and cytochrome oxidase

Table 7.4
Characteristics of Several Types of Cyclic Exercises From an Energetics Viewpoint

Type of exercise	Anaerobic	Aerobic-anaerobic		Aerobic		
Duration of exercise (min)	1–2	3–10	11–35	36–90	90–360	>360
Oxygen uptake (% $\dot{V}O_2max$)	95–100	95–100	90–95	80–95	60–85	50–60
Aerobic/anaerobic ratio	50:50	80:20	85:15	95:5	98:2	99:1
Energy expenditure (kJ/·min) kJ (total)	160 160–320	120 320–1200	110 1200–3700	105 3900–8400	80 8400–25,300	75 >27,000
Breakdown of glycogen in muscle (%)	10	30	40	60	80	95
Blood lactate (mmol/·L)	18	20	14	8	4	2
Free fatty acids in plasma (mmol/·L)	0.5	0.5	0.8	1.0	2.0	2.5

Adapted from G. Neumann 1992.

decreased in association with a pronounced drop of muscle glycogen content and hypoglycemia (see Yakovlev 1977).

During prolonged exercises after a certain time, the exercise levels of heart rate and stroke volume were elevated (Saltin and Stenberg 1964; Ekblom 1970), and arterial mean pressure (Ekblom 1970) and systolic pressure (Viru et al. 1973) declined. Accordingly, during prolonged aerobic exercises, peak oxygen uptake changes. Untrained persons were able to maintain only about half of $\dot{V}O_2max$. Training reduced the difference between $\dot{V}O_2max$ and the VO_2 level during prolonged exercise (see Åstrand and Rodahl 1986). Also hormone levels may decline during prolonged exercise (see chapter 8, p. 176). Obviously, a quality that is induced by training is increased functional stability, and it may have significance for aerobic capacity.

A possibility exists that during prolonged exercises the mechanical efficiency of muscle work may be reduced. This is a possibility because convincing evidence to the contrary is lacking.

Assessing Aerobic Capacity

Aerobic capacity has not been assessed as frequently as aerobic power. The obvious reason is that related testing procedures require use of prolonged exercises. Another problem is that assessment of aerobic capacity has practical value if the performances in test and competitive exercises are founded on the same peculiarities in muscle energetics. The tests for assessment of aerobic capacity can be divided into at least three categories.

- Capacity for aerobic-anaerobic performance
- Capacity for aerobic-glycolytic performance
- Capacity for aerobic-lipolytic performance

The testing of these categories of performance capacities requires the establishment of the total energy typically available to perform exercises corresponding to one of the three categories. In testing the aerobic-lipolytic capacity, the problem is how to find a test that is informative but does not totally exhaust the athlete.

An example of testing for aerobic-lipolytic capacity is the aerobic capacity test proposed by Boulay et al. (1984). The test consists of a 90-min nonstop exercise on a bicycle ergometer. Intensity of the exercise was adjusted by heart rate 10 beats lower than that at the ventilatory threshold. Water was available ad libitum, but no food was allowed. Room temperature was 20 to 22° C. Total work output was recorded and expressed in kilo Joules per kilogram of body mass. Boulay et al. (1984) found good reliability of the test results. A high correlation between maximal aerobic capacity and $\dot{V}O_2max$ was also established. This fact confirms that the tested aerobic capacity depended on aerobic energy production. On the other hand, from this result a question arises whether the testing of aerobic capacity is necessary if $\dot{V}O_2max$ provides the same information. Both parameters yielded a common variance in range of 81%. Thus, about 20% of the variance observed in aerobic capacity could not be accounted for by variation from $\dot{V}O_2max$ scores that yielded a standard error of approximately 9.9%, whereas prediction of $\dot{V}O_2max$ from aerobic capacity measure yielded a standard error of 8.5%. This result may be related to the fact that a great portion of ATP regenerates from fat oxidation during the aerobic capacity test, whereas in the $\dot{V}O_2max$ test carbohydrate metabolism dominates.

Because the highest intensity for "pure" aerobic performance is determined by the anaerobic threshold, for assessment of aerobic glycolytic capacity, exercises done to exhaustion at the anaerobic threshold are advisable. Aunola et al. (1990) constructed curves for expressing the relationship of exercise intensity to maximal cycling time. When exercise intensity was expressed as a percentage of the subject's own anaerobic threshold, the maximal running time predicted for an average subject was 60 min at the work rate corresponding to the anaerobic threshold. This was in accordance with the actual measure of the maximal cycling time at this work rate (Aunola et al. 1990).

The term "maximal aerobic speed" has been used for the speed at $\dot{V}O_2max$ but not for the speed at the anaerobic threshold. Actually, the speed at $\dot{V}O_2max$ is related not only to the aerobic energy processes but also to the contribution of anaerobic processes. Berthain et al. (1996) recommended an estimation of such maximal aerobic speed (MAS) with the aid of the formula MAS = ($\dot{V}O_2max$ – 0.083)/C, where $\dot{V}O_2max$ is in ml/kg/min and C is the energy cost of running (ml/kg/min).

Maximal running time at the velocity corresponding to $\dot{V}O_2max$ is indicative of aerobic-anaerobic capacity. Billat et al. (1994b, 1996) determined the running time to exhaustion at this velocity in long-distance runners. Repeating the test a week later proved the reproducibility of the time to exhaustion at $\dot{V}O_2max$. The running time to exhaustion was significantly related to the lactate threshold and the running velocity in a race for 21 km. Significant correlations of running time to exhaustion with $\dot{V}O_2max$ value or with running velocity at $\dot{V}O_2max$ were not found. Thus, the results indicated that the maximal running time at $\dot{V}O_2max$ velocity did not depend on aerobic power. Obviously, it expressed another quality, the aerobic-anaerobic capacity.

Attention should be paid to the Bruce treadmill test for prediction of the aerobic-anaerobic capacity. The test consists of treadmill running. After every 3 min, the speed and grade of the treadmill increased by 0.8 mph and 2%, respectively. Maximal duration of the treadmill performance is a test criteria (Bruce et al. 1973). It possibly depends on aerobic power and aerobic-anaerobic capacity. This idea has been used in Finland for testing top-level skiers (Kantola and Rusko 1985). The modified test characteristics are presented in table 7.5.

In conclusion, the possibilities for differential assessment of aerobic capacity exist. However, the method of a concerned test requires further evaluation. For assessment of aerobic capacity, additional means may be found in testing the functional stability of VO_2 and cardiovascular parameters during prolonged exercises and testing the transfer from carbohydrate to lipid use.

Monitoring Energy Production Mechanisms

Monitoring energy production mechanisms makes sense if it is event-specific. Therefore, the competition exercises have to be analyzed to establish the mechanism that limits the performance in competition. The value of the results obtained may be diverse, depending on the event. In this section, the experience of monitoring the energy production mechanisms in training practice is analyzed.

Lactate Diagnostics in Training Monitoring

In endurance events, so-called lactate diagnostics is an approved tool for obtaining information on changes in muscle energetics developing as a result of training. For this purpose, event-specific field tests are elaborated. The main outcome of these tests is a lactate/velocity curve, which provides the possibility for characterizing individual peculiarities of lactate patterns during running, cycling, swimming, or rowing with increasing velocities. Examples of such testing are available in publications of Mader et al. (1976), Kantola and Rusko (1985), Liesen (1985), Janssen (1987), Dickhuth et al. (1989), Raczek (1989), Heck (1990), Mader (1991), and Urhausen et al. (2000), among others.

The main point is not to extrapolate exercise intensities at 2.0, 4.0, or 8.0 mmol/L of lactate. The idea is to detect the shift of the whole curve after a training stage, period, or year. This shift is specifically related to the training exercise used.

Table 7.5
Treadmill Test for Cross-Country Skiers

Time (min)	Speed (km/h)	Incline (%)
0	5	5
3	5.5	7
6	6	9
9	6	12
12	6	15
15	6.5	17
18	6.5	20
21	6.5	23
24	6.8	25
27	7.5	25
30	8	25
33	8.5	25

Figure 7.9 shows the specific difference in the lactate/velocity curves in female and male runners at 400 and 800 m and in male runners at 1500 to 5000 m and a marathon race (Raczek 1989).

In marathon runners, aerobic training increased the anaerobic threshold and made the curve more flat than in middle-distance runners. In athletes specializing in aerobic-anaerobic events, the anaerobic threshold was lower than in marathon runners, but the curve rose promptly above the suggested aerobic threshold. In middle-distance runners, the differences from the curve of the marathon runners were pronounced. Their anaerobic threshold was lower, but the lactate increase in exercises above the anaerobic threshold was more pronounced. In 1500- to 5000-m runners, during running at 4.8 m/s, the lactate levels were lower than in 400- to 800-m runners (Raczek 1989). Possibly, middle-distance runners responded to increased running velocity more sharply, reflecting their ability to mobilize the anaerobic capacity, whereas long-distance runners exhibited a pronounced reserve for gradually increasing the lactate level.

Raczek (1989) gave two examples of training results in young middle-distance runners (figure 7.10). In the young male athlete who extensively increased the volume of training exercises in the aerobic zone, a pronounced increase of the anaerobic threshold was found, but lactate increases in exercises above the anaerobic threshold were sharp despite comparatively low running velocities. The young female athlete

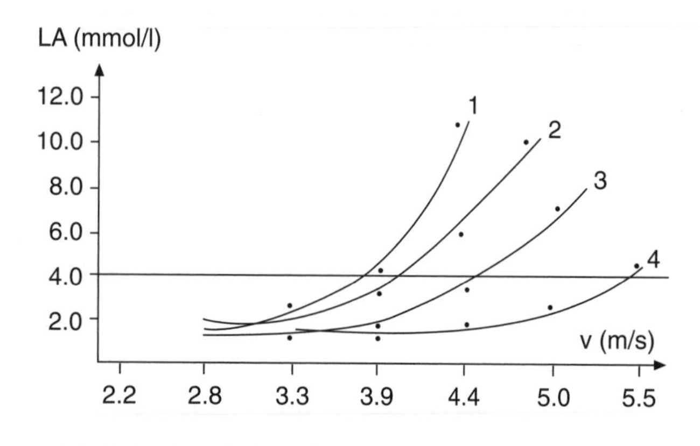

Figure 7.9 Lactate/velocity curves in female and male runners.

From Raczek 1989.

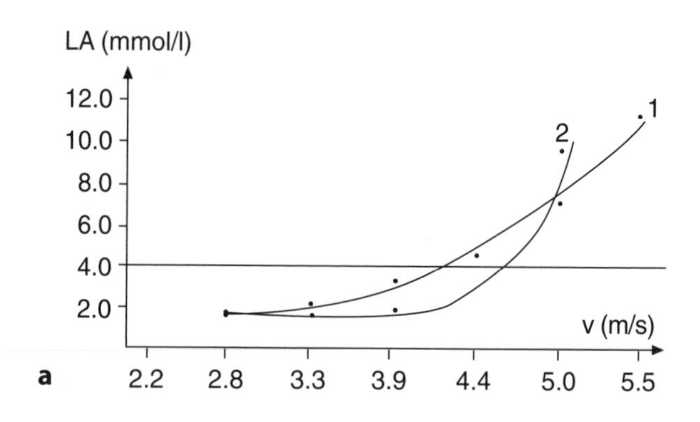

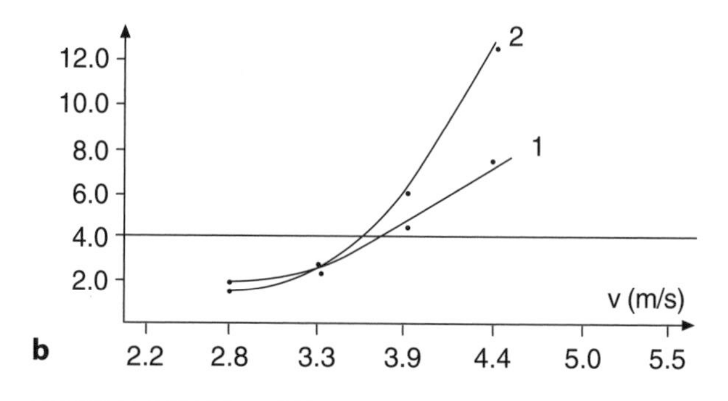

Figure 7.10 Examples of training effects on lactate/velocity curve in young middle-distance runners *(a)* before and *(b)* after a training period.

From Raczek 1989.

made an ascent to an excessive increase of the intensity of training load. He showed more pronounced lactate responses at higher running velocities than the previous runner, but the velocity that corresponded to the anaerobic threshold seemed to be reduced. These examples characterize the specificity of training effect in dependence on chosen exercises.

The data in figure 7.11 shows four-year dynamics in long-distance runners (the curves exhibit the group mean values). The data demonstrate the different positive training effects in long-distance runners and marathoners (Raczek 1989).

In turn, these analyses allowed authors to systematize training exercises according to lactate responses (Kantola and Rusko 1985; Liesen 1985; Janssen 1987; Dickhuth et al. 1989; Raczek 1989). Examples are presented in tables 7.6 and 7.7. This suggestion, although approved in training practice, is not the only possibility. Discrepancies also exist among various specialists in terminology (e.g., Janssen 1987 gives his proposal in a general manner).

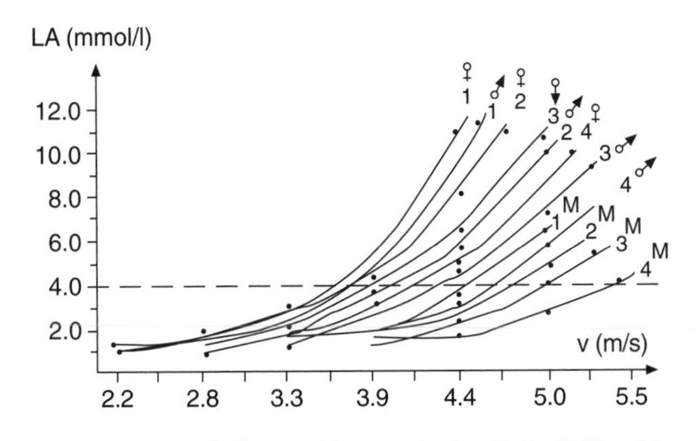

Figure 7.11 Four-year dynamics in lactate/velocity curves in long-distance runners and marathon runners (M). The year denotes the year of study.

From Raczek 1989.

1. Recovery and regeneration workout: lactate level <2 mmol/L of lactate
2. Extensive endurance workout: lactate level around 2 mmol/L
3. Intensive endurance workout: lactate level between 3 and 4 mmol/L
4. Extensive repetitions (tempo duration): lactate level between 4 and 6 mmol/L
5. Intensive repetitions (anaerobic interval training): lactate level between 6 and 12 mmol/L.

Such suggestions demonstrate the possibilities of using the results of biochemical monitoring as a training guide by coaches (see also Billat 1996).

Urhausen et al. (2000) compared the value of lactate and heart rate monitoring for estimation of exercise intensity by percentage of the individual anaerobic threshold. They established that small changes in exercise intensity up to 80% of the individual anaerobic threshold may be better assessed by controlling the heart rate because lactate remains unchanged. When exercise intensity is greater than 85% of the individual anaerobic threshold, lactate concentration better distinguishes the different intensities because the slope of heart rate changes is less than the lactate slope at these exercise intensities (table 7.8).

The superiority of athletes specializing in middle distances or other equivalent events is revealed in a specific test of anaerobic capacity or after anaerobic exercises for time trials. Lacour et al. (1990) established that blood lactate concentrations at the end of 400-m and 800-m running competitions were in direct correlation with the performances. Data obtained 3 min after semifinals or finals for 100-m and 200-m running showed a correlation between the lactate level after the 200-m race and sustained running velocity during the last 165 m. Performance in a 100-m race did not correlate with the lactate response (Hautier et al. 1994) probably because of the great contribution of the phosphocreatine mechanism in muscle energetics.

$\dot{V}O_2$max in Training Monitoring

Regular determination of aerobic power is a proved means for evaluation of the effectiveness of endurance training, particularly in athletes specializing in aerobic-anaerobic exercises. Examples of the dynamics of $\dot{V}O_2$max during the entire sports careers of two of Lithuanian's best cross-country skiers are presented in figure 7.12 (Milasius 1997). The male skier's $\dot{V}O_2$max was monitored from age 21 through 31; the female's was monitored from age 20 through 29. Results of many years of monitoring $\dot{V}O_2$max, power output at $\dot{V}O_2$max, and the anaerobic threshold in the best German (GDR) road cyclists (average values) are given in figure 7.13 (Neumann 1992).

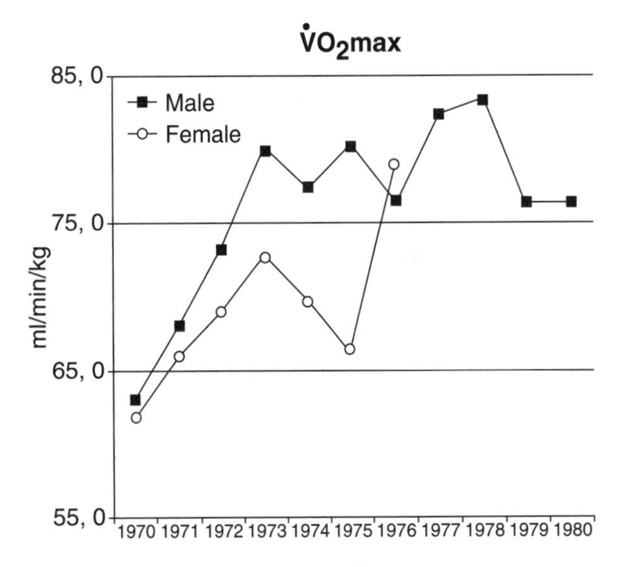

Figure 7.12 Dynamics of $\dot{V}O_2$max in two elite cross-country skiers. In 1970, they were both 20 years old.

Table 7.6
Zones of Training Workload for Young Athletes in Endurance Events

Exercise energetics	Goals	Time of main exercise (min)	Blood lactate (mmol/·L)	Heart rate (bpm)	Training mean
Aerobics	1. Regeneration, maintenance, and adaptation to long-lasting exercises and improved economy	45–120 up to 150 up to 180	1.5–2.5 1.0–2.0	130–150 100–130	Regeneration runs Extensive runs
	2. Development of aerobic power and capacity	15–45	2.5–4.0	150–180	Endurance runs Fartlek Cross-country runs
Aerobic-anaerobic	3. Intensive exercises for improved performance	8–20 1–3	4.0–7.0	170–190	Intensive endurance runs Long repetition runs Extensive interval runs
Aerobic-anaerobic	4. Critical exercises	2–8 1–3	7.0–10.0	180–200	Endurance tempo runs Intensive interval runs Time trials
Anaerobic	5. Over critical	40 s 15–40 s	>10.0	Up to 200	Speed endurance runs Tempo runs Time trials
Anaerobic	6. Maximal	Up to 15 s	Individual		Speed-development runs Short repetitions of maximal or sub-maximal velocities

Modified from Raczek 1989.

Åstrand and Rodahl (1986) presented the 8-year dynamics of oxygen uptake in a world top-level swimmer. The VO_2 level at maximal running increased during the first 4 years. Afterwards, the level remained unchanged for 4 years. However during these years, changes were found in VO_2 during swimming at maximal speed. Peak values were recorded close to the time when he won two gold medals at the Olympic Games.

The information provided by the $\dot{V}O_2max$ monitoring during a training year in athletes of aerobic-anaerobic events is essential. In 1964, Enschede and Jongblood reported that in high-level speedskaters, $\dot{V}O_2max$ increases up to the competition period. Similar data have been found in long-distance runners and road cyclists (Vasiljeva et al. 1971, 1972). However, before the Winter Olympic Games in 1964 and 1968, several cross-country skiers from the Soviet national team exhibited $\dot{V}O_2max$ levels in the range of 78 to 88 ml/kg /min the summer before the games. In October through December, aerobic power decreased in those skiers in association with an impaired performance level during competitions. Other top skiers who won Olympic medals showed a gradual increase of $\dot{V}O_2max$ up to the competition period (Ogoltsov 1968).

Table 7.7
Lactate Values of Various Running Exercises in Top-Level Marathon Runners

Exercise	Blood lactate (mmol/·L)	Exercise intensity (% of marathon velocity)
Recovery workout	1.0	<80
Extensive endurance	1.0–1.1	80–90
Intensive endurance	1.3	90–97
Tempo endurance	2.0	100
Extensive intervals (fartlek)	3.0	105
Intensive intervals	>8.0	

Adapted from H. Liesen 1985; adapted from E. Hultman et al. 1990.

Table 7.8
Means of Lactate (La) and Heart Rate (Hr) During Endurance Runs

%IAT	% $\dot{V}O_2$max	La mmol/·L	%La IAT	HR bpm	%HR IAT	%HR max
70	55	1.54	45	138	80	72
80	63	1.67	50	154	89	80
90	71	2.67	80	166	96	86
95	75	3.53	105	175	101	91
100	79	5.67	165	183	106	95

Lactate and heart rates are either absolute values or percentages of the corresponding threshold and maximal. Endurance runs are of intensities varying from 70% to 100% of individual anaerobic thresholds (IATS), corresponding to 79% of $\dot{V}O_2$max.
Adapted from Urhausen et al. 2000.

Two variants of $\dot{V}O_2$max dynamics have also been found in Finnish male and female skiers. In top-level athletes, aerobic power increased from spring to winter. In skiers of lower performance, the level of $\dot{V}O_2$max increased to autumn. In winter, a decrease followed (Kantola and Rusko 1985).

A female skier at the international level demonstrated a pronounced increase in $\dot{V}O_2$max for the most important competition (figure 7.14). At the end of the competition period, $\dot{V}O_2$max dropped (Milasius 1997).

Figure 7.15 presents the results obtained in a team of young promising cross-country skiers. At the beginning of the monitoring period, their age was 15 to 16 years. In summer, autumn, and winter, their $\dot{V}O_2$max increased in correlation with a rise in performance in the treadmill running test. After the end of the competition, the $\dot{V}O_2$max and performance index reduced transitorily (Viru M. et al. 2000b).

$\dot{V}O_2$max is not a universal indication of the performance level. Improved aerobic power is necessary, but not the main condition, for a high

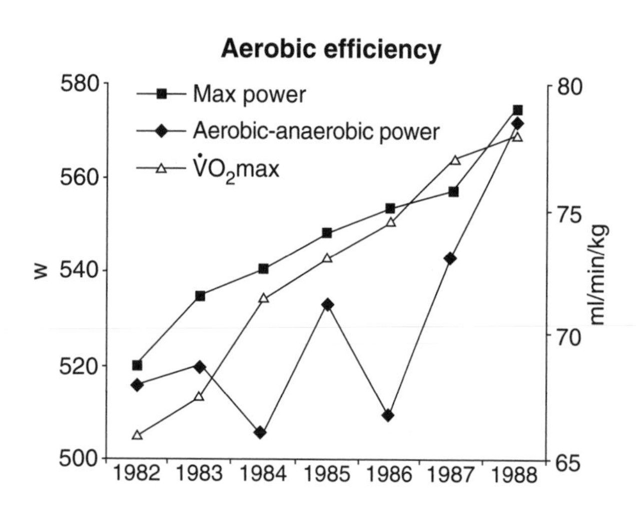

Figure 7.13 Dynamics of $\dot{V}O_2$max and power output at $\dot{V}O_2$max and the anaerobic threshold in the best German cyclists.

Reprinted from G. Neumann 1992.

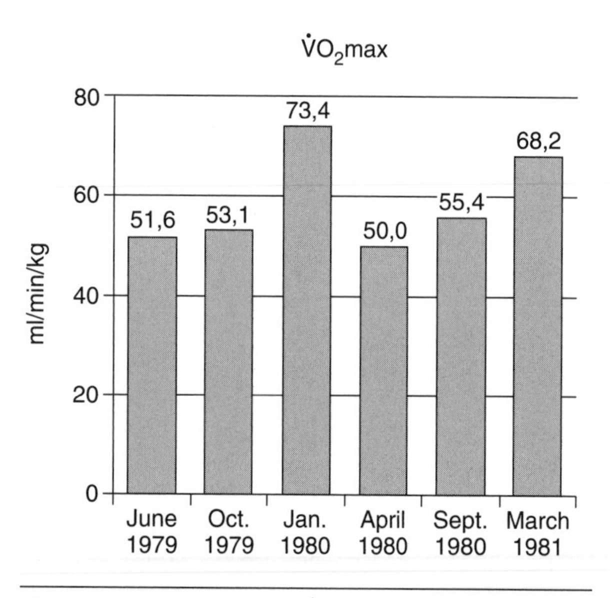

Figure 7.14 Dynamics of $\dot{V}O_2$max in an international-level female skier.

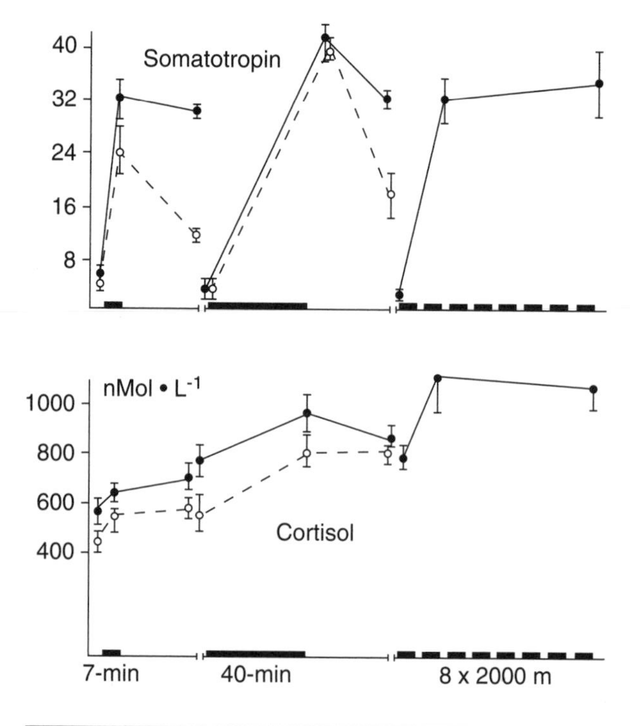

Figure 7.15 Dynamics of $\dot{V}O_2$max in young skiers over two years. Test time is the duration of incremental treadmill running.

performance level in marathon and ultramarathon events. The significance of $\dot{V}O_2$max is rather modest if at all in sprinters and power athletes. Therefore, the decrease of $\dot{V}O_2$max for the onset of the competition and during this period is a usual phenomenon. However, it is necessary to distinguish whether the decrease of $\dot{V}O_2$max is associated with the improvement of the specific performance in power and sprint events. If the performance also decreases, one should think about the exhaustion of the body's adaptivity (see chapter 9). To distinguish these two possibilities, the dynamics have to be compared with changes in event-specific factors of performance. Also, the magnitude of the decrease of $\dot{V}O_2$max has to be taken into account. Unfortunately, the quantitative standards for the evaluation of the $\dot{V}O_2$max reduction during the competition period have not been elaborated.

A specific problem of $\dot{V}O_2$max monitoring arises in events in which competitions last several hours, but the responsible brief effective movements require high power output, speed, and/or anaerobic capacity (e.g., soccer, ice hockey, basketball, decathlon, etc). In ice-hockey players, a high correlation has been found between $\dot{V}O_2$max and playing activity on the ice (Guminski et al. 1971). The explanation was that during rest minutes after a shift, restitution depends on the rate of oxidation processes and, thereby, on the supply of oxygen to muscles. In 1973, decathlonists at the international level had an average $\dot{V}O_2$max of 56.1 ± 1.2 ml/kg·min (Pärnat et al. 1975b). This level is often observed in athletes of power and sprint events. However, compared with decathlonists at the college level (48.2 ± 2.1 ml/kg·min), the superiority of international athletes was significant (Pärnat et al. 1975b). Moreover, the Soviet record holders of that time had a $\dot{V}O_2$max of 66.0 ml/kg·min (Pärnat et al. 1973). These data point out the relationship between performance and $\dot{V}O_2$max in decathlonists. Again, the explanation might be that during the 2 days of competition, recovery rate between events depended on the oxygen supply of muscles. Increased $\dot{V}O_2$max might ensure a better condition for readiness in the next effort.

In endurance athletes, the dynamics of $\dot{V}O_2$max and anaerobic threshold may not be parallel. The obvious reason is the ratio of aerobic-anaerobic and aerobic exercises in training. Training specificity is, of course, more pronounced when improvements of anaerobic capacity and aerobic power are compared. Tabata et al. (1996) demonstrated that endurance training at 70% $\dot{V}O_2$max for 6 weeks did not influence anaerobic capacity but significantly increased $\dot{V}O_2$max. High-intensity interval training improved anaerobic capacity by 28% and $\dot{V}O_2$max by only 7 ml/kg·min.

The material presented shows the significance of lactate diagnostics (including determination of the anaerobic threshold) and of repeated assessment of $\dot{V}O_2$max in the biochemical monitoring of training. It should be considered that $\dot{V}O_2$max informs mainly about the training action on the oxygen transport system, which may not correlate with the training effect on the oxidative potential of muscle fibers. Therefore, in endurance events, both parameters have to be monitored. Depending on the intensity of competition exercises, the test of anaerobic power and capacity and the test for aerobic capacity have to be added into the monitoring battery.

Assessing Other Training Effects

Training guides need more information than data about the improvement of power and capacity of the energy production systems. In several cases, an amount of satisfying information about training effects has been obtained by using motor fitness tests (see Morrow et al. 1997; Fleck and Kraemer 1997; Dintiman et al. 1998; Bosco 1999). From the aspect of metabolic adaptation, the most valuable information would be about energy stores and what happens in muscles. The latter is particularly important in high-resistive and power training. However, valid information about changes in muscles (energy stores, myofibrillar adaptation, mitochondrial changes) needs biopsy sampling. To determine adaptive changes in muscles induced by high-resistive or power training, the main possibilities are strength testing, power testing, and testing the characteristics of muscle contraction and the measurement of muscle hypertrophy with the aid of ultrasonic scanning.

For prolonged events, it is important to know the relationship between anaerobic processes and sparing of glycogen with the aid of transfer to lipid use. To check on the crossover phenomenon, levels of lactate and free fatty acids have to be measured during prolonged exercises of various intensities.

Exercise-induced hormonal responses contain essential information regarding mobilization of the body's energy reserves. However, several conditions have to be considered (see chapter 4, pp. 69-71). Otherwise, the information from hormonal changes may lead to misunderstandings. Hormonal studies are meaningful for the following purposes:

- To determine during prolonged exercises when the insulin level in blood decreases and, thereby, when lipolysis in adipose tissue releases from insulin-dependent inhibition
- To measure the magnitude of the catecholamine response in sprint exercises to determine epinephrine interference with cellular-automatic control to augment the mobilization of muscle glycogen resources
- To determine the stability of responses of the hormones controlling the mobilization of the body's energy resources by repeated measurement of the concentration of related hormones (catecholamines, cortisol, growth hormone, glucagon) during prolonged exercises
- To determine the basal level of testosterone to assess the possibility that this hormone supports the effective triggering of functions of the neuromuscular apparatus for high performance in strength and power events

In chapter 5, results were mentioned that indicate that training results in increased hormonal responses in supramaximal exercises. Increased hormonal responses were found even in highly trained rowers after a year of training in association with improvement of rowing performance. Cortisol and growth hormone responses were increased in a 7-min maximal rowing test, and the cortisol response increased at the level of anaerobic threshold (figure 7.16). Very high cortisol levels appeared in these rowers while rowing 2000 m eight times during the second year of the study (Snegovskaya and Viru 1993a). An outstanding middle-distance runner was tested with the aid of incremental treadmill running for 3 years. The progress of running performance was associated with the shift of the catecholamine/running velocity curves to the left (figure 7.17), which indicates the elevated threshold intensity for catecholamines. Another noteworthy fact was that during the last year, the running velocity and maximum level of epinephrine were the highest (Lehmann et al. 1989).

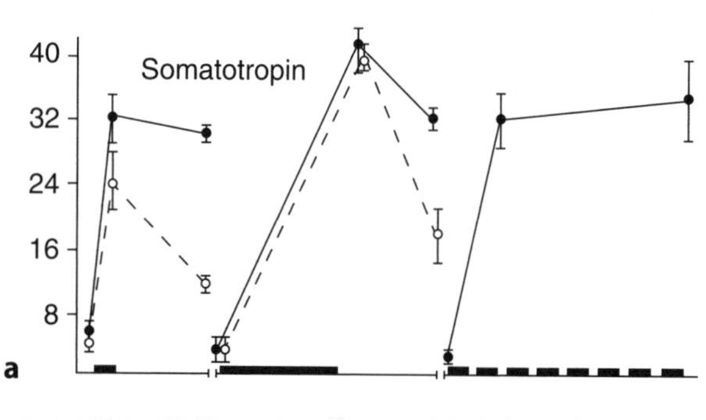

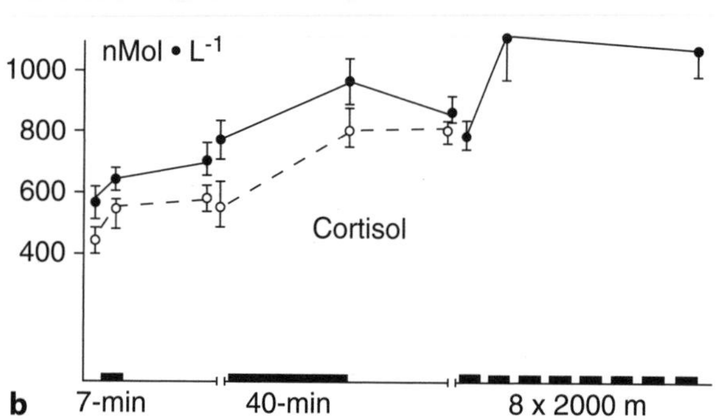

Figure 7.16 Changes in *(a)* growth hormone (somatotropin) and *(b)* cortisol during a 7-min all-out rowing test and 8 × 2000-m rowing.

Reprinted from V. Snegovskaya and A. Viru 1993.

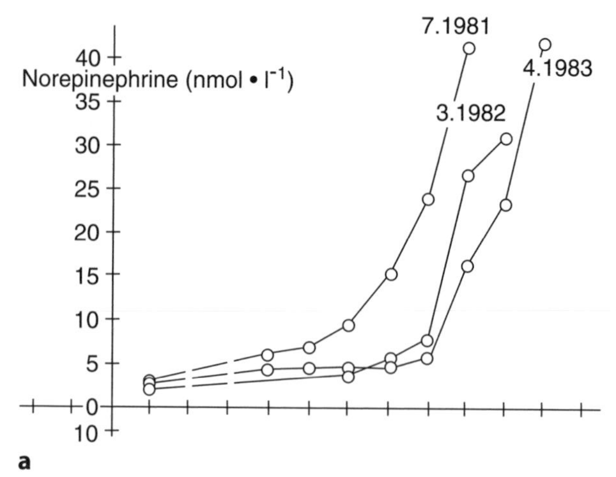

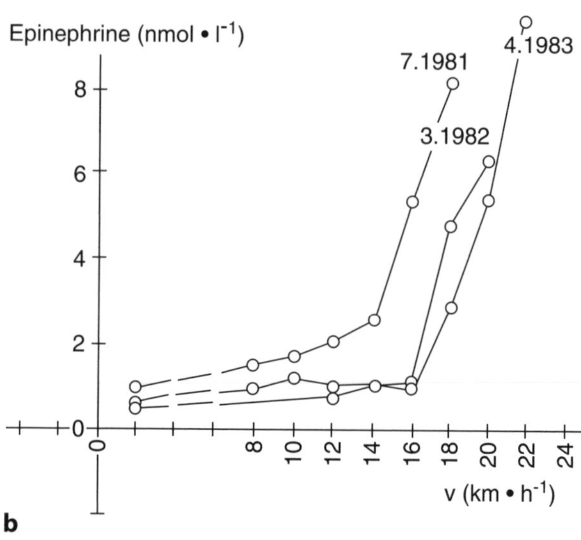

Figure 7.17 *(a)* Norepinephrine and *(b)* epinephrine during an incremental treadmill test in a runner.

Reprinted from Lehmann et al. 1989b.

In conclusion, the training effect on metabolic adaptation and endocrine function may be assessed. However, the value of the outcome of these studies is noteworthy for extensive analyses but not for ordinary monitoring. The exceptions are studies of the carbohydrate/lipid ratio in the oxidation process.

Summary

The power and capacity of the energy mechanisms represent essential criteria of performance capacity in several sports events. Therefore, their assessment is an essential task of training monitoring. In the evaluation of aerobic power, suitable approaches are determination of maximal oxygen uptake (characterizes mainly the maximal rate of oxygen transport to working muscles) and anaerobic threshold (characterizes the highest power output at the expense of oxidative phosphorylation without the additional use of anaerobic energy). A direct way to measure the anaerobic threshold consists of assessing the maximal lactate steady state. Assessment of the anaerobic threshold by lactate dynamics during incremental exercise tests is also valuable. The most reliable results are found in analysis of the whole curve of lactate increase plotted against power output.

Complications exist regarding estimation of aerobic capacity. Assessment of the total capacity of the body for oxidative phosphorylation is practically impossible. In the applied aspect, assessment of aerobic capacity for aerobic-anaerobic exercises (maximal duration of exercise performed at the intensity corresponding to $\dot{V}O_2max$), for the highest rate of exercises performed at the expense of oxidative phosphorylation (maximal duration of exercise performed at the intensity corresponding to the anaerobic threshold), and for the exercise performed at the expense of oxidation of lipids is more essential.

For the evaluation of anaerobic glycogenolysis, attention should be focused on the Wingate test, Bosco's continuous jumping test, and other similar tests consisting of exercising at the highest possible rate for 30 to 90 s. The value of the power output determination of lactate increases if the rate of lactate accumulation is estimated. Precise measurement of the rate of lactate accumulation provides the best possibility for assessment of the power of anaerobic glycogenolysis. To test the capacity of anaerobic glycogenolysis, the highest possible lactate concentration in blood should be determined. A possibility for that is provided by testing with the aid of high-intensity exercises repeated over short rest intervals (interval testing).

Indirect approaches for the assessment of the power of the phosphocreatine mechanism are the determination of the highest possible power output in short-term staircase running (Margaria's staircase test) or in maximal vertical jumping, continued during 5 s (a modification of the Bosco's test). The best method for the assessment of the capacity of the phosphocreatine mechanism is the determination of phosphocreatine loss in intensely working muscles with the aid of biopsy sampling. An indirect way is to record the dynamics of power output (e.g., movement velocity) of short-term exercises of maximal intensity.

Determination of the concentration of ammonia, uric acid, or hypoxanthine in blood plasma provides limited possibilities for characterizing the myokinase mechanism. The main significance of the determination of the products of AMP degradation consists in the possibility of getting information as to whether ATP resynthesis lagged behind ATP degradation to ADP.

Among several possibilities to test the training effects on metabolic control, in the applied aspect, the most essential is to determine the crossover point in regard to the use of lipids versus carbohydrates as oxidation substrates. In the aspect of training guidance, essential information provides simultaneous determination of lactate and free fatty acid (or glycerol) in blood plasma during prolonged exercises.

Several possibilities exist for evaluation of training effects on endocrine functions and hormonal metabolic control. These possibilities may be essential in answering questions regarding metabolic adaptation in training.

CHAPTER

8

Evaluating Training Workloads

Several tasks of biochemical monitoring of training are related to evaluating training loads. This chapter discusses the use of metabolic and hormonal parameters for assessing the intensity and volume of the workload of training sessions and microcycles. The biochemical parameters are mostly necessary for assessing the trainable effect of training workloads. The possibilities for this assessment arise from the concept of adaptive protein synthesis as the foundation of training effects at the level of cellular structures. The accumulation of protein synthesis inductors may provide a tool for evaluation of the trainable effect. Assessing training workloads and microcycles also requires the diagnostics of fatigue and monitoring in the recovery period.

Training Session Workload

Designing training for athletes consists of two groups of tasks that must be considered: training strategy and training tactics (Viru 1995). Training strategy involves ensuring the most effective use of training time in the movement from teenager to Homo Olympicus. Often, 10 to 12 years are necessary to achieve this goal. Of course, training strategy needs to take into account ontogenetic development and to use the most favorable developmental stages for inducing the necessary structural, metabolic, and functional changes. The tasks have to be distributed among years, within a year between training periods, and within a training period between mesocycles and microcycles of training.

Training tactics deal with how to act on the organism to induce the necessary changes in the body. Concerns of training tactics include the choice of exercise and training methods, design of training sessions, and microcycles. Some questions and concerns involved with training tactics follow:

- Which changes have to be induced in a specific stage of training?
- Which exercises induce the necessary changes?
- Which training methods have to be used?
- How can inductors be created for the synthesis of structural and enzyme proteins necessary for fulfilling the task?
- What sequence should various exercises in a training session follow?
- How is sufficient (trainable) workload of a training session determined?
- How are the influences of subsequent training sessions integrated?
- How are training sessions and recovery time between sessions related?

It has been hypothesized (Viru 1994, 1995) that the choice of exercises determines which structural and enzyme proteins will be synthesized, whereas the total workload of the training sessions determines the activation of endocrine function and, thereby, amplification of adaptive protein syntheses. The total load is the sum of the influences of all exercises performed during the session and depends on the intensity of exercises and rest intervals between exercises. The total workload of a training session may be

- excessive, surpassing the body's adaptivity (the functional capacity of the most responsible systems) and causing overstrain;
- a trainable load, which causes specifically directed and sufficiently intensive adaptive protein synthesis and thereby induces the desired training effect;
- a maintaining training load, which is insufficient to stimulate adaptive protein synthesis but sufficient to avoid a detraining effect;
- a restitution load, which is insufficient to avoid detraining effects but favors promotion of the recovery processes after a trainable load; and
- a useless load.

According to these levels of workloads, at least three groups of criteria are necessary for a detailed analysis of the training session influence.

1. Criteria for the highest possible training load
2. Criteria for the trainable effect
3. Criteria for the minimum load that will induce the main training effect

Studies in the field of sports medicine and the practical experience of sports physicians make it possible to identify several signs of overstrain. These signs are mostly concerned with the functional characteristics of heart overstrain as measured by an electrocardiogram or echocardiogram.

Testing the ratio between the rate of free radical oxidation and activation of antioxidant systems seems to provide an approach, but is this the most suitable index usable in field conditions of sports training for evaluating oxidation stress? This question and several others have to be answered to determine the validity of the test. Such a test is valid if it provides qualitative information on the background of the quantitative measurement of indexes of oxidation stress.

The question becomes why not assess the results of free radical oxidation on membrane function? Release of intracellular enzyme proteins into the blood may provide for this possibility. However, the activity of various intracellular enzymes (e.g., creatine kinase) increases in most vigorous exercises. What is the qualitative borderline between normal changes and overstrain? Again this criterion is yet to be determined.

In training practices workloads are frequently on the borderline between overtraining and training effective. Progress in performance capacity is based on the regeneration rate of certain organs and tissues after overstrain. In cases of rapid regeneration, overstrain does not require remediation. However, do the effects of such overstrain, repeated over time, accumulate and result in serious damage. What are the possibilities for testing these "latent" phenomena and regeneration rates?

The best test will provide information about the highest load that does not cause any overstrain. Otherwise, only negative consequences will teach athletes. Regarding criteria on the highest training load immediately preceding overstrain, some pedagogical criteria may be mentioned. For swimmers, various workloads for training levels (technique) were recommended (Platonov and Vaitsekhovski 1985). Their recommendations are presented in table 8.1. For runners, a similar proposal has been made according to the running velocity and heart rate (table 8.2). When cross-country skiers use the method of continuous training at a speed of 87% of competition velocity, a drop in skiing velocity of more than 30% is recommended for the criterion of the highest trainable load. The maintaining load constitutes 40% to 75% of that volume (distance). It has been thought that in cross-country skiers, the trainable load begins at the moment when skiing velocity can be maintained only with the aid of increasing stride frequency (Baikov 1975). These proposals were verified in training practice. However, a further checkup that uses the indexes of metabolic adaptation is necessary.

Assessing the Trainable Effect of Training Sessions

Practically speaking, the most important thing to know is whether the training session induces a trainable effect. Training effects are found in changes at the cellular level. In turn, changes at the cellular level are related to the synthesis of structure protein and augmentation of enzyme molecules catalyzing the most responsible metabolic pathways. Therefore, assessment of the trainable effect may be founded on the intracellular accumulation of metabolites or on hormonal changes during and after the session, which ensure the necessary adaptive protein synthesis (figure 8.1).

A number of problems arise regarding corresponding indexes. The most serious among methodological complications are the necessity of obtaining body fluid or tissue samples for analysis and the necessity for complicated biochemical methods. The metabolic changes controlling transcriptional and translational events are intracellular. Even with the use of a biopsy method, determining intracellular and total tissue changes in metabolic processes is not a simple task.

Metabolites

The main way to check the metabolic changes is by analysis of blood or urine. This enables the evaluation of general alterations in the metabolic status and the accumulation of metabolites causing outflow from the intracellular compartment. Regrettably, we still do not know what metabolic inductors are causing the main training effects

Table 8.1
Various Levels of Training Session Workloads in Swimmers

Load	Main characteristics	Action
Light	10%–15% of the amount of exercises up to the drop in working capacity.	Restitution load
Moderate	40%–60% of the amount of exercises up to the drop in working capacity.	Maintaining load
Heavy	60%–70% of the amount of exercises up to the drop in working capacity.	Trainable load
Very heavy	Causes pronounced fatigue (a drop in working capacity), disorders in swimming technique.	Highest trainable load

From Platonov and Vaitsekhovsky 1985.

Table 8.2
Training Workloads in Long-Distance Runners

Zones of load	Time for 1 km (min:s) during prolonged running	Heart rate (bpm)	% Running velocity in relation to the individual best result
Restitution load	4:30–5:00	130	
Maintaining load	4:00–4:30	130–150	
Trainable load	3:30–4:00	150–170	
Highly trainable load	3:00 or less	170–190	80
Sprint, acceleration running			81–95
Jumping, competition running			100

From Doroshenko 1976.

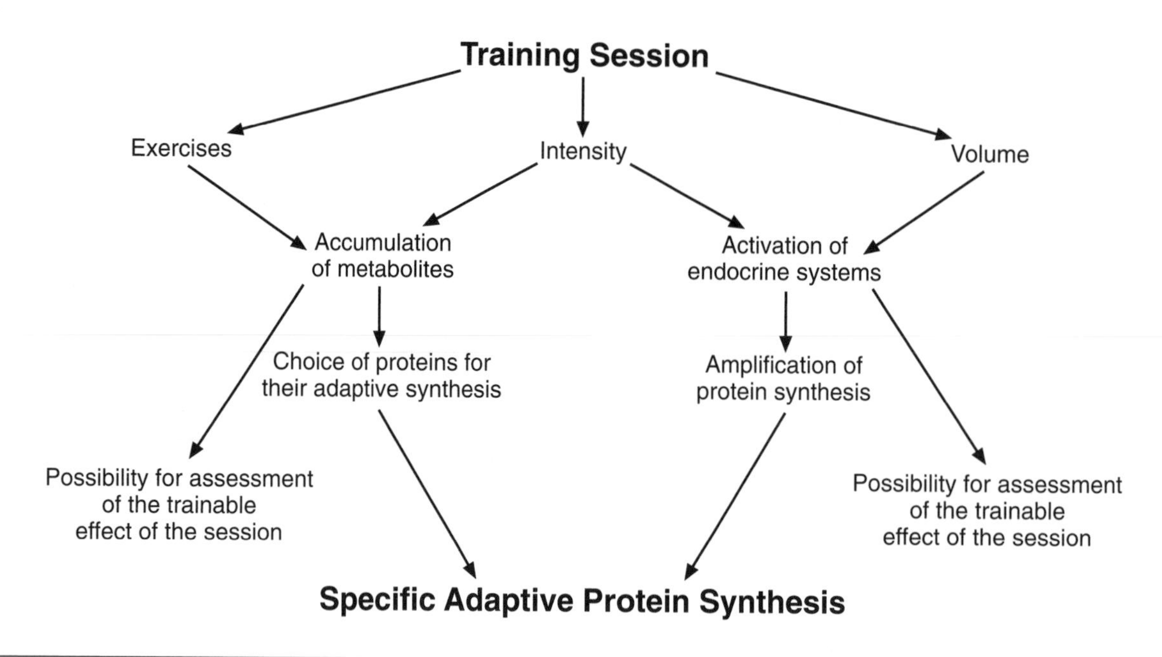

Figure 8.1 Appearance of possibilities for metabolic and hormonal criteria for the trainable effect of training sessions.

nor do we know how great the accumulation of metabolites should be. There is also the question of how great the necessary intracellular accumulation of metabolites that causes their measurable outflow from the intracellular compartment must be. Consequently, the metabolic indexes are still usable only for detecting general alterations in the metabolic status.

During strenuous anaerobic exercises, the rise in blood lactate is so tremendous that there is no reason to doubt the value of lactate for the "semiquantitative" evaluation of the anaerobic capacity used. However, we do not know exactly how high a lactate level is necessary or for how long the increased lactate levels have to be maintained to achieve an effective stimulus for improved anaerobic capacity. In issues of training methods, suggestions can be found that for minimal doses of exercises stimulating the improvement of anaerobic capacity, exercise has caused a blood lactate level greater than 4 mmol/L. For qualified athletes, the minimal effective exercise dose is thought to be characterized by rises in the blood lactate level to more than 11 mmol/L. For elite athletes, a blood lactate rise up to 19 to 22 mmol/L is supposed to be necessary. In all these cases, the training effect increases with the duration of the period during which the lactate concentration persists at the indicated levels.

One may argue whether these high lactate levels are indeed necessary for stimulating improvement of anaerobic capacity. There are three principal ways to put a large demand on anaerobic glycogenolysis.

1. When intensive exercises are performed, high lactate levels may be caused but only for a short time.
2. When the same amount of exercise is performed in parts, by use of the interval method, the final lactate level may be the same or even higher than in the first case, but the time during which the increased lactate level persisted is more prolonged.
3. A possibility also exists that in continuous aerobic-anaerobic exercises the blood lactate concentration gradually increases. The high lactate levels may persist in these cases for a longer time than in the previously mentioned cases.

Are the training effects of these variants the same? Maybe these variants have to be used by choice for improvement of different forms of anaerobic capacity (rapid use during a short time or gradual use during a prolonged time).

Biochemical investigations failed to show an inductor action of lactate on protein synthesis (Hedden and Buse 1982). Therefore, lactate seems to indicate only the situation in which actual inductors are created for stimulation of synthesis of proteins responsible for improved anaerobic capacity. However, one may not exclude the possibility that the minor amount of proteins whose synthesis is induced by lactate may not be reflected in the total rate of protein synthesis in muscles.

A widespread tendency exists to use blood urea for evaluation of the training session workload and recovery process. It is thought that a pronounced rise in the blood urea concentration indicates a strong influence of the training session and, thereby, suggests a trainable effect. Normalization of the urea level in blood is used as an index of time to perform subsequent strenuous training sessions. Nevertheless, links between exercise-induced blood urea changes and stimulation of adaptive protein synthesis have not yet been established. In rat experiments, discordance was found between blood urea changes and other indexes of protein metabolism during a short-term training cycle. Therefore, urea may not exactly express the actual alterations in the status of protein metabolism in training (Ööpik et al. 1988).

As was indicated earlier (see chapter 3, p. 38), urea production is suppressed when exercises induce high lactate levels. Therefore, the urea level provides the best information on training workloads when continuous aerobic exercises are used. In endurance athletes, distinct correlations are found between postexercise blood urea levels and the total amount of performed exercise (Haralambie and Berg 1976; Steinacker et al. 1993). A great body of results accumulated in Moscow has been used for elaboration of an evaluation scale of training workload in cases of prolonged continuous exercises (table 8.3). The next morning, urea levels less than 7.5 mmol/L have been considered the index of optimal recovery. Higher morning urea levels point to inadequate recovery. An increase of the morning urea level up to the values close to the clinically relevant upper limit (8.3 mmol/L) indicates overtraining (Lorenz and Gerber 1979; Urhausen and Kindermann 1992a). Such a dynamic was found in female rowers during a 3-week period. After a week of reduced training (blood lactate <2 mmol/L), the morning urea level declined from the high values and leveled off at 5 to 6 mmol/L for the next 5 weeks of training (Kindermann 1986).

A specific index of the catabolism of muscle contractile proteins is 3-methylhistidine excretion (see chapter 3, pp. 41-42). This metabolite may be used only in a meat-free diet or when 3-methylhistidine excretion is corrected for the 3-methylhistidine originating from the consumed meat product. Data obtained in humans (Dohm et al. 1982, 1985) and rats (Dohm et al. 1982; Varrik and Viru 1988) indicated that after exercise, 3-methylhistidine excretion increases gradually. In athletes and untrained persons, the highest excretion was observed when urine was collected 12 to 24 h after resistance- or power-training sessions (Viru and Seli 1992). Hence, if the training session takes place before noon, the urine collected during the following night expresses the most intensive 3-methylhistidine excretion caused by the training. The corrected 3-methylhistidine excretions during the night after a training session significantly correlated with the total excretion of corrected 3-methylhistidine measured during 48 h after resistance- or power-training sessions. Therefore, the use of the night 3-methylhistidine excretions is suitable for studying the effect of training on endogenous 3-methylhistidine productions.

To find a relationship between 3-methylhistidine excretion and training effects, excretion of 3-methylhistidine was measured in young men during an 8-week training period for improved power or strength. In both cases, one group of persons used training exercises 70% of 1 repetition maximum (1 RM); the other group used training exercises 50% 1 RM. Power training caused a significant improvement in the 30-m dash, vertical jump, standing long jump,

Table 8.3
Evaluation of the Workload in Prolonged Continuous Exercises by Blood Urea Levels

Load	Blood levels of urea (mmol/L)
Hard training load	9.0–10.0
Medium training load	7.5–8.5
Moderate training load	<7.5

According to Hnõkina 1983 and Tchareyeva et al. 1986a.

standing triple jump, and squat lift when the training exercise loads were 70% 1 RM (in 50% 1 RM exercise, the effects were less pronounced). Heavy resistance training improved the results in the vertical jump, standing triple jump, and most of all in the squat lift. A pronounced increase in a cross-section of the thigh muscle (evaluated by X-ray photography) resulted from heavy resistance training with 70% 1 RM exercise. In all groups, an elevation of corrected 3-methylhistidine excretions was observed in the night urine during the first 3 to 5 weeks of training (figure 8.2). The response was most pronounced and most lasting in a group performing heavy-resistance training with 70% 1 RM exercise. In other groups, the degree of muscular hypertrophy and the duration of increased 3-methylhistidine excretions were less. In power training with 70% 1 RM exercise, mean 3-methylhistidine excretion reached approximately the level observed in heavy-resistance training when two persons were excluded from the power group who did not reveal an increase in a cross-section of their thigh muscle with power training. Consequently, a relationship between training-induced muscular hypertrophy and 3-methylhistidine excretions was suggested (Viru and Seli 1992).

Increased 3-methylhistidine excretion during a training period agrees with the results of Hickson and collaborators (Pivornik et al. 1989), who found significant increases of this metabolite excretion from the third day of progressive resistance training.

At least two ways exist to understand this relationship between training effectiveness and 3-methylhistidine excretion. First, an increased 3-methylhistidine excretion during postexercise recovery is assumed to express an enhanced turnover of contractile proteins. An enhanced protein turnover is an indispensable condition for muscular growth. Thus, increased 3-methylhistidine excretions are related to an overall condition of muscle protein anabolism. On the other hand, an increased 3-methylhistidine production may describe a condition in which inductor metabolites for the synthesis of myofibrillar proteins accumulate. Either way, a high concentration of 3-methylhistidine excretion may be used as an index of training effectiveness in the stimulation of muscular hypertrophy.

During the last weeks of the 8-week training, increased 3-methylhistidine excretion disappeared (see figure 8.2). Most likely, this was related to adaptation to the given exercise stimulus. If this is the case, the reduced 3-methylhistidine excretion indicated that the training stimulus had to be increased.

Although 3-methylhistidine is derived from contractile proteins, its estimation may have significance in assessment of the trainable effect in heavy-resistance or power exercises acting on the myofibrillar size.

Hormones

Regarding hormone inductors of protein synthesis, the most attention should be paid to the exercise-induced changes in testosterone and thyroxin + triiodothyronine levels in a training session for improved strength or endurance, respectively (see chapter 2, p. 16). The reason is that these hormones exert a strong inductor influence eliciting intensive synthesis of myofibrillar proteins (both hormones) and mitochondrial proteins (thyroid hormones).

For synthesis of myofibrillar proteins, the most essential factor is the dynamics of androgens during the recovery period. A general characteristic of testosterone dynamics is its low level during the several hours after exercises (Kuoppasalmi 1980; Kuoppasalmi et al. 1980; Viru et al. 1999). After a triathlon, a low level of testosterone persisted for 4 days (Urhausen and Kindermann 1987). However, apart from endurance exercises, a tendency toward increased production of testosterone follows strength exercises after a decrease during the first 1 to 3 h. In rats, this change is associated with the augmentation of testosterone and androstenedione content and an increase in the number of androgen-binding sites in skeletal muscles (Tchaikovsky et al. 1986). In humans, increased protein synthesis in muscles during the recovery period after resistance exercises (Chesley et al. 1992) and the testosterone effect on protein synthesis in muscles (Urban et al. 1995) have been confirmed. Statistically significant correlations were found between testosterone concentration or testosterone/cortisol ratio and changes in strength and power during training periods up to a year (Häkkinen et al. 1987). Despite the low testosterone level in women, a correlation was found between serum testosterone level and an individual change in

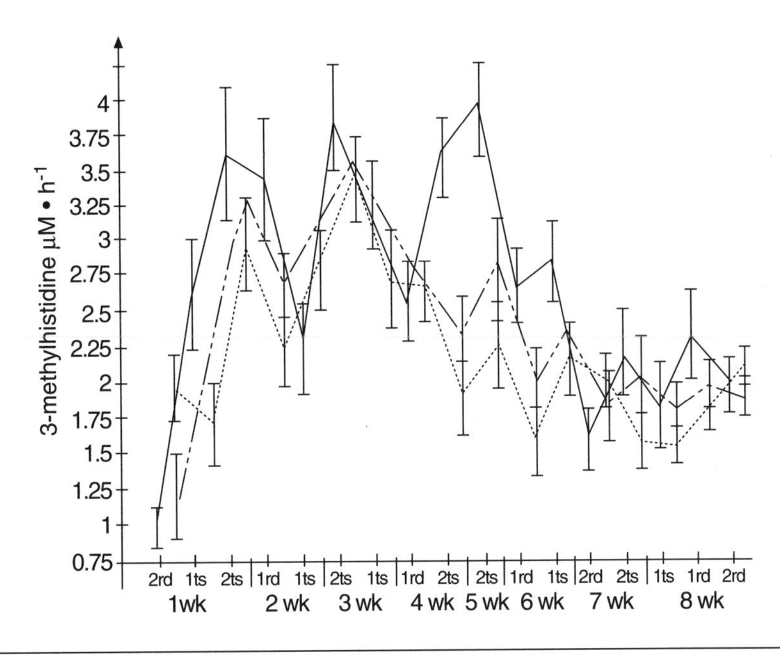

Figure 8.2 Excretion of 3-methylhistidine (μmol/h)during a training period for improved strength.
Reprinted from A. Viru et al. 1992.

maximal force (Häkkinen et al. 1990). This correlation might be founded on the estradiol-dependent increase of sensitivity to the anabolic effect of testosterone in the female body (Danhaive and Rousseau 1988).

After endurance exercises, the main locus of the increased rate of protein synthesis is the mitochondria of fast oxidative glycolytic (FOG) and slow oxidative (SO) fibers. In rats, the highest rate of synthesis of mitochondrial and myofibrillar proteins was found in FOG and SO fibers 24 h after 30 min of running. In FG fibers, protein synthesis remained suppressed at least 48 h after an endurance exercise (Viru and Ööpik 1989). The main result of this study was confirmed and extended with the aid of an autoradiographic study. A postexercise rise of the blood level of triiodothyronine and thyroxin coincided with an increased incorporation of ^{3}H-tyrosine in all types of muscle fibers in normal rats. The most pronounced increase of label incorporation was found 24 h after exercise. In hypothyroid rats, no increase was found in label incorporation during a 48-h recovery period after 30 min of running. In these rats, a low level of label was found in the mitochondria and in all regions of sarcoplasm and myofibrils during the recovery period (Konovalova et al. 1997).

A stimulatory effect on protein synthesis (probably on the translational level) is produced by insulin and somatotropin (Ullman and Oldfors 1986; Balon et al. 1990).

These results together suggest that information on the trainable effect may be obtained by responses in testosterone, thyroid hormones, growth hormone, and insulin during and after training sessions. In rats, postexercise increases in concentrations of testosterone (Tchaikovsky et al. 1986) and thyroid hormones (Konovalova et al. 1997) have been found. According to Hackney and Gulledge (1994), in men a high thyroxin level persisted 8 h after 1-h anaerobic or aerobic interval exercises. A postexercise increase of insulin has been demonstrated in men (Pruett 1985; Viru et al. 1992a). Increase in the growth hormone level has been found the night after exercise (Adamson et al. 1974), but this result was not confirmed later (Hackney et al. 1989). During the night after daytime exercise (90 min of cycling), prolactin and thyroxin increased but not cortisol, testosterone, or growth hormone (Hackney et al. 1989). After resistance exercise, a nocturnal rise of testosterone and thyroxin was observed (McMurray et al. 1995). Daytime exercise sessions produced suppressed cortisol levels at night, whereas the magnitude of this effect depended on the intensity at which the daytime exercise was performed (Hackney and Viru 1999).

Thus, for monitoring training not only are responses significant during the session but also in the later dynamics. However, frequent hormone determinations, which disturb the athlete because of repeated vein punctures, are necessary to establish the dynamics and to find the increased hormone level.

Assessing the Intensity of Training Sessions

In sport practice, the workload of training sessions is usually evaluated with the aid of two parameters—volume and intensity. Training volume is a quantitative characteristic. It may be measured in kilometers of total distance covered during the session, total weight of resistance exercises, number of exercise repetitions, total time of exercising, and so on. Intensity of a workload is the volume per unit of time. Intensity may be evaluated by the relative intensity of performed exercises (in percentage to $\dot{V}O_2max$, maximal 1-repetition weight, velocity). An account of the time (percentage of the total time) expended during exercises with certain intensity is popular. Each sport event has its own characteristics for workload volume and intensity.

According to a number of articles, blood lactate assay does not always give a dependable assessment of training intensity or current performance ability (Stegmann et al. 1981; Busse et al. 1989, Hopkins 1991). Negative results might be related to substantial contributions of the diet effect (Fröhlich et al. 1989; Yoshida 1989) and to other conditions. Assessment of the intensity of a training session by the lactate level depends on

- the duration of relief intervals between highly intensive exercise bouts,
- the time interval between the last intensive exercise and blood sampling, and
- the fact that in short-term sprint exercises (races up to 100 to 200 m) the lactate response is not the highest because of ATP resynthesis caused by phosphocreatine breakdowns.

Nevertheless, in cyclic events, determination of lactate with rapid express methods is useful to control the actual running, swimming, cycling, or rowing intensity during the session. This approach is meaningful if the exercise intensity is regulated according to the anaerobic (or aerobic) threshold (see Janssen 1987). However, by adjusting exercise intensity to the anaerobic threshold, the latter should be measured by the individual lactate curve (individual anaerobic threshold) or by the maximal lactate steady state. Lactate use values of 2.0 (2.5) or 4.0 mmol/L ignore individual variability and will not guarantee a precise result.

Sprint exercises are strongly related to the accumulation of ammonia in blood. The threshold for ammonia in running for 300 m at velocity is 87.5% of the individual maximum. At this velocity, the lactate level increased up to 8 mmol/L. Four repetitions of 300-m running bouts with velocities less than the threshold velocity did not evoke ammonia responses, although they increased the lactate response. The authors concluded that increase in ammonia is a primary consequence of exercise intensity but not of volume, whereas the lactate response is related to the intensity and the volume (Schlicht et al. 1990).

In high-resistance and power exercises, intensity is determined by the ratio between performed exercises and the highest possible strength or power in corresponding conditions. When only 1 repetition is possible (maximal strength), this exercise level is expressed as 1 RM. When 5, 8, or 12 repetitions are possible, the expressions are 5 RM, 8 RM, and 12 RM, respectively.

In strength and power exercises, total workload is the sum of products of repetition numbers and weights for each exercise.

$$\text{Workload volume} = (\text{number of repetitions} \times \text{mean weight in exercise } E_1) + (\text{number of repetitions} \times \text{mean weight in } E_2)\ldots + (\text{number of repetitions} \times \text{mean weight in } E_n).$$

Workload intensity may be calculated according to the following formula:

$$\text{Workload intensity} = \frac{\text{Workload volume in high-load exercises (90\% to 100\%)}}{\text{Total workload}}$$

Bosco (1992, 1997) recommended recording the muscle power dynamics during the main exercises in a training session. An electronic apparatus has been constructed for this purpose (Bosco et al. 1995). The apparatus consists of an electronic measuring system that can be adapted to any muscular machine that uses gravitational force (e.g., leg press, leg extension) as external resistance. Whenever a muscular activity is performed by lowering a load, the apparatus records displacement as a function of time. This information can be used to compute velocity, work, and power of the movement. Through visual feedback, the person being tested is informed about the quality and quantity of the muscular effect performed. In this way, the apparatus allows immediate feedback on the actual power output and its decline during repetitions. When the set of repetitions has been designed to reduce power output or movement velocity until a certain level is reached, the apparatus provides that information and avoids useless work. In subsequent sessions, the optimal number of repetitions can then be used until the next recording is taken. The precise information on power output and strength used is also necessary to know which types of muscle fibers are recruited and where the stimulus for improvement reaches (Tihanyi 1989, 1997).

No doubt exists that recordings of mechanical parameters ensure valuable information for the guiding training. Nevertheless, this approach ignores the significance of the duration of rest pauses between repetitions and series. In this regard, the necessity for metabolic and hormonal studies arises not only in endurance but also in strength and power training. The main consideration is whether volume and intensity are sufficient to elicit the trainable effect. Essential information may be obtained with the aid of metabolic and hormonal studies. Fleck and Kraemer (1997), summarizing the results of their investigations (Kraemer et al. 1990b, 1993a), demonstrated that when a higher strength (5 RM) was used for 8 repetitions over a 3-min rest, the lactate response was less pronounced than in 8 repetitions of 10 RM strength over a 1-min rest. Summarizing another set of studies (Keul et al., 1978; Gettman and Pollock 1981; Kraemer et al. 1987; Kraemer et al. 1990b), Fleck and Kraemer pointed out high lactate responses in short rest intervals between workouts in bodybuilders and in high-intensity circuit weight training. Lactate responses were modest in workouts with prolonged rest intervals (power lifting, Olympic weightlifting).

The duration of rest pauses also influences the hormonal responses during a session. The highest growth hormone, β-endorphin, and cortisol contractions appeared when short-rest (1-min) exercises were reported in 3 sets using only 10 RM strength. Testosterone responses appeared in a high-intensity (5-RM), long-rest (3-min) exercise protocol and in a 10-RM short-rest protocol (Kraemer et al. 1990b; Kraemer et al. 1991b; Kraemer et al. 1993a). In weightlifters, testosterone responses were detected after squat lifts at 90% to 95% or 60% to 65% of 6 RM (Schwab et al. 1993). A bench-press protocol (5 sets to failure using a 10-RM load) increased testosterone concentrations by 7.49% and a jump squat protocol (15 sets of 10 repetitions using 30% 1-RM squats) by 15.1%, whereas cortisol levels did not change (Volek et al. 1997). A weightlifting protocol caused an increase of testosterone concentration in 17-year-old juniors with training experience of more than 2 years but not in subjects with training experience less than 2 years. All juniors exhibited cortisol, β-endorphin, and growth hormone responses (Kraemer et al. 1992). In elite athletes, after 20 sets of squat lifts at 1 RM, total and free testosterone, cortisol, and growth hormone levels did not change. Concentrations of all three hormones increased significantly after 10 sets of 10 repetitions of 10% 10 RM (Häkkinen and Pakarinen 1993).

Vanhelder et al. (1984b) found a growth hormone response after 7 series of squat lifts at 85% of maximal leg strength but not after 7 series of 21 squat lifts at 30% of the previous load. A 3-set heavy-resistance exercise protocol resulted in greater increases of growth hormone and testosterone than a 1-set exercise protocol, whereas the cortisol response was almost the same (Cotshalk et al. 1997).

Discrepancies among the results of various studies are obviously related to the simultaneous action of several factors. Fleck and Kraemer (1997) pointed out that the hormonal response to a resistance exercise workout depends on muscle mass recruited, intensity of the workout, amount of rest between sets and exercises, volume of total work, and training level of the individual. Of course, the simultaneous interplay of various factors complicates the use of hormonal changes for establishing the action of one of these factors. However, if we consider hormonal responses as an integral part of the action of volume, intensity, and regime of training exercises and of other possible factors, we get an understanding of their general influence on the endocrine system. The problem is that we do not know anything about the reception of hormones in muscle fibers. Therefore, we have to keep in mind that hormonal responses allow us to make many suggestions but not reach a final conclusion. We also need to know the dynamics of hormone levels during the recovery period. All in all, hormonal responses provide an understanding of the activities of the endocrine systems that are essential in adaptive protein synthesis but not about the actualization of considered endocrine influences.

Bosco and coauthors (2000) studied the influence of a different power-training session on hormone levels. A large number of repetitions with low power output in bodybuilders decreased testosterone and increased growth hormone levels, whereas a high number of repetitions with power output in weightlifters increased testosterone concentrations without changes in growth hormone level. When power output remained close to maximum and the application of force increased (number of series decreased), no significant hormone changes were found in weightlifters. Sprinters performed exercises at maximal power with a force of 80% 1 RM. Although the number of series was modest, the total workload was obviously high. In men, blood concentrations of lutropins, testosterone, and cortisol decreased in compared with pre-session values. It was possible to suggest that the reversed hormonal responses were related to pronounced fatigue that developed during the session. This possibility was confirmed by a significant decrease of average power in full squats and by an increase of the ratio of electromyogram to power in full squats. In female sprinters, lutropin, testosterone, and cortisol levels did not change. This situation expressed the same trend that was found in men, but the rate of fatigue development might be lower in women, and, therefore, the hormone levels did not decrease to values below initial ones.

Further studies are necessary to establish the expression of hormonal response during power exercises in the actualization of specific manifestations of adaptive protein synthesis.

Training Microcycles

An important element of training organization is the training microcycle. It includes a limited number of training and rest days to obtain an adequate influence on the body. The organization of microcycles has to

- concern the action of subsequent training sessions,
- determine the ratio between training time and rest hours, and
- ensure complete restitution before the beginning of the next microcycle.

The time for the actualization of adaptive protein synthesis (the reconstructive function of the recovery period) and supercompensation of energy stores is determined by the organization of the training microcycles. Although training exercises determine the specificity of adaptive protein synthesis and the workload of training sessions ensures hormonal alterations for its amplifications, the training microcycle summarizes and interrelates various influences on protein turnover. For that purpose, microcycles can be categorized as follows:

- Developmental microcycle—ensures the desired results of training: (a) ordinary microcycle (the difference between the total workload of this microcycle and the previous microcycle is moderate) or (b) "blow" microcycle (total workload of the microcycle is increased to a great extent compared with the previous microcycles)
- Applied microcycle—adjusts the athlete's body to training at the beginning of a training period or to new training conditions (transition from outdoor to indoor conditions or vice versa, or from running to skiing, etc.) or ensures readiness for competition
- Competition microcycle—last day(s) before competition and days of competition
- Restitution microcycle—relief days or weeks just after the competition microcycle or after the "blow" microcycle

In regard to ordinary microcycles, the ratio of training workloads to restitution time between sessions is the most decisive. From a physiological point of view, ordinary microcycles have to be integrated into microcycles with complete restitution before the next training set and into microcycles with summation of workloads. The first type exists in the physical education of adolescents and in the training of beginners (Atha 1981). In advanced athletes, especially in qualified athletes, this variant of the ordinary microcycle is a waste of time (Kraemer et al. 1987; Hoffman et al. 1990a). The summation of workloads within a microcycle is caused by the daily repetition of vigorous training sessions.

Principally, microcycles with summation of workloads may produce three different results.

1. Summation of workloads causes general fatigue during the last training day(s); during subsequent rest day(s), recovery processes ensure repletion of energy stores and functions, and a moderate stimulus for further adaptation arises.
2. Summation of workloads causes a drop in the body's energy until the borderline to dangerous exhaustion is reached; this situation is a strong stimulus for adaptation processes: pronounced structural, metabolic, and functional improvements may be achieved for the beginning of the next microcycle.
3. Summation of workloads creates such a demanding situation that dangerous exhaustion (characterized by overstrain) develops; restitution days are necessary for alleviation of overstrain, and the next microcycle begins from the level of decreased work capacity.

In the latter case, a restitution microcycle has to follow to avoid overtraining.

In monitoring training, the main task is to establish the borderline between dangerous exhaustion and optimal stimulation of adaptation processes. Summation of the influences on training sessions performed on several subsequent days has been studied in athletes and experimental animals.

When 30 min of swimming was repeated daily in rats for 5 days, on the first day, adrenocortical activity increased in response to exercise. During further repetitions of the exercise, a generally high level of adrenocortical activity persisted in the resting state and after exercise. After 3 or 4 days of exercising, a decreased content of corticosterone in the adrenal glands is associated with a low level of corticosterone production in vitro. On the fifth day of the training week, a restoration of the normal level of adrenocortical activity was observed without any response to the same exercise workload. Activation of the adrenocortical function was observed again when the workload of the training exercise was augmented. The same dynamics reappeared during the next training weeks (Viru et al. 1988). These data suggest that adaptation to a specific level of exercise workload is achieved through a phase of subtotal depletion of the resources of the adrenal cortex. Obviously, an improvement in the apparatus of corticosteroid biosynthesis was necessary for achievement of adaptation. In the third week of training, a high level of corticosterone production in vitro was observed. This suggests that the glucocorticoid biosynthesis mechanism had reached a new augmented level.

This experiment was repeated with 90-min swims instead of 30-min swims (Ööpik et al. 1991). These results also indicated that daily repetitions of exercises for 5 days caused a subtotal exhaustion of the adrenal reserve. For comparison with the results obtained in overtrained athletes, see chapter 9, page 213. It is important to point out that adrenal responsiveness to corticotropin persisted, although the magnitude of response decreased.

Rat experiments showed that the cortisol response to exercise is necessary to obtain improved work capacity (Viru 1976b). The cortisol response indicates the activation of the mechanism of general adaptation that is reasonably required for the transition from acute adaptation to stable continuous adaptation.

Determination of 17-hydroxycorticoid (cortisol and its metabolites) in the urine is comparatively simple. In female students, 8 weeks of experimental training on a bicycle ergometer induced an increase in peak work capacity (PWC_{170}). Simultaneously, increased 17-hydroxycorticoid excretion was recorded as a response to training sets in the first and last week of training (Viru E. et al. 1979). In 15- to 17-year-old skiers, the training sessions caused increased 17-hydroxycorticoid excretions when they skied at the velocity of 87% to 90% of competition velocity. When the skiing velocity used in training sessions corresponded to 81% to 83%

of competition velocity, increased 17-hydroxycorticoid excretions were found only at the end of the training microcycle. Improvement of performance was detected only in those skiers who used the higher velocity of skiing (Alev and Viru 1982).

In international- and national-level basketball players, 17-hydroxycorticoid excretion was determined on the first and fourth days of the training microcycle. Four variants of response were found.

1. Activation of the adrenocortical function only at the end of a 4-day microcycle
2. Activation on the first and last day of the microcycle
3. Activation on the first but suppression on the last day
4. Suppression of adrenocortical function throughout the microcycle

In basketball players exhibiting adrenocortical activation throughout the microcycle, $\dot{V}O_2max$, PWC_{170}, and playing effectiveness increased during the corresponding training stage. In those who did not exhibit adrenocortical activation, detraining manifestations were found: $\dot{V}O_2max$ and PWC_{170} decreased. In cases of persisting suppression of adrenocortical activity, $\dot{V}O_2max$ did not change, PWC_{170} decreased, and ST-depression and isoelectric T waves were observed in electrocardiograms (Jalak and Viru 1983).

In chapter 4 (see p. 62), methodological problems were indicated in regard to investigations of hormone excretion with urine. Therefore, although this is a convenient method, the result has to be checked with determination of hormones in the blood. The results suggest that increased adrenocortical activity associated with the trainable effect and the decreased adrenocortical activity developing during a microcycle indicate the use of too many exhaustive training workloads. Attention should also be paid to individual differences in the responses to basketball. In team events, the creative ability of the coach is required for individualization of workloads to have a trainable effect for each player.

Results that were in agreement were obtained in a group of middle- and long-distance runners (figure 8.3). Strenuous daily training first caused a pronounced elevation of the cortisol resting level in conjunction with unusually high responses of cortisol and growth hormone to training exercises. Afterwards, a reversed cortisol response accompanied a high resting level of blood cortisol after exercise (Viru et al. 1988).

A 1-week intensive strength-training microcycle with two training sessions per day was studied in eight elite weightlifters. A significant increase in total and free serum testosterone levels was found during the afternoon sessions but not in the morning sessions. Cortisol concentration decreased after the morning sessions but increased after the afternoon sessions. Growth hormone level increased in both sessions; free testosterone declined from the fourth day. Morning levels of cortisol and growth hormone were constant during the week (Häkkinen et al. 1988a).

A demonstrative example of the metabolic manifestation of summation of the daily workload has been provided by dynamics of muscle glycogen (figure 8.4) (Kantola and Rusko 1985). Similar results are available in several articles (Hultman 1967; Costill et al. 1971, 1988; Kirwan et al. 1988b). Because a decreased glycogen level induces peculiarities in metabolic regulation, one must take into consideration that at the end of a summation microcycle metabolic adaptation may be altered; decreased lactate and increased ammonia productions are possible (Broberg and Sahlin 1988). After 10 days of summation of daily workloads, a decrease in muscle glycogen concentration was associated with reduced glycogen and enhanced lipid use during exercise (Kots et al. 1986).

Various dynamics of blood urea during different microcycles are presented in figure 8.5. These data showed that summation of training workload or insufficient time for recovery may lead to a gradual increase of blood urea or its leveling off at a high level (Neumann and Schüler 1989).

Daily summation of workloads may lead to a persistent overall decrease of the protein synthesis rate in skeletal muscles. In rats, daily swimming for 90 min for 5 days resulted in a pronounced and long-lasting suppression of the rate of synthesis of sarcoplasmic and myofibrillar proteins in the soleus, gastrocnemius, and red and white portions of the quadriceps muscle. The decreased rate of protein synthesis also persisted during the 2 subsequent rest days. The blood urea

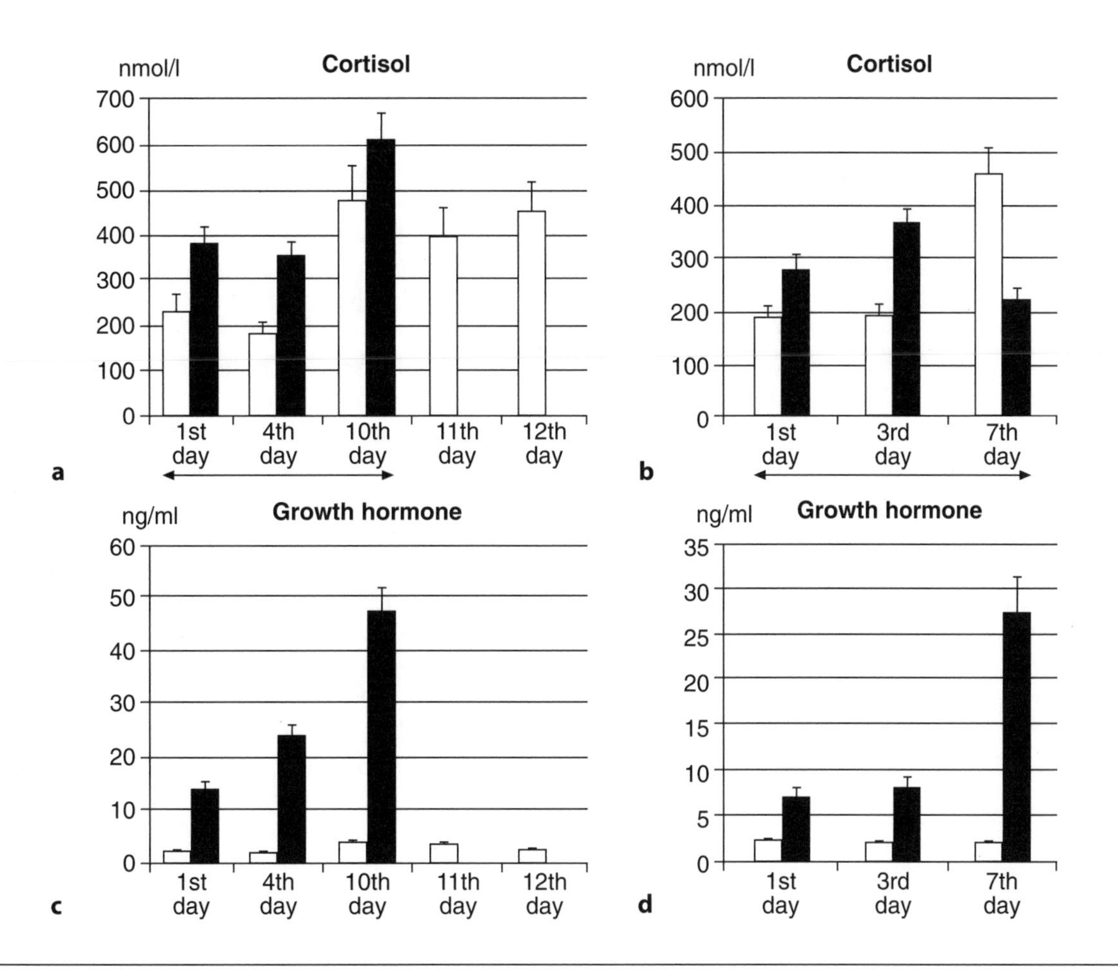

Figure 8.3 Changes in blood levels of *(a, b)* cortisol and *(c, d)* growth hormone (somatotropin) in female middle-distance (graphs on the left) and long-distance runners (graphs on the right) during daily high-volume training. The training-stage (indicated by horizontal arrow) lasted 10 days in middle-distance runners and 7 days in long-distance runners followed by rest days. The morning hormone levels are indicated by white columns, and the hormone levels 5 min after training are indicated by shaded columns.

From Viru et al. 1988.

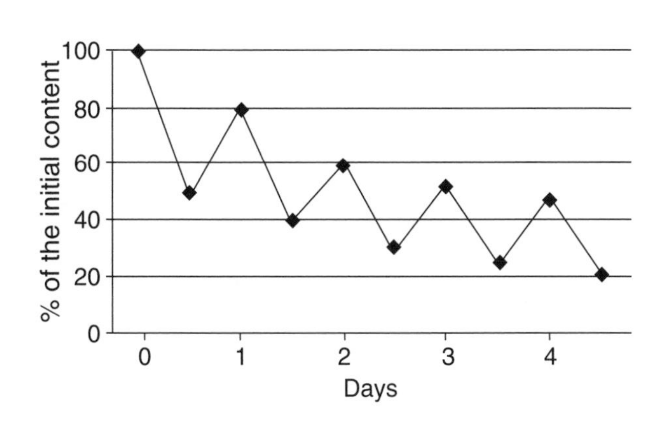

Figure 8.4 Dynamics of glycogen content in vastus lateralis muscle during a summating microcycle.

From Kantola and Rusko 1985.

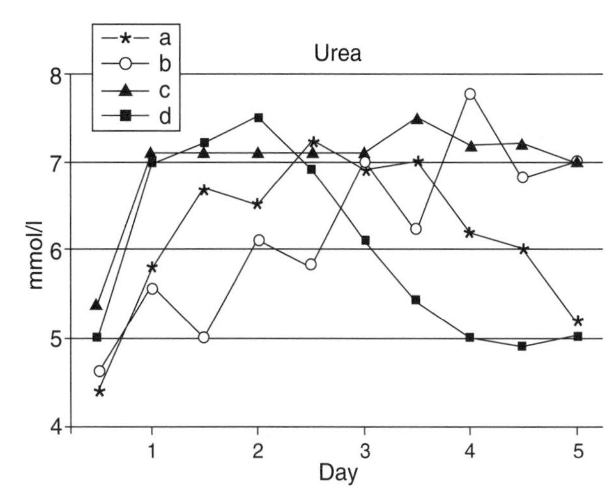

Figure 8.5 Urea dynamics during various microcycles.

Reprinted from G. Newman 1992.

level was elevated during exercise days. However, 32 h after the last swimming set, it normalized, although the diminished rate of protein synthesis persisted in the muscles. The level of free tyrosine increased in the muscles during swimming sets. In muscles of predominantly red fibers, the free tyrosine level usually normalized after 8 h of every exercise bout, and in muscles of predominantly white fibers, the elevated content of free tyrosine persisted 24 h or more after exercise (Ööpik and Viru 1988; Ööpik et al. 1988).

In another series of experiments, swimming duration was reduced by 60 min after the first 2 days of every week. In this case, protein synthesis was suppressed only on exercise days, and it intensified to a level above control values during the days of recovery (Ööpik and Viru 1988). During 4 weeks of training (at the beginning of each week, swimming duration was increased by 15 min), physical work capacity (maximal duration of swimming), glycogen reserves, and activity of succinate dehydrogenase in red muscles and dry weight of the adrenal glands increased in both training regimes (figure 8.6) (Ööpik and Viru 1992).

Hence, the overall suppression of protein synthesis in skeletal muscles did not exclude an increase in work capacity and other training effects. The increased activity of the mitochondrial enzyme indicated that at least in regard to these enzymes, adaptive protein synthesis took place. This agrees with the previously noted fact that after an endurance exercise, the main locus of elevated protein synthesis is mitochondrial proteins in the red fibers. Training effects were even more pronounced in cases of persisting suppression of protein synthesis than in a less strenuous training regime, leading to an elevated rate of protein synthesis during recovery after 5 days of training. This fact led us to the assumption that the overall suppression of protein synthesis excluded the competition between the syntheses of various proteins for "building materials" and helped to concentrate

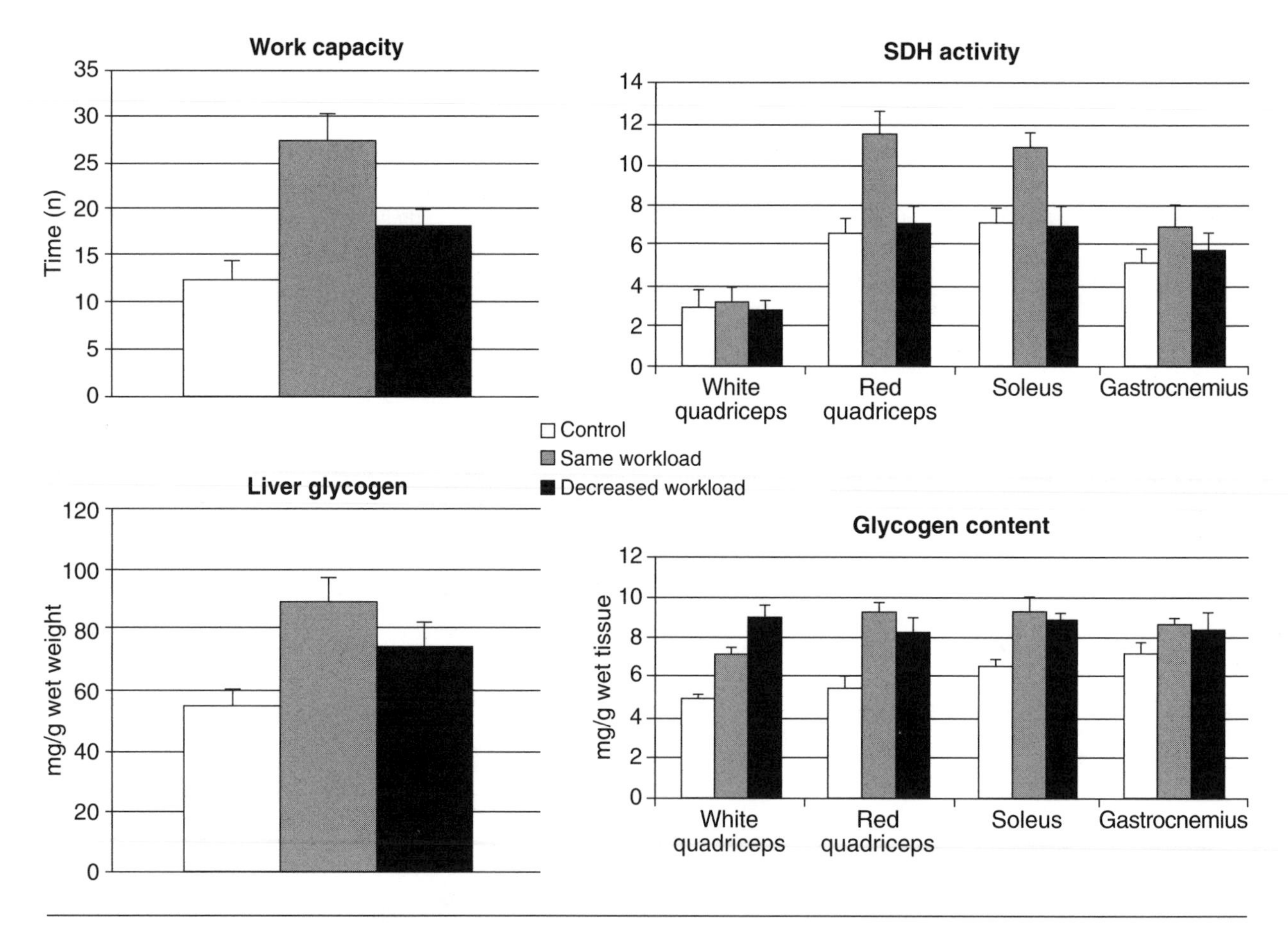

Figure 8.6 Changes induced by 4 weeks of training in maximal duration of swimming, glycogen stores, and activity of succinate dehydrogenase in various muscles of rats.

From Ööpik and Viru 1992.

adaptive protein synthesis for the most responsible proteins.

The last microcycle before competitions has to ensure the repletion and supercompensation of energy stores. Several variants of tapering in combination with a specially altered diet have been used (e.g., Eynde 1983). A special problem is how to monitor the effectiveness of the precompetition microcycle. An example is provided by Hooper et al. (1999). The authors studied the effect of 2 weeks of tapering for a major competition. The Profile of Mood Status (POMS) indicated a decrease of fatigue and an increase of vigor. Significant changes were not found in a 100-m swim time and peak force, as well as various hematological and cardiovascular parameters. Trends (nonsignificant) were for an increase of plasma creatine kinase and cortisol, and the decline of norepinephrine and epinephrine was noticed.

Diagnosing Fatigue

Evaluation of the workload of a training session, and particularly assessment of the training microcycle arrangement, is related to diagnosing the degree of fatigue. An indication of fatigue is a decrease in work capacity, mainly an inability to maintain the performance quality. However, to know the location of fatigue or its primary reason, biochemical testing of fatigue may be useful. The significance of biochemical testing of fatigue increases to detect the "latent fatigue" preceding the actual decrease in performance.

Fatigue

Fatigue represents a complicated problem. Various approaches have been used to define the phenomenon and to penetrate into its essence. According to the most widely used approach, fatigue is a failure to maintain a required or expected force or power output (Edwards 1981, 1983). Knuttgen et al. (1983) defined fatigue as the inability of a physiological process to continue functioning at a particular level and/or the inability of the total organism to maintain a predetermined exercise intensity. The latter definition emphasizes that fatigue may be a phenomenon either of a functional system or of the whole organism. Manifestations of fatigue are located not only in the neuromuscular apparatus, but they may also be related to the inability of the various physiological processes "to continue functioning at a particular level." In a principle in accordance with Knuttgen's definition is the definition that fatigue is a physiological state developing as a result of intense or prolonged activity and manifested by the decrease of work capacity, a feeling of tiredness, and discordance of functions (Zhimkin 1975). In this regard, a period of surmounted fatigue (latent fatigue) was discerned during prolonged exercise. A specific manifestation of latent fatigue in endurance events is the compensation of decreased force output by an increased movement frequency to maintain the movement velocity at a constant level (Farfel et al. 1972). This compensation is effective only for a short time and may be related to an additional intensification of anaerobic energy production. The metabolic consequences of the latter speed up the decrease of possibilities to maintain the required level of power output. Another example of compensation of decreased force is recruitment of additional motor unit and muscle groups. The result is decreased mechanical efficiency and, thereby, increased demands on the oxygen transport system. In cardiovascular function, fatigue development is related to decreased stroke volume that is compensated by increased heart rate. Thus, manifestations of fatigue express not only the failure but also the compensatory processes in the neuromuscular apparatus and in various attendant functions and metabolic processes. The failure related to fatigue may have various locations (figure 8.7) (Kirkendall, 2000).

A catastrophe theory (Edwards 1983) and a defense theory (Zhimkin 1975; Viru 1975) have been used to understand the essence of fatigue. The basis of the catastrophe theory of fatigue indicates that fatigue may be due to breaking any of the links in the chain of command in muscular contraction (figure 8.8). Edwards (1983) emphasizes that the theory illustrates the principle, whereas the more complex forms could pertain to intermediate fatigue states with impaired energy supply and excitation. The catastrophe theory may properly describe the common final path of cellular functions leading to impaired performance.

The defense theory is based on discerning fatigue from exhaustion. The main points of this theory are that

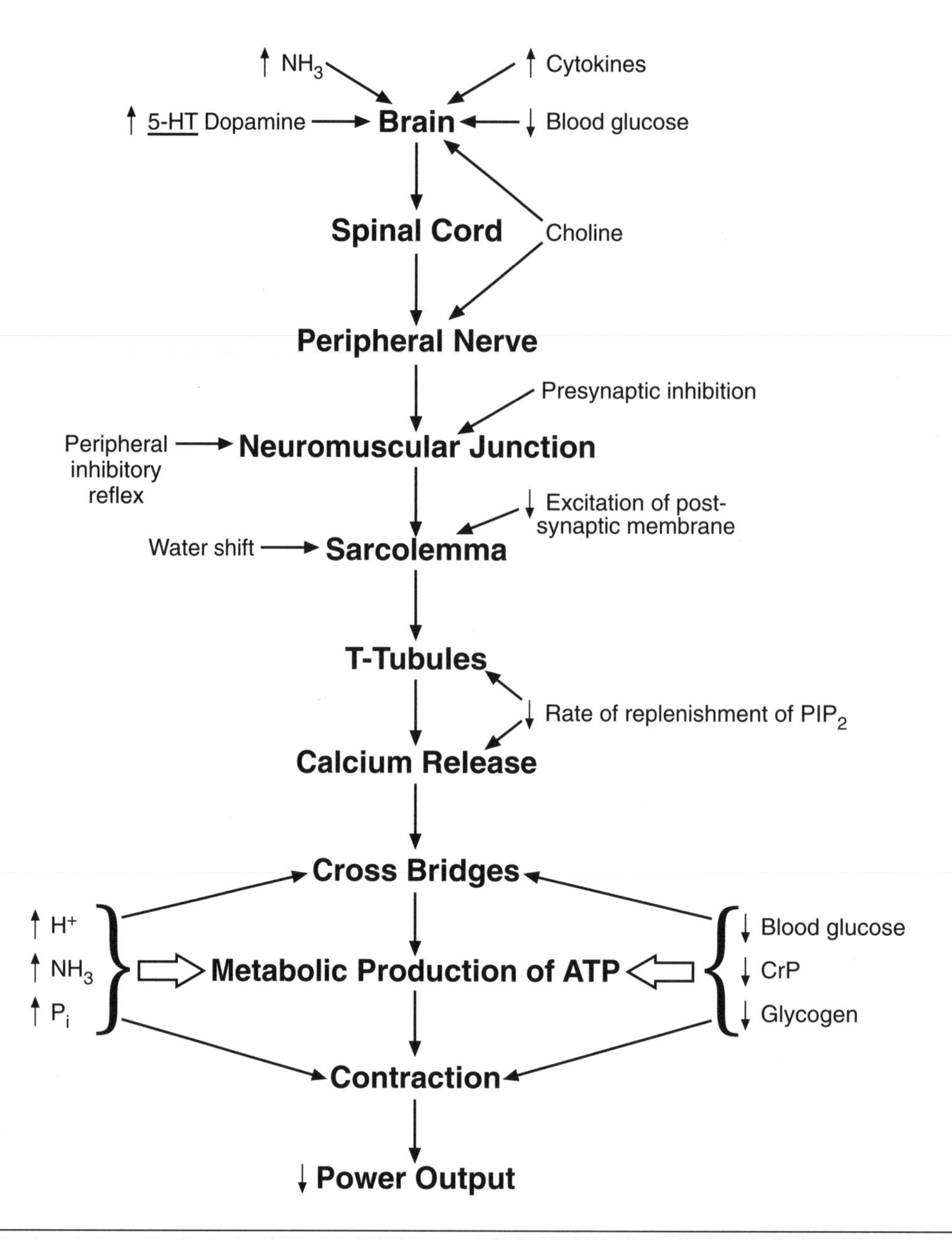

Figure 8.7 Neuromuscular pathway leading to contractions and probable mechanisms of fatigue.
Reprinted from V. Oopik and A. Viru 1992.

- the body never uses up all its reserves,
- the use of those "reserve forces" is protected by a special barrier deeply connected with the fatigue processes, and
- fatigue precedes the exhaustion of body resources and results in the termination of work (exercise) before pronounced exhaustion sets in.

Therefore, the function of fatigue is to avoid a fatal depletion of resources in the body, organs, and cells. It has been suggested that the protective role of fatigue is accomplished by three types of defense reactions.

1. Influence of the feeling of tiredness
2. Direct disconnection of motor unit activation by protective inhibition of the nerve cells or by cumulative changes in muscle fibers
3. Inhibition of the mechanism of mobilization of metabolic resources (Viru 1975a)

To compare the defense and catastrophe theories, the term "defense reaction" has to be defined. The defense reaction is a centrally or-

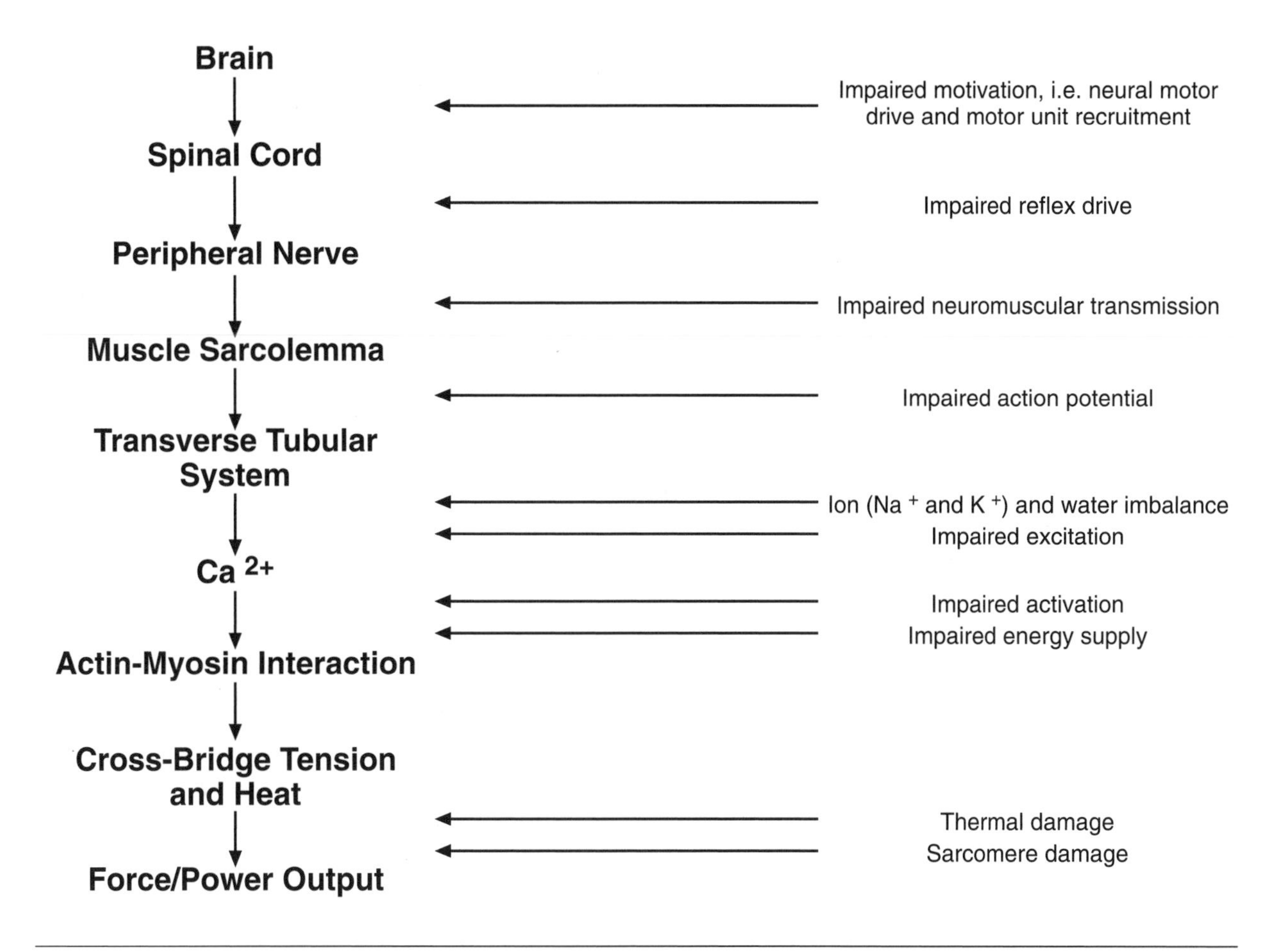

Figure 8.8 Chain of events in muscle contraction and possible locations of failure underlying fatigue.
Reprinted from R.H.T. Edwards 1983.

ganized change in regulative functions leading to inhibition of mechanisms responsible for the mobilization of an organism's resources.

The central inhibition of mobilization of the body's resources is possibly related to serotonergic neurons located in the hippocampus. After electrocoagulation of the hippocampus area, the maximal duration of swimming was slightly increased, but the decrease in blood corticosterone concentration, observed in sham-operated rats after swimming until exhaustion, was not observed (Viru 1975a). Consequently, the subnormal adrenocortical activity was the result of a regulatory mechanism. This regulatory mechanism is, obviously, related to increased activity of the hippocampal serotonergic neurons. These neurons are known to cause inhibition of hypothalamic neurosecretory cells producing corticoliberin (corticotropin-releasing factor). The resulting subnormal adrenocortical activity limits the catabolic action of glucocorticoids during prolonged exercise. Simultaneously, it decreases epinephrine synthesis (Matlina et al. 1978) and, thereby, reduces glycogenolysis in skeletal muscles and lipolysis in adipose tissue. It is noteworthy that inhibition of the hypothalamic neurosecretory cells is related to activity in the serotonergic neurons. Production of their neurotransmitter, serotonin, is generally considered to be responsible in the central mechanism of fatigue (see chapter 3, p. 48).

Edwards (1983) agrees that fatigue is a manifestation of one or more of the "fail-safe" mechanisms in the organism, which call for moderation before damage occurs. Nevertheless, the significance of the defense mechanism(s) does not exclude the fact that several functional inabilities are related directly to the local exhaustion of resources without their link to the central mechanisms.

Assessing Fatigue in Biochemical Monitoring of Training

In monitoring training, the main point is to detect the critical reduction of performance capacity and energy stores for the end of a training microcycle. This means the assessment of "critical fatigue." In this context, "critical" denotes the lack of possibility for complete recovery during the 1 or 2 rest days at the end of the microcycle. Complete recovery assumes the restoration of exercise performance, creating change in the central neuron structures ensuring a psychic readiness to begin the next microcycle with trainable workloads and replenishment of energy resources. When the actual fatigue is over the "critical" phase, the next microcycle should be a restitution microcycle instead of a developmental microcycle, or an additional day(s) is necessary before the onset of the next developmental microcycle.

In several cases, the performance level at the end of the final session of a microcycle provides valuable information on whether the fatigue is over the "critical" level. If these observations are founded on certain objective parameters, they express the manifestation of fatigue related to the event-specific performance in an athlete. In power events, mostly, these observations may be sufficient as a successful guide to training. In endurance and game events, the possibilities for such an observation are limited, although they exist. In several cases, biochemical methods are more convenient for diagnostics of "critical" fatigue compared with performance observations.

Metabolic changes are related to either peripheral or central fatigue (see Edwards 1983; Hultman et al. 1990; Sjøgaard 1990; Newsholme et al. 1992; Davis 1994; Brooks et al. 1996; Krieder 1998; Kirkendall, 2000) (see figure 8.8). Most of these changes are specifically related to the fatigue process in a certain link of the neuromuscular apparatus. The stability of a parameter does not exclude the possibility of critical changes in structures, the state of which is expressed by other parameters. Consequently, the structure most responsible for event-specific performance (or specific demands of the training session) has to be determined. It is not a simple task. The solution requires the comparison of various changes associated with a pronounced drop in the specific performance.

Another problem is that complicated methods are necessary for detecting changes. Methodologically, a more or less realistic pathway is the determination of the tryptophan/branched-chain amino acid ratio (see chapter 3, p. 48). In this way, certain characteristics are possible to determine from a condition favoring central fatigue. For peripheral fatigue, attention should be paid to the Na^+-K^+ pump function. This function is critically related to the shift of H_2O into the cell and decreased excitation of the postsynaptic membrane. Earlier, it was mentioned that failure of the Na^+-K^+ pump function is expressed by an increased potassium level in plasma (see chapter 6, p. 135). Therefore, determination of the K^+ concentration in plasma (taking into account possible hemoconcentration) is a suitable method for fatigue assessment, as suggested by Sjøgaard (1990, 1991).

Furthermore, in several cases, information may be an exaggerated accumulation of urea, ammonia, lactate, and creatine kinase in blood, as well as hypoglycemia and extreme reduction of blood pH. Regarding these indexes, the significance may have not only the magnitude of responses but also the time during which the altered levels persisted in the postexercise period.

All mentioned metabolic changes might appear without pronounced fatigue. Therefore, their use requires an individual comparison of changes obtained after an event-specific workload (e.g., at the end of the training microcycle) with changes in performance and the rate of recovery. Those changes that are associated with recovery of more than 2 days have to be considered as indexes of the "critical" fatigue.

Earlier (see this chapter, p. 179) data were presented that a high-load microcycle may induce a pronounced decrease in the excretion of cortisol metabolites. In these athletes, progress in performance was not obvious, and in several cases overstrain signs appeared. Thus, a possibility appears to use decreased adrenocortical activity for fatigue diagnostics. Of course, the information is more valid when the blood hormone level is measured. However, the study of hormone excretion disturbs an athlete to a lesser degree.

The link between decreased excretion of corticosteroid (mainly glucocorticoid) and fatigue was found in the 1950s (Rivoire et al. 1953) and confirmed later in several studies (Bugard et al. 1961; Viru 1977; Kassil et al. 1978). Measurement of corticosteroids in blood with the color reaction specific for 17-hydroxysteroids (Staehelin et al. 1955) or fluorometric assay (Viru et al. 1973; Keibel 1974) demonstrated that after an initial increase of the corticosteroid level, a decrease below initial values followed during prolonged exercise. In the 1970s, radioimmuno assays provided highly specific and valid methods for hormone determinations in the blood. The use of radioimmunoassay methods with high cortisol levels was found at the finish of a marathon race (Maron et al. 1975; Dessypris et al. 1976; Weicker et al. 1981), 100-km running (Keul et al. 1981), or cross-country skiing for 70 km (Sundsfjord et al. 1975), as well as during triathlon competition (Feldmann et al. 1988; Jürimäe et al. 1990b; Gastmann et al. 1998), 24-h skiing (Zuliani et al. 1984) or 24-h running (Feldmann et al. 1988). Nevertheless, results were published indicating a possible link between fatigue and a low level of cortisol in the blood. Dessypris et al. (1976) reported that difference from other persons; a low cortisol level was found in a marathon runner who collapsed after he ran 15 km. A male rower who collapsed during a regatta had a remarkably low cortisol concentration the day before the competition (Urhausen et al. 1987). In both cases, low cortisol levels were associated with low levels of testosterone. Feldmann et al. (1992) studied a 6-day period of a Nordic ski race (distance per day, 44 to 61.5 km). In the first 2 days, skiing increased the blood level of cortisol in association with high levels of corticotropin. From the third day, blood samples obtained after skiing did not show a cortisol response.

In a study of the fatigue effect on hormone responses, a 10-min test exercise at 70% $\dot{V}O_2max$ was used before and after a 2-h training session consisting of continuous aerobic exercise. The participants were endurance athletes. Before the prolonged aerobic exercise, the test exercise induced a significant increase of cortisol and growth hormone concentrations, whereas the testosterone response was not significant. After the long-lasting aerobic workout, the cortisol response to the test exercise allowed us to categorize persons into two groups: in four persons, the magnitude of the cortisol response increased in combination with high preexercise and postexercise growth hormone concentrations. In these persons, the long-lasting aerobic workout did not decrease performance in a 1-min anaerobic test. In the other eight persons, the cortisol response decreased or the inverse response appeared. In these persons, no increase of growth hormone level was found in the test exercise after a prolonged aerobic workload in association with decreased anaerobic performance. No common changes were found in testosterone concentration. Thus, two fatigue effects appeared: either an increased mobilization of glucocorticoid and somatotropic function or a suppression of these responses (Viru et al., 2000). Odagiri et al. (1996) reported two types of hormonal responses during a triathlon. A group of persons, possessing high vigor regardless of high fatigue score (tested with the Profile of Mood Status), had high postcompetition levels of corticotropin and β-endorphin. Other persons showing low vigor and high fatigue exhibited suppressed responses of corticotropin and β-endorphin. Cortisol responses were similar in both groups.

The results that were discussed confirm that exercise-induced fatigue may alter the pituitary-adrenocortical responses to exercise. The stronger the fatigue, the higher the possibility for suppressed pituitary-adrenocortical function. In these cases, suppression may extend to the somatotropic function. Consequently, a hypothesis may be proposed, according to which signs of suppressed pituitary-adrenocortical activity, blunted pituitary somatotropic function, and a decreased blood testosterone level may appear as indications of critical fatigue.

Monitoring Recovery

Several tasks can be solved with the aid of recovery monitoring. Such tasks may arise to determine optimal resting intervals in training sessions and during competitions. Another group of tasks is related to the design of training microcycles.

In the interval method of training, the usual rest (relief) intervals are too short for complete equilibrium between lactate concentrations in working muscles and blood. Therefore, measurement of blood lactate at the onset and end of the rest intervals is meaningless. Experiences from the 1960s taught athletes that useful information can be provided by the change in heart rate during the rest intervals. Nevertheless, monitoring lactate dynamics during an interval training session is meaningful. The lactate rise depends on the intensity of exercise bouts and the duration of rest intervals. Thus, changing these two components, it is possible to speed up or slow down lactate accumulation in the blood. After taking measurements of the lactate levels during rest intervals after certain numbers of repetitions of exercise bouts, the coach or athlete will know what the actual influence of an interval training regime is. An increase in the duration of rest intervals within certain limits ensures prolongation of the time during which an athlete has to perform intensive exercises in conditions of increasing lactate concentrations. Another effect is achieved by shortening rest intervals. Then, repeated intensive exercise prepares the body for rapid mobilization of anaerobic energy production. In weight or power events, the best information for the optimization of rest intervals between series and between repetitions can be obtained by the dynamics of the characteristics of force and power implications. However, as was noted earlier (this chapter, p. 174), lactate monitoring may be useful in several cases. In these training sessions, the pronounced lactate accumulation suggests that a condition develops that speeds up the loss of the possibility for further repetition of strength or power exercise bouts. This condition is mostly related to short rest intervals.

In game events, it is usual to substitute players on the field or court. There are tactical, psychological, and other reasons for that. In several cases, the reason is to reduce the development of manifestations of fatigue with the aid of rest for a couple of minutes (e.g., after a high-intensity shift, ice-hockey players sit passively for 4 to 6 min on the bench waiting for their next on-ice shift [see Montgomery, 2000]). During a training session, a modeling experiment may be organized to find out the optimal times for on-ice shifts for each player by way of lactate dynamics.

After the training session, the return of blood lactate to its initial level does not last more than 30 to 45 min. Therefore, lactate is not a means for monitoring postsession recovery. However, postexercise monitoring of lactate provides information on the dynamics of the training effect on various energy production pathways. An example has been provided by Pelayo and coauthors (1996). In elite 200-m freestyle swimmers, blood lactate was determined 3 and 12 min after a maximal anaerobic lactate test (four all-out 50-m swims interspersed with 10-s rest intervals). The test was performed six times during a 21-week season. The percentage of lactate recovery between 3 and 12 postexercise minutes increased from week 2 to week 10 with aerobic training and decreased from week 10 to week 21 with anaerobic training. Test performance improved continuously throughout the season. Competition performance improved during the competitions on the first, seventh, and thirteenth weeks but declined on the twenty-third week, coinciding with the lowest percentage of lactate recovery after test exercise.

The most popular index for monitoring postsession recovery is urea dynamics. Urea concentration has been mostly determined just after the session and the next morning (or only the next morning). According to a simplified approach, the high urea level the next morning (see this chapter, p. 175) indicates the necessity of a restitution or maintaining workload but not a trainable workload for the next training session. Depending on the design of the microcycle, the urea dynamics might be different (see figure 8.5).

Information about recovery and supercompensation of energy stores should be provided by monitoring postsession recovery. Regrettably, the biopsy method is necessary to ensure reliability of results.

Earlier, several metabolic parameters were discussed in regard to their use for fatigue diagnostics (p. 185 of this chapter). It was noted that the significance might have not only the magnitude of responses but also the rate of their return to initial values. Regarding training design, the values in the morning of the beginning of a new training microcycle are the most important.

A problem is the use of hormonal changes in recovery monitoring. Hormones may exhibit altered levels for hours and days after a heavy-load workout. For instance, Fry et al. (1991) found after intensive anaerobic interval training sessions decreases of cortisol and testosterone concentrations within 2 h to levels less than the initial levels. Testosterone remained at a decreased level for at least 24 h. During this period, urea, uric acid, and creatine kinase persisted at increased levels. It is possible to suggest that delayed hormonal changes are essential for control of the recovery processes and particularly for control of adaptive protein synthesis. However, the related hormonal studies are complicated by their cost and methods and by the necessity for repeated blood sampling of athletes. When possibilities exist to overcome the complications, attention should be paid to hormones (1) for the study of rapid recovery—insulin and cortisol, which contribute to the control of the replacement of carbohydrate stores, and (2) for the study of delayed recovery—testosterone, cortisol, thyroid hormone, growth hormone, and insulin, which contribute to the control of protein turnover and adaptive protein synthesis.

Summary

Assessment of the training design requires evaluation of training sessions and microcycles. The most necessary information is provided by assessing the trainable effect. The possibility for that arises from the knowledge of the essential role of the induction of adaptive protein synthesis as the foundation of the main training effect. Related inductors are metabolites and hormones. Because metabolites, which may have an inductor effect, accumulate within cells, assessing them is complicated. A possible solution is to assess the catabolic effect of training sessions (e.g., with the aid of excretion of 3-methylhistidine), taking into consideration the fact that there should also be an inductor(s) of adaptive protein synthesis among the catabolites. Thus, the greater the metabolite outflow from cells into the blood plasma or urine, the higher should be the possibility that the specific inductors accumulated in the cell.

Several problems also exist in ensuring that the rise of hormone concentration is directly related to its inductor effect because it depends on attaining hormone receptors. Thus, the blood hormone responses give us only an approximation for suggesting the actual inductor effect of the hormone. Another limitation lies in the fact that more important than hormone changes during the training sessions may be the pattern of hormone availability for cells during the recovery period. However, the more intense the hormone response, the greater the possibility of the inductor action of this hormone.

The assessment of training workload is related to the problem of biochemical diagnostics of fatigue. In monitoring training, attention should be focused on the accumulation of potassium in blood serum, hypoglycemia, reduction of blood pH, and the rate of recovery processes (e.g., postexercise dynamics of urea in blood).

CHAPTER

9

Assessing Changes in Adaptivity for Optimizing Training Strategies

Intensive and voluminous training in high-level athletes causes certain changes that are not directly related to the performance level but influence the effectiveness of training. It seems that something happens during the course of training that results in a training efficiency breakdown. Continuation of training leads to an overtraining state. This breaking point is frequently associated with reaching peak performance. Mostly, it appears after the achievement of high results, when hard training continues. Peak performance is also associated with increased susceptibility to various viral diseases. Thus, the hypothetical change in the body must be general and not limited to structures ensuring neuromuscular performance. The consequence of this change is, obviously, altered adaptational possibilities. This is referred to as the body's adaptivity.

In this chapter, we discuss training-induced adaptive changes. Attention is focused on saturation phenomenon, changes in immunoactivities, and possibilities for assessment of the adaptive pattern in training with hormonal responses in exercise. This approach makes it necessary to also analyze the peak performance state and overtraining in light of biochemical monitoring of these phenomena of sports training.

Changes in Adaptivity in Training

Exercises performed in training sessions or competitions trigger acute adaptation processes necessary to adjust body functions to the corresponding level of elevated energy metabolism. Adjustments are also necessary to avoid harmful alterations in the internal milieu of the body. In turn, all these adjustments enable exercise performance. Systematic repetitions of exercise induce long-term stable adaptation that is founded on structural and metabolic changes, making increased functional capacities possible. Accordingly, exercise training is founded on the body's capacity for adaptation.

The fundamental studies of Selye (1950) established that in chronic influences of strong factors (stressors), adaptation processes constitute the general adaptation syndrome. The stages of the syndrome are alarm, resistance, and exhaustion. To explain the pattern of the adaptation processes, Selye (1950) made a suggestion about adaptation energy. Adaptation energy expresses the ability of the body to acquire resistance to changes in an internal and external medium. Too much use of adaptation energy is the reason that the stage of resistance is followed by the stage of exhaustion. Selye defined the stage of exhaustion as representing the sum of all nonspecific systemic reactions, which ultimately develop as a result of prolonged overexposure to stimuli to which adaptation has been developed but could no longer be maintained. In the stage of exhaustion, the body becomes highly susceptible to harmful influences of stressors.

Soon after the publication of Selye's comprehensive monograph on stress (1950), Delanne (1952) and Mitolo (1951) argued that the general adaptation syndrome took place in exercise training. Increased training workload evokes the alarm response. After a number of repetitions of this workload, the alarm response disappears, meaning that the body achieved resistance to this exercise level. This new level of adaptation represents an improvement in general and/or specific physical fitness. Overtraining was considered an expression of the stage of exhaustion.

In 1959, Prokop pointed out the dependence of performance dynamics on exhaustion of adaptation abilities in athletes. According to him, an athlete has to exhaust a great part of his or her adaptation energy to reach top performance. Thereafter, a decrease in the performance level will follow. In this situation, continuing the training with high loads unduly magnifies the drop in performance, whereas reduced training helps to overcome the decrease in performance and ensures improvement (figure 9.1). Prokop (1959) did not have the opportunity to provide experimental evidence for his concept, which was founded on the analysis of experiences accumulated in the training practices of athletes. His work made an impact because he transferred the experience of practice into a biological concept using Selye's (1950) theory of the general adaptation syndrome.

Following Selye (1950), proper use of the term "adaptation energy" is necessary to understand the appearance of overtraining compared to extensive training. To avoid any search for energy in physical meaning, it seems better to speak about "adaptability" or "adaptivity" instead of "adaptation energy." Adaptivity can be defined as the ability of the body to adequately use the adaptation processes, ensuring

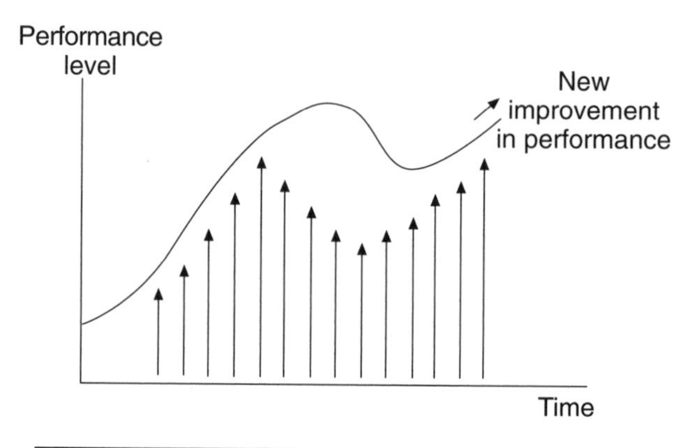

Figure 9.1 Two variants of the dynamics of performance level in high-level athletes.

- normal life activities despite changes within the body or in the external environment, and
- adaptive alterations in cellular structures and the amount of enzyme molecules aimed at achieving a stable adaptation (resistance) to the influence of chronically acting factors (Viru and Viru 1997a, 1997b).

The studies of Zhimkin and Korobkov (see Zhimkin 1968) showed that previously trained rats are more resistant to the influences of hypoxia, irritation, high or low environmental temperatures, and the action of various toxins compared with sedentary rats. However, with a forced training regime, the resistance of experimental animals to various stressors decreased to lower levels than it was in sedentary rats. Thus, training induced an increase in adaptivity, which disappeared when the training regime was too difficult. Obviously, in the latter case, training itself exhausted a part of the adaptivity.

In mice, swimming training in water 32° C increased the percentage of survived animals for

22 h at a temperature of 6 to 7° C, as well as after injection of formalin. Swimming training in water 18° C improved resistance to stressors when daily swimming duration was moderate (60% maximal) but was impaired at daily swimming for maximal duration. Thus, the combination of muscular activity with the influence of cold (swimming in 18° C) did not increase resistance to stressors any more and caused an obvious drop in adaptivity. The drop in adaptivity also influenced the training-induced increase in maximal swimming duration. Swimming duration improved in mice that swam at 32° C either with moderate or maximal exercise, as well as in mice that swam at 18° C with moderate exercise. In mice that swam at 18° C with maximal exercise, no significant improvement of swimming duration was found (Viru 1976a). These results support changes in adaptivity dependent on the muscular activity regime and environmental conditions.

Much data evidence the positive effect of exercising on health (see Bouchard et al. 1993). It has been suggested that the background for the health benefit of training is composed of elimination of bad consequences of hypokinesia, specific preventive effect of exercises in regard to certain pathophenomena, and changes in the body that simultaneously are essential for improved exercise performance and accomplishment of the general mechanism of adaptation (Viru and Smirnova 1995). In this way, increased adaptivity as a training outcome contributes significantly to the health benefit. The significance of increased adaptivity is emphasized by the fact that after a rest (transitory period of training) followed by a competition season, athletes are able to increase training workload to higher levels compared with the previous year.

Saturation Phenomenon

The saturation phenomenon is expressed by the loss of training effects after a certain total volume of training workload is done despite further increase of volume and intensity of exercises used.

Biochemical studies on rats indicated that a rapid increase in muscle glycogen and phosphocreatine in the early stage of training slows afterwards. Later, despite continued training, energy stores level off (Yakovlev 1977). Similar dynamics were found in the $\dot{V}O_2$max of men (Henriksson and Reitman 1977). In these cases, it is possible to suggest that the increase in exercise load (stimulus for improvement) was insufficient. However, a rat study of the dependence of an increase in muscle succinate dehydrogenase activity on training volume indicated a direct dependence of the enzyme activity increase on training volume only up to a certain volume of training (called saturation threshold). Further increase in the volume induces a trend toward a reduction in enzyme activity (figure 9.2). Strenuous training that used the interval method resulted in the saturation phenomenon at a lower total training volume than the continuous steady training method. However, the highest levels of enzyme activity obtained were almost the same (Volkov 1974, 1990).

In athletes, the saturation phenomenon was revealed by plotting the increase in oxygen supply of the body to the total time expended for training by the interval method in middle-distance runners. It was found that the appearance of the saturation phenomenon depends on the previous fitness level; the higher the performance level, the larger the training volume needed to induce saturation (figure 9.3) (Volkov 1974, 1990).

Actually, the saturation phenomenon has also been revealed in the results of a comprehensive

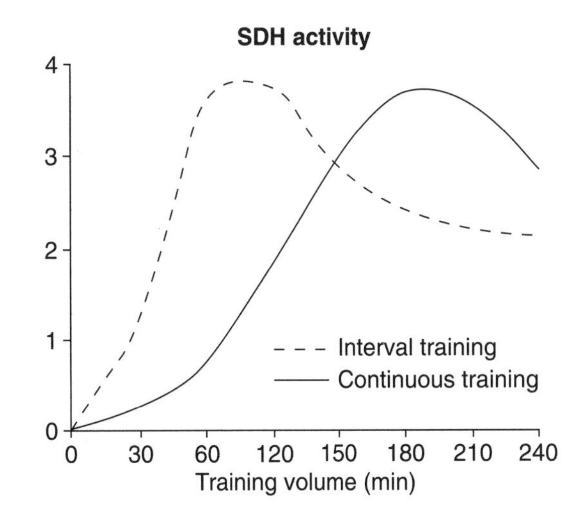

Figure 9.2 Changes in succinate dehydrogenase activity in the quadriceps muscle of rats.
Modified from Volkov 1974.

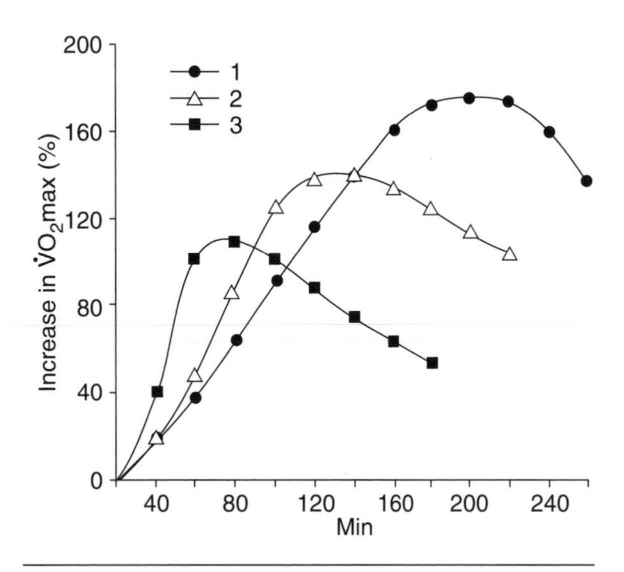

Figure 9.3 Increases in $\dot{V}O_2$max depending on the volume of interval training.

Modified from Volkov 1974.

study done by Dudley et al. (1982). In rats, when the running duration was 60 min, a further prolongation of exercise did not enhance an increase in cytochrome c concentration any more. Thus, a saturation response was found.

The following results seem to confirm the existence of the saturation phenomenon in athletes. It has been found that swimming training, 3 to 4 h per day, 6 days eper week, does not provide any greater benefits than when training is limited to only 1 to 1.5 h per day. In more difficult training, a significant decrease in muscle strength and sprint swimming performance was observed (Costill et al. 1992). The swimmers who trained twice per day for 6 weeks did not show any additional improvement over those who trained only once per day. The benefits were found neither immediately after the 6-week period nor during 14 weeks of later training. To determine the influence of long-term excessive training, performance improvements in swimmers who trained twice daily for a total distance of more than 10,000 m per day were compared with improvements in those swimmers who swam approximately half that distance in a single session each day. Changes in performance time for the 100-yd front crawl were identical in both groups over a 4-year period (Costill et al. 1992).

Intensive swimming training (daily swimming distance was increased from 4.266 to 8.790 m, during 10 successive days, swimming intensity was at 94% of $\dot{V}O_2$max) did not cause a progressive change in swimming power, sprinting and endurance performance, aerobic capacity, and deltoid citrate synthase activity in eight highly trained male swimmers. Four swimmers were unable to tolerate this training. Together with decreased swimming performance, the four swimmers had significantly reduced muscle glycogen values (Costill et al. 1988).

It is plausible that the appearance of the saturation phenomenon is related to a temporary exhaustion of the body's adaptivity.

Cyclic Training Design

An investigation of the pattern of actual training loads in high-level athletes showed undulating alterations in the intensity and volume of the exercises used. Three groups of waves of training workloads were distinguished in qualified athletes. Duration of the waves of the first group was approximately a week (training microcycles, see chapter 8, p. 180). The waves of the second group summed up three to six waves of the first group, respective microcycles. The secondary waves meant that the microcycles are not repeated in the same way, but secondary waves included microcycles with both increased and reduced total workload. A group of microcycles with increasing workload was usually found near a microcycle of reduced workload; this constituted a training mesocycle. The total workload of mesocycles increased from the beginning of the preparatory period and throughout this period and most of the competition period. Reduced training followed the competition season, constituting the transitory period and terminating the training year. Thus, a macrocycle has been formed with the duration of a year or half a year. Against this background, Matveyev (1964) formulated the principle of cyclic organization of training. In several issues on training methods, we can find suggestions related to actualization of the principle of cyclic organization of training (Counsilman 1968; Harre 1973; Thayer 1983; Verkhoshanski 1985; Platonov 1986; Kantola 1989; Rowbottom 2000).

The cyclic organization of training is indispensably connected to the necessity for relief time. It is possible to find sessions of maintaining loads or sessions of restitution loads between trainable sessions during training microcycles. Restitution sessions and rest day(s) are, obviously, necessary for the recovery processes and for complete elimination of manifestations of fatigue. However, what is the benefit of training sessions with maintaining loads that are insufficient to stimulate the development of training? Possibly, they are necessary to provide a rest for the molecular mechanism, ensuring training effects.

Sports practice showed that relief times are also necessary in the training mesocycle. After one or two (in exclusive cases three) trainable microcycles, particularly after a blow microcycle, a relief microcycle is necessary. It is also necessary after high-strain competition. Reduced training has been used between two or three important and, therefore, stressful competitions. Moscow's ice-hockey players used three to six restitution microcycles after the first important competition. During the following three to six microcycles, the training workload increased gradually to achieve the same level of training intensity that existed before the first competition (Klimov and Koloskov 1982). Accordingly, in female rowers on the Dutch national team after a training camp and the Olympic qualification race, a decrease was found in training intensity, with a following increase in the training workload (Vermulst et al. 1991).

A typical manifestation of reduced training verified by athletes' experiences is the transition or relief period after the competition period.

Why are reduced training stages necessary? Of course, several functions and body resources may remain incompletely recovered in a high-strain training regime. Possibly, among them are those that determine the body's adaptivity. Accordingly, it seems to be justifiable to suggest that the reduced training stages avoid the untimely exhaustion of the body's adaptivity.

The use of reduced training has to account for the possibility of a detraining effect. After cessation of exercising, results of the previous training disappear within a certain time. In most cases, detraining manifestations occur in exercise performance within weeks (see Neufer 1989). However, several articles reported that training-induced adaptation was maintained for several weeks when training was maintained at a reduced level (e.g., Hickson et al. 1985; Neufer et al. 1987; Neufer 1989; Houmard et al. 1990a). In jumpers, concentrated strength training for 4 weeks caused a pronounced decrease in the results of strength tests. Then the direction of training was changed, and strength exercises were not used. During the following two or three microcycles, muscle strength increased to a higher level than it had been before the strength training (Verkhoshanski 1985). Similarly, Jeukendrup et al. (1992) found impairment of cycling velocity in cyclists after weeks of increased training loads, but a pronounced improvement occurred when 2 weeks of reduced training followed.

To compete at their peak, many athletes reduce their training intensity before a major competition to give their bodies and their minds a break from the rigors of intense training. The taper period, during which training intensity is reduced, should provide adequate time for healing of tissue damage caused by intense training and time for the body's energy reserve to be fully replenished (Wilmore and Costill 1994, p. 308). However, it is justifiable to ask whether the taper period also favors the restoration of the athlete's adaptivity.

In conclusion, exercise training induced changes in the body, which increased the possibilities of the mechanism of general adaptation. The result is enhanced adaptivity. At the same time, training of athletes with a high load requires use of the body's adaptivity to a great extent. Because of this, achieving top performance is related to a drop in the body's adaptivity. Therefore, the organization of training has to provide for restoration of the body's adaptivity. Tools for accomplishing this goal are as follows:

- Inclusion of maintenance and restitution workloads (with trainable workloads in microcycles) into training sessions
- Use of restitution microcycles after high-volume/high-intensity microcycles, after blocks of concentrated unidirectional training, and after competitions
- Following a competition period with a transition period

Theoretical Considerations

It was indicated earlier that training results in two opposite outcomes. On the one hand, training seems to increase the body's adaptivity. On the other hand, an intensive training regime may exhaust adaptivity. Figure 9.4 shows a multiyear dynamic of possibilities for the generation of high power output for short efforts in a jumper (Verkhoshanski and Viru 1990). The main idea that at the end of each training year, progress stops. During the following transition period (detraining/reduced training), power output declined. However, during the transition periods, the bodies of athletes restored their opportunities to respond to training with a further increase in power output. In this way, from year to year, the performance level of an athlete improves.

In chapter 7 (p. 163), a gradual increase of $\dot{V}O_2$max throughout many years of training was documented. Consequently, before the beginning of the next training year, the body's susceptibility to training influences is restored, and athletes possess the opportunity to increase motor capacities. Thus, exhaustion of the body's adaptivity during a training year is a temporary phenomenon. After a rest period, adaptivity reaches a new higher level.

It is justifiable to suggest that the increase of adaptivity is founded on improvement of the possibilities of the mechanism of general adaptation (Viru 1995). By analyzing the health promotion effect of exercise training, we indicated that training-induced changes are simultaneously essential for increased performance and for effectiveness of the mechanism of general adaptation (Viru and Smirnova 1995).

The reason for transitory exhaustion of the body' s adaptivity in training is not clear. One possibility is that a change occurs at the level of gene induction and/or expression that allows us to suggest the appearance of fatigue in the cellular genetic apparatus (Viru 1995). It is not excluded that a regulatory mechanism exists that eliminates uniform adaptive protein synthesis after stimulation for a while. An alternative possibility is that the disappearance of training efficiency is related to the exhaustion processes in the central nervous or the endocrine systems. If this suggestion is correct, the alteration in endocrine function may indicate possibilities for monitoring adaptivity in training. Because the temporary loss of adaptivity is associated with increased susceptibility of athletes for pathogenetic agents, another opportunity for monitoring adaptivity should be provided by alterations in immune activities.

Hormonal and Metabolic Changes During a Training Year

The hypothesis of an expression of the temporary loss of the body's adaptivity in endocrine function presupposes hormonal changes during a training year. Essential information may provide not only altered basal levels of hormones in the blood, but also exercise-induced hormonal responses. This section will analyze hormonal changes in athletes during stressful training. In

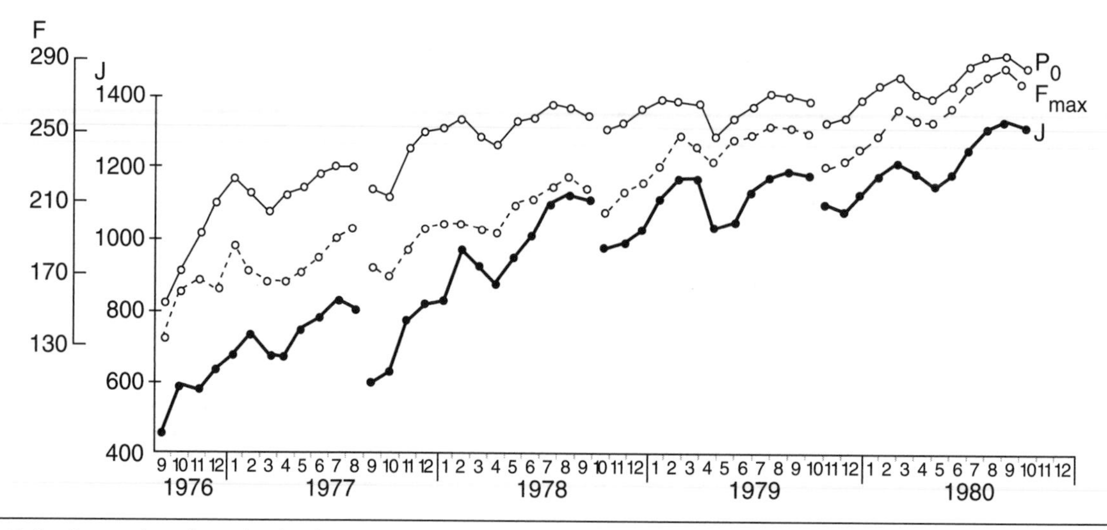

Figure 9.4 Dynamics of maximal voluntary strength, maximal explosive strength, and velocity of force development in a highly qualified triple jumper.

From Verkhoshanski and Viru 1990.

the first part of this section, hormonal changes induced by short-term stages of increased training intensity and duration are discussed. In the second part, the discussion pertains to the pattern of the basal level of hormones and exercise-induced responses during prolonged training, particularly during the competition period.

Effects of Short-Term Intensive or Voluminous Training Stage

Several studies show that the basal (morning) level of cortisol increases in response to a pronounced rise in the intensity or volume of training workload (tables 9.1 to 9.5). The results obtained are variable. Approximately half of the articles did not report any significant change in cortisol basal level. A reason may be that the increase of training intensity or duration was insufficient to evoke a strain in adaptation processes and thereby increased the cortisol level. Another possibility is that in some athletes, the cortisol level increased and in others it decreased in relationship to the individual strain in training. Actually, the second group of articles indicated possibilities for both an increased and decreased cortisol level. The changes in basal testosterone level were also variable, whereas in some articles the direction of changes was parallel, in others it was opposite. In a limited number of articles, hormone changes were assessed together with alterations in performance. An increase in performance as a result of a hard training stage might be associated with either an increased, decreased, or unchanged cortisol level. Also, the ratio of testosterone to cortisol does not give a clear-cut relationship to performance. The results of Urhausen et al. (1987), Vervoorn et al. (1991), and Hoogeveen and Zonderland (1996) are important according to which decreased ratio of testosterone to cortisol may coincide with increased performance. Thus, an increase of this ratio does not necessarily provide evidence of the loss of performance capacity.

Accordingly, Urhausen et al. (1995) affirm that the behavior of (free) testosterone and cortisol is a physiological indicator of the training influence but does not necessarily indicate overtraining.

What is the meaning of a decreased basal cortisol level? It may be related to dysfunction of the pituitary-adrenocortical system, as it has been suggested in regard to the overtraining state (see p. 214 of this chapter). However, in endurance athletes, the cortisol level tends to be decreased (Bosco and Viru 1998). Therefore, a low basal cortisol level after a hard stage of endurance training may be related to general adjustment to prolonged exercises.

Banfi et al. (1993) studied elite speedskaters from the end of June to the beginning of January (the preparatory period and the onset of the competition period). After altitude training, the cortisol level decreased significantly in men and women. At the same time, testosterone concentration increased in men but not in women. Another temporary decrease of blood cortisol level appeared at the end of the preparatory period (end of November) in association with a trend to increased testosterone concentrations. The authors considered the cortisol decrease a reflection of adaptation. An alternative explanation is that training altered the interrelations between corticotropin and lutropin secretion. Significance might have also altered secretions of related hypothalamic neurohormones. A question remains whether the shift in hormonal interrelations is caused by adaptation or dysadaptation. Mackinnon et al. (1997) did not find differences in concentrations of plasma norepinephrine, cortisol, and testosterone between overreached (decrements in swim performance and persistent high ratings of fatigue) and well-adapted swimmers during 4 weeks of training with progressively increasing volume. Only lower urinary excretion of norepinephrine in overreached swimmers distinguished the two groups.

All in all, instead of a basal hormone level, exercise-induced responses may provide better possibilities for assessment of the changes in adaptivity.

In endurance athletes, hard training stages caused increased cortisol levels before and after the test exercise (Kirwan et al. 1988a; Snegovskaya and Viru 1992; Hoogeveen and Zonderland 1996). In qualified rowers, parallel increases of cortisol levels were found during the competition period (Snegovskaya and Viru 1993b). However, variability also exists in cortisol responses to exercise.

Apart from the result obtained in endurance athletes, the cortisol level tends to decrease in most cases during strenuous strength or power training.

Table 9.1
Effects of Hard Endurance Training on Cortisol and Testosterone Levels in Male Rowers

		Cortisol		Testosterone		
Reference	**Brief remarks on the study**	**Basal**	**Response in exercise**	**Basal**	**Response in exercise**	**Performance**
	Rowers					
Urhausen et al. 1987	Rowers were studied during the competition period. T decreased particularly in weeks of competition.	=		–		
Vervoorn et al. 1991	Members of Dutch national team were studied during a 9-mo period before Olympic Games. Study included monitoring of a 2-wk intense training period.	+		–		+
Snegovskaya and Viru 1992	A national-level crew of an eight-oar bout was studied during 6 consecutive micro-cycles (3 training d + 1 rest d). Exercise testing was on the 2nd d of each microcycle. The 5th microcycle was a high-volume microcycle.	=	=			+
Snegovskaya and Viru 1993b	National-level rowers. Repeated testing with the aid of 7-min all-out rowing tests during the competition period.	+	+			+
Steinacker et al. 1993	Juniors. 16 d of voluminous training at the lactate threshold followed by 10 d of taper. Performance level was evaluated by power output as well as by individual anaerobic threshold.	=		=		
Steinacker et al. 1999	Juniors. 5-wk training camp before the World championship. The highest workload on the 2nd and 3rd wk. Afterwards, taper. The highest workload at an early stage during the training camp. Later, the load reduced.	+				

d=day; wk=week; mo=month; +=increase; ==no change; –=decrease; T–testosterone.
Empty cells denote the lack of information in the paper.

Table 9.2
Effects of Hard Endurance Training on Cortisol and Testosterone Levels in Male Swimmers

		Cortisol		Testosterone		
Reference	**Brief remarks on the study**	**Basal**	**Response in exercise**	**Basal**	**Response in exercise**	**Performance**
	Swimmers					
Kirwan et al. 1988a	11d of intensive training. Collegiate swimmers. Cortisol response was recorded in front crawl for 365.8 m at 95% $\dot{V}O_2max$	+	+	+		
Häkkinen et al. 1989	Elite swimmers. A training year was monitored. First intense training stage was studied.	+		=		
Flynn et al. 1994	21-wk period before championship. Measurements before and after a wk of intense training as well as before and after taper at the end of the study period.	–		–		–
Mujika et al.1996a	A taper stage was studied in the competition period (between two competitions). Swimming velocity increased in part of swimmers together with a moderate decrease in cortisol. In other swimmers performance did not change, cortisol increased.	– +				+ =
Mujika et al.1996b	The pool contingent of the previous study.	=				=
Mackinnon et al. 1997	Elite swimmers progressively increased the volume training up to high fatigue ratings.	=		=		–

d=day; wk=week; mo=month; +=increase; ==no change; –=decrease.
Empty cells denote the lack of information in the paper.

Two-week tapering for the National Championship of Australia caused interindividual variability in changes of cortisol, testosterone, and catecholamine basal levels. Multiregressional analysis on the results obtained showed that the performance change with tapering was predicted by reduction of the norepinephrine level (Hooper et al. 1999).

Reduced training after a stay of highly intensive endurance training normalized the cortisol basal level when it was increased in the strenuous training stage (Häkkinen et al. 1989, Tabata et al. 1989). After a hard regime of heavy resistance training, a low cortisol level persisted for several weeks of detraining (Häkkinen et al. 1985). During reduced training that followed the

Table 9.3
Effects of Hard Endurance Training on Cortisol and Testosterone Levels in Male Distance Runners, Orientieers and Cross-Country Skiers, and Cyclists

		Cortisol		Testosterone		
Reference	**Brief remarks on the study**	**Basal**	**Response in exercise**	**Basal**	**Response in exercise**	**Performance**
	Runners					
Verde et al. 1992	In distance runners a 38% increment in training for for 3 wk. Sustained fatigue and decreased vigor was indicated by POMS without clinical picture of overtraining. The increase of cortisol level, normally found during 30 min of submaximal exercise, disappeared.	=	–			=
Flynn et al. 1994	Collegiate runners. 10-wk period before championship. The stage of hard training was for 3 wk. Performance was assessed with the aid of running time to exhaustion at 110% $\dot{V}O_2$max.	=		=		=
Fry et al. 1992	A 10-d stage of interval training twice a day.	–		=		
	Orienteers and Cross-Country Skiers					
Tsai et al. 1991	Elite orienteers and cross-country skiers were studied during the competition period. Cortisol decreased at the onset of the competition period; afterwards, no change.	–		+		
	Cyclists					
Hoogeveen and Zonderland 1996	Professional cyclists. 10 d hard training in January (6 h a day). Blood sampling before and after an incremental exercise. Performance was tested with the aid of maximal power output.	+	+	–	=	+

d=day; wk=week; mo=month; +=increase; ==no change; –=decrease.
Empty cells denote the lack of information in the paper.

Table 9.4
Effects of Hard Endurance Training on Cortisol and Testosterone Levels in Female Rowers and Orienteers and Cross-Country Skiers

		Cortisol		Testosterone		
Reference	**Brief remarks on the study**	**Basal**	**Response in exercise**	**Basal**	**Response in exercise**	**Performance**
	Rowers					
Urhausen et al. 1987	Rowers were studied during the competition period. A trend for testosterone decrease was throughout the competition period.	=		–		
Vervoorn et al. 1991	Members of Dutch National rowing team during a 9-mo period before Olympic Games. Testosterone concentration decreased gradually; maximal power and output increased slightly. An intense training stage temporarily increased T and decreased C.	– or +		–		+
	Orienteers and Cross-Country Skiers					
Tsai et al. 1991	Orienteers and cross-country skiers. Cortisol increased at the onset of the competition season and during the season.	+		=		

d=day; wk=week; mo=month; +=increase; ==no change; –=decrease; T=testisterone; C=cortisol.
Empty cells denote the lack of information in the paper.

hard training stage, the testosterone basal level normalized or did not change (Häkkinen et al. 1985, 1989; Urhausen et al.1987; Steinacker et al. 1993; Mujika et al. 1996). Reduced training was accompanied by a decline in urea level and creatine kinase activity (Steinacker et al. 1993). In power athletes, 2 weeks of detraining led to an increase of plasma growth hormone and testosterone, whereas plasma cortisol and creatine kinase levels decreased (Hortobagyi et al. 1993). In distance runners, 3 weeks of reduced training after normal baseline training did not cause any changes in testosterone and cortisol levels in combination with further improvement of performance (Houmard et al. 1990).

Earlier, the results obtained in male and female runners during the high-volume training stage (volume of a training session 105% to 120% of the usual for each athlete) were mentioned (chapter 8, p. 183, figure 8.3). Individual analysis of these data showed four variants of hormonal changes.

1. A moderate increase of blood cortisol and growth hormone levels during training sessions without any change in basal hormone levels
2. An elevation of cortisol level in the resting state together with great rises of cortisol and growth hormone concentration during the training session
3. A pronounced elevation of cortisol resting level in association with a decrease of cortisol concentration and a sharp rise of

Table 9.5
Effects of Hard Resistance/Power Training on Cortisol and Testosterone Levels in Male Athletes

		Cortisol		Testosterone		
Reference	**Brief remarks on the study**	**Basal**	**Response in exercise**	**Basal**	**Response in exercise**	**Performance**
Häkkinen et al. 1985	Elite weightlifters were studied during 24 wk, followed by detraining. Group A—heavy resistance training Group B—moderate strength and jumping training	– – –		– + =		+ + +
Häkkinen et al. 1987	Elite weightlifters were studied by 2-wk stages: stressful training, "normal" training, reduced training.	=		–		
Busso et al. 1990	Elite weightlifters were monitored 51 wk. Three times intensive training periods were followed by taper.	–		+		
Busso et al. 1992	Elite weightlifters were studied during a 6-wk period before a primary competition. Four wk of this period was a stage of intense training and last 2 wk were a taper.			–		
Fry et al. 1994	Male junior weightlifters (17.6 ± 0.3 years) were tested before and after 1 wk of increased training volumes. The overreaching wk caused attenuated exercise-induced T increase in the 1st year. A year later, T increase did not change.	1st year + 2nd year +	= =	– =	=	
Fry et al. 1998	Weight-trained men. Daily training at 100% 1-RM intensity for 2 wk. Hormones were determined before and after a test exercise.	=	–	+	+	–

d=day; wk=week; mo=month; +=increase; ==no change; –=decrease; T=testisterone.
Empty cells denote the lack of information in the paper.

growth hormone concentration during the training session

4. A high basal level of cortisol together with low levels of cortisol and growth hormone after training sessions

The first variant was found in most cases on the first and second days, whereas the fourth variant appeared in few cases at the end of a 2-week period. Meanwhile, either the second or the third variant appeared during training days.

Suppression of adrenocortical activity during heavy training over a 3-week period was indicated by the fact that a 30-min submaximal exercise, which usually induced an increase of blood cortisol level, failed to cause a change after heavy training (Verde et al. 1992). We found a reversed cortisol response to a heavy incremental treadmill running test after 3 weeks of altitude training in elite skiers in combination with feelings of fatigue, decreased performance, low testosterone levels, and suppressed growth hormone responses (figure 9.5). After 2 weeks of reduced training, fatigue disappeared, performance normalized, and hormone responses to the treadmill test were again pronounced (Viru M. et al., 2000a).

Lehmann et al. (1991) reported decreased excretion of cortisol in middle-distance and long-distance runners during the heavy training stage.

In conclusion, training stages, characterized by pronounced increases in volume or intensity of training exercises, may induce variable changes in cortisol and testosterone concentrations. Obviously, diverse changes in the hormone basal level are related to individual strain on the body caused by altered training. However, no strict evidence exists that increased cortisol and decreased testosterone levels are necessarily associated with loss of performance capacity. Compared with basal hormone levels, exercise-induced cortisol responses are more informative. The influence of the hard-training stage is mostly reflected by a high cortisol level in the test exercises when performance capacity is enhanced at the same time. In the case of further hard training, the cortisol response in the training session or test exercises becomes suppressed or reversed.

Increased volume or intensity of strength and power exercises usually induces a decreased cortisol level in conjunction with variability of testosterone changes (mostly testosterone level increases). Reduced training tends to normalize hormonal levels.

Monitoring Hormones During a Training Year

In elite rowers, plasma cortisol and growth hormone basal levels and their responses to a 7-min all-out test on a rowing apparatus were recorded three times in the preparatory period and competition period. Increased performance (power output during rowing test) was common during the period of observation together with an elevated basal level of cortisol and a gradual

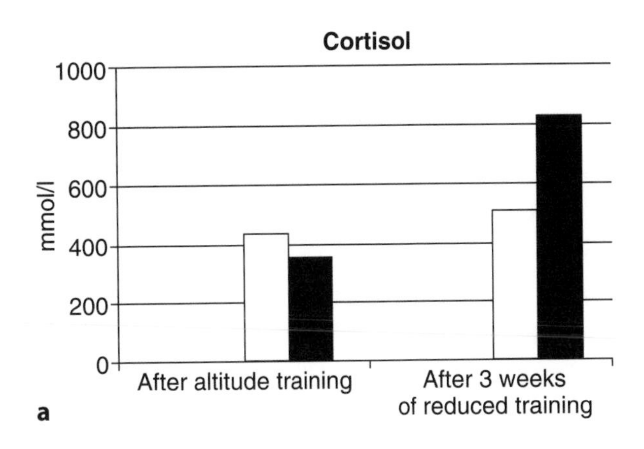

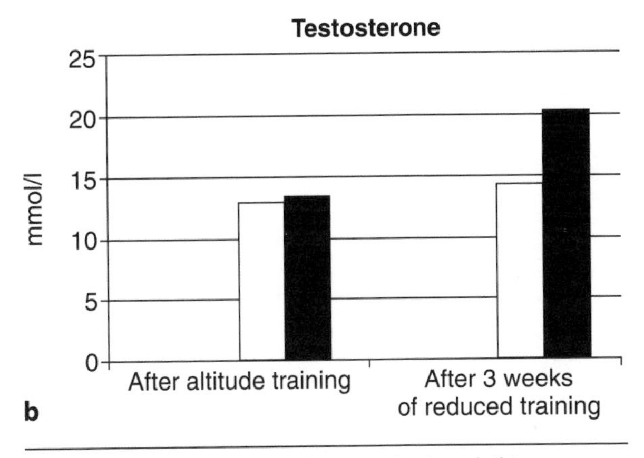

Figure 9.5 Responses of *(a)* cortisol and *(b)* testosterone in an incremental treadmill test after altitude training and after 3 weeks of reduced training.

rise of cortisol and growth hormone responses to the rowing test (figure 9.6) (Snegovskaya and Viru 1993b). In runners, the blood cortisol and somatotropin responses to training exercises increased from June to July (Kostina et al. 1986). In swimmers, the cortisol concentration varied during the competition period on higher levels than during the preparatory period (Port and Viru 1987). The same was found in female but not in male orienteers (Tsai et al. 1991). Häkkinen et al. (1987, 1989a) found an elevation of the basal cortisol level for the competition season in swimmers but not in strength and power athletes (Häkkinen et al. 1987, 1989). Testosterone levels remained constant (Häkkinen et al. 1987, 1989).

Two-year monitoring in elite weightlifters showed gradual increases of testosterone, lutropins, and follitropin blood levels, whereas cortisol remained constant (Häkkinen et al. 1988c). A group of rowers was studied in two subsequent years. When they reached the national

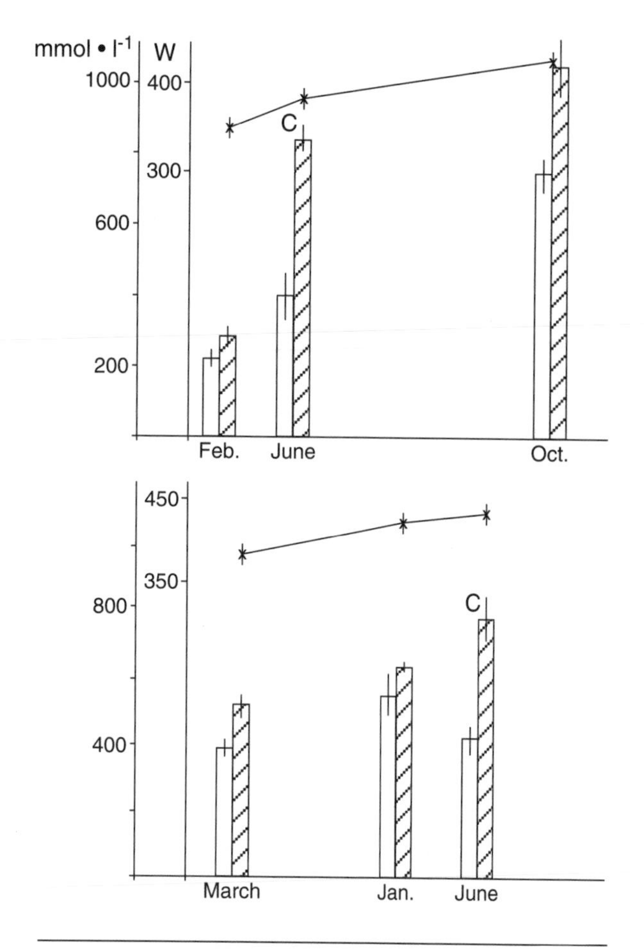

Figure 9.6 Power output and cortisol responses in 7-min rowing test performed at the highest possible rate or in a 2000-m competition race in two groups of qualified rowers.

Reprinted from V. Snegovskaya and A. Viru 1993b.

level, the increase in performance was associated with increased cortisol responses to a 7-min all-out rowing test and 40 min of rowing at the anaerobic threshold. The growth hormone response increased in the 7-min test (Snegovskaya and Viru 1993a). A 9-month monitoring was performed with Dutch rowers (Vervoorn et al. 1991, 1992). Episodes of increased cortisol and decreased testosterone concentrations were found in conjunction with temporary rises in training volume or intensity. The results obtained by Flynn et al. (1994) in college athletes were similar. In runners, cortisol and testosterone basal levels remained constant independent of changes in training intensity or volume, as well as during the precompetition taper. In swimmers, an increase of training intensity was associated with an insignificant reduction of cortisol and a significant decrease of testosterone. Afterwards, the average cortisol concentration returned to the initial level. Precompetition taper increased the testosterone level. The free testosterone concentrations changed in a manner similar to total testosterone concentration, but the changes in free testosterone were more pronounced.

Thus, two types of results were obtained. One type demonstrated increased cortisol levels before and after test exercises in association with reaching top performance. Another type showed no relation of hormonal changes with increases in training workload. Reaching peak performance is associated with a pronounced rise in training intensity. On the other hand, the effect of high-intensity training might remain masked because of the effect on precompetition tapering. The results of the training studies can only be compared if the time points for hormone determinations were similar, keeping in mind the training characteristics and performance dynamic.

Another important point in the testing of endocrine function is that responses to exercises are more essential than the basal level. We analyzed the databank of own laboratory (results obtained in 122 endurance-trained athletes, 20 athletes of acute events, and 115 untrained persons). Statistical analysis of the material and comparison of hormonal changes with metabolic indexes, $\dot{V}O_2max$, and performance characteristics allowed us to conclude that first, for evaluation of the athlete's status, exercise-induced changes of cortisol, growth hormone, and testosterone are most informative, and second, the changes of these hormones have to be recorded at least before and after a test exercise, requiring a high degree of mobilization of the endocrine system for adaptation to strenuous exercises (Viru et al. 1999).

A Finnish treadmill test of skiers was adjusted for hormonal studies. Longitudinal studies for several years were performed in elite cross-country skiers, including athletes at the international level. The individual analysis of the data confirmed the prolonged aftereffect of the altitude training. Nine to 14 days after return from the medium-altitude training camp, the basal levels of cortisol and testosterone were increasing in the athletes, but the cortisol responses to the test exercise were reduced or suppressed. Growth hormone responses seemed exaggerated in some athletes, and in others, suppressed. Testosterone basal level was increased in most athletes. Before the onset of the competition period, an increase of cortisol basal level and response to exercise appeared. However, at the end of the competition period, reversed cortisol responses from the high basal level appeared with low testosterone levels and suppressed growth hormone responses to the test exercise (Viru M. et al. 2000a).

Figure 9.7 reproduces the result of monitoring training in two international levels of female skiers over several years (Viru M. et al. 2000). The following list provides background information about the testing:

- An inversed cortisol response appeared in September 1995, 9 days after altitude training camp.
- An inversed or blunted cortisol response appeared in May 1996 after an exhaustive competition season.
- An exaggerated cortisol response appeared in October 1997 in skier 1, but there was an inhibited response in skier 2, despite high levels of VO_2 in both skiers. After an intensive and voluminous training stage for skier 2, reduced training was recommended before the final stage for the Winter Olympic Games. She achieved the pace among the first 20 in all distances (her first success at the international level). A serious trauma prevented skier 1 from competing.
- At the end of the competition period (April 1998), both skiers exhibited pronounced cortisol responses, suggesting good adaptivity. In skier 2, the exhaustion of adaptivity was avoided by a timed change in training workload; in skier 1, the exhaustion of adaptivity was avoided by a constrained rest caused by trauma.
- Pronounced cortisol responses appeared at the end of the next competition season (April 1999).
- At the beginning of the preparation period (June 1999), the cortisol response was modest in skier 1 and inverse.' in skier 2.
- After a stage of intensive and voluminous training (October 1999), both skiers showed a pronounced cortisol response.
- Patterns of cortisol responses and $\dot{V}O_2max$ were different.

In conclusion, the results suggest that cortisol, testosterone, and growth hormone levels and responses to the test exercise are indicative of the resources of adaptivity. The high cortisol basal level and exaggerated cortisol responses suggest the high activity of the adaptation process, whereas the suppression of cortisol and growth hormone responses, and particularly the appearance of the reversed cortisol response, are indications of significant reductions in the reserves for adaptive.

Are there possibilities to use metabolic parameters for long-term monitoring of metabolic situations in the body? Figure 9.8 shows urea dynamics in blood serum in top-level rowers. An increase of urea in morning blood samples is possible in various stages of training and even over the clinically relevant urea level of 8.3 mmol/L

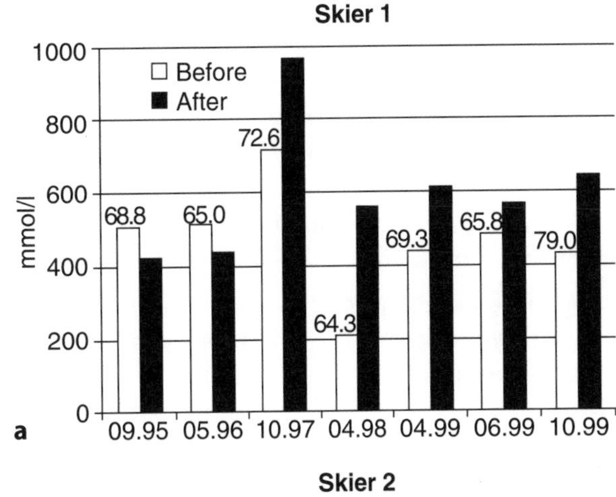

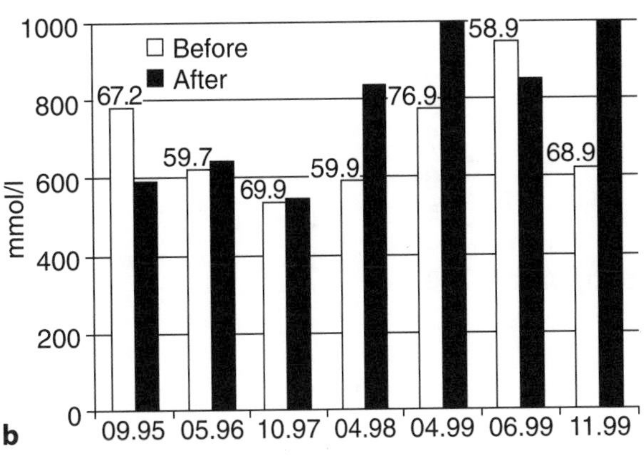

Figure 9.7 Blood cortisol concentrations before and after exhaustive treadmill running tests in two female skiers. The numbers above the columns indicate $\dot{V}O_2max$.

(Kindermann 1986). However, the situation has to be considered critical when the high urea level persists for several weeks. In the same way, results of the repeated assessment of plasma enzyme activities have to be evaluated. If there are methodological possibilities, the ratio between activities of oxidant and antioxidant systems may also provide promising indications.

Altering Immune Activities During a Training Year

Several studies (for a review, see Roberts 1986; Shephard 1986; Mackinnon 1992; Nieman 1997a, 2000) indicate an elevated susceptibility of athletes to various viral diseases (hepatitis, skin and upper respiratory tract infections, and other miscellaneous infections). Various reasons may favor the admission of viral pathogens (Nieman, 2000).

- Athletes perform in environments in which certain pathogenic microorganisms are particularly widespread.
- Athletes are prone to abrasions or other tissue injury, allowing the transfer of microbiological agents.
- Athletes are vulnerable to cross-infection from others with whom they are in close contact.
- Athletes are exposed to alien environmental pathogens during foreign travel and lack specific immunity.

In addition, there is a potential for immunosuppression from psychological and physiological strain that can arise during the periods of hard training and competition (Nieman and Nehlsen-Cannarella 1991; Nieman 1997). A J-shaped relation exists between various exercises performed in training and the risk of upper respiratory tract infection; in moderate training volume, the risk is lower than in sedentary persons. In athletes, the risk is highest when the training volume is great (Nieman 1997a).

Increased risk of viral diseases has been found mainly in athletes of prolonged endurance events. In those athletes, episodes of upper respiratory tract infections appear to be related to intensive preparation for competitions and the influence of the competition itself. Thus, episodes of upper respiratory tract infections were found mostly before and after marathon races (see Nieman 2000).

Studies on athletes showed a diversity of changes of various indexes of immune activities in training. The activity of natural killer cells has been found in marathon runners to be more than 57% greater than in sedentary persons (Nieman et al. 1995). This result is in accordance with data obtained previously by Pedersen et al. (1989). In junior cyclists, natural killer cell activity was increased in summer in association with the competition period compared with winter when training workloads were less (Tvede et al. 1991). In professional football play-

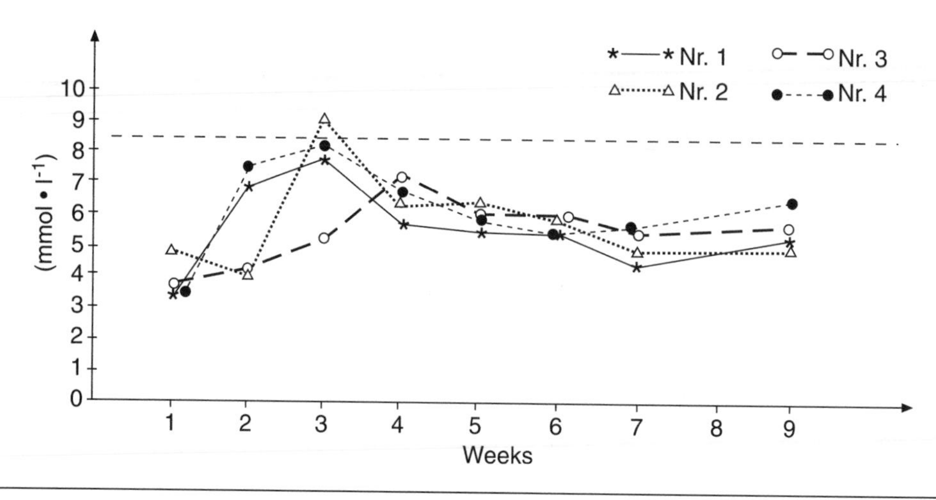

Figure 9.8 Urea concentrations in blood (the morning level) of four top-level rowers during nine weeks of the training stage in the competition period. The interrupted horizontal line indicates the clinically accepted highest urea level.
Reprinted from W. Kindermann 1986.

ers, training and competitions did not induce changes in the number of natural killer cells or in the cytotoxic activity of natural killer cells (Bury et al. 1998).

Assessment of neutrophil function by phagocytosis and the release of immunomodulatory cytokines failed to show increased activity in endurance athletes (see Nieman 2000). However, when endurance athletes were studied at different training periods, neutrophil function was suppressed in high-intensity training, but it was similar to sedentary people in the low training period (Hack et al. 1994). In junior cyclists stimulated chemiluminescence of neutrophils diminished for the end of a training and competition season (Baj et al. 1994). Significant suppression of neutrophil function was also found in professional soccer players during the training and competition season (Bury et al. 1998).

In swimmers, neutrophil oxidative activity was significantly lower before national-level competition compared with age- and sex-matched sedentary persons (Pyne et al. 1995).

Lymphocyte function was assessed with the aid of determination of the proliferative response of human lymphocytes to stimulation with various mitogens in vitro. Mostly, no significant difference was established between athletes and nonathletes (see Nieman 2000), as well as junior cyclists studied at the beginning of a training season and after 6 months of intensive training and the racing season. Soccer players exhibited a moderate but significant suppression of lymphocyte proliferation at the end of a competitive season (Bury et al. 1998). In cyclists, the training and competition season was associated with a decrease in absolute numbers of $CD3^+$ and $CD4^+$ cells and diminished interleukin-2 generation (Baj et al. 1994). Tvede et al. (1989) did not find any difference when junior cyclists and nonathletes were compared during low- and high-training periods.

Exercise training may enhance specific antibody formation in response to immunogenic challenge, even after marathon-race runners exhibited a slightly higher titer of serum immunoglobulin (Ig) specific to injected tetanus toxin than nonathletic persons (Eskola et al. 1978). However, another picture has been found in top-level athletes during the competition period. In most athletes, the concentration of IgA, IgG, and IgM decreases in blood and saliva during the competition period together with suppression of the formation of serum antibodies to tetanus, diphtheria, staphylococcus, and typhus (Wit 1984; Chogovadze et al. 1988; Levando et al. 1988, Pershin et al. 1988). Levando et al. (1990) reported that during the competition period, a decrease in Ig and suppression of specific antibody formation took place mainly in high-level athletes falling ill during the preparation for competitions. A study of 250 athletes allowed the authors to distinguish four stages of immune activity (activation, compensation, decompensation, and recovery) during the training year (figure 9.9) (Pershin et al. 1988). Thus, the suppression of immunoactivity found in highly qualified athletes in association with a high performance level is a transitory phenomenon. Probably this phenomenon is related to a partial exhaustion of adaptivity when athletes use high training workloads to achieve top performance.

Gotovtseva et al. (1998) found decreased levels of IgG and IgM but not of IgA in cross-country skiers after 4 weeks of intensive training. The Ig response was in conjunction with decreased lymphocyte proliferation and drops in cytokine production (interferon α and γ). In long-distance runners interferon γ and interleukin-2 production,

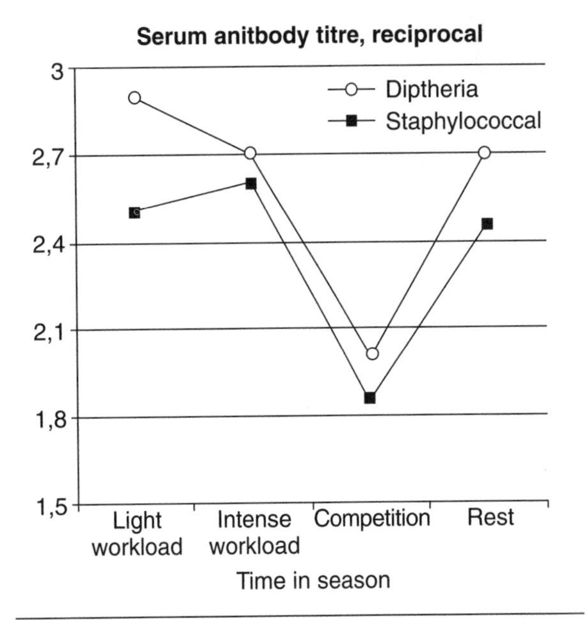

Figure 9.9 Serum antibodies specific to diptheria and staphylococcus after the administration of specific antiviral serums.

Adapted from B.B. Pershin et al. 1988.

adjusted to lymphocyte number, decreased with heavy training and more with competition.

The low level of salivary IgA in elite cross-country skiers, found by Tomasi et al. (1982), confirms that described changes in immune activity can be detected by saliva analysis. Tharp and Barnes (1990) studied university swimmers over a 4-month season. They found a decrease in the resting level of salivary IgA with an increase in training intensity.

In conclusion, highly intensive training and competition may suppress immune activities in athletes. These are reflected in neutrophil functions, levels of Ig (including salivary Ig), and the formation of cytokines and specific antibodies. An analysis of training-induced changes in various immune indexes led to the conclusion that the most promising immunological marker of excessive training is a decrease in salivary IgA concentration (Shephard and Shek 1998). The significance of this index has been emphasized by the fact that mucosal secretory IgA provides the first line of defense against inhaled pathogens. A decrease in the mucosal concentration of IgA should be related to a decrease in resistance to infection (Shephard and Shek 1998).

Suppression of immune activities in athletes during the competition period is a transitory phenomenon. It may be considered a reflection of the exhaustion of adaptivity caused by high-volume and high-intensity training and competitions. Possibly, these changes explain, at least in part, the susceptibility of top-level athletes to viral and other similar diseases.

At the same time, another explanation exists for morbidity in athletes. For 3 to 12 h after strenuous prolonged exercises, the immune activities are suppressed, which makes an "open window" for the harmful effects of infections (Pedersen and Brunsgaard 1995). In regard to this hypothesis, attention should be paid to the study of Muns (1994). Muns studied the respiratory tract with the aid of the determination of the number of polymorphonuclear neutrophils recovered by nasal lavage. The percentage of phagocytizing neutrophils decreased during the last training week before a 20-km race. The acute effects of the 20-km race were an increased number of polynuclear neutrophils in nasal lavage and pronounced drops in the percentage of phagocytizing neutrophils and ingested *Escherichia coli* per phagocytizing neutrophil.

Special Phenomena of Top-Level Sport

Peak performance and the overtraining syndrome are phenomena deeply related to the body's adaptivity. Hard training for reaching peak performance exhausts much of the body's adaptivity. The overtraining syndrome is the result of critically exhausted adaptivity. Peak performance is the aim of training but only at the right time. Therefore, assessment of achieving peak performance is important in training guidance. Early detection of overtraining affords a possibility for its rapid treatment.

Difficulty of Assessing Peak Performance

Training strategy is directed at a high performance level during the competition season. However, there are only a limited number of days (or 1 to 2 weeks) during the competition season when an athlete can achieve his or her best achievements. Consequently, only a short period (in some cases two or three such periods) is characterized as peak performance. Unfortunately, we know very little about the biological nature of peak performance and the metabolic background of the concerned state in the athlete's body.

Peak performance is founded on well-designed event-specific training effects. However, a principal question is whether the sum of concerned changes induced by training is enough for the highest possible performance of an athlete. This question is justified by the fact that training-induced changes mainly persist for a longer time than a short-term period of peak performance.

The ergogenic action of emotions shows that usually a person does not completely mobilize his or her performance capacities. Certain reserves, called "reserve forces," always remain for external situations. Physiologically, the essential role of fear and rage and of pain in the mobilization of bodily functions for fight was analyzed by Cannon (1925). Performance in sport competitions is unavoidably connected with stimulating or suppressing actions exerted by emotions. However important, it is not only the ergogenic action of emotions. "The records are the apotheosis of mind and volition of a stronger among the strongest" (Kuznetsov 1980).

Obviously the performance of an athlete depends to a certain degree on the ability to mobilize of body motor capacities. Accordingly, two tasks of training should be distinguished: generating motor potential and achieving opportunities for mobilizing motor potential up to the possible maximum (figure 9.10) (Viru 1993).

On one hand, mobilization of the motor potential has to be event-specific (e.g., power and sprint events require rapid mobilization of the maximum potential, whereas in endurance events success is guaranteed when the resources are used most efficiently and optimally distributed for the whole distance). The possibility of resisting the physiological and psychological manifestations of developing fatigue is also significant. Mobilization of the motor potential is possibly related to similar changes in the body that are under the influence of strong ergogenic emotions. The two main components of this influence are (1) an arousal effect of the activating system of brain stem reticular formation and thalamic unspecific nuclei, which increases the excitability and lability of nervous centers, and (2) exaggerated production of hormones, ensuring the purposeful mobilization of metabolic resources (see Viru 1993). Naturally, the arousal effect and exaggerated hormone production cannot persist continuously for a couple of weeks. Peak performance is related to the possibility of a rapid and pronounced arousal effect and hormonal changes. Thus, certain preconditioning should constitute the background for peak performance.

An example of increased metabolic mobilization obtained for the competition is that the same interval-training session resulted in an elevation of blood lactate up to 10 to 12 mmol/L during the preparatory period but up to 14 to 16 mmol/L during the competition period (Tchagovets et al. 1983). Lehmann et al. (1989) demonstrated that in cross-country skiers, nocturnal norepinephrine and dopamine excretions increased for the competition period. Epinephrine excretion increased promptly in response to altitude training, decreased during the next 3 weeks, and increased again afterwards (figure 9.11). Monitoring of training of a champion swimmer before the 1988 Olympic Games showed a parallel increase in swimming speed and blood lactate concentration after the 200-m time trial (figure 9.12) (Spikermann 1989).

Earlier, a possible precondition for the performance in power events was suggested (chapter 5, p. 106). In this regard, the significance of an increased testosterone level was indicated. Among tennis players (Booth et al. 1989) and judo fighters (Elias 1981), the winners had higher testosterone levels than the losers. In the winners, testosterone concentration was also elevated before the next match. Losers went to the next match with decreased testosterone concentration in blood and saliva.

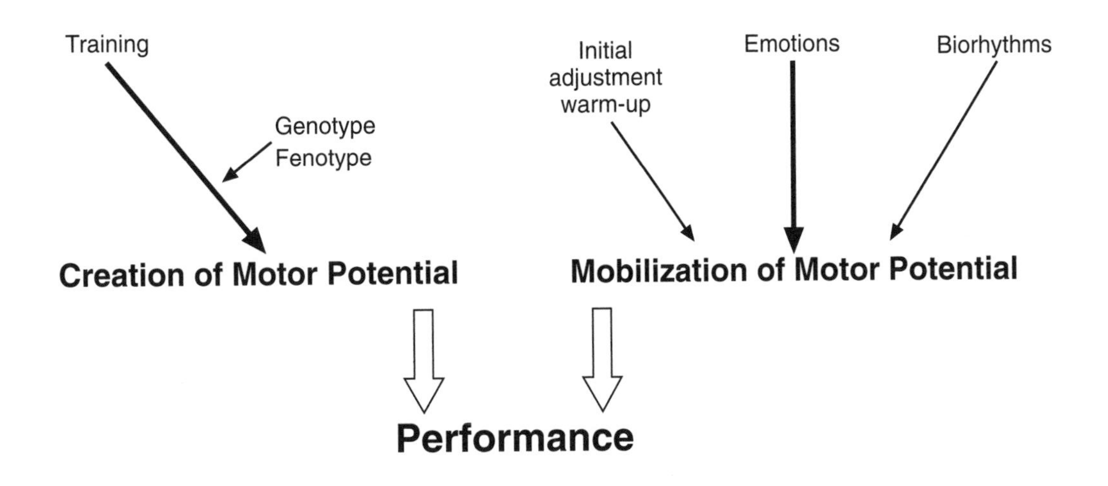

Figure 9.10 Creation of the motor potential and capacity for mobilization of the motor potential determine performance.

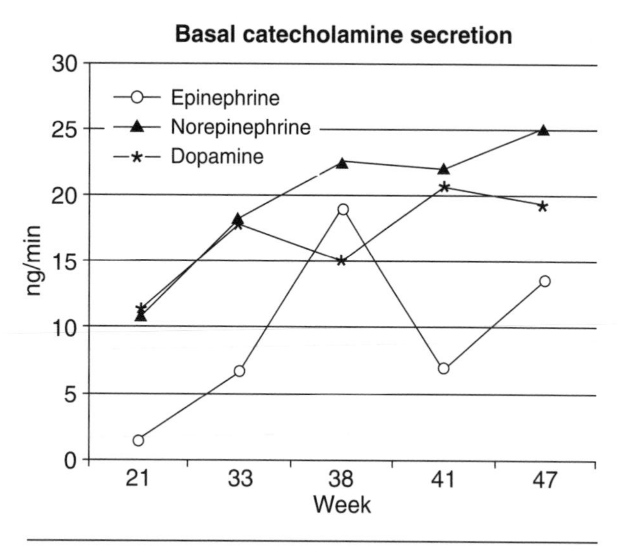

Figure 9.11 Increase of nocturnal excretion of catecholamines in cross-country skiers during the preparatory period.

Reprinted from Lehmann et al. 1989

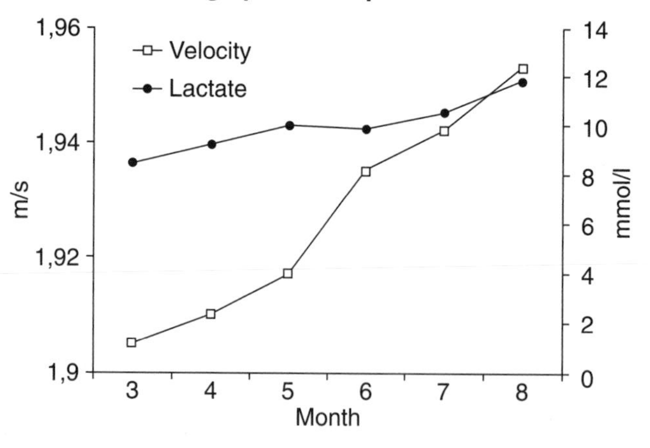

Figure 9.12 Swimming velocity in 200-m swimming and blood lactate level after the time trials in an Olympic champion during preparation for the Olympic Games.

From Spikerman 1985

We found a low testosterone concentration immediately before an important competition while monitoring the training in a discus thrower. His performance was much lower than expected. Three weeks after the competition, the testosterone concentration normalized (unpublished results).

Preconditioning should also be related to the possibilities for activation of the sympathoadrenal system. An enhanced reactivity of this system has been pointed to by the excretion of epinephrine and norepinephrine in response to specific exercises (repeated sprints for 20 to 80 m at a speed of 95% to 110% of maximal) in sprinters. At the time of the competition, rises in excretion and catecholamines were exaggerated compared with the responses at the onset of the competition period (Kassil and Mekhrikadze 1985).

Earlier, data were presented on the rise of cortisol level and responses to exercise during the competition period (pp. 205-206 of this chapter). The significance of these changes for peak performance has not been evidenced. However, in monitoring decathletes, results were obtained that a certain link exists between performance in competition and adrenocortical activity. A study of 17-hydroxycorticoid excretion in these athletes revealed three variants of changes before and during competition: (1) activation of adrenocortical function before and at the beginning of competition, followed by a modest decline in activity with a new activation at the end of the competition; (2) decreased adrenocortical activity before and during competition; and (3) no substantial change in 17-hydroxycorticoids excretion during the day of competition. The average result in the competition was 7510 ± 9 points in variant 1 and 7517 ± 41 points in variant 3. Variant 2 was associated with significantly lower results, on average 7112 ± 51. In some cases, the excretion of 17-hydroxycorticoids increased promptly on the eve of the competition day. In these cases, the excretion of 17-hydroxycorticoids decreased to low levels during the competition and performance was unexpectedly poor (Savi and Viru 1975).

Of course, a direct link between adrenocortical activity and performance of decathletes has not been evidenced. Obviously, the increased adrenocortical activity reflected an opportunity for the general mobilization of body resources, which created favorable conditions for performance in the decathlon.

In a more general manner, it is possible to suggest that during peak performance, the functions that directly determine the result must be highly developed; however, the body's reactive and mobilization capacity have to be at the highest level (figure 9.13).

In conclusion, hormone studies together with detection of blood level of neurotransmitters and some metabolites may open a door for biochemical monitoring of achieving peak performance. Metabolic indexes should demonstrate either an extreme mobilization or a particularly high economy of the use of metabolic resources.

Overtraining

After the body's adaptivity is lost, further training may lead to overtraining. This possibility becomes reality if training continues using of high-volume and/or high-intensity exercises.

Among several definitions of overtraining, the most acceptable for us is the one proposed by Lehmann et al. (1999a, p. 1) ". . . overtraining can be defined as stress > recovery (regeneration) imbalance, that is too much stress combined with too little time for regeneration." In this context, stress summarizes all individual training, nontraining, and competition-dependent stress factors. Undoubtedly, nutritional factors, travel, and genetic predisposition exacerbate the risk of overtraining in a completely individual manner. From the point of phenomenology, this definition has been completed by Mackinnon and Hooper (2000, p. 487): "Overtraining is a process of excessive training in high-performance athletes that may lead to persistent fatigue, performance decrements, neuroendocrine changes, alterations in mood states, and frequent illness, especially upper respiratory tract infection (URTI)." They considered overtraining syndrome as a neuroendocrine disorder that may result from accumulated fatigue during periods of excessive training with inadequate recovery.

Different from overtraining, overreaching is characterized by a transient performance incompetence that is reversible within a short-term recovery period of 1 to 2 weeks and "can be rewarded by a state of supercompensation" (Lehmann et al. 1999a). An example of typical overreaching has been provided by Jeukendrup et al. (1992). They found significant decreases of maximal power output; $\dot{V}O_2max$; maximal heart rate in an incremental exercise; blood lactate concentration at 200 W, 250 W, and 300 W; maximal lactate concentration in incremental exercise; power output at 4 mmol/L of lactate; and cycling velocity and heart rate in a time trial for 8.5 km in professional cyclists after 2 weeks of intensified training. Sleeping heart rate increased. After 2 weeks of reduced training, all changes disappeared, and performance indexes (maximal power output and cycling velocity in time trial) improved.

By analysis of the training of national- and international-level swimmers during a 44-week period, Mujika et al. (1996a) found taper stages repeatedly after 10 to 12 weeks of high workload

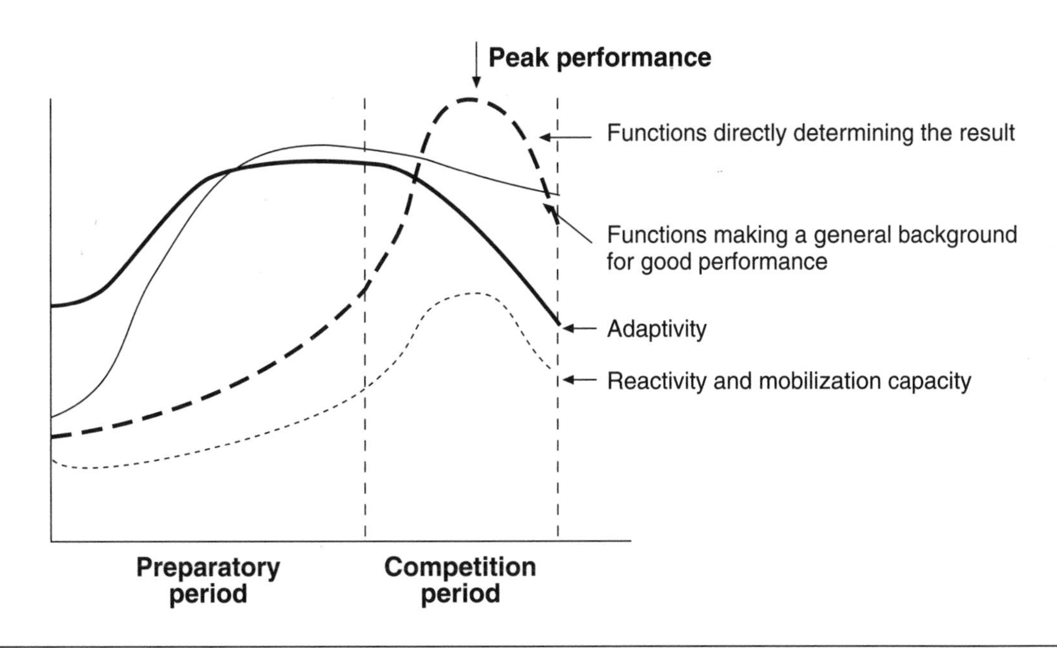

Figure 9.13 Peak performance is founded on a high reactive and mobilization capacity together with a high level of functional capabilities that directly determine the result.

training. The taper stages were associated with improvements in performance. Obviously, tapering helps to avoid overtraining.

Keizer (1998) emphasized that overreaching and overtraining are characterized by an array of events over time. If the balance between exercise-induced fatigue and time devoted to recovery is incorrect, adaptation will fail, and a situation of chronic fatigue will occur (Kuipers and Keizer 1988; Keizer 1998).

From a practical point of view, the main difference between overreaching and overtraining is that for top athletes, overreaching is evoked voluntarily according to the training plan and aimed at ensuring further progress of performance. Overtraining is an undesirable phenomenon, a misfortune related to exhaustion of the body's adaptivity. Once the overtraining syndrome appears, physical performance can decline so severely that the athlete's competition season may be over (Hackney et al. 1990).

In principle, the link between the loss of adaptivity and overtraining is confirmed by the fact that the onset of persistent performance incompetence is reflected in a period during which athletes fail to equalize or improve their personal records (Lehmann et al. 1997). Connecting overtraining to exhaustion of the body's adaptivity makes the term "overreaching" preferrable to the term "short-term overtraining."

Hormones

Based on his background of practical contact with overtrained athletes and analysis of published data, Israel (1958, 1976) assumed that overtrained athletes reveal clinical picture of either an "Addisonoid" or a "Basedovoid" syndrome. His assumption has been accepted (Kindermann 1986; Kuipers and Keizer 1988; Lehmann et al. 1993a). At the same time, it was recognized that the Addison type of overtraining syndrome resembles a state that is characterized by predominance in vagal tone, and the Basedow type of overtraining resembles a hyperadrenergic state (Kuipers and Keizer 1988; Lehmann et al. 1992a, 1997, 1998b; Urhausen et al. 1995). However, these types may be related to the manifestation of adrenocortical subactivity or thyroid hyperfunction, respectively (Hooper et al. 1995b). Parasympathetic ("Addisonoid") overtraining is the most frequent type of overtraining syndrome (Lehmann 1992a, 1992b, 1999a; Hooper et al. 1995b).

Clinical pathophenomena related to either adrenal cortex or thyroid gland disorders have never been evidenced in overtrained athletes (Foster and Lehmann 1997). Nonetheless, dysfunction of the pituitary-adrenocortical system is usual.

Several articles indicated that a high workload training stage for 3 to 6 weeks induced an increased cortisol basal level. Authors of some of these articles consider this change a response to hard training (tables 9.1 to 9.5). The others thought the increased cortisol level was an indication of overtraining (Stray-Gundersen et al. 1986; Roberts et al. 1993). In these studies, the training situation resembles more overreaching than overtraining. When overtraining is reflected in a prolonged drop in performance (for more than 3 to 4 weeks) and pronounced changes in mood characteristics or when the training analyses point out chronically exhausted adaptivity, the overtraining is associated with a decreased cortisol basal level (Lehmann et al. 1993b; Snyder et al. 1995) and/or reduced or reversed cortisol responses either in exercises (Lehmann et al. 1992a; Urhausen and Kindermann 1994; Urhausen et al. 1998) or after other influences (Barron et al. 1985).

Suppressed adrenocortical activity obviously reflects a general imbalance induced by overtraining either on the level of hypothalamic neurosecretion or in the function of the anterior pituitary gland. In overtrained distance runners, the responses of cortisol, corticotropin, growth hormone, and prolactin to insulin-hypoglycemia were reduced (Barron et al. 1985). Using these results, integral hormonal responses (hormone level above the basal values during a 90-min period after intravenous insulin injections) were calculated (Hackney 1999). This approach convincingly showed that in overtrained marathon runners the integral responses of growth hormone, corticotropin, and cortisol were lower than in the same athletes after 4 weeks of rest as in control marathon runners.

Keizer and collaborators (1991) found that exhaustive running training reduced the β-endorphin response to incremental treadmill running and also blunted β-endorphin response to human corticotropin-releasing hormone.

The possibility of imbalance in endocrine control was supported by the fact that after an intensive training stage, the release of corticotropin, caused by exogenous administration of corticotropin-releasing hormone (corticoliberin), was 60% higher, but the increase of blood cortisol level was 30% lower than before the strenuous training stage (Lehmann et al. 1993b). Similar results were also obtained in junior cyclists (Gastmann and Lehmann 1999). The impaired cortisol release combined with the increased growth hormone response to somatoliberin (Lehmann et al. 1993b).

Keizer (1998) studied lutropin secretion in female amenorrheic athletes before and after a test exercise. When training volume or intensity was promptly increased during an ovarian-menstrual cycle, lutropin pulse frequency was not altered, but the amplitude of secretory bursts was decreased. These results support the suggestion that menstrual disorders in endurance female athletes are caused by overload training (Kuipers and Keizer 1988).

Keizer (1998, pp. 158-159) assumes that "overtraining originates at the level of the hypothalamus and higher brain centers. Changes in neurotransmitter content and receptor sensitivity might play an important role in adaptation to training as well as in maladaptations." According to Lehmann and coworkers (1997, 1998a, 1999b), the first step to overload training, inducing overtraining, is depressed adrenal responsiveness to corticotropin, obviously caused by downregulation of corticotropin receptors on adrenocortical cells. They interpreted this change as a protective mechanism against chronic overload (Lehmann et al. 1999b). The positive feedback influence or other mechanisms increase the secretion of corticotropin. An increased corticotropin level has been found to be influenced by overload exercise (Wittert et al. 1996). However, the increased corticotropin release fails to overcome the depressed adrenal responsiveness, and cortisol secretion decreases, particularly during an exercise. In an advanced stage of the development of overtraining, the pituitary corticotropin and growth hormone responses reduce (Barron et al. 1985). Lehmann et al. (1999b) suggest that a decreased secretion of corticotropin and growth hormone is connected with the mechanism of central fatigue amplifying the metabolic system in the competence of affected athletes. This suggestion is supported by the role of serotonin in the mechanism of central fatigue (see chapter 3, p. 48). Hippocampal serotonergic neurons send inhibitory influences to hypothalamic neurons secreting corticoliberin (see chapter 8, p. 187). This nervous mechanism possibly contributes to the dysfunction of the endocrine system.

Other changes, typical of overtraining, are decreased β-adrenoreceptor density and, thereby, diminished sensitivity to catecholamines and decreased intrinsic sympathetic activity (for a review, see Lehmann et al. 1997, 1998a, 1999b). These changes are also considered a part of the mentioned protective mechanism against chronic overload (Lehmann et al. 1999b). Altered autonomic balance is expressed by low nocturnal catecholamine excretion (Lehmann et al. 1992b). In male middle-distance and long-distance runners, an exhaustive training stage increased norepinephrine levels in rest and during submaximal but not during maximal exercise; epinephrine and dopamine levels did not change (Lehmann et al. 1992c). Uusitalo et al. (1998) found a decreased epinephrine level in the blood during maximal exercises and a decreased norepinephrine level during submaximal exercises in overtrained female athletes. However, the resting catecholamine level and exercise-induced responses were found to be increased in overtrained athletes in several other studies (see Lehmann et al. 1998b). An increased norepinephrine level was confirmed in overtrained swimmers, whereas a high norepinephrine level remained when usual training was substituted by tapering (Hooper et al. 1995a).

A typical overtraining effect is a low testosterone level in the blood (Lehmann et al. 1992a; Roberts et al. 1993; Urhausen et al. 1995; Kraemer and Nindl 1998). Some authors pointed out an associated increase of cortisol level. Therefore, the testosterone/cortisol ratio is also low. Adlercreutz and coauthors (1986) recommended use of the ratio between free testosterone and cortisol as an indication of overtraining if the ratio decreases more than 30% or if the ratio is less than 0.35×10^{-3}. In several articles, the authors forgot the quantitative measure proposed by Adlercreutz and coworkers. They considered any decrease in the ratio (as well as in the ratio between total testosterone and cortisol) a bad indication. However, even if the ratio decreases more than 30% or it is less than 0.35×10^{-3}, there may no indication of overtraining (Kuipers and Keizer 1988; Vervoorn et al. 1991, 1992; Urhausen et al. 1995). Another essential point is the fact that the physiological meaning of the altered ratio is different if it is caused by a decrease of testosterone together with a lack of cortisol increase, by a less pronounced increase of testosterone than increase of cortisol, or by a more pronounced decrease of cortisol than of testosterone. Instead of the expected testosterone decrease and cortisol increase, Urhausen et al. (1987) found a decreased testosterone/cortisol ratio caused by a reduction in testosterone concentration, while the cortisol level remained unchanged.

Earlier (pp. 214-215 of this chapter), the overtraining-induced suppression of cortisol release was discussed. Thus, in overtraining, particularly in advanced overtraining, a cortisol increase appears as an exception. Therefore, the use of the ratio of testosterone to cortisol for overtraining diagnostics is meaningless. Mostly, a decrease of the testosterone level in conjunction with an increase of the cortisol level takes place in strenuous training stages (tables 9.1 to 9.5). Obviously, these changes indicate overreaching but not overtraining. The most correct conclusion is to join that of Urhausen and coauthors (1995), as did Keizer (1998), that the behavior of testosterone and cortisol is a physiological indicator of the current training workload, but it does not necessarily indicate the overtraining syndrome.

In conclusion, the main hormonal indicators of overtraining are suppressed production of cortisol, testosterone, and growth hormone and nocturnal excretion of catecholamines and decreased tissue sensitivity to hormones. Decreased sensitivity to catecholamines and altered responsiveness of various links in the pituitary-adrenocortical system have been demonstrated. An increased catecholamine level may point out the hyperadrenergic type of overtraining. An exaggerated drop of insulin during an exercise has been linked with this type of overtraining (Urhausen and Kindermann 1994).

Earlier (chapter 5, p. 98) the relative stability and low-rate changes of thyroid activity were mentioned. The plausible reason is, as was indicated, the long life span of thyroid hormones and their low turnover rate. However, during the postexercise recovery period, prolonged changes appear in thyroid activity (see chapter 5, p. 98). The hypothalamic-pituitary imbalance in overtraining may also influence the function of the pituitary-thyroid system. This problem is still not studied.

Foster (1998) affirmed that simple methods of monitoring the characteristics of training may allow an athlete to achieve the goal of training, while minimizing a undesired training outcome. Hormonal indicators of overtraining have strength only in conjunction with a decrease in performance level and mood changes. When hormonal indicators appear without performance and mood alterations, they inform about the high risk of the appearance of overtraining. The detection of this situation gives great value to hormonal monitoring because it informs trainers in a timely manner about the necessity to decrease the training workload and increase the time for recovery. This situation requires special attention if it appears in association with stagnation in training progress.

From the other side, if the stagnation of performance progress is associated with a lack of hormonal indications of overtraining, the training may continue without restrictions.

Hormonal studies in overtraining are essential to recognize the actual change in the metabolic situation of the exhausted athlete and to judge by the hormonal situation that the reparation process is necessary.

An example of monitoring overtraining by a complex of means has been provided by Keizer (1998). Male runners were studied repeatedly over 8-day intervals. During the stage of exhaustive training, the testosterone/cortisol ratio increased because the plasma cortisol level decreased significantly. Four of 12 subjects revealed no training desire, inability to perform the training workload, sleep disturbances, lethargy, and a constant feeling of tiredness. These athletes were considered to be in an overreaching state. One runner was obviously overtrained. His mood changes were extremely pronounced. His normal performance was restored after as much as 5 weeks of light training. In overreaching athletes, a blunted β-endorphin response to standard treadmill running appeared. The overtrained person showed decreased endorphin release.

In trained runners, 10 days of twice-daily interval training sessions resulted in a significant fall in performance and several personality and mood changes that persisted for at least 5 days. They felt tired and depressed, had sore muscles, had a feeling of nausea appear after light running, and had trouble getting to sleep. Their cortisol levels began to decrease at the end of the 10-day exercise period, and the low levels persisted during the 5 recovery days (minimum 3 days after the training stage). Testosterone, lutropin, and follitropin levels did not change significantly. A high level of creatine kinase activity was detected from the sixth day of exercise up to the third postexercise day. The blood urea level did not change (Fry et al. 1992).

Overtrained weightlifters exhibited a moderate increase in testosterone response and a pronounced decrease in cortisol response to the test exercise in conjunction with 1-RM strength decrements (Fry et al. 1998).

Metabolic Changes

In overtraining, metabolic changes express exhaustion of the body and contraregulatory processes aimed at preventing fatal exhaustion. Depletion of glycogen stores is common for overreaching and overtraining (Snyder 1998). Although reduced muscle glycogen is a factor of decreased performance in several events, the decreased glycogen store is not the main cause of overtraining (Snyder et al. 1995; Snyder 1998). When glycogen reduction was avoided with the aid of a carbohydrate supplement, the overreaching symptoms appeared in cyclists (Snyder et al. 1995).

The consequences of muscle glycogen depletion are reduced anaerobic work capacity and diminished lactate responses in exercise (Lehmann et al. 1991, 1992c, 1998b; Mackinnon and Hooper 2000). However, when glycogen depletion was avoided with the aid a carbohydrate supplement, a decreased maximal work capacity and a reduced ratio of lactate to rating of perceived exertion appeared in the overreaching state (Snyder et al. 1995).

Figure 9.14 shows that in middle-distance runners $V.O_2max$ and running time in an anaerobic test decreased in conjunction with a less pronounced increase of lactate concentration in response to an aerobic-anaerobic exercise or anaerobic test (Kindermann 1986).

A result of liver glycogen depletion is a lower glucose level at submaximal exercise in the overreaching state (Lehmann et al. 1991, 1992c, 1998b). Overreaching also decreases the free fatty acid level during submaximal exercise (Lehmann et al. 1992c, 1998a). Thus, mobilization of the lipid reserve is also impaired.

Decreased carbohydrate availability should exaggerate the responses of hormones controlling energy metabolism (see chapter 3, p. 52).

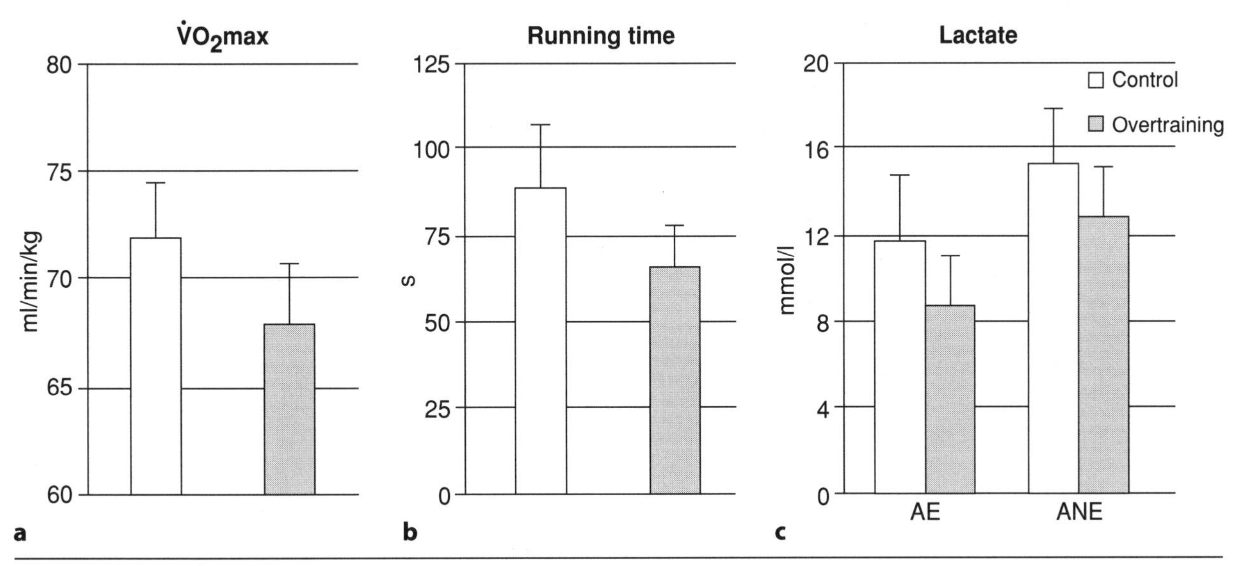

Figure 9.14 *(a)* $\dot{V}O_2max$, *(b)* running time during an anaerobic test, and *(c)* lactate levels during incremental exercise at $\dot{V}O_2max$ (AE) and after an anaerobic test (ANE) in middle-distance runners in good performance and in overtraining.

Reprinted from W. Kindermann 1986.

However, this is not the case in overtraining. Thus, in overtraining, the activity of the endocrine glands is controlled by a stronger influence than the glucostatic regulation.

Urea, creatine, and uric acid levels remained constant in blood during 4 weeks of running training with unusually high volume. Plasma activities of creatine kinase increased, but alanine-aminotransferase and glutamine-aminotransferase did not change (Lehmann et al. 1991). The blood amino acid concentration varied in runners during the exhaustive training stage, but no strict relationship with overtraining development was found (Lehmann et al. 1996).

With regard to monitoring overtraining, attention should be paid to exercise-induced changes of glucose and lactate, as well as the glucose basal level. Lactate diagnostics of overtraining must be founded on repeated assessments over several weeks and months by the use of strictly controlled exercise, diet, and environmental conditions (Mackinnon and Hooper 2000). A decrease in the blood lactate concentration at maximal work rate or the ratio of lactate response to rating of perceived exertion is most closely related to overtraining (Hooper and Mackinnon 1995; Snyder et al. 1995; Pelayo et al. 1996; Mackinnon and Hooper 2000). However, the significance of lactate diagnostics is different for sprinters and athletes in power events and for middle-distance and long-distance runners. Even for marathon runners the significance of lactate diagnostics is different compared with long-distance runners (see chapter 7, pp. 161-163). All in all, the overtraining markers have to consider event-specific differences (Hooper et al. 1995a; Hooper and Mackinnon 1995, Mackinnon and Hooper 2000).

Plasma creatine kinase, serum ferritin concentration, and erythrocyte and leukocyte counts are not reliable markers of overreaching and overtraining (Hooper et al. 1995b; Rowbottom et al. 1995; Mackinnon et al. 1997). Rowbottom et al. (1998) consider depressed concentration of plasma glutamine an essential marker of overtraining. Another point of view has been published by Mackinnon and Hooper (2000); although plasma glutamine concentration may be lower in overtrained or overreached athletes compared with well-trained athletes, no direct link has been established between plasma glutamine level and overtraining symptoms, prognosis, and immune functions.

The metabolic situation in overtraining may be related to oxidation stress (Tiidus 1998). Still, no convincing evidence has been obtained about changes of the ratio between the production of oxidants and activities of the antioxidant system in overtraining. Therefore, it is impossible to propose quantitative measures characteristic for overreaching and overtraining. For analyzing the events in the central nervous system during overtraining, attention should be paid to brain neurotransmissions (Meeusen 1999).

When the monitoring has a special aim to assess the contribution of central fatigue, the determination of the branched-chain amino acid/tryptophan ratio and serotonin and dopamine levels in blood plasma will be essential (for the metabolic parameter of central fatigue, see Meeusen and De Meirlleir 1995). However, Tanaka et al. (1997) did not find in trained endurance runners changes in plasma-free and total tryptophan, branched-chain amino acid concentration, or tryptophan/branched-chain amino acid ratio during the high-volume training stage, resulting in overreaching.

Immune Activity

Several review articles indicate that training with a high workload suppresses immune activity and promotes frequent illness (see chapter 6, p. 129). As extrapolation, these characteristics are considered common for overtraining (Fitzgerald 1991; Fry et al. 1991; Lehmann et al. 1993a). Mackinnon and Hooper (2000) indicated that the susceptibility to illness has been investigated in an overtraining situation in only two articles (Pyne et al. 1995; Mackinnon and Hooper 1996). Surprisingly, the risk of upper respiratory tract infections was not greater in overtrained swimmers compared with intensively trained athletes. Earlier, the effects of intensive training, including overreaching, were discussed (pp. 208-210 of this chapter). There are not many results to add that are obtained in overtrained athletes, keeping in mind long-lasting performance incompetence and mood changes. In overtrained female swimmers (investigated at the end of a 6-month competition season), saliva IgA concentration was lower than in those swimmers who finished the season in good condition (Mackinnon and Hooper 1994; Hooper et al. 1995b).

Noteworthy is a prospective longitudinal study of 19 ± 3 months conducted by Gabriel and coauthors (1998). The studied persons were 12 cyclists and 3 triathletes. In 15 of 85 examinations, overtraining was diagnosed (reduction of results at recent competition, unexpectedly premature interruption of training or competition, decreased subjective performance capacity, and early fatigue with training with vegetative symptoms). Neither at physical rest nor after an incremental cycle ergometer exercise at 110% of individual anaerobic threshold did counts of neutrophils, T-cells and B-cells, and number of natural killer cells differ between overtraining cases and cases of good performance capacity. Eosinophils were lower, and activated T-cells increased slightly during overtraining. Differences were not found in most membrane antigen contents, but overtraining increased the percentge of T-cell subsets $CD3^{+}CD16/CD56^{+}$, $CD3^{+}HLA\text{-}DR^{+}$, $CD4^{+}CD45RO^{+}$, $CDB^{+}CD45RO^{+}$, and $CD45RO^{+}$. The authors concluded that overtraining does not lead to clinically relevant alterations of immunophenotypes in peripheral blood. However, a moderate activation of T-cells and an upregulation of CD45RO or T-cells indicated an enhanced functional state of T-cells. Exercise-induced mobilization and redistribution patterns of leukocyte and lymphocyte subpopulations remained unaltered.

In the overtraining state, information on alterations of immune activities may be obtained measuring saliva IgA and blood T-cell activity and the glutamine level in blood plasma. However, there are no markers for the justification of the overtraining state. According to Shephard and Shek (1998), the most promising immunological marker of excessive training seems to be a decrease in salivary IgA concentration. They also emphasized that no single change occurred with sufficient consistency to identify a competitor who is at risk of overtraining on the background of immune activity alterations.

In conclusion, although several possibilities for biochemical monitoring of overtraining were indicated, we still have to agree with Kuipers (1998) that reliable, specific, and sensitive parameters for overreaching and overtraining have not been found.

Summary

An essential, although not widely recognized, task for training monitoring is the assessment of the body's adaptivity. The actual level of adaptivity determines the possibility for adaptation, including prolonged continuous adaptation founded on adaptive protein synthesis. The saturation phenomenon expresses the limitation of the possibilities to have training effects during a long-lasting period. Limits to adaptivity also provide opportunities to increase and maintain peak performance. The obvious reason is that although increased adaptivity is a general benefit of exercise training, hard training regimes exhaust a certain part of the body's adaptivity. When adaptivity decreases, instead of a positive training effect, overtraining manifestations develop. Decrease of adaptation is a temporary phenomenon. A rest (transitory) period after the competition season ensures recovery of adaptivity, and from the beginning of the next training year, the body is again capable of adaptation in hard training and increases of training workloads.

Assessment of adaptivity is a problem because the mechanism of adaptivity reduction has not been established. Several monitoring studies showed typical hormone and immune activity changes during hard training. Therefore, decreased adaptivity may be related to changes in endocrine function in conjunction with alterations in several immune parameters. It has been hypothesized that increased cortisol and decreased testosterone responses to serious exercises together with an increased basal level of cortisol express a high strain of adaptation processes in the body. A suppressed or reversed cortisol response in exercise together with decreased cortisol and testosterone basal levels and blunted growth hormone responses in exercise appears when the adaptivity of an athlete decreases. In the overtraining state, usually low cortisol levels and dysregulation of the pituitary-adrenocortical and sympathoadrenal systems has been found.

The physiological aspect of peak performance is insufficiently treated. Therefore, the criteria for monitoring peak performance are yet to be determined. It is known that a hard strain on the processes of adaptivity expressed by exaggerated cortisol responses characterizes the achievement of peak performance. At the same time changes in immune activities result in increased susceptibility of athletes to viral diseases. Thus, a high performance level accompanies reduced adaptivity.

General Conclusion to Part III

The arrangement of sport training is effective if the factors influencing the outcome are understood and considered. Therefore, feedback information on training-induced effects is essential (chapter 7). Biochemical monitoring provides possibilities for evaluation of power and capacity of energy mechanisms so that important training effects can be established (chapter 8). Biochemical monitoring may also provide information about accomplishment of metabolic control and endocrine function. Knowledge of the dynamics of carbohydrates and lipids during prolonged exercise is most essential.

The benefits that arise from biochemical monitoring are whether the design of training sessions and microcycles and mesocycles is suitable. It is essential to know whether the adaptation possibilities of the athletes' bodies (adaptivity) are used appropriately during various stages of training. Possibilities for monitoring adaptivity arise in recording hormonal responses to test exercise or training sessions (the most information seems to be attained with cortisol responses) or in assessing parameters of immune activities. In this way, the desired state of peak performance and the undesired state of overtraining may be assessed if these changes are compared to performance competence.

Concluding Remarks

During the 20th century clinics were supplied with an extended arsenal of biochemical methods for improved diagnosis. Wide use of these methods promoted the recognition that an essential background for pathophenomena consists in failure of metabolic processes and/or their regulation. Over the years, especially during the last quarter of the 20th century, opportunities were created for a similar approach in understanding training effects on athletes' performance. Knowledge of metabolic foundations and hormonal regulation of metabolic processes in exercise was essential not only for theoretical considerations but also for a more scientific and more justified pathway for training guidance. Trainers had extended opportunities to assess training effects and evaluate training design. The value of related feedback information rose significantly because of the use of data about metabolic events in an athlete's body, including reflections of metabolic events at the cellular level. Thus, biochemical monitoring of training appeared and was approved in practice.

The monitoring of training is aimed at helping trainers. It does not intend to make any kind of biochemical, hormonal, or immunological measurements in athletes. These studies can be considered as purposeful training monitoring if the results obtained provide necessary and direct information for improved guidance and design of practical training in athletes.

Biochemical studies are not necessary if their practical outcome supplies the trainers with the same information as other simpler and less costly methods. However, in the practice of training monitoring, several questions arise that can be answered only by penetrating into the foundations of training. Metabolic and hormonal studies are indispensable in analyzing these mechanisms. Our purpose in this book has been to indicate the possibilities for metabolic and hormonal studies and discuss their significance and limitations in training monitoring. The idea is as follows: when simplified methods do not help trainers, metabolic and hormonal studies may be beneficial.

Biochemical monitoring will be successful only in conjunction with other methods, particularly with well-designed recording of the training process and activities during competition. If this requirement is satisfied, (1) the results of biochemical studies can be used for direct analyses of training in an athlete and (2) trainers will begin to understand the actual essence of training. The main outcome will be that traditional, frequently "blind," exercising will be replaced with the conscious choice of exercises, training methods, work/relief regimes, and training workloads.

Any result of biochemical monitoring or any metabolic or hormonal parameter provides useful information only if the essence of the concerned metabolic process and its regulation is understood. It is not enough to know about lactate values or about "good" and "bad" hormonal changes without knowing what the lactate accumulation actually means or what the function of the hormone is. Moreover, it is necessary to keep in mind that metabolic processes are interrelated and controlled by a multifactorial system. Any simplification may lead to misunderstandings. Therefore, in this book, great attention has been paid to a discussion of the essence of metabolites or hormones used in training monitoring.

We started writing this book with the belief that the essence of sport training is inducing changes in the athlete's body. At the conclusion of this book, we want to emphasize once more that this approach provides an opportunity for scientifically founded management of training. However, this approach works only if the necessary information on the changes in training effects is available and understood by those who guide the training.

References

Adams, G. A. 1998. Role of insulin-like growth factor I in the regulation of skeletal muscle adaptation to increased loading. *Exercise and Sport Science Reviews* 26:31-60.

Adamson, L., W. M. Hunter, O. O. Ogurremi, I. Oswald, and I. W. Percy-Robb. 1974. Growth hormone increase during sleep after daytime exercise. *Journal of Endocrinology* 62:473-78.

Adlercreutz, H., K. Härkönen, K. Kuoppasalmi, H. Näveri, H. Huthaniani, H. Timsanen, K. Remes, A. Dessypris, and J. Karvonen. 1986. Effect of training on plasma anabolic and catabolic steroid hormones and their response during physical exercise. *International Journal of Sports Medicine* 7(Suppl.):27-8.

Adlercreutz, H., M. Härkönen, K. Kuoppasalmi, H. Näveri, and S. Rehunen. 1976. Physical activity and hormones. *Advances in Cardiology* 18:144-57.

Ahlborg, B., J. Bergström, L.-G. Ekelund, and E. Hultman. 1967. Muscle glycogen and muscle electrolytes during prolonged physical activity. *Acta Physiologica Scandinavica* 70:129-40.

Ahlborg, G., and P. Felig. 1976. Influence of glucose ingestion on fuel-hormone response during prolonged exercise. *Journal of Applied Physiology* 41:683-88.

Ahlborg, G., and P. Felig. 1977. Substrate utilization during prolonged exercise preceded by ingestion of glucose. *American Journal of Physiology* 233:E188-94.

Ahlborg, G., P. Felig, L. Hagenfeldt, R. Hendler, and E. J. Wahren. 1974. Substrate turnover during prolonged exercise in man: splanchnic and leg metabolism of glucose, free fatty acids and amino acids. *Journal of Clinical Investigations* 53:1080-90.

Ahlman, K., and M. J. Karvonen. 1962. Weight reduction by sweating in wrestlers and its effect on physical fitness. *Journal of Sports Medicine and Physical Fitness* 1:58-62.

Alessio, H. M. 1993. Exercise-induced oxidative stress. *Medicine and Science in Sports and Exercise* 25:218-24.

Alessio, H., and A. Goldfarb. 1988. Lipid peroxidation and scavenger enzymes during exercise: adaptive response to training. *Journal of Applied Physiology* 64:1333-6.

Alev, M., and A. Viru. 1982. A study of the state of mechanisms of general adaptation by excretion of 17-hydroxycorticoids in young skiers [in Russian]. *Teoria i Praktika Fizcheskoi Kulturõ (Moscow)*, 12:16-8.

Allen, D. G., C. D. Balnave, E. R. Chin, H. Westerblood. 1999. Failure of calcium release in muscle fatigue. In *Biochemistry of Exercise X,* ed. Hargreaves, M., and M. Thompson. 135-146. Champaign, IL: Human Kinetics.

Aloia, J. F., P. Rasulo, L. J. Deftas, A. Voswani, and J. K. Yeh. 1985. Exercise-induced hypercalcemia and calciotropic hormones. *Journal of Laboratory and Clinical Medicine* 106:229-32.

Altenkirch, H. U., R. Gerzer, K. A. Kirsch, J. Weil, B. Heyduck, I. Schultes, and L. Röcker. 1990. Effect of prolonged physical exercise on fluid regulating hormones. *European Journal of Applied Physiology* 61:209-13.

Ament, W. J. R. Huizenga, E. Kort, T. W. van der Mark, R. G. Grevnik, and G. J. Verhorke. 1999. Respiratory ammonia output and blood ammonia concentration during incremental exercise. *International Journal of Sports Physiology* 20:71-7.

Andersen, P., and B. Saltin. 1985. Maximal perfusion of skeletal muscle in man. *Journal of Physiology* 366:233-49.

Andersen, P., and J. Henriksson. 1977. Training induced changes in the subgroups of human type II skeletal muscle fibers. *Acta Physiologica Scandinavica* 99:123-5.

Apple, F. S., and M. A. Rogers. 1986. Skeletal muscle lactate dehydrogenase isozyme alterations in men and women marathon runners. *Journal of Applied Physiology* 61:477-81.

Apple, F. S., and P. A. Tesch. 1989. CK and LD isozymes in human single muscle fibers in trained athletes. *Journal of Applied Physiology* 66:2717-20.

Ardawi, M. S., and Y. S. Jamal. 1990. Glutamine metabolism in skeletal muscle of glucocorticoid-treated rats. *Clinical Sciences* 79:139-47.

Armstrong, L. D., D. L. Costill, and W. J. Fink. 1985. Influence of diuretic-induced dehydration on competitive running performance. *Medicine and Science in Sports and Exercise* 17:456-61.

Armstrong, R. B. 1986. Mechanisms of exercise-induced delayed onset of muscular soreness and training. *Clinical Sports Medicine* 5:605-14.

Armstrong, R. B. 1990. Initial events in exercise-induced muscular injury. *Medicine and Science in Sports and Exercise* 22:429-35.

Armstrong, R. B., R. W. Ogilvie, and J. A. Swane. 1983. Eccentric exercise-induced injury to rat skeletal muscle. *Journal of Applied Physiology* 54:80-93.

Arnall, D. A., J. C. Marker, R. K. Conlee, and W. W. Winder. 1986. Effect of infusing epinephrine on liver and muscle glycogenolysis during exercise in rats. *American Journal of Physiology* 250:E641-9.

Arner, P., J. Bolinder, and J. Ostman. 1983. Glucose stimulation of the antilipolytic effect of insulin in humans. *Science* 220:1057-9.

Arner, P., E. Kriegholm, P. Engfeldt, and J. Bolinder. 1990. Adrenergic regulation of lipolysis in situ at rest and during exercise. *Journal of Clinical Investigation* 85:893-8.

Åstrand, P.-O., and K. Rodahl. 1986. *Textbook of work physiology*. 3d ed. New York, McGraw-Hill.

Åstrand, P. O., and B. Saltin. 1964. Plasma and red cell volume after prolonged severe exercise. *Journal of Applied Physiology* 19:829-32.

Atha, J. 1981. Strengthening muscle. *Exercise and Sport Science Reviews* 9:1-73.

Aunola, S., E. Alanus, J. Mariniemi, and H. Rusko. 1990. The relation between cycling time to exhaustion and anaerobic threshold. *Ergonomics* 33:1027-42.

Aunola, S., J. Marniemi, E. Alanen, M. Mantyla, M. Saraste, and H. Rusko. 1988. Muscle metabolic profile and oxygen transport capacity on determinant of aerobic and anaerobic thresholds. *European Journal of Applied Physiology* 50:405-11.

Aunola, S., and H. Rusko. 1984. Reproducibility of aerobic and anaerobic thresholds in 20-50 year old men. *European Journal of Applied Physiology* 53:260-6.

Aunola, S., and H. Rusko. 1992. Does anaerobic threshold correlate with maximal lactate steady-state? *Journal of Sports Sciences* 10:309-23.

Aurell, M., and P. Vikgren. 1971. Plasma renin activity in supine muscular exercise. *Journal of Applied Physiology* 31:339-41.

Babij, P., S. M. Matthews, and M. J. Rennie. 1983. Changes in blood ammonia, lactate and amino acids in relation to workload during bicycle ergometer exercise in man. *European Journal of Applied Physiology* 50:405-11.

Baikov, V. M. 1975. Experimental foundation of regimes of training loads for improved endurance in cross-country skiers [in Russian]. Thesis acad. diss., Tartu University, Tartu.

Bailey, S. P., J. M. Davis, and E. N. Ahlborn. 1993a. Neuroendocrine and substrate responses to altered brain 5-HT activity during prolonged exercise to fatigue. *Journal of Applied Physiology* 74:3006-12.

Bailey, S. P., J. M. Davis, and E. N. Ahlborn. 1993b. Brain serotogenic activity affects endurance performance in the rat. *International Journal of Sports Medicine* 6:330-3.

Baj, Z., J. Kantorski, E. Majewska, K. Zeman, L. Pokoca, E. Fornalczyk, H. Tchorzewski, Z. Sulowska, and R. Lewicki. 1994. Immunological status of competitive cyclists before and after the training season. *International Journal of Sports Medicine* 15:319-24.

Balon, T. W., A. Zorzano, J. L. Treadway, M. N. Goodman, and N. B. Ruderman. 1990. Effects of insulin on protein synthesis and degradation in skeletal muscle after exercise. *American Journal of Physiology* 258:E92-7.

Banfi, G., M. Marineii, G. S. Roi, and V. Agape. 1993. Usefulness of free testosterone/cortisol ratio during a season of elite speed skating athletes. *International Journal of Applied Physiology* 14:373-9.

Bang, O. 1936. The lactate content of the blood during and after muscular exercise in man. *Scandinavische Archiv für die Physio*logie 74(10):51-82.

Bangsbo, J., P. D. Gollnick, T. E. Graham, C. Juel, B. Kiens, M. Mizuno, B. Saltin. 1990. Anaerobic energy production and O_2-deficit-debt relationship during exhaustive exercise in humans. *Journal of Physiology* 422:439-559.

Bangsbo, J., P. D. Gollnick, T. E. Graham, and B. Saltin. 1991. Substrates for muscle glycogen synthesis in recovery from intense exercise in man. *Journal of Physiology* 434:423-40.

Bannister, E. W., and B. J. C. Cameron. 1990. Exercise-induced hyperammonemia: peripheral and central effects. *International Journal of Sports Medicine* 11 (Suppl 2):S129-42.

Banister, E. W., A. K. Singh, and T. W. Calvert. 1980. The effect of training on the formation of catecholamines, ammonia, glutamic acid and glutamine during exhaustive exercise. *Medicine and Science in Sports and Exercise* 12:84 (abstract).

Barach, J. H. 1910. Physiological and pathological effects of severe exertion. The Marathon Race on the circulatory and renal systems. *Archives of International Medicine* 5:382-405.

Barcroft, J., and J. G. Stephens. 1927. Observations upon the size of the spleen. *Journal of Physiology* 64:1-22.

Bar-Or, O. 1981. La test anaerobie de Wingate. *Symbioses* 13:157-72.

Bar-Or, O. 1987. The Wingate anaerobic test. An update on methodology, reliability and validity. *Sports Medicine* 4:381-94.

Barron, G. L., T. D. Noakes, W. Levy, C. Smith, and R. P. Millar. 1985. Hypothalamic dysfunction in overtrained athletes. *Journal of Clinical Endocrinology and Metabolism* 60:803-6.

Bärtsch, P., B. Welsch, M. Albert, B. Friedmann, M. Levi, and E. K. O. Kruithof. 1995. Balanced activation of coagulation and fibrinolysis after a 2-h triathlon. *Medicine and Science in Sports and Exercise* 27:1465-70.

Barwich, D., A. Rettenmeier, and H. Weicker. 1982. Serum levels of so-called "stress-hormones" in athletes after short-term consecutive exercise. *International Journal of Sports Medicine* 3(Suppl):8.

Barwich, D., H. Hägele, M. Weiss, and H. Weicker. 1981. Hormonal and metabolic adjustment in patients with central Cushing's disease after adrenalectomy. *International Journal of Sports Medicine* 2:220-7.

Beckmann, R. P., L. E. Mizzen, and W. J. Welch. 1990. Interaction of HSP 70 with newly synthesized proteins: implications for protein folding and assembly. *Science* 248:850-4.

Bell, G. J., and H. A. Wenger. 1988. The effect of one-legged sprint training on intramuscular pH and non-bicarbonate buffering capacity. *European Journal of Applied Physiology* 58:158-64.

Bell, P. A., and T. R. Jones. 1979. Interactions of glucocorticoid agonists and antagonists with cellular receptors. In *Antihormones,* ed. Agarwall M. K., 35-50. Amsterdam: Elsevier/North-Holland Biomedical Press.

Beneke, R., and S. P. van Duvillard. 1996. Determination of maximal lactate steady state in selected sports events. *Medicine and Science in Sports and Exercise* 28:241-46.

Bennett, N. B., C. M. Oyston, and D. Ogsten. 1968. The effect of prolonged exercise on the components of the blood fibrinolytic enzyme. *Journal of Physiology* 198:479-85.

Berg, A., and G. Harlambie. 1978. Changes in serum creatine kinase and hexose phosphate isomerase activity with exercise duration. *European Journal of Applied Physiology* 39:191-201.

Berg, A., and J. Keul. 1980. Serum alanine during long-lasting physical exercise. *International Journal of Sports Medicine* 1:199-202.

Bergh, U., H. Hartley, L. Landsberg, and B. Ekblom. 1979. Plasma norepinephrine concentration during submaximal and maximal exercise at lowered skin and core temperatures. *Acta Physiologica Scandinavica* 106:383-4.

Berglund, B. O., G. Birgegård, and P. Hemmingsson. 1988. Serum erythropoietin in cross-country skiers. *Medicine and Science in Sports and Exercise* 20:208-9.

Bergström, J. 1962. Muscle electrolytes in man. Determined by neutron activation analysis in needle biopsy specimens. A study on normal subjects, kidney patients, and with chronic diarrhea. *Scandinavian Journal of Clinical and Laboratories Investigation* 14(Suppl 68).

Bergström, J., P. Furst, and E. Hultman. 1985. Free amino acids in muscle tissue and plasma during exercise in man. *Clinical Physiology* 5:155-60.

Bergström, J., G. Gaurnieri, and E. Hultman. 1973. Changes in muscle water and electrolytes during exercise. In *Limiting factors of physical performance,* ed. Keul J., 173-178. Stuttgart: G. Thieme.

Bergström, J., L. Hermansen, E. Hultman, and B. Saltin. 1967. Diet muscle glycogen and physical performance. *Acta Physiologica Scandinavica* 71:140-50.

Bergström, J., and E. Hultman. 1966a. The effect of exercise on muscle glycogen and electrolytes in normals. *Scandinavian Journal of Clinical and Laboratories Investigation* 18:16-20.

Bergström, J., and E. Hultman. 1966b. The muscle glycogen synthesis after exercise: An enhancing factor localized to the muscle cell in man. *Nature* 210:309-10.

Bergström, J., E. Hultman, L. Jorfeldt, B. Pernow, and J. Wahren. 1969. Effect of nicotinic acid on physical working capacity and on metabolism of muscle glycogen in man. *Journal of Applied Physiology* 26:170-6.

Berk, L. S., D. C. Nieman, W. S. Youngberg, K. Arabatzis, M. Simpson-Westerberg, J. W. Lee, S. A. Tan, and W. C. Eby. 1990. The effect of long endurance running on natural killer cells in marathoners. *Medicine and Science in Sports and Exercise* 22:207-12.

Bernet, F., and J. Denimal. 1974. Evolution de la response adrenosympatheque á l'exercise au cours de l'entrainement chez le rat. *European Journal of Applied Physiology* 33:57-70.

Berthain, S., P. Pelayo, G. Lousel-Corbeil, H. Robin, and M. Gerbeanin. 1996. Comparison of maximal aerobic speed on assessed with laboratory and field measurements in moderately trained subjects. *International Journal of Sports Medicine* 17:525-9.

Bevegård, B. S., and J. T. Shepherd. 1967. Regulation of the circulation during exercise in man. *Physiological Reviews* 47:178-213.

Bichler, K. H., E. Lachmann, and F. Porzsolt. 1972. Untersuchungen zur mechanischen Hämolyse bei

Langstreckenläufern. *Sportarzt und Sportmedizin* 23:9-14.

Bieger, W. P., M. Weiss, G. Michel, and H. Weicker. 1980. Exercise-induced monocytosis and modulation of monocyte function. *International Journal of Sports Medicine* 1:30-6.

Billat, L. V. 1996. Use of blood lactate measurements for prediction of exercise performance and for control of training. Recommendations for long-distance runners. *Sports Medicine* 22:157-75.

Billat, V. L., D. W. Hill, J. Pinoteau, B. Petit, and J. P. Korolsztein. 1996. Effect of protocol on determination of velocity at $\dot{V}O_2$max and on its time to exhaustion. *Archives of Physiology and Biochemistry* 104:313-21.

Billat, V., F. Dalmay, M. T. Antonini, and A. P. Chassain. 1994a. A method for determining the maximal steady state of blood lactate concentration from two levels of submaximal exercise. *European Journal of Applied Physiology* 69:196-202.

Billat, V., J. C. Renoux, J. Pinoteau, B. Petit, and J. P. Koralsztein. 1994b. Reproducibility of running time to exhaustion at $\dot{V}O_2$max in elite runners. *Medicine and Science in Sports and Exercise* 26:254-7.

Bishop, P., and M. Martino. 1993. Blood lactate measurement in recovery as an adjunct to training. *Sports Medicine* 16:5-13.

Björkman, O., P. Felig, L. Hagenfeldt, and J. Wahren. 1981. Influence of hypogluconemia on splanchnic glucose output during leg exercise in man. *Clinical Physiology* 1:43-57.

Björntorp, P., M. Fahlen, J. Holm, T. Schersten, and V. Szostak. 1970. Determination of succinic oxidase activity in human skeletal muscle. *Scandinavian Journal of Clinical and Laboratories Investigation* 26:145-52.

Blair, S. N., H. W. Kohl, R. S. Paffenbarger, D. G. Clark, K. H. Cooper, and L. W. Gibbons. 1989. Physical fitness and all-cause mortality: A prospective study of healthy men and women. *JAMA* 262:1395-2401.

Blake, M. J., E. A. Stein, and A. Vomachka. 1984. Effects of exercise training on brain opioid peptides and serum LH in female rats. *Peptides* 5:953-8.

Blomstrand, E., F. Celsing, and N.A. Newsholme. 1988. Changes in plasma concentration of aromatic and branched-chain amino acids during sustained exercise in man and their possible role in fatigue. *Acta Physiologica Scandinavica* 133:115-22.

Blomstrand, E., and B. Ekblom. 1982. The needle biopsy technique for fibre type determination in human skeletal muscle. A methodological study. *Acta Physiologica Scandinavica* 116:437-42.

Blomstrand, E., D. Perrett, M. Parry-Billings, and E. A. Newsholme. 1989. Effect of sustained exercise on plasma amino acid concentration and on 5-hydroxytryptamine metabolism in six different brain regions in the rat. *Acta Physiologica Scandinavica* 136:473-81.

Bloom, S. R., R. H. Johnson, D. M. Park, M. J. Rennie, and W. R. Sulaiman. 1976. Differences in the metabolic and hormonal responses to exercise between racing cyclists and untrained individuals. *Journal of Physiology* 258:1-18.

Bock, A. V., C. Vancoulert, D. B. Dill, A. Fölling, and L. M. Hurthal. 1928. Dynamic changes occurring in man at work. *Journal of Physiology* 66:136-61.

Bonen, A. A. N. Belcastro, K. MacIntyre, and J. Gardner. 1977. Hormonal responses during rest and exercise with glucose. *Medicine and Science in Sports* 9:64.

Bonen, A. K. MacIntyre, A. N. Belcastro, and G. Piarce. 1981. Effect of reduced hepatic and glycogen depots on substrate and endocrine response during menstrual cycle of teenage athletes. *Journal of Applied Physiology* 50:545-51.

Bonen, A., F. Haynes, W. Watson-Wright, M. M. Sopper, G. N. Pierce, M. P. Low, and E. Graham. 1983. Effects of menstrual cycle on metabolic responses to exercise. *Journal of Applied Physiology* 55:1506-13.

Bonen, A., W. Y. Ling, K. P. MacIntyre, R. Neil, J. C. McGroil, and A. N. Belcastro. 1979. Effects of exercise on the serum concentrations of FSH, LH, progesterone and estradiol. *European Journal of Applied Physiology* 49:15-23.

Bonen, A., M. H. Tan, P. Clune, and R. O. Kirby. 1985. Effects of exercise on insulin binding to human muscles. *American Journal of Physiology* 248:E403-8.

Bonifazi, M., E. Bela, G. Carli, L. Lodi, C. Lupo, E. Maioli, G. Murtelli, M. Paghi, A. Ruspetti, and A. Vitti. 1994. Responses of atrial natriuretic peptide and other fluid regulating hormones to long distance swimming in the sea. *European Journal of Applied Physiology* 68:504-7.

Boobis, L. H., C. Williams, and S. A. Wooton. 1983. Influence of sprint training on muscle metabolism during brief maximal exercise in man. *Journal Physiology* 342:36P-7P.

Booth, A., G. Shelley, A. Mazur, G. Thorp, and R. Kittok. 1989. Testosterone, and winning and losing in human competition. *Hormones and Behaviour* 23:556-71.

Booth, F. W. 1988. Perspectives on molecular and cellular exercise physiology. *Journal of Applied Physiology* 65:1461-71.

Booth, F. W., and D. B. Thomason. 1991. Molecular and cellular adaptation of muscle in response to exercise: perspectives of various models. *Physiological Review* 71:541-85.

Bosco, C. 1992. Eine neu Methodik zur Einschätzung und Programmierung der Trainings. *Leistungsports* 5:21-8.

Bosco, C. 1997. A new training method with Ergopower—Bosco system. *Acta Academia Olympiquae Estoniae* 5:24-34.

Bosco, C. 1999. *Strength assessment with the Bosco's Test*. Rome: Italian Society of Sport Science.

Bosco, C., A. Belli, M. Astrua, J. Tihanyi, R. Pozzo, S. Kellis, O. Tsarpela, C. Foti, R. Manno, and C. Tranquilli. 1995. Dynamometer for evaluation of dynamic muscle work. *European Journal of Applied Physiology* 70:379-86.

Bosco, C., R. Colli, R. Bonomi, S. P. von Duvillard, and A. Viru. 2000. Monitoring of strength training. Neuromuscular and hormonal profile. *Medicine and Science in Sports and Exercise* 32:202-8.

Bosco, C., P. Luhtanen, and P. V. Komi. 1983. A simple method for measurement of mechanical power in jumping. *European Journal of Applied Physiology* 50:273-82.

Bosco, C., J. Tihanyi, L. Rivalta, G. Parlato, C. Tranquilli, G. Pulverenti, C. Foti, M. Viru, and A. Viru. 1996a. Hormonal responses in strenuous jumping effort. *Japanese Journal of Physiology* 46:93-8.

Bosco, C., J. Tihanyi, and A. Viru. 1996b. Relationships between field fitness test and basal serum testosterone and cortisol levels in soccer players. *Clinical Physiology* 16:317-22.

Bosco, C., and A. Viru. 1998. Testosterone and cortisol levels in blood of male sprinters, soccer players and cross-country skiers. *Biology of Sport* 15:3-8.

Bouchard, C., A. W. Taylor, J. A. Simoneau, and S. Dulac. 1991. Testing anaerobic power and capacity. In *Physiological testing of the high performance athlete*, 2nd ed., ed. MacDougal, J. O., H. A. Wenger, and H. J. Green., 175-221. Champaign, IL: Human Kinetics.

Bouchard, C., R. J. Shephard, and T. Stephens. 1993. *Physical activity, fitness, and health*. Champaign, IL: Human Kinetics.

Bouhuys, A., J. Pool, and R. A. Binkhorst. 1966. Metabolic acidosis of exercise in healthy males. *Journal of Applied Physiology* 21:1040-6.

Bouissou, P., F. Perronet, G. Brisson, R. Helie, and M. Ledoux. 1987. Fluid-electrolyte shift and renin-aldosterone responses to exercise under hypoxia. *Hormone and Metabolic Research* 19:285-344.

Boulay, M. R., P. Hamel, J. A. Simoneau, G. Lortie, D. Prud'homme, and C. Bouchard. 1984. A test of aerobic capacity: description and reliability. *Canadian Journal of Applied Sports Sciences* 9:122-6.

Boulay, M. R., G. Lortie, J. A. Simoneau, P. Hamel, C. Leblanc, and C. Bouchard. 1985. Specificity of aerobic and anaerobic work capacities and power. *International Journal of Sports Medicine* 6:325-8.

Brahmi, Z., J. E. Thomas, M. Park, and I. R. G. Dowdeswell. 1985. The effect of acute exercise on natural killer-cell activity of trained and sedentary human subjects. *Journal of Clinical Immunology* 5:321-8.

Brandenberger, C., V. Canadas, M. Follenius, J. P. Liebert, and J. M. Kahn. 1986. Vascular fluid shifts and endocrine responses to exercise in the heat. *European Journal of Applied Physiology* 55:123-9.

Brandenberger, G., M. Follenius, B. Hietter, B. Reinhardt, and M. Simeoni. 1982. Feedback from meal-related peaks determines diurnal changes in cortisol response to exercise. *Journal of Clinical Endocrinology* 54:592-6.

Brandenberger, G., R. Schnedecker, K. Spiegel, B. Mettauer, B. Geny, J. Lampert, and J. Londsdorfer. 1995. Parathyroid function in cardiac transplant patients: evaluation during physical exercise. *European Journal of Applied Physiology* 70:401-6.

Broberg, S., and K. Sahlin. 1988. Hyperammonemia during prolonged exercise: an effect of glycogen depletion? *Journal of Applied Physiology* 65:2475-7.

Brodan, V., E. Kuhn, J. Pechard, and D. Tomnova. 1976. Changes of free amino acids in plasma in healthy subjects induced by physical exercise. *European Journal of Applied Physiology* 35:69-77.

Brooks, G. A. 1985. Anaerobic threshold: Review of the concept and directions for future research. *Medicine and Science in Sports and Exercise* 17:22-31.

Brooks, G. A. 1987. Amino acid and protein metabolism during exercise and recovery. *Medicine and Sciences of Exercise Sport* 19:S150-6.

Brooks, G. A. 2000. Intra- and extra-cellular lactate shuttles [review]. *Medicine and Science in Sports and Exercise* 32:790-9.

Brooks, G. A., T. D. Fahey, and T. P. White. 1996. *Exercise physiology. Human energetics and its applications*. Mountain View, London, Toronto: Mayfield.

Brooks, G. A., and J. Mercier. 1994. The balance of carbohydrate and lipid utilization during exercise: the crossover concept. *Journal of Applied Physiology* 76:2253-64.

Brooks, S., J. Burrin, M. E. Cheetham, G. M. Hall, T. Yeo, and C. Williams. 1988. The responses of the catecholamines and ß-endorphin to brief maximal exercise in man. *European Journal of Applied Physiology* 57:230-4.

Brothershood, J., B. Brozovic, and C. G. C. Pugh. 1975. Hematological status of middle and long-distance runners. *Clinical Sciences and Molecular Medicine* 48:139-45.

Bruce, R. A., F. Kusumi, and D. Hosmer. 1973. Maximal oxygen intake and nomographic assessment of functional aerobic impairment in cardiovascular disease. *American Heart Journal* 85:546-62.

Buckler, J. M. 1972. Exercise as a screening test for growth hormone release. *Acta Endocrinology* 69:219-29.

Buckler, J. M. H. 1973. The relationship between changes in plasma growth hormone levels and body temperature occurring with exercise in man. *Biomedicine* 19:193-7.

Buetiner, G. R. 1993. The peaking order of free radicals and antioxidants: lipid peroxidation, alpha-tocopherol, and ascorbate. *Archives of Biochemistry and Biophysics* 300:535-43.

Bugard, P., M. Henry, F. Plas, and P. Chailley-Bert. 1961. Les corticoids et l'aldosterone dans l'effort prolonge du sportif. Indication avec les metabolismes. *Review Pathologique Generale et Physiologique Clincale* 61:159-4.

Bullen, B. A., G. S. Surinar, I. Z. Beitins, D. B. Carr, S. M. Reppert, C. O. Dotson, M. DeFencl, E. V. Gervino, and J. W. McArthur. 1984. Endurance training effects on plasma hormonal responsiveness and sex hormone excretion. *Journal of Applied Physiology* 56:1453-63.

Bülow, J. 1988. Lipid mobilization and utilization. In *Principles of exercise biochemistry,* ed. Poortmans, J. R., 140-163. Basel: Karger.

Bülow, J., and J. Madsen. 1978. Human adipose tissue blood flow during prolonged exercise. *Pflügers Archive* 376:41-5.

Bülow, J., and J. Madsen. 1981. Influence of blood flow on fatty acid mobilization from lipolytically active adipose tissue. *Pflügers Archive* 390:169-74.

Buono, M. J., and J. E. Yeager. 1991. Increases in aldosterone precede those of cortisol during graded exercise. *Journal of Sports Medicine and Physical Fitness* 31:48-51.

Buono, M. J., J. E. Yeager, and J. A. Hodgdon. 1986. Plasma adrenocorticotropin and cortisol responses to brief high-intensity exercises in humans. *Journal of Applied Physiology* 61:1337-9.

Buresova, M., E. Gutmann, and M. Klicpera. 1969. Effect of tension upon rate of incorporation of amino acids into protein of cross-striated muscles. *Experimentia* 25:144-5.

Burge, C. M., and S. L. Skinnar. 1995. Determination of hemoglobin mass and blood volume with CO: evaluation and application of a method. *Journal of Applied Physiology* 79:623-31.

Burton, G. W., K. U. Ingold. 1989. Vitamin E as an in vitro and in vivo antioxidant. *Annals of New York Academy of Sciences* 570:7-22.

Burton, R. F. 1978. Intracellular buffering. *Respiratory Physiology* 33:51-8.

Bury, T., R. Marechal, P. Mahieu, and F. Pirnay. 1998. Immunological status of competitive football players during the training season. *International Journal of Sports Medicine* 19:364-8.

Buse, M. G. 1981. In vivo effects of branched amino acids on muscle protein synthesis in fasted rats. *Hormonal and Metabolic Research* 13:502-5.

Busse, M., N. Maassen, and K. M. Braumann. 1989. Interpretation of lactate values. *Moderne Athlete and Coach* 27:26-8.

Busso, T., K. Häkkinen, A. Pakarinen, C. Carasso, J. R. Lacour, P. V. Komi, and H. Kauhanen. 1990. A systems model of training responses and its relationship with hormonal responses in elite weightlifters. *European Journal of Applied Physiology* 61:48-58.

Busso, T., K. Häkkinen, A. Pakarinen, H. Kauhanen, P. V. Komi, and J. R. Lacour. 1992. Hormonal adaptations and modelled responses in elite weightlifters during 6 weeks of training. *European Journal of Applied Physiology* 64:381-6.

Bylund, P., E. Eriksson, E. Jansson, and L. Nordberg. 1981. A new biopsy needle for percutaneous biopsies of small skeletal muscles and erector spinae muscles. *International Journal of Sports Medicine* 2:119-20.

Bylund-Fellenius, A. C., T. Bjurö, G. Cederblad, J. Holm, K. Lundholm, M. Sjöström, K. A. Ängquist, and T. Schersten. 1977. Physical training in man. Skeletal muscle metabolism in relation to muscle morphology and running ability. *European Journal of Applied Physiology* 36:151-69.

Bylund-Fellenius, A. C., M. Davidsson, A. Arvidsson, A. Elander, and T. Schersten. 1982. Optimal conditions for assay of cytochrome-c-oxidase activity in human skeletal tissue. *Clinical Physiology* 2:71-9.

Byrnes, W. C., and P. M. Clarkson. 1986. Delayed onset of muscle soreness and training. *Clinical Sports Medicine* 5:605-14.

Byrnes, W. C., P. M. Clarkson, J. P. White, S. S. Hsieh, P. N. Frykman, and R. J. Maughan. 1985. Delayed onset muscle soreness following repeated bouts of downhill running. *Journal of Applied Physiology* 59:710-5.

Caiozzo, V. J., and F. Haddad. 1996. Thyroid hormone: modulation of muscle structure, function, and adaptive responses to mechanical loading. *Exercise and Sport Sciences Reviews* 24:321-61.

Calvo, F., J. L. Chichardo, F. Bandres, A. Lucia, M. Perez, J. Alvarez, L. L. Mojares, A. F. Vaquero, and J. C. Legido. 1997. Anaerobic threshold determination with analysis of salivary amylase. *Canadian Journal of Applied Physiology* 22:553-61.

Cannon, J. G., R. A. Fielding, M. A. Fiatarone, S. F. Orencole, C. A. Dinarello, and W. J. Evans. 1989. Increased interleukin 1 beta in human skeletal muscle after exercise. *American Journal of Physiology* 26:R451-5.

Cannon, J. G., and M. J. Kluger. 1983. Endogenous pyrogen activity in human plasma after exercise. *Science* 220:617-9.

Cannon, W. B. 1925. *Bodily changes in pain, hunger, fear and rage*. New York, London: D. Appleton.

Carlson, K. I., J. C. Marker, D. A. Arnall, M. L. Terry, H. L. Young, L. G. Lindsay, M. E. Bracken, and W. W. Winder. 1985. Epinephrine is unessential for stimulation of liver glycogenolysis during exercise. *Journal of Applied Physiology* 58:544-8.

Carlson, L. A., L.-G. Ekelund, and S. Fröberg. 1971. Concentration of triglycerides, phospholipids and glycogen in skeletal muscle and of free fatty acids and ß-hydroxybutyric acid in blood in man in response to exercise. *European Journal of Clinical Investigation* 1:248-54.

Carlson, M. G., W. L. Snead, J. O. Hill, N. Nurjahan, and P. J. Campbell. 1991. Glucose regulation of lipid metabolism in humans. *American Journal of Physiology* 261:E815-20.

Carlsten, A., B. Hallgren, R. Jagenburg, A. Svanborg, and L. Werkö. 1962. Arterial concentrations of free fatty acids and free amino acids in healthy human individuals at rest and at different work loads. *Scandinavian Journal of Clinical and Laboratory Investigations* 14:185-91.

Carlsten, A., B. Hallgren, R. Jagenburg, and L. Svan Werkö. 1961. Myocardial metabolism of glucose, la amino acids and fatty acids in healthy hum at rest and at different work load. *Scandinavian Journal Investigation* 13:418-28.

Caro, J. F., M. K. Sinha, J. W. Kolaczynski, P. L. Zhang, and P. V. Considine. 1996. Leptin: a tale of an obesity gene. *Diabetes* 45:1455-62.

Carr, D. B., B. A. Bullen, G. S. Surinar, M. A. Arnold, M. Rosenblatt, I. Z. Beitins, J. B. Martin, and J. M. McArthur. 1981. Physical conditioning facilitates the exercise-induced secretion of beta-endorphin and beta-lipotropin in women. *New England Journal of Medicine* 305:560-3.

Carraro, F., T. D. Kimbrough, and R. R. Wolfe. 1993. Urea kinetics in humans at two levels of exercise intensity. *Journal of Applied Physiology* 75:1180-5.

Carraro, F., C. A. Stuart, W. H. Harte, J. Rosenblatt, and R. R. Wolfe. 1990. Effect of exercise and recovery on muscle protein synthesis in human subjects. *American Journal of Physiology* 259:E470-6.

Carson, J. A. 1997. The regulation of gene expression in hypertrophying skeletal muscle. *Exercise and Sport Sciences Reviews* 25:301-20.

Carson, J. A., and F. W. Booth. 2000. Molecular biology of exercise. In *Exercise and sport sciences,* ed. Garrett, W. E., and D. T. Kirkendall, 251-264. Philadelphia: Lippincott Williams & Wilkins.

Casoni, I., C. Barsetto, A. Cavicchi, S. Martinelli, and F. Conconi. 1985. Reduced hemoglobin concentration and red cells hemoglobinization in Italian marathon and ultramarathon runners. *International Journal of Sports Medicine* 6:176-9.

Castenfors, J., F. Mossfeldt, and M. Piscator. 1967. Effect of prolonged heavy exercise on renal function and urinary protein excretion. *Acta Physiologica Scandinavica* 70:194-206.

Catapano, A. L. 1987. *High density lipoproteins. Physiopathological aspects and clinical significance.* New York, Raven Press.

Cathcart, E. P. 1925. The influence of muscle work on protein metabolism. *Physiological Review* 5:225-43.

Cazalets, J. R., Y. Sqalli-Houssaini, and F. Clarac. 1994. GABAergic inactivation of the central pattern generators for locomotion in isolated neonatal rat spinal cord. *Journal of Physiology* 474:173-81.

Cerny, F. 1975. Protein metabolism during two hour ergometer exercise. In *Metabolic adaptation to prolonged physical exercise*, ed. Howald, H., and J. R. Poortmans, 232-237. Basel: Birkhäuser.

Cerretelli, P., D. Pendergost, C. Marcow, and J. Piiper. 1986. Blood flow in exercising muscles. *International Journal of Sports Medicine* 7 (Suppl.):29-33.

Chailley-Bert, P., F. Plas, M. A. Henry, and P. Bugard. 1961. Les modifications metaboliques au cours d'effort prolonges chez le sportif. *Revue de Pathologie generale et de Physiologie clinique* 61:143-57.

Chalmers, R. J., S. R. Bloom, G. Duncan, R. H. Johnson, and W. R. Sulaimen. 1979. The effect of somatostatin on metabolic and hormonal changes during and after exercise. *Clinical Endocrinology* 10:451-8.

Chance, B., H. Sies, and A. Boveris. 1979. Hydroperoxide metabolism in mammalian organs. *Physiological Reviews* 59:527-605.

Chang, T. W., and A. L. Goldberg. 1978. The metabolic fates of amino acids and the formation of glutamine in skeletal muscle. *Journal of Biology Chemistry* 253:3685-95.

Chaouloff, F. 1993. Physiolopharmacological interactions between stress hormones and central serotonergic systems. *Brain Research Reviews* 18:1-32.

Chaouloff, F., J. L. Elgohozi, Y. Guezennec, and D. Laude. 1985. Effects of conditioned running on plasma, liver and brain tryptophan and on brain 5-hydroxytryptamine metabolism of the rat. *British Journal of Pharmacology* 86:33-41.

Chaouloff, F., G. A. Kennett, B. Serrurier, D. Merina, and G. Curson. 1986. Amino acid analysis demonstrates that increased plasma free tryptophan causes the increase of brain tryptophan during exercise in the rat. *Journal of Neurochemistry* 46:1647-50.

Chasiotis, D. 1983. The regulation of glycogen phosphorylase and glycogen breakdown in human skeletal muscle. *Acta Physiologica Scandinavica* (Suppl) 518:1-68.

Cheetham, M. E., C. Williams, and H. K. A. Lakomes. 1985. A laboratory sprint running test: metabolic responses of endurance and sprint trained athletes. *British Journal of Sports Medicine* 19:81-4.

Chesley, A., J. D. MacDougall, M. A. Tarnopolsky, S. A. Atkinson, and K. Smith. 1992. Changes in human muscle protein synthesis after resistance exercise. *Journal of Applied Physiology* 73:1383-8.

Chichardo, J. L., Legido, J. C., J. Avarez, L. Serrotosa, F. Bamdres, and C. Camella. 1994. Saliva electrolytes an useful tool for anaerobic threshold determination. *European Journal of Applied Physiology* 68:214-8.

Chogovadze, A. V., Y. I. Smirnova, and A. N. Sakrebo. 1988. Immunological reactivity of swimmers during the preparatory and competition period. *Sports Training, Medicine and Rehabilitation* 1:41-3.

Christie, M. J. 1982. Regional specificity of changes in (^{3}H)-leuenkephalin binding associated with warm water swimming in mice. *Neuroscience Letters* 33:197-202.

Christie, M. J., and G. B. Chester. 1983. (^{3}H)-leu-enkephalin binding following chronic swim-stress in mice. *Neuroscience Letters* 36:323-8.

Chwalbinska-Moneta J., H. Kaciuba-Uscilko, H. Krysztofiak, A. Ziemba, K. Krzeminski, B. Kruk, and K. Nazar. 1998. Relationship between EMG, blood lactate and plasma catecholamine threshold during graded exercise in men. *Journal of Physiology and Pharmacology* 49:433-4.

Chwalbinska-Moneta, J., H. Krysztofiak, A. Ziemba, K. Nazar, and H. Kaciuba-Uscilko. 1996. Threshold increases in plasma growth hormone in relation to plasma catecholamine and blood lactate concentrations during progressive exercise in endurance-trained athletes. *European Journal of Applied Physiology* 73:117-20.

Clarkson, P. M., F. S. Apple, W. C. Byrnes, K. McCarnick, and K. M. Triffletti. 1987. Creatine kinase isoforms following isometric exercise. *Muscle and Nerve* 10:41-4.

Clarkson, P. M., W. C. Byrnes, K. M. McCormick, L. P. Turcotte, and J. S. White. 1986. Muscle soreness and serum creatine kinase activity following isometric eccentric and concentric exercise. *International Journal of Sports Medicine* 7:152-5.

Clarkson, P. M., P. Litchfield, J. Graves, J. Kirwan, and W. Byrnes. 1985. Serum creatine kinase activity following forearm flexion isometric exercise. *European Journal of Applied Physiology* 53:368-71.

Clarkson, P. M., K. Nosaka, and B. Braun. 1992. Muscle function after exercise-induced muscle damage and rapid adaptation. *Medicine and Science in Sports and Exercise* 24:512-20.

Clausen, J. P. 1977. Effect of physical training on cardiovascular adjustments to exercise in man. *Physiological Reviews* 57:779-815.

Clausen, T., S. L. V. Andersen, and J. A. Flatman. 1993. Na^+-K^+ pump stimulation elicits recovery of contractility in K^+-paralysed rat muscle. *Journal of Physiology* 472:521-36.

Clausen, T., and M. E. Everts. 1991. K^+-induced inhibition of contractile force in rat skeletal muscle: role of active Na^+-K^+ transport. *American Journal of Physiology* 261:C799-807.

Clausen, T., O. B. Nielsen, A. P. Harrison, J. A. Flatman, and K. Overgaard. 1998. The Na^+-K^+ pump and muscle excitability. *Acta Physiologica Scandinavica* 162:183-90.

Clement, D. B., and L. L. Sanichuk. 1984. Iron status and sports performance. *Sports Medicine* 1:67-74.

Coggan, A. R., D. L. Habash, L. A. Mendenhall, S. C. Swanson, and C. L. Kien. 1993. Isotopic estimation of CO_2 production during exercise before and after endurance training. *Journal of Applied Physiology* 75:70-5.

Coggan, A. R., and B. D. Williams. 1995. Metabolic adaptations to endurance training: substrate metabolism during exercise. In *Exercise metabolism,* ed. Hargreaves, M., 177-210. Champaign, IL: Human Kinetics.

Colt, E. W. D., S. L. Wardlaw, and A. G. Frantz. 1981. The effect of running on plasma beta-endorphin. *Life Sciences* 28:1637-40.

Conlay, L. A., R. J. Wurthman, G. Lopez, I. Coviella, J. K. Blusztajn, C. A. Vacanti, M. Logue, M. During, B. Caballero, T. J. Maher, and G. Evoniuk. 1989. Effect of running the Boston Marathon on plasma concentrations of large neural amino acids. *Journal of Neural Transmiss* 76:65-71.

Conley, D. L., and G. S. Krahenbuhl. 1980. Running economy and distance running performance. *Medicine Sciences in Sports Exercise* 12:357-60.

Convertino, V. A., L. C. Keil, E. M. Bernauer, and J. E. Greenleaf. 1981. Plasma volume, osmolality, vasopressin, and renin activity during graded exercise in man. *Journal of Applied Physiology* 50:123-8.

Convertino, V. A., L. C. Keil, and J. E. Greenleaf. 1983. Plasma volume, renin and vasopressin responses to graded exercise after training. *Journal of Applied Physiology* 54:508-14.

Conzalez-Alonso, J. 1998. Separate and combined influences of dehydration and hyperthermia on cardiovascular responses to exercise. *International Journal of Sports Medicine* 19(Suppl. 2):S111-4.

Cook, N. J., A. Ng, G. F. Read, B. Harris, D. Riad-Fahmy. 1987. Salivary cortisol for monitoring adrenal activity during marathon runs. *Hormone Research* 25:18-23.

Coppack, S. W., M. D. Jensen, and J. N. Miles. 1994. In vivo regulation of lipolysis in humans. *Journal of Lipid Research* 35:177-93.

Core, C. J., A. G. Hahn, C. M. Burge, and R. D. Telford. 1997. $\dot{V}O_2$max and haemeglobin mass of trained athletes during high intensity training. *International Journal of Sports Medicine* 18:477-82.

Cornet, F., G. Heynen, A. Cession-Fossion, A. Adams, and J. M. Hoaft. 1978. Effects de l'exercise musculaire sur la calcémie, la clairance du calcium et la sécrétion d'hormone parathyroidienne. *Comptes Rendes Société de Biologie* 172:1245-9.

Costill, D. L. 1984. Energy supply in endurance activities. *International Journal of Sports Medicine* 5(Suppl.):19-21.

Costill, D. L. 1986. Muscle metabolism and electrolyte balance during heat acclimation. *Acta Physiologica Scandinavica* 128(Suppl. 556):111-8.

Costill, D. L., R. Bowers, G. Graham, and K. Sporks. 1971. Muscle glycogen utilization during prolonged exercise on successive days. *Journal of Applied Physiology* 31:834-8.

Costill, D. L., G. Branam, W. Fink, and R. Nelson. 1976b. Exercise-induced sodium conservation: changes in plasma renin and aldosterone. *Medicine and Science in Sports* 8:209-13.

Costill, D. L., R. Cote, and W. J. Fink. 1976c. Muscle water and electrolytes following varied levels of dehydration in man. *Journal of Applied Physiology* 40:6-11.

Costill, D. L., R. Cote, W. J. Fink, and P. Van Handel. 1981. Muscle water and electrolyte distribution during prolonged exercise. *International Journal of Sports Medicine* 2:130-4.

Costill, D. L., E. Coyle, G. Dalsny, W. Evans, W. Fink, and D. Hooper. 1977. Effects of elevated plasma FFA and insulin on muscle glycogen usage during exercise. *Journal of Applied Physiology* 43:695-9.

Costill, D. L., E. F. Coyle, W. F. Fink, G. R. Lesmes, and F. A. Witzman. 1979. Adaptation in skeletal muscle following strength training. *Journal of Applied Physiology* 46:96-9.

Costill, D. L., J. Daniels, W. Evans, W. Fink, G. Krahenbuhl, and B. Saltin. 1976a. Skeletal muscle enzymes and fiber composition in male and female track athletes. *Journal of Applied Physiology* 40:149-54.

Costill, D. L., W. J. Fink, L. H. Getchell, J. L. Ivy, and F. A. Witzmann. 1979. Lipid metabolism in skeletal muscle of endurance-trained males and females. *Journal of Applied Physiology* 47:787-91.

Costill, D. L., M. G. Flynn, J. P. Kirwan, J. A. Horward, R. Thomas, J. Mitchell, and S. H. Park. 1988. Effects of repeated days of intensified training on muscle glycogen and swimming performance. *Medicine and Science in Sports and Exercise* 20:249-54.

Costill, D. L., P. D. Gollnick, E. Jansson, B. Saltin, and E. M. Stein. 1973a. Glycogen depletion patterns in human muscle fibers during distance running. *Acta Physiologica Scandinavica* 89:374-83.

Costill, D. L., E. W. Maglischo, and A. B. Richardson. 1992. *Swimming. Handbook of Sports Medicine and Science*. London: Blackwell Scientific.

Costill, D. L., H. Thomason, and B. Roberts. 1973b. Fractional utilization of the aerobic capacity during distance running. *Medicine and Sciences in Sports* 5:248-52.

Cotshalk, L. A., C. C. Loebel, M. Nindl, Putukian, W. J. Sebastianelli, R. U. Newton, K. Häkkinen, and W. J. Kraemer. 1997. Hormonal responses of multiset, versus singleset heavy-resistance exercise protocols. *Canadian Journal of Applied Physiology* 22:244-55.

Counsilman, J. E. 1968. *The science of swimming*. Englewood Cliffs, NJ: Prentice Hall.

Cousineau, D., R. J. Fergusson, J. de Champlain, P. Gauthier, P. Cote, and M. Bourassa. 1977. Catecholamines in coronary sinus during exercise in man before and after training. *Journal of Applied Physiology* 43:801-8.

Coyle, E. F. 1998. Cardiovascular drift during prolonged exercise and the effects of dehydration. *International Journal of Sports Medicine* 19 (Suppl. 2):S121-4.

Crampes, F., M. Beaville, D. Riviere, and M. Garrigues. 1986. Effect of physical training in humans on the response of isolated fat cells to epinephrine. *Journal of Applied Physiology* 61:25-9.

Crim, M. C., D. H. Calloway, and S. Margen. 1975. Creatine metabolism in man: urinary creatine excretion with creatine feeding. *Journal of Nutrition* 105:428-38.

Critz, J. B., and A. W. Merrick. 1962. Serum glutamic-oxalacetic transaminase levels after exercise in men. *Proceedings for Society of Experimental Biology* 109:608-10.

Cumming, D. C., and R. W. Rebar. 1985. Hormonal changes with acute exercise and with training in women. *Seminar Reproductive Endocrinology* 3:55-64.

Cuneo, R. C., E. A. Espiner, M. G. Nichalls, and T. G. Yemdle. 1988. Exercise induced increase in plasma natriuretic peptide and effect of sodium loading in normal man. *Hormone and Metabolic Research* 20:115-7.

Czerwinski, S. M., T. G. Kurowski, T. M. O'Neil, and R. C. Hickson. 1987. Initiating regular exercise protects against muscle hypertrophy from glucocorticoids. *Journal of Applied Physiology* 63:1504-10.

Danhaive, P. A., and G. G. Rousseau. 1988. Evidence for sex-dependent anabolic response to androgenic steroids mediated by glucocorticoid receptor in rat. *Journal of Steroid Biochemistry* 29:275-81.

Darmaun, D., and P. Dechelotte. 1991. Role of leucine as a precursor of glutamine α-amino nitrogen in humans. *American Journal of Physiology* 260:E326-9.

Darmaun, D., D. E. Matthews, and D. M. Bier. 1988. Physiological hypercortisolemia increases proteolysis, glutamine and alanine production. *American Journal of Physiology* 225:E366-73.

Davidson, R. J. L., L. D. Robertson, G. Galea, and R. J. Maughan. 1987. Hematological changes associated with marathon running. *International Journal of Sports Medicine* 8:19-25.

Davies, C. T. M., and J. D. Few. 1973. Effect of exercise on adrenocortical function. *Journal of Applied Physiology* 35:887-91.

Davies, C. T. M., and J. D. Few. 1976. Effect of hypoxia on adrenocortical response to exercise. *Journal of Endocrinology* 71:157-8.

Davies, C. T. M., J. Few, K. G. Foster, and A. J. Sargent. 1974. Plasma catecholamine concentration during dynamic exercise involving different muscle groups. *European Journal of Applied Physiology* 32:195-206.

Davies, K. J., T. Quantanilla, G. A. Brooks, and L. Packer. 1982. Free radicals and tissue damage produced by exercise. *Biochemistry and Biophysics Research Communication* 107:1198-205.

Davis, J. A. 1995. Direct determination of aerobic power. In *Physiological assessment of human fitness,* ed. Maud, and C. Foster, 9-17. Champaign, IL: Human Kinetics.

Davis, J. A., M. H. Frank, B. J. Whipp, and K. Wasserman. 1979. Anaerobic threshold alterations caused by endurance training in middle-aged men. *Journal of Applied Physiology* 46:1039-46.

Davis, J. M. 1994. Nutritional influences on central mechanisms of fatigue involving serotonin. In *Biochemistry of exercise IX,* ed. Maughan, R. J., and S. M. Shirreffs, 445-455. Champaign, IL: Human Kinetics.

Davis, J. M., S. P. Bailey, J. A. Woods, F. J. Galiano, M. T. Hamilton, and W. P. Bartoli. 1992. Effects of carbohydrate feedings on plasma free tryptophan and branched-chain amino acids during prolonged cycling. *European Journal of Applied Physiology* 65:513-9.

Dawson, A. A., and A. Ogston. 1969. Exercise-induced thrombocytosis. *Acta Haemotologica* 42:241-6.

Decombaz, J., P. Reinhardt, K. Anantharaman, G. V. Glutz, and J. R. Poortmans. 1979. Biochemical changes in a 100km run: free amino acids, urea and creatinine. *European Journal of Applied Physiology* 41:61-72.

Delanne, R. 1952. Stress, adaptation et exercise musculaire. *Medicine, éducation physique et sport* 26(3):7-21.

Demos, M. A., E. L. Gitin, and L. J. Kagen. 1974. Exercise myoglobinemia and acute exertional rhabdomyolysis. *Archives of Internal Medicine* 134:669-73.

De Paoli Vitali, E., C. Guglielmini, I. Casoni, M. Vedovato, P. Gilli, A. Farinelli, G. Saliatorelli, and F. Conconi. 1988. Serum erythropoietin in cross-country skiers. *International Journal of Sports Medicine* 9:99-101.

De Souza, M. J., C. M. Maresh, M. S. Maguire, W. J. Kraemer, G. Flora-Ginter, and K. L. Goetz. 1989. Menstrual status and plasma vasopressin, renin activity, and aldosterone exercise responses. *Journal of Applied Physiology* 67:736-43.

Despres, J. P., C. Bouchard, R. Savard, A. Tremblay, M. Marcotte, and M. Theriault. 1984. Effect of exercise training and detraining on fat cell lipolysis in men and women. *European Journal of Applied Physiology* 53:25-30.

Dessypris, A., K. Kuoppasalmi, and H. Adlercreutz. 1976. Plasma cortisol, testosterone, androstenedione and luteinizing hormone (LH) in a noncompetitive marathon run. *Journal of Steroid Biochemistry* 7:33-7.

Dessypris, A., G. Wägar, F. Fyhrquist, T. Mäkiness, M. G. Welin, and B. A. Lamberg. 1980. Marathon run: effects on blood cortisol—ACTH, iodothyronines—TSH and vasopressin. *Acta Endocrinologica* 95:151-7.

Devol, D. L., P. Rotwein, J. L. Sadow, J. Novanofski, and P. J. Bechtel. 1990. Activation of insulin-like growth factor gene expression during work-induced skeletal muscle growth. *American Journal of Physiology* 259:E89-95.

DeVries, H. A. 1974. *Physiology of exercise for physical education and athletics.* 2d ed. Dubuque, Iowa: C. Brown.

Diamond, P., G. R. Brisson, B. Candas, and F. Peronnet. 1989. Trait anxiety, submaximal physical exercise and blood androgens. *European Journal of Applied Physiology* 58:699-704.

DiCarlo, S. E., C.-Y. Chan, and H. L. Collins. 1996. Onset of exercise increases lumbar sympathetic nerve activity in rats. *Medicine and Science of Sports and Exercise* 28:677-84.

Dickhuth, H. H., W. Aufenanger, P. Schmidt, G. Simon, M. Huonker, and J. Keul. 1989. Möglichkeiten und Grenzen der Leistungsdiagnostes und Trainingssteuerung im Mittel und Lang strechenlauf. *Leistungssport* 19(4):21-4.

Dickson, D. N., R. L. Wilkinson, and T. D. Noaks. 1982. Effects of ultramarathon training and racing on hematologic parameters and serum ferritin levels in well-trained athletes. *International Journal of Sports Medicine* 3:111-7.

Dill, D.B., and D. L. Costill. 1974. Calculation of percentage changes in volumes of blood, plasma and red cells in dehydration. *Journal of Applied Physiology* 2:247-48.

Dill, D. B., J. H. Talbot, and H. T. Edwards. 1930. Studies in muscular activity. VI. Response of several individuals to a fixed task. *Journal of Physiology* 69:267-305.

Dillard, C. J., R. E. Litov, W. M. Savin, E. E. Dumelin, and A. L. Tappel. 1978. Effects of exercise, vitamin E, and ozone in pulmonary function and lipid peroxidation. *Journal of Applied Physiology* 45:927-32.

Dintiman, G., B. Ward, and T. Tellez. 1998. *Sports speed.* 2nd ed. Champaign, IL: Human Kinetics.

Di Prampero, P. E. 1981. Energetics of muscular exercise. *Review of Physiology, Biochemistry and Pharmacology* 89:144-222.

Dohm, G. A. 1986. Protein as a fuel for endurance exercise. *Exercise Sports Sciences Review* 14:143-73.

Dohm, G. L., G. R. Beecher, R. Q. Warren, and R. T. Williams. 1981. Influence of exercise on free amino acid concentration in rat tissues. *Journal of Applied Physiology* 50:41-4.

Dohm, G. L., A. L. Hecker, W. E. Brawn, G. J. Klain, F. R. Puente, E. W. Askew, and G. R. Beecher. 1977. Adaptation of protein metabolism to endurance training. *Biochemical Journal* 164:705-8.

Dohm, G. L., R. G. Israel, R. L. Breedlove, B. L. Williams, and E. W. Askew. 1985. Biphasic changes in 3-methylhistidine excretion in humans after exercise. *American Journal of Physiology* 248:588-92.

Dohm, G. L., M. K. Sinha, and J. F. Caro. 1987. Insulin receptor binding and protein kinase activity in muscles of trained rats. *American Journal of Physiology* 252:E170-5.

Dohm, G. L., R. T. Williams, G. J. Kasperek, and A. M. Rij. 1982. Increased excretion of urea and N^{t}-methylhistidine by rats and humans after a bout of exercise. *Journal of Applied Physiology* 52:27-33.

Dolny, D. G., and P. W. R. Lemon. 1988. Effect of ambient temperature on protein breakdown during prolonged exercise. *Journal of Applied Physiology* 64:550-5.

Donath 1970. Enzymologie in der Sportmedizin. *Medizina und Sport* 10:2-8.

Donath, R., C. Clauswitzer, W. Rockstrach, and S. Israel. 1969a. Die Ausscheidung der unkonjugierten 11-Hydroxykortikoide im Harn bei extremen Ausbelastungen. *Medizin und Sport* 9:117-23.

Donath, R., C. Clauswitzer, and K.-P. Schüler. 1969b. Zur Bewertung des Blutlaktatverhalten in der sportmedizinischen Funktiondiagnostiks. *Medizin und Sport* 9:355-9.

Dons, B., K. Bollerup, F. Bonde-Peterson, and S. Hanke. 1978. The effect of weight-lifting exercises related to muscle fiber composition and muscle cross-sectional area in humans. *European Journal of Applied Physiology* 40:95-106.

Doroshenko, N. I. 1976. Study of training and competition loads in runners for middle and long distances [in Russian]. Thesis acad. diss. Moscow: Central Institute Physical Culture.

Dotan, R., and O. Bar-Or. 1983. Load optimization from the Wingate anaerobic test. *European Journal of Applied Physiology* 51:409-17.

Dudley, G. A., W. M. Abraham, and R. L. Terjung. 1982. Influence of exercise intensity and duration on biomechanical adaptations in skeletal muscle. *Journal of Applied Physiology* 53:844-50.

Dudley, G. A., R. S. Staron, T. F. Murray, F. C. Hagerman, and A. Luginbuhl. 1983. Muscle fibre composition and blood ammonia levels after intense exercise in humans. *Journal of Applied Physiology* 54:582-6.

Dufaux, B., G. Assmann, and W. Hollmann. 1982. Plasma lipoproteins and physical activity: a review. *International Journal of Sports Medicine* 3:123-36.

Dufaux, B., G. Assmann, V. Order, A. Hoederath, and W. Hollmann. 1981a. Plasma lipoproteins, hormones, and energy substrates during the first days after prolonged exercise. *International Journal of Sports Medicine* 2:256-60.

Dufaux, B., A. Hoederath, L. Streitberger, W. Hollmann, and G. Assmann. 1981b. Serum ferritin, transferrin, haptoglobin and iron in middle and long-distance runners, elite rowers and professional racing cyclists. *International Journal of Sports Medicine* 2:43-6.

Dux, L., E. Dux, and F. Guba. 1982. Further data on the androgenic dependence of the skeletal musculature: the effect of pubertal castration on the structural development of the skeletal muscle. *Hormonal and Metabolic Research* 14:191-4.

Ebbeling, C. B., and P. M. Clarkson. 1989. Exercise-induced damage and adaptation. *Sports Medicine* 7:207-34.

Eckardt, K. U., U. Bouttellier, A. Kurtz, M. Schopen, E. P. Kollen, and C. Bauer. 1989. Rate of erythropoietin formation in human in response to acute hypobaric hypoxia. *Journal of Applied Physiology* 66:1785-8.

Eckardt, K. U., A. Kurtz, and C. Bauer. 1990. Triggering of erythropoietin product2ion by hypoxia is inhibited by respiratory and metabolic acidosis. *American Journal of Physiology* 258:R678-83.

Edgerton, V. R., B. Essen, B. Saltin, and D. R. Simpson. 1975. Glycogen depletion in specific types of human skeletal muscle fibers in intermittent and continuous exercise. In *Metabolic adaptation to prolonged physical exercise,* ed. Howald, H., and J. R. Poortmans, 402-416. Basel: Birkhäuser.

Edwards, R. H. T. 1981. Human muscle function and fatigue. In *Human muscle fatigue: physiological mechanisms,* ed. Porter, R., and J. Whelen, 1-18. Ciba Foundation Symposium. No 82, London: Pitman Medical Books.

Edwards, R. H. T. 1983. Biochemical bases of fatigue in exercise performance: catastrophe theory of muscular fatigue. In *Biochemistry of exercise,* ed. Knuttgen, H. G., J. A. Vogel, and J. Poortmans, 3-28. Champaign, IL: Human Kinetics.

Edwards, R. H. T., J. M. Round, and D. A. Jones. 1983. Needle biopsy of skeletal muscle: a review of 10 years experience. *Muscle and Nerve* 6:676-83.

Egoroff, A. 1924. Die Veränderung der Blutbilder während Muskelarbeit bei Gesunden. *Zeitschift für die klinische Medizin* 100:485-97.

Eichner, E. R. 1985. Runner's macrocytosis: a clue to footstrike hemolysis. *American Journal of Medicine* 78:321-5.

Ekblom, B. 1970. Effects of physical training on circulation during prolonged severe exercise. *Acta Physiologica Scandinavica* 78:145-58.

Elder, G. C. B., K. Bradbury, and R. Roberts. 1982. Variability of fiber type distributions within human muscles. *Journal of Applied Physiology* 53:1473-80.

Elias, A. N., A. F. Wilson, M. R. Pandian, G. Chune, A. Utsuki, R. Kayaleh, and S. C. Stone. 1991. Corticotropin releasing hormone and gonadotropin secretion in physically active males after acute exercise. *European Journal of Applied Physiology* 62:171-4.

Elias, M. 1981. Serum cortisol, testosterone, and testosterone-binding globulin response to competitive fighting in human males. *Aggressive Behaviour* 7:215-24.

Elia, M., A. Schlatmann, A. Goren, and S. Austin. 1989. Amino acid metabolism in muscles and in the whole body of man before and after ingestion of a single mixed meal. *American Journal of Clinical Nutrition* 49:1203-10.

Eller, A., and A. Viru. 1983. Alterations of the content of free amino acids in skeletal muscle during prolonged exercise. In *Biochemistry of exercise,* ed. Knuttgen, H. G., J. A. Vogel, and J. Poortmans, 363-366. Champaign, IL: Human Kinetics.

El-Sayed, M. S., B. Davies, and D. B. Morgan. 1990. Vasopressin and plasma volume responses to submaximal and maximal exercise in man. *Journal of Sports Medicine and Physical Fitness* 30:420-5.

Engfred, K., M. Kjaer, N. H. Secher, D. B. Friedman, B. Hanel, O. J. Nielsen, F. W. Bach, H. Galbo, and B. D. Levine. 1994. Hypoxia and training-induced adaptation of hormonal responses to exercise in humans. *European Journal of Applied Physiology* 68:303-9.

Enschede, F. A., and A. I. Jongblood. 1964. The physical condition of top-skaters during training. *Internationale Zeitschift für die angevante Physiologie* 20:252-7.

Eränkö, O., M. J. Karvonen, and L. Räisänen. 1962. Long-term effects of muscular work on the adrenal medulla of the rat. *Acta Endocrinologica* 39:285-7.

Eriksson, B. O., P. D. Gollnick, and B. Saltin. 1973. Muscle metabolism and enzyme activities after training in boys 11-13 years old. *Acta Physiologica Scandinavica* 87:485-97.

Eriksson, L. S., S. Broberg, O. Björkman, and J. Wahren. 1985. Ammonia metabolism during exercise in man. *Clinical Physiology* 5:325-36.

Eskola, J., O. Ruuskanen, E. Sopp, M. K. Viljanen, M. Jarvinen, H. Toivonen, and K. Koualainen. 1978. Effect of sport stress on lymphocyte transformation and antibody formation. *Clinical Experimental Immunology* 32:339-45.

Espersen, G. T., A. Elbaek, E. Ernst, E. Toft, S. Kaalund, C. Jersild, and N. Grunnet. 1990. Effect of physical exercise on cytokines and lymphocyte subpopulations in human peripheral blood. *Acta Pathologica Microbiologica et Immunologica Scandinavica* 98:395-400.

Essig, D. A. 1996. Contractile activity-induced mitochondrial biogenesis in skeletal muscle. *Exercise and Sport Sciences Reviews* 24:289-319.

Evans, M. J., C. N. Meredith, J. C. Cannon, C. A. Dinarello, W. R. Fronters, V. A. Hughes, B. H. Jones, and H. G. Knuttgen. 1986. Metabolic changes following eccentric exercise in trained and untrained men. *Journal of Applied Physiology* 61:1864-8.

Evans, W. J., S. D. Phinney, and V. R. Young. 1982. Suction applied to a muscle biopsy maximizes sample size. *Medicine and Science in Sports and Exercise* 14:101-2.

Exton, J. H., N. Friedmann, E. H. Wong, J. P. Brineaux, J. D. Corbin, and C. R. Park. 1972. Interaction of glucocorticoids with glucagon and epinephrine in the control of gluconeogenesis and glycogenolysis in liver and lipolysis in adipose tissue. *Journal of Biological Chemistry* 247:3579-88.

Eynde, E. V. 1983. Training for long distance runners. *Modern Athlete and Coach* 21(2):35-7.

Fagard, R., A. Amery, T. Reybrouck, P. Lijnen, L. Billiet, M. Bogaert, E. Moerman, and A. De Schaepdryver. 1978. Effects of angiotensin antagonism at rest and during exercise in sodium-deplete man. *Journal of Applied Physiology* 45:403-7.

Fagard, R., A. Amery, T. Reybrouck, P. Lijnen, B. Moerman, M. Bagaert, and A. De Schaepdryver. 1977. Effects of angiotensin antagonism on hemodynamics, renin and catecholamines during exercise. *Journal of Applied Physiology* 43:440-4.

Fagraeus, L., J. Häggendahl, and D. Linnarsson. 1973. Heart rate, arterial blood pressure and noradrenaline levels during exercise with hyperbaric oxygen and nitrogen. *Swedish Journal of Defence Medicine* 9:265-70.

Fain, J. N. 1979. Inhibition of glucose transport in rat cells and activation of lipolysis by glucocorticoids. In *Glucocorticoid hormone action,* ed. Baxter, J. D., and G. G. Rousseau, 547-560. Berlin, Heidelberg, New York: Springer.

Fain, J. N., V. P. Kovacev, and R. O. Scow. 1965. Effect of growth hormone and dexamethasone on lipolysis and metabolism in isolated fat cells of the rat. *Journal of Biological Chemistry* 240:3522-9.

Fain, J. N., V. P. Kovacev, and R. O. Scow. 1966. Antilipolytic effect on insulin in isolated fat cells of the rat. *Endocrinology* 78:773-8.

Fain, J. N., R. O. Snow, and S. S. Chornick. 1963. Effects of glucocorticoids on metabolism of adipose tissue in vitro. *Journal of Biological Chemistry* 238:54-8.

Farfel, V. S., M. S. Zakharov, M. A. Kurakin, and N. N. Nikolayeva. 1972. Motor manifestations of fatigue in long distances. In *Physiological characteristics and methods of assessment of endurance in sports* [in Russian], ed. Zimkin, 68-81. Moscow: FiS.

Farjanel, J., C. Denis, J. C. Chatard, and A. Geyssant. 1997. An accurate method of plasma volume measurement by direct analysis of Evans blue spectra in plasma without dye extraction: origins of albumin-space variations during maximal exercise. *European Journal of Applied Physiology* 75:75-82.

Farrell, P. A., W. K. Gates, M. Mausnal, and W. P. Morgan. 1982. Increases in plasma ß-endorphin, ß-lipotropin immunoreactivity after treadmill running in humans. *Journal of Applied Physiology* 52:1245-9.

Farrell, P. A., M. Kjaer, F. W. Bach, and H. Galbo. 1987. Beta-endorphin and adrenocortical response to supramaximal exercise. *Acta Physiologica Scandinavica* 130:619-25.

Farrell, P. A., J. H. Wilmore, E. F. Coyle, J. E. Billing, and D. L. Costill. 1979. Plasma lactate accumulation and distance running performance. *Medicine and Science in Sports* 11:338-44.

Fehr, H.-G., H. Lotzerich, and H. Michna. 1989. Human macrophage function and physical exercise: phagocytic and histochemical studies. *European Journal of Applied Physiology* 58:613-7.

Feldmann, N., M. Bedu, G. Boudet, M. Mage, M. Sagnol, J.-M. Pequignot, B. Claustrat, B. Jocelyne, L. Peyrin, and J. Coudert. 1992. Inter-relationships between pituitary-adrenal hormones and catecholamines during a 6-day Nordic ski race. *European Journal of Applied Physiology* 64:258-65.

Feldmann, N., M. Sagnol, M. Bedu, G. Falgairette, E. Van Pragh, G. Gaillard, P. Jouanel, and J. Coudert. 1988. Enzymatic and hormonal responses following a 24-h endurance race and a 10-h triathlon race. *European Journal of Applied Physiology* 57:545-53.

Felig, P. 1973. The glucose-alanine cycle. *Metabolism* 22:179-207.

Felig, P. 1977. Amino acid metabolism in exercise. *Annals of New York Academy of Sciences* 301:56-63.

Felig, P., J. D. Baxter, A. E. Broadus, and L. A. Frohman. 1987. *Endocrinology and metabolism.* 2d ed. New York: McGraw-Hill.

Felig, P., A. Cherif, A. Minagawa, and J. Wahren. 1982. Hypoglycemia during prolonged exercise in normal men. *New England Journal of Medicine* 306:895-900.

Felig, P., and E. J. Wahren. 1971. Amino acid metabolism in exercising man. *Journal of Clinical Investigation* 50:2703-14.

Felig, P., and J. Wahren. 1979. The role of insulin and glucagon in the regulation of hepatic glucose production during exercise. *Diabetes* 28(Suppl 1):71-5.

Fell, J. W., J. M. Rayfield, J. P. Gulbin, and P. T. Gaffney. 1998. Evaluation of the Accusport Lactate Analyzer. *International Journal of Sports Medicine* 19:199-204.

Fenn, W. O. 1936. Electrolytes in muscle. *Physiological Reviews* 450-87.

Fenn, W. O. 1939. The deposition of potassium and phosphate with glycogen in rat liver. *Journal of Biological Chemistry* 128:297-307.

Fentem, P. H., I. A. Macdonald, B. Munoz, and S. A. Watson. 1985. Catecholamine responses to brief maximum exercise in man. *Journal of Physiology* 361:89P.

Fergusson, D. B., D. A. Price, and S. Wallace. 1980. Effects of physiological variables on the concentration of cortisol in human saliva. *Advances of Physiological Sciences* 28:301-11.

Fernstrom, J. 1983. Role of precursor availability in control of monoamine biosynthesis in brain. *Physiological Reviews* 63:484-546.

Ferry, A., F. Picard, A. Duvallet, B. Weill, and M. Rieu. 1990. Changes in blood leucocyte populations induced by acute maximal and chronic submaximal exercise. *European Journal of Applied Physiology* 59:435-42.

Few, J. D., G. C. Cashmore, and G. Turton. 1980. Adrenocortical response to one-leg and two-leg exercise on a bicycle ergometer. *European Journal of Applied Physiology* 44:167-74.

Few, J. D., F. J. Imms, and J. C. Weiner. 1975. Pituitary-adrenal response to static exercise in men. *Clinical Science and Medicine* 49:201-6.

Few, J. D., and D. E. Worsley. 1975. Human pituitary-adrenal response to hypothermia. *Journal of Endocrinology* 66:141-2.

Fick, A., and J. Wislicenus. 1866. On the orgin of muscle power. *Philosophical Magazine* 31:485-503.

Fimbel, S., A. Abdelmaki, B. Mayet, R. Sempore, and R. J. Favier. 1991. Exercise training fails to prevent glucocorticoid-induced atrophy from glucocorticoids [abstract]. Eighth International Biochemistry Exercise Conference. Nagoya, 146.

Fischer, E. H., M. G. Heilmeyer, and R. H. Haschke. 1971. Phosphorylase and the control of glycogen degradation. *Current Topics of Cell Regulation* 3:211-51.

Fitts, R. H. 1994. Cellular mechanisms of muscle fatigue. *Physiological Reviews* 74:49-94.

Fitzgerald, L. 1991. Overtraining increases the susceptibility to infection. *International Journal of Sports Medicine* 12(Suppl.):S5-S8.

Fleck, S. J., and W. J. Kraemer.1997. *Designing resistance programs*. Champaign, IL: Human Kinetics.

Flynn, M. G., D. L. Costill, J. A. Hawley, W. J. Fink, P. D. Neifer, R. A. Fielding, and M. D. Sleeper. 1987. Influence of selected carbohydrate drinks on cycling performance and glycogen use. *Medicine and Science in Sports and Exercise* 19:37-40.

Flynn, M. G., F. X. Pizza, J. B. Boone, F. F. Andres, T. A. Michaud, and J. R. Rodriguez-Zayas. 1994. Indices of training stress during competitive running and swimming seasons. *International Journal of Sports Medicine* 15:21-6.

Follenius, M., and G. Brandenberger. 1988. Increase in atrial natriuretic peptide in response to physical exercise. *European Journal of Applied Physiology* 57:159-82.

Follenius, M., V. Candas, B. Bothorel, and G. Brandenberger. 1989. Effect of rehydration on atrial natriuretic peptide release during exercise in the heat. *Journal of Applied Physiology* 66:2516-21.

Foster, C. 1998. Monitoring training in athletes with reference to overtraining syndrome. *Medicine and Science in Sports and Exercise* 30:1164-8.

Foster, C., L. L. Hector, K. S. McDonald, and A. C. Snyder. 1995. Measurement of anaerobic power and capacity. In *Physiological assessment of human fitness,* ed. Maud, P. J., and C. Foster, 73-85. Champaign, IL: Human Kinetics.

Foster, C., and M. Lehmann. 1997. Overtraining syndrome. In *Running injuries,* ed. Guten, 173-188. Philadelphia: Saunders.

Fowler, W. M., S. R. Chowdburg, L. M. Pearson, G. Gardner, and R. Bratton. 1962. Changes in serum enzyme levels after exercise in trained and untrained subjects. *Journal of Applied Physiology* 17:943-6.

Fowler, W. M., G. W. Gardner, H. H. Kazerunian, and W. A. Lanostad. 1968. The effect of exercise on serum enzymes. *Archives of Physical Medicine and Rehabilitation* 49:554-65.

Fox, E. L. 1975. Differences in metabolic alterations with sprint versus endurance interval training programs. In *Metabolic adaptation to prolonged physical exercise,* ed. Howald, H., and J.R. Poortmans, 119-126. Basel: Birkhäuser Verlag.

Fox, E. L., and D. Mathews. 1974. *Interval training: Conditioning for sports and general fitness.* Philadelphia: Saunders.

Foxdal, P., B. Sjödin, H. Rudstam, C. Östman, B. Östman, and G. C. Hedenstierna. 1990. Lactate concentration differences in plasma, whole blood, capillary finger blood and erythrocytes during submaximal graded exercise in humans. *European Journal of Applied Physiology* 61:218-22.

Fraioli, F., C. Moretti, D. Paolucci, E. Aicicco, F. Crescenzi, and G. Fortunio. 1980. Physical exercise stimulates marked concomitant release of ß-endorphin and adrenocorticotropic hormone (ACTH) in peripheral blood in man. *Experientia* 36:987-9.

Francis, K. T. 1979. Effect of water and electrolyte replacement during exercise in the heat on biochemical indices of stress and performance. *Aviation, Space and Environmental Medicine* 50:115-9.

Francis, K. T., and R. MacGregor. 1978. The effect of exercise in the heat on plasma renin and aldosterone with either water or a potassium-rich electrolyte solution. *Aviation, Space and Environmental Medicine* 49:461-5.

Frewin, D. B., A. G. Frantz, and J. A. Downey. 1976. The effect of ambient temperature on the growth hormone and prolactin response to exercise. *Australian Journal of Experimental Biology and Medical Sciences* 54:97-101.

Friden, J., M. Sjöström, and B. Ekblom. 1983. Myofibrillar damage following intense eccentric exercise in man. *International Journal of Sports Medicine* 4:170-6.

Friedman, J. M. 2000. Obesity in the new millennium. *Nature* 404:632-34.

Friedmann, B., and W. Kindermann. 1989. Energy metabolism and regulatory hormones in women and men during endurance exercise. *European Journal of Physiology* 59:1-9.

Fröhlich, J., A. Urhausen, U. Seul, and W. Kindermann. 1989. Beinflussung der individuellen anaeroben Schwelle durch kohlenhydratarme und reiche Ernährung. *Leistungssport* 19(4):18-20.

Fry, A. C., W. J. Kraemer, and C. T. Ransay. 1998. Pituitary-adrenal-gonadal responses to high-intensity resistance exercise overtraining. *Journal of Applied Physiology* 85:2352-9.

Fry, R. W., A. R. Morton, P. Garcia-Webb, G. P. M. Crawford, and D. Keart. 1992. Biological responses to overload training in endurance sports. *European Journal of Applied Physiology* 64:335-44.

Fry, R. W., A. R. Morton, P. Garcia-Webb, and D. Heast. 1991. Monitoring exercise stress by changes in metabolic and hormonal response over a 24-h period. *European Journal of Applied Physiology* 63:228-34.

Fry, A. C., W. J. Kraemer, M. H. Stone, B. J. Warren, S. J. Fleck, J. T. Kearney, and S. E. Gordon. 1994. Endocrine responses to overreaching before and after 1 year of weightlifting. *Canadian Journal of Applied Physiology* 19:400-10.

Fryburg, D. A., R. A. Gelfand, and E. J. Barrett. 1991. Growth hormone acutely stimulates forearm muscle protein synthesis in normal humans. *American Journal of Physiology* 260:E499-504.

Fulks, R. M., J. B. Li, and A. L. Goldberg. 1975. Effect of insulin, glucose, and amino acids on protein turnover in rat diaphragm. *Journal Biological Chemistry* 250:290-8.

Gabriel, H., and W. Kindermann. 1995. Flow cytometry. Principles and applications in exercise immunology. *Sports Medicine* 20:302-20.

Gabriel, H., H.-J. Muller, L. Brechted, A. Urhausen, and W. Kindermann. 1992b. Increased phagocytic capacity of the blood, but decreased phagocytic activity per individual circulating neutrophil after an ultradistance run. *European Journal of Applied Physiology* 71:281-4.

Gabriel, H., L. Schwarz, P. Born, and W. Kindermann. 1991b. Differential mobilization of leucocyte and lymphocyte subpopulation into the circulation during exercise to exhaustion. *European Journal of Applied Physiology* 63:449-57.

Gabriel, H., L. Schwarz, G. Steffens, and W. Kindermann. 1992a. Immunoregulatory hormones circulating leucocytes and lymphocyte subpopulations before and after endurance exercise of different intensities. *International Journal of Sports Medicine* 13:359-66.

Gabriel, H., A. Urhausen, and W. Kindermann. 1991a. Circulating leukocyte and lymphocyte subpopulations before and after intense endurance exercise to exhaustion. *European Journal of Applied Physiology* 63:449-57.

Gabriel, H. H. W., A. Urhausen, G. Valet, U. Heidelbach, and W. Kindermann. 1998. Overtraining and immune system: a prospective longitudinal study in endurance athletes. *Medicine and Science in Exercise and Sports* 30:1151-7.

Galbo, H. 1983. *Hormonal and metabolic adaptation to exercise*. Stuttgart: Thieme.

Galbo, H., H. J. Christensen, and J. J. Holst. 1977a. Glucose-induced decrease in glucagon and epinephrine responses to exercise in man. *Journal of Applied Physiology* 42:525-30.

Galbo, H., C. J. Hedeskov, K. Capito, and J. Vinter. 1981b. The effect of physical training on insulin secretion of rat pancreatic islets. *Acta Physiologica Scandinavica* 111:75-9.

Galbo, H., J. J. Holst, and N. J. Christensen. 1975. Glucagon and plasma catecholamine responses to graded and prolonged exercise in man. *Journal of Applied Physiology* 38:70-6.

Galbo, H., J. J. Holst, and N. J. Christensen. 1979a. The effect of different diets and of insulin on the hormonal response to prolonged exercise. *Acta Physiologica Scandinavica* 107:19-32.

Galbo, H., J. J. Holst, N. J. Christensen, and J. Hilsted. 1976. Glucagon and plasma catecholamines during beta-receptor blockade in exercising man. *Journal of Applied Physiology* 40:855-63.

Galbo, H., L. Hummer, I. B. Petersen, N. J. Christensen, and N. Bie. 1977c. Thyroid and testicular hormone responses to graded and prolonged exercise in man. *European Journal of Applied Physiology* 36:101-6.

Galbo, H., M. E. Houston, N. J. Christensen, J. J. Holst, B. Nielsen, E. Nygaard, and J. Suzuki. 1979b. The effect of water temperature on the hormonal response to prolonged swimming. *Acta Physiologica Scandinavica* 105:326-37.

Galbo, H., M. Kjaer, and N. H. Secher. 1987. Cardiovascular, ventilatory and catecholamine responses to maximal exercise in partially curarized man. *Journal of Physiology* 389:557-68.

Galbo, H., I. B. Peterson, K. Mikenes, B. Sonne, J. Hilsted, C. Hagen, and J. Fahrensurug. 1981a. The effect of fasting on the hormonal response to graded exercise. *Journal of Clinical Endocrinology and Metabolism* 52:1106-12.

Galbo, H., E. A. Richter, J. Hilsted, J. J. Holst, N. J. Christensen, and J. Hendriksson. 1977b. Hormonal regulation during prolonged exercise. *Annals of the New York Academy of Sciences* 301:72-80.

Galliven, E. A., A. Singh, D. Michelson, S. Bina, P. W. Gold, and P. A. Deuster. 1997. Hormonal and menstrual responses to exercise across time of day and menstrual cycle phase. *Journal of Applied Physiology* 83:1822-31.

Gambert, S. R., T. L. Garthwaite, C. H. Pontzer, E. E. Cook, F. E. Tristani, E. H. Duthie, D. R. Martinson, T. C. Hagen, and D. J. McCarty. 1981. Running elevates plasma beta-endorphin immunoreactivity and ACTH in untrained human subjects. *Proceedings of the Society of Experimental Biology and Medicine* 168:1-4.

Garber, A. J., J. E. Karl, and D. M. Kipnis. 1976. Alanine and glutamine synthesis and release from skeletal muscle. I. Glycolysis and amino acid release. *Journal of Biological Chemistry* 251:826-35.

Garrett, W. E., and D. T. Kirkendall (eds.). 2000. *Exercise and sport science.* Philadelphia: Lippincott Williams & Wilkins.

Gastin, P. B. 1994. Quantification of anaerobic capacity. *Scandinavian Journal of Medicine and Science in Sports* 4:91-112.

Gastmann, U., F. Dimeo, M. Huonker, J. Bocker, J. M. Steinacker, K. G. Petersen, H. Wieland, J. Keul, and M. Lehmann. 1998. Ultra-triathlon-related blood-chemical and endocrinological responses in nine athletes. *Journal of Sports Medicine and Physical Fitness* 38:18-23.

Gastmann, U. A. L., and M. J. Lehmann. 1999. Monitoring overload and regeneration in cyclists. In *Overload, performance incompetence, and regeneration in sport,* ed. Lehmann, M., C. Foster, U. Gastmann, H. Keizer, and J. M. Steinacker, 131-137. New York: Kluwer Academic/Plenum.

Gerber, G., and W. Roth. 1969. Über das Verhalten von anorganischen Phosphate Glucose, Lactate und Hämoglobin in menschlichen Blut vor, während und nach Fahrradergometerbelastung. *Medicin und Sport* 9:218-22.

Gettman, L. R., and M. L. Pollock. 1981. Circuit weight training: A critical review of its physiological benefits. *Physician and Sportmedicine* 9:44-60.

Geyssant, A., D. C. Geelen, A. M. Allevard, M. Vincent, E. Jarsaillon, C. A. Bizollon, J. P. Lacour, and C. Gharib. 1981. Plasma vasopressin, renin activity, and aldosterone: effect of exercise and training. *European Journal of Applied Physiology* 46:21-30.

Glenmark, B., G. Hedberg, and E. Jansson. 1992. Changes in muscle fibre type from adolescence to adulthood in women and men. *Acta Physiologica Scandinavica* 146:251-9.

Godsen, R., J. Smith, and J. Kime. 1991. There is essentially no order effect associated with fingertip blood sampling. *International Journal of Sports Medicine* 12:250-1.

Goldberg, A. K., and H. M. Goodman. 1969. Relationship between growth hormone and muscular work in determining muscle size. *Journal of Physiology* 200:655-65.

Goldberg, A. L. 1968. Protein synthesis during work-induced growth of skeletal muscle. *Journal of Cell Biology* 36:653-8.

Goldberg, N. D. 1985. Changes of activity and isozyme spectra of hexokinase of skeletal muscles and brain in adaptation to intensive physical exercises. *Ukrainskij Biokhimicheskij Zhurnal* 57(2):46-51.

Gollnick, P. D., R. B. Armstrong, B. Saltin, C. W. Sauber, W. L. Semberovich, and E. R. Shephard. 1973a. Effect of training on enzyme activity and fiber composition of human skeletal muscle. *Journal of Applied Physiology* 34:107-11.

Gollnick, P. D., R. B. Armstrong, W. L. Sembrowich, R. E. Shephard, and B. Saltin. 1973b. Glycogen depletion pattern in human muscle fibers after heavy exercise. *Journal of Applied Physiology* 34:615-8.

Gollnick, P. D., and L. Hermansen. 1973. Biochemical adaptations to exercise: anaerobic metabolism. *Exercise and Sport Science Reviews* 1:1-43.

Gollnick, P. D., L. Hermansen, and B. Saltin. 1980. The muscle biopsy: Still a research tool. *Physician Sportsmed* 8(1):1-7.

Gollnick, P. D., and King, D. W. 1969. Effect of exercise and training on mitochondria of rest skeletal muscle. *American Journal of Physiology* 216:1502-9.

Gollnick, P. D., P. Kõrge, J. Karpakka, and B. Saltin. 1991. Elongation of skeletal muscle relaxation during exercise is linked to reduced calcium uptake by the sarcoplasmic reticulum in man. *Acta Physiologica Scandinavica* 142:135-6.

Gollnick, P. D., M. Riedy, J. J. Quintinskie, and L. A. Bertocci. 1985. Differences in metabolic potential of skeletal muscle fibers and their significance for metabolic control. *Journal of Experimental Biology* 115:153-63.

Gollnick, P. D., and B. Saltin. 1982. Significance of skeletal oxidative enzyme enhancement with endurance training. *Clinical Physiology* 2:1-12.

Gontzea, I., S. Dumitrache, and P. Schutzescu. 1961. Untersuchungen über den Mechanismus der vermehrten Stickstoffausscheidung durch den Harn unter der Einwirkung von Muskelarbeit. *International Zeitschrift für angewante Physiologie* 19:7-17.

Gordon, B., L. A. Kohn, S. A. Levine, M. Matton, W. M. Seriver, and W. B. Whiting. 1925. Sugar content of the blood in runners following a marathon race, with especial reference to the prevention of hypoglycemia: Further observations. *Journal of American Medicine and Association* 85:508-9.

Gorokhov, A. L. 1969. Action of muscular activity on catecholamine content in tissues of trained and untrained white rats. *Sechenov Physiological Journal of the USSR* 55:1411-5.

Gorokhov, A., A. Krasnova, and N. Yakovlev. 1973. Dynamics of blood urea content and urinary excretion of catecholamines in sportsmen during physical exercise of various character [in Russian]. *Teoria i Praktika Fizicheskoi Kulturõ* 9:34-8.

Gorski, J., K. Lesczak, and J. Woiceszak. 1985. Urea excretion in sweat during short-term efforts of high intensity. *European Journal of Applied Physiology* 54:416-9.

Gorski, J., M. Nowacka, Z. Namiot, and T. Kiryluk. 1987. Effect of exercise on energy substrates metabolism in tissues of adrenalectomized rats. *Acta Physiologica Polonia* 38:331-7.

Gotovtseva, E. P., I. D. Surkina, and P. N. Uchakin. 1998. Potential interventions to prevent immunosuppression during training. In *Overtraining in sport,* ed. Kreider, R. B., A. C. Fry, and M. L. O'Toole, 243-272. Champaign IL: Human Kinetics.

Graham, T. E., J. Bangsbo, P. D. Gollnick, C. Juel, and B. Saltin. 1990. Ammonia metabolism during intense dynamic exercise and recovery in humans. *American Journal of Physiology* 259:E170-6.

Graham, T. E., J. W. E. Rush, and D. A. MacLean. 1995. Skeletal muscle amino acid metabolism and ammonia production during exercise. In *Exercise metabolism,* ed. Hargreaves, M., 131-175. Champaign, IL: Human Kinetics.

Granner, D. K. 1979. The role of glucocorticoid hormones as biological amplifiers. In *Glucocorticoid hormone action,* ed. Baxter, J. D., and G. G. Rousseau, 593-611. Berlin, Heidelberg, New York: Springer-Verlag.

Grassi, B., V. Quaresima, C. Marconi, M. Ferrari, and P. Cerretelli. 1999. Blood lactate accumulation and muscle deoxygenation during incremental exercise. *Journal of Applied Physiology* 87:348-55.

Green, H. J. 1998. Cation pumps in skeletal muscle: potential role in muscle fatigue. *Acta Physiologica Scandinavica* 182:201-13.

Green, H. J., E. R. Chin, M. Ball-Burnett, and D. Ranney. 1993. Increases in human skeletal muscle Na^+-K^+ ATPase concentration with short term training. *American Journal of Physiology* 264:C1538-41.

Green, H. J., S. Jones, M. Ball-Burnett, and I. Fraser. 1991. Early adaptation to in-blood substrates, metabolites, and hormones to prolonged exercise training in man. *Canadian Journal of Physiology and Pharmacology* 69:1222-3.

Green, H. J., H. M. Houston, J. A. Thomson, J. R. Sutton, and P. D. Gollnick. 1979. Metabolic consequences of supramaximal arm work performed during prolonged submaximal leg work. *Journal of Applied Physiology* 46:249-55.

Green, S. 1994. A definition and systems view of anaerobic capacity. *European Journal of Applied Physiology* 69:168-73.

Greenhaff, P. L., J.-M. Ren, K. Söderlund, and E. Hultman. 1991. Energy metabolism in single human muscle fibers during intense contraction without and with adrenaline infusion. *American Journal of Physiology* 260:E713-8.

Greenhaff, P. L., and J. A. Timmons. 1998. Interaction between aerobic and anaerobic metabolism during intense muscle contraction. *Exercise Sport Science Review* 26:1-30.

Greenleaf, J. E., V. A. Convertino, R. W. Stremel, E. M. Bernauer, W. C. Adams, S. R. Vignau, and P. J. Brock. 1977. Plasma [Na^+], [Ca^{2+}], and volume shifts and thermoregulation during exercise in man. *Journal of Applied Physiology* 43:1026-32.

Guezennec, C. Y., A. Abdelmalki, B. Serrurier, D. Merino, X. Bigard, M. Berthelot, C. Pierard, and M. Peres. 1998. Effects of prolonged exercise on brain ammonia and amino acids. *International Journal of Sports Medicine* 19:323-7.

Guezennec, C. Y., P. Ferre, B. Serrurier, D. Merino, M. Aymonod, and P. C. Pesquies. 1984. Metabolic effects of testosterone during prolonged physical exercise and fasting. *European Journal of Applied Physiology* 52(3):300-4.

Guezennec, C. Y., E. Fournier, F. X. Galen, M. Lartiques, F. Louisy, and J. Gutowska. 1989. Effects of physical exercise and anti-G suit inflation on atrial natriuretic factor plasma level. *European Journal of Applied Physiology* 58:500-7.

Guezennec, C. Y., L. Legar, F. Lhoste, M. Aymonod, and P. C. Pesquies. 1986. Hormones and metabolite responses to weight-lifting training sessions. *International Journal of Sports Medicine* 7:100-5.

Guglielmini, C., A. R. Paolini, and F. Conconi. 1984. Variation of serum testosterone concentration after physical exercise of different duration. *International Journal of Sports Medicine* 5:246-9.

Guminski, A. A., A. V. Tarassov, O. S. Jelizarova, and O. A. Samsonov. 1971. Assessment of aerobic and anaerobic work capacity in young and adult hockey-players [in Russian]. *Teoria i Praktika Fizicheskoi Kulturõ* 11:39-41.

Gyntelberg, F., M. J. Rennie, R. C. Hickson, and J. O. Holloszy. 1977. Effect of training on the response of plasma glucagon to exercise. *Journal of Applied Physiology* 43:302-5.

Haahr, P. M., B. K. Pedersen, A. Fomsgaard, N. Tvede, M. Diamant, K. Klarlund, J. Halkjaer-Kristensen, and K. Bendtzen. 1991. Effect of physical exercise on *in vitro* production of interleukin 1, interleukin

6, tumour necrosis factor-α, interleukin 2, and interferon-γ. *International Journal of Sports Medicine* 12:223-7.

Hack, V., G. Strobel, M. Weiss, and H. Weicker. 1994. PMN cell counts and phagocytotic activity of highly trained athletes depend on training period. *Journal of Applied Physiology* 77:1731-5.

Hackney, A. C. 1989. Endurance training and testosterone. *Sports Medicine* 8:117-27.

Hackney, A. C. 1996. The male reproductive system and endurance exercise. *Medicine and Sciences in Sports and Exercise* 28:180-9.

Hackney, A. C. 1999. Neuroendocrine system. Exercise overload and regeneration. In *Overload, performance incompetency, and regeneration in sport,* ed. Lehmann, M., C. Foster, U. Gastmann, H. Keizer, J. M. Steinacker, 173-186. New York: Kluwer Academic/Plenum.

Hackney, A. C., C. S. Curley, and B. J. Nicklas. 1991. Physiological responses to submaximal exercise at the mid-follicular, ovulatory and mid-luteal phases of the menstrual cycle. *Scandinavian Journal of Medicine and Science in Sport* 1:94-8.

Hackney, A. C., and T. Gulledge. 1994. Thyroid hormone responses during an 8-hour period following aerobic and anaerobic exercise. *Physiological Research* 43:1-5.

Hackney, A. C., R. J. Ness, and A. Schrieber. 1989. Effects of endurance exercise on nocturnal concentrations in males. *Chronobiology International* 6:341-6.

Hackney, A. C., S. N. Pearman, and J. M. Nowacki. 1990. Physiological profiles of overtrained and stale athletes: a review. *Applied Sport Psychology* 2:21-33.

Hackney, A. C., M. C. Premo, and R. G. McMurray. 1995. Influence of aerobic versus anaerobic exercise on the relationship between reproductive hormones in man. *Journal of Sports Sciences* 13:305-11.

Hackney, A. C., C. L. Fahrner, and R. Stupnicki. 1997. Reproductive hormonal responses to maximal exercise in endurance trained men with low testosterone levels. *Experimental and Clinical Endocrinology and Diabetes* 105:291-5.

Hackney, A. C., and A. Viru. 1999. Twenty-four-hour cortisol response to multiple daily exercise session of moderate and high intensity. *Clinical Physiology* 19:178-82.

Hagberg, J. M., and E. F. Coyle. 1983. Physiological determinants of endurance performance as studied in competitive racewalkers. *Medicine and Science in Sports and Exercise* 15:287-9.

Hageloch, W., S. Schneider, and H. Weicker. 1990. Blood ammonia determination in a specific field test as a method supporting talent selection in runners. *International Journal of Sport Medicine* 11(Suppl 2):S56-61.

Hagenfeldt, L., and J. Wahren. 1968. Human forearm muscle metabolism during exercise. II. Uptake, release and oxidation of individual FFA and glycerol. *Scandinavian Journal of Clinical Investigations* 21:263-76.

Hagg, S. A., E. L. Morse, and S. A. Adibi. 1982. Effect of exercise on rates of oxidation turnover, and plasma clearance of leucine in human subjects. *American Journal of Physiology* 282;242:E407-10.

Häggendal, J., L. H. Hartley, and B. Saltin. 1970. Arterial noradrenaline concentration during exercise in relation to the relative work levels. *Scandinavian Journal of Clinical Laboratory Investigation* 26:337-42.

Haibach, H., and M. W. Hosler. 1985. Serum creatine kinase in marathon runners. *Experientia* 41:39-40.

Haier, R. J., K. Quaid, and J. S. C. Mills. 1981. Naloxone alters pain perception after jogging. *Psychiatric Research* 5:231-2.

Häkkinen, K., K. L. Keskinen, M. Alén, P. V. Komi, and H. Kauhanen. 1985. Serum hormone concentrations during prolonged training in elite endurance-trained and strength-trained athletes. *European Journal of Applied Physiology and Occupational Physiology* 59(3):233-8.

Häkkinen, K., and A. Pakarinen. 1993. Acute hormonal responses to two different fatiguing heavy-resistance protocols in male athletes. *Journal of Applied Physiology* 74:882-7.

Häkkinen, K., A. Pakarinen, M. Alén, and P. V. Komi. 1985. Serum hormones during prolonged training of neuromuscular performance. *European Journal of Applied Physiology* 53:287-93.

Häkkinen, K., A. Pakarinen, M. Alén, H. Kauhanen, and P. V. Komi. 1988a. Daily hormonal and neuromuscular responses to intensive strength training in 1 week. *International Journal of Sports Medicine* 9:422-8.

Häkkinen, K., A. Pakarinen, M. Alén, H. Kauhanen, and P. V. Komi. 1988b. Neuromuscular and hormonal responses in elite athletes to two successive strength training sessions in one day. *European Journal of Physiology* 57:133-9.

Häkkinen, K., A. Pakarinen, M. Alén, H. Kauhanen, and P. V. Komi. 1987. Relationships between training volume, physical performance capacity, and serum hormone concentrations during prolonged training in elite weight lifters. *International Journal of Sports Medicine* 8(Suppl.):61-5.

Häkkinen, K., A. Pakarinen, A. Alén, H. Kauhanen, and P. V. Komi. 1988c. Neuromuscular and hormonal adaptations in athletes to strength training in two years. *Journal of Applied Physiology* 65:2406-12.

Häkkinen, K., K. L. Keskinen, M. Alén, P. V. Komi, and H. Kauhanen. 1989. Serum hormone concentrations during prolonged training in elite endurance-trained and strength-trained athletes. *European Journal of Applied Physiology* 55:233-8.

Häkkinen, K., A. Pakarinen, H. Kyrölainen, S. Chang, D. H. Kim, and V. P. Komi. 1990. Neuromuscular adaptation and serum hormones in females during prolonged power training. *International Journal of Sports Medicine* 11:91-8.

Hales, C. N., J. B. Luzio, and K. Siddle. 1978. Hormonal control of adipose-tissue lipolysis. *Biochemical Society Symposium* 43:97-135.

Halkjaer-Kristensen, J., and T. Ingemann-Hansen. 1981. Variations in single fibre area and in fiber composition in needle biopsies from the quadriceps muscle in man. *Scandinavian Journal of Clinical and Laboratory Investigations* 41:391-6.

Halliwell, B., and Gutteridge. 1989. *Free radicals in biology and medicine*. Oxford: Clarendon Press.

Halonen, P. I., and A. Konttinen. 1962. Effect of physical exercise on some enzymes in serum. *Nature* 193:942-4.

Hamosh, M., Lesch, M., Baron, J., and S. Kaufman. 1967. Enhanced protein synthesis in a cell-free system from hypertrophied skeletal muscle. *Science* 157:935-7.

Han, J. W., R. Hieleczek, M. Varsanyi, and L. M. Heilmeyer. 1992. Compartmentalized ATP synthesis in skeletal muscle trials. *Biochemistry* 31:377-84.

Hanson, P. G., and D. K. Flaherty. 1981. Immunological responses to training in conditioned runners. *Clinical Science* 60:225-8.

Haralambie, G. 1962. Die Veränderungen des Säure-Basen-Gleichgewichts als Kontrolmittel der Anpassung bei körperliches Belastung. *Medicina und Sport* 2:58-63.

Haralambie, G. 1964b. L'elimination de la tyrosine pendant l'effort physique. *Medicine, Education Physique et Sport* 38:325-31.

Haralambie, G. 1964a. La valeur de certains constantes biochimiques du serum chez les sportifs en regime d'entrainement intense. *Acta Biologica et Medica Germanica* 13:30-9.

Haralambie, G., and A. Berg. 1976. Serum urea and amino nitrogen changes with exercise duration. *European Journal of Applied Physiology* 36:39-48.

Hardman, A. E., R. Mayes, and C. Williams. 1987. Onset of blood lactate accumulation and endurance performance in endurance-trained and sprint-trained athletes. *Journal of Physiology* 394:10P.

Hargreaves, M. (ed.). 1995. *Exercise metabolism*. Champaign, IL: Human Kinetics.

Hargreaves, M., B. Kiens, and E. A. Richter. 1991. Effect of increased plasma free fatty acid concentration on muscle metabolism in exercising men. *Journal of Applied Physiology* 70:194-201.

Harre, D. 1973. *Traininglehre*. Berlin: Sportverlag.

Harris, R. C., E. Hultman, and L. O. Nordesjo. 1974. Glycogen, glycolytic intermediates and high-energy phosphates determinated in biopsy samples of musculus quadriceps femoris of man at rest. Methods and variance of values. *Scandinavian Journal of Clinical and Laboratory Investigations* 33:109-20.

Harro, J., H. Rimm, M. Harro, M. Grauber, K. Karelson, and A. Viru. 1999. Association of depressiveness with blunted growth hormone responses to maximal physical exercise in young healthy men. *Psychoneuroendocrinology* 24:505-17.

Hartley, L. H., J. W. Mason, R. P. Hogdan, L. G. Jones, T. A. Kotchen, E. H. Mougly, F. E. Wherry, L. L. Pennington, and P. T. Rickets. 1972a. Multiple hormonal responses to graded exercise in relation to physical training. *Journal of Applied Physiology* 33:602-6.

Hartley, L. H., J. W. Mason, R. P. Hogdan, L. G. Jones, T. A. Kotchen, E. H. Mougley, F. E. Wherry, L. L. Pennington, and P. T. Rickets. 1972b. Multiple hormonal responses to prolonged exercise in relation to physical training. *Journal of Applied Physiology* 33:607-10.

Hartog, M., R. J. Havel, G. Copinschi, J. M. Earl, and B. C. Ritchie. 1967. The relationship between changes in serum levels of growth hormone and mobilization of fat during exercise in man. *Quarterly Journal of Experimental Physiology* 52:86-96.

Harvey, W. D., G. R. Faloon, and R. H. Unger. 1974. The effect of adrenergic blockade on exercise-induced hyperglucagonemia. *Endocrinology* 94:1254-8.

Hatfaludy, S., Shansky, J., and H. H. Vanderburgh. 1989. Metabolic alteration induced in cultured skeletal muscle by stretch-relaxation activity. *American Journal of Physiology* 256:C175-81.

Hautier, C. A., D. Wouassi, L. M. Arsac, E. Bitanga, P. Thiriet, and J. R. Lacour. 1994. Relationship between postcompetition blood lactate concentration and average running velocity over 100-m and 200-m races. *European Journal of Applied Physiology* 68:508-13.

Havas, E., J. Komalainen, and V. Vihko. 1997. Exercise-induced increase in serum creatine kinase is modified by subsequent bed rest. *International Journal of Sports Medicine* 18:578-82.

Havel, R. J., B. Pernow, and N. L. Jones. 1967. Uptake and release of free fatty acids and other metabolites in the legs of exercising men. *Journal of Applied Physiology* 23:90-9.

Hawley, J. A., and W. G. Hopkins. 1995. Aerobic glycolytic and aerobic lipolytic power systems. *Sports Medicine* 19:240-50.

Haymes, E. M., and J. J. Lamanca. 1989. Iron loss in runners during exercise. Implications and recommendations. *Sports Medicine* 7:277-85.

Heck, H. 1990. *Laktat in der Leistungdiagnostik.* Schorndorf: Verlag K. Hofmann.

Heck, H., A. Mader, G. Hess, S. Mucke, R. Muller, and W. Hollmann. 1985. Justification of the 4 mmol/l lactate threshold. *International Journal of Sports Medicine* 6:117-30.

Hedden, M., and M. G. Buse. 1982. Effects of glucose, pyruvate, lactate and amino acids in muscle protein synthesis. *American Journal of Physiology* 242:E184-92.

Hedfors, E., G. Holm, M. Ivansen, and J. Wahren. 1983. Physiological variation of blood lymphocyte reactivity: T-cell subsets. Immunoglobulin production, and mixed-lymphocyte reactivity. *Clinical Immunology and Immunopathology* 27:9-14.

Heitcamp, H. C., M. Holdt, and K. Scheib. 1991. The reproducibility of the 4 mmol/l lactate threshold in trained and untrained women. *International Journal of Sports Medicine* 12:363-8.

Heitkamp, H.-C., K. Schmid, and K. Scheib. 1993. ß-endorphin and adrenocorticotropin hormone production during marathon and incremental exercise. *European Journal of Applied Physiology* 66:269-74.

Hellestein, Y., F. Apple, and B. Sjodin. 1996. Effect of sprint cycle training on activities of antioxidant enzymes in human skeletal muscle. *Journal of Applied Physiology* 81:1484-7.

Henderson, S. A., A. L. Black, and G. A. Brooks. 1985. Leucine turnover and oxidation in trained rats during exercise. *American Journal of Physiology* 249:E137-44.

Henriksson, J. 1977. Training-induced adaptation of skeletal muscle and metabolism during submaximal exercise. *Journal of Physiology* 270:677-90.

Henriksson, J., M. M. Chi, C. S. Hintz, D. A. Young, K. K. Kaiser, S. Salmons, O. H. Lowry. 1986. Chronic stimulation of mammalian muscle: changes in enzymes of six metabolic pathways. *American Journal of Physiology* 251:C614-32.

Henriksson, J., and J. S. Reitman. 1977. The time course of changes in human skeletal muscle succinate dehydrogenase and cytochrome oxidase activities and maximal oxygen uptake with physical activity and inactivity. *Acta Physiologica Scandinavica* 99:91-7.

Henriksson, K. G. 1979. "Semi-open" muscle biopsy technique. A simple outpatient procedure. *Acta Neurology Scandinavica* 59:317-23.

Hermansen, L., A. Orheim, and O. M. Sejersted. 1984. Metabolic acidosis and changes in water and electrolyte balance in relation to fatigue during maximal exercise of short duration. *International Journal of Sports Medicine* 5(Suppl.):110-15.

Hermansen, L., and J.-B. Osnes. 1972. Blood and muscle pH after maximal exercise in man. *Journal of Applied Physiology* 32:304-8.

Hermansen, L., and O. Vaage. 1977. Lactate disappearance and glycogen synthesis in human muscle after maximal exercise. *American Journal of Physiology* 233:E422-9.

Hermansen, L., E. D. R. Pruett, J. B. Osnes, and F. A. Giere. 1970. Blood glucose and plasma insulin in response to maximal exercise and glucose infusion. *Journal of Applied Physiology* 29:13-6.

Herxheimer, H. 1933. *Grundriss der Sportmedizin für Arzte und Studierende.* Leipzig: Thieme.

Hesse, B., I.-L. Kanstrup, N. J. Christensen, T. Ingemann-Hansen, J. F. Hansen, J. Halkjaer-Kristensen, and F. B. Petersen. 1981. Reduced norepinephrine response to dynamic exercise in human subjects during O_2 breathing. *Journal of Applied Physiology* 51:176-8.

Heymsfield, S. B., C. Arteaga, C. McManus, J. Smith, and S. Moffitt. 1983. Measurement of muscle mass in humans: validity of the 24-hour urinary creatinine method. *American Journal of Clinical Nutrition* 37:478-94.

Hickey, M. S., R. V. Considine, R. G. Israel, T. L. Mahar, M. R. McCammon, G. L. Tyndall, J. A. Houmard, and J. F. Caro. 1996. Leptin is related to body fat content in male distance runners. *American Journal of Physiology* 271(5 Pt 1):E938-40.

Hickey, M. S., J. A. Houmard, R. V. Considine, G. L. Tyndall, J. B. Midgette, K. E. Gavigan, M. L. Weidner, M. L. McCammon, R. G. Israel, and J. F. Caro. 1997. Gender-dependent effects of exercise training on serum leptin levels in humans. *American Journal of Physiology* 272(4 Pt 1):E562-6.

Hickson, R. C., H. A. Bomze, and J. O. Holloszy. 1978. Faster adjustment of O_2 uptake to the energy requirement of exercise in the trained state. *Journal of Applied Physiology* 44:877-81.

Hickson, R. C., and J. R. Davis. 1981. Partial prevention of glucocorticoid-induced muscle atrophy by endurance training. *American Journal of Physiology* 241:E226-32.

Hickson, R. C., C. Forter, M. L. Pollok, T. M. Gallassi, and S. Rich. 1985. Reduced training intensities and loss of aerobic power, endurance, and cardiac growth. *Journal of Applied Physiology* 58:492-99.

Hickson, R. C., J. M. Hagberg, R. K. Conlee, D. A. Jones, A. A. Ehsani, and W. W. Winder. 1979. Effect of training on hormonal responses to exercise in competitive swimmers. *European Journal of Applied Physiology* 41:211-9.

Hill, A. V. 1925. *Muscular activity*. Baltimore: Williams & Wilkins.

Hill, R., F. C. Goetz, H. M. Fox, B. J. Murawski, L. J. Kranauer, R. W. Reffenstein, S. Gray, W. J. Reddy, S. E. Hedberg, J. R. St. Marc, and G. W. Thorn. 1956. Studies on adrenocortical and psychological response to stress in man. *Archives of Internal Medicine* 97:269-98.

Hilsted, J., H. Galbo, B. Somme, T. Schwartz, J. Fahrenkrug, O. B. Schaffalitzky de Muckadell, K. B. Lauritsen, and B. Tronier. 1980. Gastroenteropancreatic hormonal changes during exercise. *American Journal of Physiology* 239:G136-40.

Hnõkina, A. 1983. *System of biochemical control in cross-country skiing, biathlon and Nordic skiing* [in Russian]. Moscow: Research Institute of Sport.

Hochacka, P. W., and G. N. Somero. 1984. *Biochemical adaptation*. Princeton: Princeton University Press.

Hoelzer, D. R., G. P. Dalsky, W. E. Clutter, S. D. Shah, J. O. Holloszy, and P. E. Cryer. 1986. Glucoregulation during exercise. Hypoglycemia is prevented by redundant glucoregulatory systems, sympathochromaffin activation and changes in islet hormone secretion. *Journal of Clinical Investigation* 71:212-21.

Hoffman, J. R., W. J. Kraemer, A. C. Fry, M. Deschenes, and M. Kemp. 1990a. The effects of self-selection for frequency of training in a winter conditioning program for football. *Journal of Applied Sport Science Research* 4:76-82.

Hoffmann, P., L. Terenius, and P. Thoren. 1990b. Cerebrospinal fluid immunoreactive beta-endorphin concentration is increased by long-lasting voluntary exercise in the spontaneously hypertensive rat. *Regulatory Peptides* 28:233-9.

Hollmann, W., and T. Hettinger. 1976. *Sportmedizin—Arbeits- und Trainingsgrundlagen*. Stuttgart: Schauttauer.

Holloszy, J. O. 1973. Biochemical adaptations to exercise: aerobic metabolism. *Exercise and Sport Sciences Reviews* 1:45-71.

Holloszy, J. O., and E. F. Coyle. 1984. Adaptation of skeletal muscle to endurance exercise and their metabolic consequences. *Journal of Applied Physiology* 56:831-8.

Holz, G., G. Gerber, R. Lorenz, G. Neuman, and H. G. Schuster. 1979. Veränderungen der Konzentration der Aminosäuren im Blutplasma bei körperlichen Ausdauerbelastung. *Medicina und Sport* 19:231-5.

Hood, D. A., and R. L. Terjung. 1987a. Effect of endurance training on leucine metabolism in perfused rat skeletal muscle. *American Journal of Physiology* 253:E648-56.

Hood, D. A., and R. L. Terjung. 1987b. Leucine metabolism in perfused rat skeletal muscle during contractions. *American Journal of Physiology* 253:E636-47.

Hoogeveen, A. R., J. Hoogsteen, and G. Schep. 1997. The maximal lactate steady state in elite endurance athletes. *Japanese Journal of Physiology* 47:481-5.

Hoogeveen, A. R., and M. L. Zonderland. 1996. Relationship between testosterone, cortisol and performance in professional cyclists. *International Journal of Sports Medicine* 17:423-8.

Hooper, S., and L. T. Mackinnon. 1995. Monitoring overtraining in athletes: recommendations. *Sports Medicine* 20:321-7.

Hooper, S. L., L. T. Mackinnon, A. Howard, R. D. Gordon, and A. W. Bachmann. 1995a. Markers for monitoring overtraining and recovery. *Medicine and Science in Sports and Exercise* 27:106-12.

Hooper, S. L., L. T. Mackinnon, and A. Howard. 1999. Physiological and psychometric variables for monitoring recovery during tapering for major competition. *Medicine and Science in Sports and Exercise* 31:1205-10.

Hooper, S. L., L. T. Mackinnon, R. D. Gordon, and A. W. Bachmann. 1995b. Hormonal responses of elite swimmers to overtraining. *Medicine and Science in Sports and Exercise* 27:106-12.

Hopkins, W. G. 1991. Quantification of training in competitive sports. *Sports Medicine* 12:161-83.

Hoppeler, H., H. Howald, K. E. Conley, S. L. Lindstedt, H. Claasen, P. Vook, and E. R. Weibel. 1985. Endurance training in humans: Aerobic capacity and structure of skeletal muscle. *Journal of Applied Physiology* 59:320-7.

Hort, W. 1951. Morphologische und physiologische Untersuchungen an Ratten während eines Lauftrainings und nach dem Training. *Virchows Archiv* 320:197-237.

Hortobagyi, T., and T. Denahan. 1989. Variability in creatine kinase: methodological, exercise, and clinically related factors. *International Journal of Sports Medicine* 10:69-80.

Hortobagyi, T., J. A. Houmard, J. R. Stevenson, D. D. Fraser, R. A. Johns, and R. G. Israel. 1993. The effects of detraining on power athletes. *Medicine and Science in Sports and Exercise* 25:929-35.

Houmard, J. A., D. L. Costill, J. B. Mitchell, S. H. Park, W. J. Fink, and J. M. Burns. 1990a. Testosterone, cortisol, and creatine kinase levels in male distance runners during reduced training. *International Journal of Sports Medicine* 11:41-5.

Houmard, J. A., D. L. Costill, J. B. Mitchell, S. H. Park, R. C. Hicker, and J. N. Roemmish. 1990. Reduced training maintains performance in runners. *International Journal of Sports Medicine* 11:46-51.

Howlett, T. A., S. Tomlin, L. Ngahfoong, L. H. Rees, B. A. Bullen, G. S. Skrinar, and J. W. McArthur. 1984. Release of ß-endorphin and met-enkephalin during exercise in normal women: response to training. *British Medical Journal* 288:1950-2.

Hultman, E. 1967. Studies in muscle metabolism of glycogen and active phosphates in man with special reference to exercise and diet. *Scandinavian Journal of Clinical and Laboratory Investigations* 19(Suppl 1):1-63.

Hultman, E. 1971. Muscle glycogen stores and prolonged exercise. In *Frontiers of fitness,* ed. Shephard, R. J., 37-60. Springfield IL: Charles C Thomas.

Hultman, E., and J. Bergström. 1973. Local energy-supplying substrates as limiting factors in different types of leg muscle work in normal man. In *Limiting factors of physical performance,* ed. Keul, J., 113-125. Stuttgart: G. Thieme Verlag.

Hultman, E., M. Bergström, L. L. Spriet, and K. Söderlund. 1990. Energy metabolism and fatigue. In *Biochemistry of exercise VII,* ed. Taylor, A. W. et al., 73-92. Champaign, IL: Human Kinetics.

Hultman, E., and R. C. Harris. 1988. Carbohydrate metabolism. In *Principles of exercise biochemistry,* ed. by Poortman, J. R., 78-119. Basel: Karger.

Hultman, E., and L. Nilsson. 1973. Liver glycogen as a glucose-supplying source during exercise. In *Limiting factors of physical performance,* ed. Keul, J. 179-189. Stuttgart: G. Thieme.

Hultman, E., and K. Sahlin. 1980. Acid-base balance during exercise. *Exercise and Sports Science Reviews* 7:41-128.

Hultman, E., and H. Sjöholm. 1983. Energy metabolism and contraction force of human skeletal muscle in situ during electrical stimulation. *Journal of Physiology* 345:525-32.

Hultman, E., and L. L. Spriet. 1988. Dietary intake prior to and during exercise. In *Exercise, nutrition, and energy metabolism,* ed. Horton, E. S., and R. L. Terjung, 132-149. New York: Macmillan.

Hunding, A., R. Jordal, and P. E. Paulev. 1981. Runner's anemia and iron deficiency. *Acta Medica Scandinavica* 209:149-55.

Hunter, W. M., and M. Y. Sukkar. 1968. Changes in plasma insulin levels during steady-state exercise. *Journal of Physiology* 196:110P-2P.

Hunter, W. M., C. C. Fonseka, and R. Passmore. 1965. The role of growth hormone in the mobilization of fuel for muscular exercise. *Quarterly Journal of Experimental Physiology* 50:406-16.

Hutter, M. M., R. E. Sievers, V. Barbosa, and C. L. Wolfe. 1994. Heat-shock protein induction in rat hearts: a direct correlation between the amount of heat-shock protein induced and the degree of myocardial protection. *Circulation* 89:355-60.

Hyatt, J.-P. K., and P. M. Clarkson. 1998. Creatine kinase release and clearance using MM variants following repeated bouts of eccentric exercise. *Medicine and Science in Sports and Exercise* 30:1059-65.

Inbar, O., O. Bar-Or, and J. S. Skinner. 1996. *The Wingate Anaerobic Test.* Champaign, IL: Human Kinetics.

Ingjer, F. 1979. Effects of endurance training on muscle fibre ATP-ase activity, capillary supply and mitochondrial content in man. *Journal of Physiology* 249:419-32.

Ingle, D. J. 1952. The role of the adrenal cortex in homeostesis. *Journal of Endocrinology* 8:xxii-xxxvii.

Ingle, D. J., and J. E. Nezamis. 1949. Work performance of adrenally insufficient rats given cortex extract by continuous intravenous injection. *American Journal of Physiology* 156:365-7.

Ingle, D. J., J. E. Nezamis, and E. H. Morley. 1952. The comparative values of cortisone, 17-hydroxycorticosterone and adrenal cortex extract given by continuous intravenous injection in sustaining the ability of the adrenalectomized rat to work. *Endocrinology* 50:1-4.

Ingwall, J. S., F. Morales, and F. W. Stockdale. 1972. Creatine and the control of myosin synthesis in differentiating skeletal muscle. *Proceedings of the National Academy of Sciences USA* 69:2250-3.

Ingwall, J. S., C. D. Weiner, F. Morales, E. Davis, and F. E. Stockdale. 1974. Specificity of creatine in the control of muscle protein synthesis. *Journal of Cell Biology* 63:145-51.

Ingwall, J. S., and K. Wildenthal. 1976. Role of creatine in the regulation of cardiac protein synthesis. *Journal of Cell Biology* 68:159-63.

Inoue, K., Yamasaki, S., Fushiki, T., Okada, Y., and E. Sugimoto. 1994. Androgen receptor agonist suppresses exercise-induced hypertrophy of skeletal muscle. *European Journal of Applied Physiology* 69:88-91.

Israel S. 1969. Die C-17-ketosteroid-Auscheidung bei extreme Ausdauerbelastung. *Medizin und Sport* 9:81-6.

Israel, S. 1958. Die Erscheinungsformen des Übertrainings. *Sportmedizin* 9:207-9.

Israel, S. 1976. Zur Problematik der Übertrainings auf internistischer and leistungsphysiologisher Sicht. *Medicina und Sport* 16:1-12.

Issekutz, B., W. M. Bortz, H. I. Miller, and P. Paul. 1967. Turnover rate of plasma FFA in humans and in dogs. *Metabolism* 16:1001-9.

Issekutz, B., and H. Miller. 1962. Plasma free fatty acids during exercise and the effect of lactic acid.

Proceedings of Society for Experimental Biology and Medicine 110:237-9.

Issekutz, B., W. A. Shaw, and T. B. Issekutz. 1975. Effect of lactate on FFA and glycerol turnover in resting and exercising dogs. *Journal of Applied Physiology* 39:349-53.

Issekutz, B., and M. Vranic. 1980. The role of glucagon in regulation of glucose production in exercising dogs. *American Journal of Physiology* 238:E13-20.

Itoh, H., and T. Ohkuwa. 1991. Ammonia and lactate in the blood after short-term sprint exercise. *European Journal of Applied Physiology* 62:22-5.

Ivy, J. L., R. T. Withers, P. J. VanHandel, D. H. Elger, and D. L. Costill. 1980. Muscle respiratory capacity and fiber types as determinants of the lactate threshold. *Journal of Applied Physiology* 48:523-27.

Jackson, M. 1998. Free radical mechanism in exercise-induced skeletal muscle fibre injury. In *Oxidative stress skeletal muscle,* ed. Reznik, A., 75-86. Basel: Birkhäuser Verlag.

Jacob, R., X. Hu, D. Neiderstock, and S. Hasan. 1996. IGF-I stimulation of muscle protein synthesis in the awake rat: permissive role of insulin and amino acids. *American Journal of Physiology* 270:E60-6.

Jacobs, I. 1980. The effect of thermal dehydration on performance of the Wingate Anaerobic Test. *International Journal of Sports Medicine* 1:21-4.

Jacobs, I. 1986. Blood lactate implications for training and sports performance. *Sports Medicine* 3:10-25.

Jacobs, I., O. Bar-Or, J. Karlsson, R. Dotam, P. Tesch, P. Kaiser, and O. Inbar. 1982. Changes in muscle metabolism in females with 30-s exhaustive exercise. *Medicine and Science in Sports and Exercise* 14:457-60.

Jacobs, I., P. A. Tesch, O. Bar-Or, J. Karlsson, and R. Dotan. 1983. Lactate in human skeletal muscle after 10 and 30s of supramaximal exercise. *Journal of Applied Physiology* 55:356-67.

Jalak, R., and A. Viru. 1983. Adrenocortical activity in many times repeated physical exercises in a day. *Fiziologia cheloveka (Moscow)* 9:418-24.

Janal, M. N., E. W. D. Colt, W. C. Clark, and M. Glusman. 1984. Pain sensitivity, mood and plasma endocrine levels in man following long-distance running: effects of naloxone. *Pain* 19:13-25.

Jansen, P. G. J. M. 1987. *Training, lactate, pulse-rate.* Oulu: Polar Electro Oy.

Jansson, E. 1980. Diet and muscle metabolism in man. *Acta Physiologica Scandinavica* Suppl. 487:1-24.

Jansson, E. 1994. Methodology and actual perspectives of the evaluation of muscular enzymes in skeletal muscle by biopsy during rest, exercise and detraining. An overview. *Medicina dello Sport* 47:377-83.

Jansson, E., Hjemdahl, P., and L. Kaijser. 1982. Diet induced changes in sympatho-adrenal activity during submaximal exercise in relation to substrate utilization in man. *Acta Physiologica Scandinavica* 114:171-8.

Jansson, E., and L. Kaijser. 1987. Substrate utilization and enzymes in skeletal muscle of extremity endurance-trained men. *Journal of Applied Physiology* 62:955-1005.

Jansson, E., B. Sjödin, and P. Tesch. 1978. Changes in muscle fibre type distribution in man after physical training. *Acta Physiologica Scandinavica* 104:235-7.

Jansson, E., and C. Sylven. 1985. Creatine kinase MB and citrate synthetase in type I and type II muscle fibers in trained and untrained men. *European Journal of Applied Physiology* 54:207-9.

Järhult, J., and J. Holst. 1979. The role of the adrenergic innervation to the pancreatic islets in the control of insulin release during exercise in man. *Pflügers Archiv für die gesamte Physiologie* 383:41-5.

Jenkins, A. B., D. J. Chisholm, D. E. James, K. Y. Ho, and E. W. Kraeger. 1985. Exercise-induced hepatic glucose output is precisely sensitive to the rate of systemic glucose supply. *Metabolism* 34:431-6.

Jenkins, A. B., S. M. Furler, D. L. Chisholm, and E. W. Kraeger. 1986. Regulation of hepatic glucose output during exercise by circulating glucose and insulin in humans. *American Journal of Physiology* 250:R411-7.

Jenkins, R. R. 1988. Free radical chemistry: relationship to exercise. *Sport Medicine* 5:156-70.

Jensen-Urstad, M., J. Svedenhag, and K. Saltin. 1994. Effect of muscle mass on lactate formation during exercise in humans. *European Journal of Applied Physiology* 69:189-95.

Jeukendrup, A. E., M. C. K. Hesselink, A. C. Snyder, H. Kuipers, and H. A. Keizer. 1992. Physiological changes in male competitive cyclists after two weeks of intensified training. *International Journal of Sports Medicine* 13:534-41.

Jeukendrup, A. E., W. H. M. Saris, and A. J. M. Wagenmakers. 1998. Fat metabolism during exercise: Part 1: Fatty acid mobilization and muscle metabolism. *International Journal of Sports Medicine* 19:231-44.

Jezova, D., M. Viga, P. Tatar, R. Kventnansky, K. Nazar, H. Kaciuba-Uscilko, and S. Kozlowski. 1985. Plasma testosterone and catecholamine responses to physical exercises of different intensities in men. *European Journal of Applied Physiology* 54:62-6.

Ji, L. L. 1995. Exercise and oxidative stress: role of cellular antioxidant systems. *Sport Science Reviews* 23:135-66.

Ji, L. L., A. Katz, R. G. Fu, M. Parchert, and M. Spencer. 1992. Blood glutathione status during exercise: effect of carbohydrate supplementation. *Journal of Applied Physiology* 74:788-92.

Johansen, K., and O. Munck. 1979. The relationship between maximal oxygen uptake and glucose tolerance/insulin response ratio in normal young men. *Hormone and Metabolic Research* 11:424-7.

Johansen, M. A., J. Polgar, D. Weightman, and D. Appleton. 1973. Data on the distribution of fibre types in thirty-six human muscles. An autopsy study. *Journal of Neurological Sciences* 18:111-29.

Jones, D. A., D. J. Newham, J. M. Round, and S. E. Tolfree. 1986. Experimental human muscle damage: morphological changes in relation to other indices of damage. *Journal of Physiology* 375:435-48.

Jones, N. L., and R. E. Ehrsam. 1982. The anaerobic threshold. *Exercise and Sports Science Review* 10:49-83.

Jones, N. L., and N. McCartney. 1986. Influence of muscle power on aerobic performance and the effects of training. *Acta Medica Scandinavica* 711:115-22.

Joye, H., and J. Poortmans. 1970. Hematocrit and serum proteins during arm exercise. *Medicine and Science in Sports* 2:187-90.

Juel, C., J. Bangsbo, T. Graham, and B. Saltin. 1990. Lactate and potassium fluxes from human skeletal muscle during and after intense, dynamic knee extensor exercise. *Acta Physiologica Scandinavica* 140:147-59.

Jürimäe, T., K. Karelson, T. Smirnova, and A. Viru. 1990b. The effect of a single-circuit weight-training session on the blood biochemistry of untrained university students. *European Journal of Applied Physiology* 61:344-48.

Jürimäe, T., A. Viru, K. Karelson, and T. Smirnova. 1990a. Biochemical changes in blood during the long and short triathlon competition. *Journal of Sports Medicine and Physical Fitness* 29:305-9.

Jurkowski, J. E., N. I. Jones, W. C. Walker, E. V. Younglai, and J. R. Sutton. 1978. Ovarian hormonal responses to exercise. *European Journal of Applied Physiology* 44:109-14.

Kacl, K. 1932. Der Einfluss der Muskelarbeit auf den Kreatin—und Kreatiningehalt im Blut normaler Menschen. *Biochemical Zhurnal* 245:452-8.

Kaiser, V., G. M. E. Hanssen, and J. W. J. Wersch. 1989. Effect of training on red cell parameters and plasma ferritin: a transverse and longitudinal approach. *International Journal of Sports Medicine* 10:S169-75.

Kaltreider, N. L., and C. R. Meneely. 1940. The effect of exercise on the volume of the blood. *Journal of Clinical Investigation* 19:627-34.

Kantola, H. 1989. *Suomalainen valmennusoppi Harjoittelu.* Jyväskylä: Gummerus Kirjapaino OY.

Kantola, H., and H. Rusko. 1985. *Sykettä ladulle.* Jyväskylä: Valmennuskirjad Oy.

Karagiorgos, A., J. F. Garcia, and G. A. Brooks. 1979. Growth hormone response to continuous and intermittent exercise. *Medicine and Science in Sports* 11:302-7.

Karelson, K., T. Smirnova, and A. Viru. 1994. Interrelations between plasma ACTH and cortisol levels during exercise in men. *Biology of Sports* 11:75-82.

Kargotich, S., C. Goodman, D. Keast, R. W. Fry, P. Gracia-Webb, P. M. Crawford, and A. R. Morton. 1997. Influence of exercise-induced plasma volume changes on the interpretation of biochemical data following high-intensity exercise. *Clinical Journal of Sport Medicine* 7:185-91.

Karlsson, J. 1971. Lactate and phosphagen concentrations in working muscle of man. *Acta Physiologica Scandinavica* 81(Suppl 358):1-72.

Karlsson, J., B. Diamant, and B. Saltin. 1971. Muscle metabolites during submaximal and maximal exercise in man. *Scandinavian Journal of Clinical and Laboratory Investigations* 26:385-94.

Karlsson, J., and I. Jacobs. 1982. Onset of blood lactate accumulation during muscular exercise. *International Journal of Sports Medicine* 3:189-201.

Karlsson, J., B. Sjödin, A. Thorstensson, B. Hulten, and K. Frith. 1975. LDH isoenzymes in skeletal muscles of endurance and strength trained athletes. *Acta Physiologica Scandinavica* 93:150-6.

Karlsson, J., L. O. Nordesjö, and B. Saltin. 1974. Muscle glycogen utilization during exercise after physical training. *Acta Physiologica Scandinavica* 90:210-7.

Karlsson, J., L. O. Nordesjö, L. Jorfeldt, and B. Saltin. 1972. Muscle lactate, ATP and CP levels during exercise and after physical training in man. *Journal of Applied Physiology* 33:199-203.

Kasperek, G. J. 1989. Regulation of branched-chain 2-oxo acid dehydrogenase activity during exercise. *American Journal of Physiology* 256:E186-90.

Kasperek, G. J., G. L. Dohm, E. B. Tapscott, and T. Powell. 1980. Effect of exercise on liver protein loss and lyosomal enzyme levels in fed and fasted rats. *Proceedings for Society of Experimental Biology and Medicine* 164:430-4.

Kassil, G. N., and V. V. Mekhrikadze. 1985. Status of sympatho-adrenal system in sprinters in various kinds of training exercises [in Russian]. *Teoria i Praktika Fizicheskoi Kulturõ* 6:18-9.

Kassil, G. N., I. L. Vaisfeld, E. S. Matlina, and G. L.

Schreiberg. 1978. *Humoral-hormonal mechanisms of regulation of functions in sports activities* [in Russian]. Moscow: Nauka.

Katch, V., A. Weltman, R. Martin, and L. Gray. 1977. Optimal test characteristics for maximal anaerobic work on the bicycle ergometer. *Research Quarterly* 263:276.

Katz, A., S. Broberg, K. Sahlin, and J. Wahren. 1986. Muscle ammonia and amino acid metabolism during dynamic exercise in man. *Clinical Physiology* 6:365-79.

Kavanagh, M., and I. Jacobs. 1988. Breath-by-breath oxygen consumption during performance of the Wingate test. *Canadian Journal of Sports Sciences* 13:91-3.

Keast, D., K. Cammeron, and A. R. Morton. 1988. Exercise and the immune response. *Sports Medicine* 5:248-67.

Keibel, D. 1974. Nebennierenrinden-Hormone und sportliche Leistung. *Medizine und Sport* 14:65-76.

Keiser, A., J. Poortmans, and S. J. Bunnik. 1980. Influence of physical exercise on sex hormone metabolism. *Journal of Applied Physiology* 48:767-9.

Keizer, H. A. 1998. Neuroendocrine aspects of overtraining. In *Overtraining in Sport,* ed. Kreider, R. B., A. C. Fry, and M. L. O'Toole, 145-167. Champaign IL: Human Kinetics.

Keizer, H. A., E. Beckers, J. de Haan, G. M. E. Janssen, H. Kuipers, and G. van Kranenburg. 1987b. Exercise-induced changes in the percentage of free testosterone and estradiol in trained and untrained women. *International Journal of Sports Medicine* 8(Suppl. 3):151-3.

Keizer, H. A., H. Kuipers, J. de Haan, E. Beckers, and L. Habets. 1987a. Multiple hormone responses to physical exercise in eumenorrheic trained and untrained women. *International Journal of Sports Medicine* 8(Suppl. 3):139-50.

Keizer, H. A., P. Platen, H. Kuoppeschaar, W. R. de Vries, C. Vervoorn, P. Geurten, and G. von Kraneburg. 1991. Blunted ß-endorphin response to corticotropin releasing hormone and exercise after exhaustive training. (abstract) *International Journal of Sports Medicine* 12:97.

Keul, J., E. Doll, and D. Kepler. 1972. *Energy metabolism in human muscle*. Basel: Karger.

Keul, J., E. Doll, H. Stein, U. Singer, and H. Reindell. 1964. Über den Stoffwechsel der menschlien Herzens: das Verhalten der arteriokoronarvenösen Differences der Aminosäuren und des Ammoniaks beim gesunden menschlichen Herzen in Ruhe, während und nach körperlicher Arbeit. *Deutche Archiv for Klinische Medizin* 209:717-25.

Keul, J., G. Haralambie, M. Bruder, and H. J. Gottstein. 1978. The effect of weight lifting exercise on heart rate and metabolism in experienced lifters. *Medicine and Science in Sports and Exercise* 10:13-5.

Keul, J., B. Kohler, G. Glutz, U., Berg, A., and H. Howald. 1981. Biochemical changes in a 100 km run: carbohydrates, lipids and hormones in serum. *European Journal of Applied Physiology* 47:181-9.

Keul, J., G. Simon, A. Berg, H.-H. Dickhuth, I. Goertler, and R. Kuebel. 1979. Bestimmung der individuellen aeroben Schwelle zur Leistungsbewertung und Trainingsgestaltung. *Deutche Zeitschrift für die Sportmedizin* 7:212-8.

Khouhar, A. M., C. W. H. Harvard, and J. D. H. Slater. 1979. Effect of the inhibition of angiotensin converting enzyme on renin release mediated by exercise in man. *Journal of Endocrinology* 83:37P.

Kiens, B., and B. Saltin. 1989. Endurance training of man decreases muscle potassium loss during exercise (abstract). *Acta Physiologica Scandinavica* 126:P5.

Kindermann, W. 1986. Das Übertraining—Ausdruck einer vegetativen Fehlsteuerung. *Deutche Zeitschrft für Sportmedizin* 37:238-45.

Kindermann, W., and J. Keul. 1977. *Anaerobic Energiebereitstellung im Hochleistungssport.* Schorndorf: Verlag Karl Hofmann.

Kindermann, W., and W. M. Schmitt. 1985. Verhalten von Testosteron im Blutserum bei Körperarbeit unterschiedlicher Dauer und Intesität. *Deutsche Zeitschrift für Sportmedizin* 36:99-104.

Kindermann, W. A., A. Schnabel, W. M. Schmidt, G. Bird, J. Cassens, and F. Weber. 1982. Catecholamines, growth hormone, cortisol, insulin and sex hormones in anaerobic and aerobic exercise. *European Journal of Applied Physiology* 49:389-99.

Kindermann, W., G. Simon, and J. Keul. 1979. The significance of the aerobic-anaerobic transition for the determination of work load intensities during endurance training. *European Journal of Applied Physiology* 42:25-34.

Kirkendall, D. T. 2000. Fatigue from voluntary motor activity. In *Exercise and sport science,* ed. Garrett, W. G., and D. T. Kirkendall, 97-104. Philadelphia: Lippincott Williams & Wilkins.

Kirsch, K., W.-D. Risch, U. Mund, L. Röcker, and H. Stobby. 1975. Low pressure system and blood volume regulating hormones after prolonged exercise. In *Metabolic adaptation to prolonged physical exercise,* ed. Howald, H., and J. R. Poortmans, 315-321. Basel: Birkhäuser Verlag.

Kirwan, J. P., D. L. Costill, M. G. Flynn, J. B. Mitchell, M. G. Fink, J. B. Mitchell, W. J. Fink, P. D. Neufer, and J. A. Houmard. 1988a. Physiological responses to successive days of intensive training in competitive swimmers. *Medicine and Science in Sports and Exercise* 20:255-9.

Kirwan, J. P., D. L. Costill, J. B. Mitchell, J. A. Houmard, M. G. Flynn, W. J. Fink, and J. D. Bettz. 1988b. Carbohydrate balance in competitive runners during successive days of intense training. *Journal of Applied Physiology* 65:2601-6.

Kjaer, M. 1992. Regulation of hormonal and metabolic responses during exercise in humans. *Exercise and Sports Sciences Reviews* 20:161-84.

Kjaer, M., and H. Galbo. 1988. Effect of physical training on the capacity to secrete epinephrine. *Journal of Applied Physiology* 64:11-6.

Kjaer, M., J. Bangsbo, G. Lortie, and H. Galbo. 1988. Hormonal responses to exercise in humans: influence of hypoxia and physical training. *American Journal of Physiology* 254:R197-203.

Kjaer, M., N. J. Christensen, B. Sonne, E. A. Richter, and H. Galbo. 1985. Effect of exercise on epinephrine turnover in trained and untrained male subjects. *Journal of Applied Physiology* 59:1061-67.

Kjaer, M., P. A. Farrell, N. J. Christensen, and H. Galbo. 1986. Increased epinephrine response and inaccurate glucoregulation in exercising athlete. *Journal of Applied Physiology* 61:1693-1700.

Kjaer, M., K. J. Mikines, J. Christensen, B. Tronier, J. Vinten, B. Sonne, E. A. Richter, and H. Galbo. 1984. Glucose turnover and hormonal changes during insulin-induced hypoglycemia in trained humans. *Journal of Applied Physiology* 57:21-7.

Kjaer, M., N. H. Secher, F. W. Bach, and H. Galbo. 1987. Role of motor center activity for hormonal changes and substrate mobilization in humans. *American Journal of Physiology* 253:R687-97.

Kjaer, M., N. H. Secher, F. W. Bach, S. Skeikh, and H. Galbo. 1989. Hormonal and metabolic responses to exercise in humans: Effect of sensory nervous blockade. *American Journal of Physiology* 257:E95-101.

Kjellberg, S. R., V. Rudhe, and T. Sjöstrand. 1949. Increase of the amount of hemoglobin and blood volume in connection with physical training. *Acta Physiologica Scandinavica* 19:146-51.

Klimov, V. M., and V. I. Koloskov. 1982. *Guidance of the training in ice hockey players* [in Russian]. Moscow: FiS.

Klitgaard, H., and T. Clausen. 1989. Increased total concentration of Na, K pump in vastus lateralis muscle of old trained human subjects. *Journal of Applied Physiology* 67:2491-4.

Knochel, J. P., L. N. Dotin, and R. J. Hamburger. 1972. Pathophysiology of intense physical conditioning in a hot climate. I. Mechanism of potassium depletion. *Journal of Clinical Investigation* 51:242-55.

Knopik, J., C. Meredith, B. Jones, R. Fielding, V. Young, and W. Evans. 1991. Leucine metabolism during fasting and exercise. *Journal of Applied Physiology* 70:43-7.

Knowlton, R. G., D. D. Brown, R. K. Hetzler, and L. M. Sikora. 1990. Venous and fingertip blood to calculate plasma volume shift following exercise. *Medicine and Science in Sports and Exercise* 22:854-7.

Knuttgen, H. G., J. A. Vogel, and J. Poortmans. 1983. Glossary of exercise terminology. In *Biochemistry of exercise,* ed. Knuttgen, H. G., J. A. Vogel, and J. Poortmans, XXIV-XXV. Champaign IL: Human Kinetics.

Kohl, H. W., R. E. LaPorte, and S. N. Blair. 1988. Physical activity and cancer: An epidemiological perspective. *Sports Medicine* 6:222-37.

Koivisto, V., R. Hendler, E. Nagel, and P. Felig. 1982. Influence of physical training on the fuel-hormone response to prolonged low intensity exercise. *Metabolism* 31:192-7.

Koivisto, V. A., S. L. Karonen, and E. A. Nikkilä. 1981. Carbohydrate ingestion before exercise: comparison of glucose, fructose, and sweat placebo. *Journal of Clinical Physiology* 26:277-85.

Koivisto, V. A., and H. Yki-Järvinen. 1987. Effect of exercise on insulin binding and glucose transport in adipocytes of normal humans. *Journal of Applied Physiology* 63:1319-23.

Komi, P. V., and J. Karlsson. 1978. Skeletal muscle fibre types, enzyme activities and physical performance in young males and females. *Acta Physiologica Scandinavica* 103:210-8.

Komi, P., J. H. T. Viitasalo, M. Ham, A. Thorstensson, B. Sjödin, and J. Karlsson. 1977. Skeletal muscle fibers and muscle enzyme activities in monozygous and dizygous twins of both sexes. *Acta Physiologica Scandinavica* 100:385-92.

Konovalova, G., R. Masso, V. Ööpik, and A. Viru. 1997. Significance of thyroid hormones in post-exercise incorporation of amino acids into muscle fibers in rats. An autoradiographic study. *Endocrinology and Metabolism* 4:25-31.

Kõrge, P. 1999. Factors limiting ATPase activity in skeletal muscle. In *Biochemistry of exercise X,* ed. Hargreaves, M., and M. Thompson, 125-134. Champaign IL: Human Kinetics.

Kõrge, P., and K. B. Campbell. 1994. Local ATP regeneration is important for sarcoplasmic reticulum Ca^{2+} pump function. *American Journal of Physiology* 267:C357-66.

Kõrge, P., and K. B. Campbell. 1995. The importance of ATPase microenvironment in muscle fatigue: a hypothesis. *International Journal of Sports Medicine* 172-9.

Kõrge, P., A. Eller, S. Timpmann, and E. Seppet. 1982. The role of glucocorticoids in the regulation of postexercise glycogen repletion and the mechanism of their action. *Sechenov Physiological Journal of the USSR* 68:1431-6.

Kõrge, P., and S. Roosson. 1975. The importance of adrenal glands in the improved adaptation of trained animals to physical exertion. *Endokrinologie* 64:232-8.

Kõrge, P., S. Roosson, and M. Oks. 1974a. Heart adaptation to physical excretion in relation to work duration. *Acta Cardiology* 29:303-20.

Kõrge, P., and A. Viru. 1971a. Water and electrolyte metabolism in skeletal muscle of exercising rats. *Journal of Applied Physiology* 31:1-4.

Kõrge, P., and A. Viru. 1971b. Water and electrolyte metabolism in myocardium of exercising rats. *Journal of Applied Physiology* 31:5-7.

Kõrge, P., A. Viru, and S. Roosson. 1974b. The effect of chronic overload on skeletal muscle metabolism and adrenocortical activity. *Acta Physiologica Academiae Scientiarum Hungaricae* 45:41-51.

Kostina, L. V., L. L. Zhurkina, and N. S. Dudov. 1986. Peculiarities of alterations of hormonal responses to competition in long-distance runners in course of improvement of special fitness. In *Biochemical criteria of improvement of special fitness,* ed. Tchareyeva, A. A., 56-67. Moscow: Allunion Research Institute of Physical Culture.

Kosunen, K. J., and A. J. Pakarinen. 1976. Plasma renin, angiotensin II, and plasma urinary aldosterone in running exercise. *Journal of Applied Physiology* 41:26-9.

Kotchen, T. A., L. H. Hartley, T. W. Rice, E. H. Mongly, L. R. G. Jones, and J. W. Mason. 1971. Renin, norepinephrine, and epinephrine responses to graded exercise. *Journal of Applied Physiology* 31:178-84.

Kots, Y. M., O. L. Vinogradova, K. Mamadu, and F. D. Danicheva. 1986. Redistribution in utilization of energy substrates in daily intensive exercises. *Teoria i Praktika Fizicheskoi Kulturõ* 4:22-4.

Kozlowski, S., Z. Brzezinska, K. Nazar, and E. Turlejska. 1981. Carbohydrate availability for the brain and muscles as a factor modifying sympathetic activity during exercise in dog. In *Biochemistry of exercise*, ed. Poortmans, J., and G. Niset, IV, 54-62. Baltimore: University Park Press.

Kozlowski, S., J. Chwalbinska-Moneta, M. Viga, H. Kaciuba-Uscilko, and K. Nazar. 1983. Greater serum GH response to arm than to leg exercise performed at equivalent oxygen uptake. *European Journal of Applied Physiology* 52:131-5.

Kozlowski, S., and B. Saltin. 1964. Effect of sweat loss on body fluids. *Journal of Applied Physiology* 19:1119-24.

Kraemer, W. J., B. A. Aguilera, M. Terada, R. U. Newton, J. M. Lynch, G. Rosendaal, J. M. McBride, S. E. Gordon, and K. Häkkinen. 1995a. Responses of IGF-I to endogenous increase in growth hormone after heavy-resistance exercise. *Journal of Applied Physiology* 79:1310-5.

Kraemer, W. J., L. E. Armstrong, R. W. Hubbard, L. J. Marchitelli, N. Leva, P. B. Rock, and J. E. Dziados. 1988. Responses of plasma human atrial natriuretic factor to high intensity submaximal exercise in heat. *European Journal of Applied Physiology* 57:399-403.

Kraemer, R. R., M. S. Blair, R. M. Caferty, and V. D. Castracane. 1993. Running-induced alteration in growth hormone, prolactin, triiodothyronine, and thyroxine concentration in trained and untrained men and women. *Research Quarterly for Exercise and Sport* 64:69-74.

Kraemer, W. J., J. Dziados, S. E. Gordon, L. J. Marchitelli, A. C. Fry, and K. L. Reynolds. 1990a. The effects of graded exercise on plasma proenkephalin peptide F and catecholamine responses at sea level. *European Journal of Applied Physiology* 61:214-7.

Kraemer, W. J., J. E. Dziados, L. J. Marchitelli, S. E. Gordon, E. A. Harman, R. Mello, S. J. Fleck, P. N. Frykman, and N. T. Triplett. 1993b. Effects of different heavy-resistance exercise protocols on plasma beta-endorphin concentrations. *Journal of Applied Physiology* 74:450-9.

Kraemer, W. J., S. J. Fleck, R. Callister, M. Shealy, G. A. Dudley, C. M. Maresh, L. Marchitelli, C. Crutbirds, T. Murray, and J. E. Falkel. 1989a. Training responses of plasma beta-endorphin, adrenocorticotropin, and cortisol. *Medicine and Science in Sports and Exercise* 21:146-53.

Kraemer, W. J., S. J. Fleck, J. E. Dziadosr, E. A. Harman, L. J. Marchitelli, S. E. Gordon, R. Mello, P. N. Frykman, L. P. Kozirs, and N. T. Triplett. 1993a. Changes in hormonal concentrations after different heavy-resistance exercise protocols in women. *Journal of Applied Physiology* 75:594-604.

Kraemer, W. J., S. J. Fleck, and W. J. Evans. 1996. Strength and power training: physiological mechanisms of adaptation. *Exercise and Sport Sciences Reviews* 24:363-97.

Kraemer, W. J., A. C. Fry, B. J. Warren, M. H. Stone, S. J. Fleck, J. T. Kearkey, B. P. Conroy, C. M. Maresh, C. A. Weseman, N. T. Triplett, and S. E. Gordon. 1992. Actual hormonal responses in elite junior weightlifters. *International Journal of Sports Medicine* 13:103-9.

Kraemer, W. J., S. E. Gordon, S. J. Fleck, L. J. Marchitelli, R. Mello, J. E. Dziados, K. Friedl, E. Harman, C. Maresh, and A. C. Fry. 1991b. Endogenous anabolic hormonal and growth factor responses to heavy resistance exercise in males and females. *International Journal of Sports Medicine* 21:228-35.

Kraemer, W. J., L. Marchitelli, S. E. Gordon, E. Harman, J. E. Dziados, R. Mello, P. Frykman, D. McCurry, and S. J. Fleck. 1990b. Hormonal and growth factor responses to heavy resistance exercise protocols. *Journal of Applied Physiology* 69:1442-50.

Kraemer, W. J., and B. C. Nindl. 1998. Factors involved with overtraining for strength and power. In *Overtraining in sport,* ed. Kreider, R. B., A. C. Fry, and M. L. O'Toole, 69-86. Champaign IL: Human Kinetics.

Kraemer, W. J., B. J. Noble, M. J . Clark, and B. W. Culver. 1987. Physiologic responses to heavy resistance exercise with very short rest periods. *International Journal of Sports Medicine* 8:247-52.

Kraemer, W. J., B. Noble, B. Cluver, and R. V. Lewis. 1985. Changes in plasma proenkephalin peptide F and catecholamine levels during graded exercise in men. *Proceedings of the National Academy of Science of USA* 18:6349-54.

Kraemer, W. J., J. F. Patton, H. G. Knuttgen, C. J. Hannan, T. Kettler, E. G. Scott, J. E. Dziados, A. C. Fry, P. N. Frykman, and E. A. Harman. 1991a. Effects of high-intensity cycle exercise on sympathoadrenal-medullary response pattern. *Journal of Applied Physiology* 70:8-14.

Kraemer, W. J., J. F. Patton, H. G. Knuttgen, L. J. Marchitelli, C. Cruthirts, A. Damonash, E. Harman, P. Frykman, and J. E. Dziados. 1989b. Hypothalamic-pituitary-adrenal responses to short-duration high-intensity cycle exercise. *Journal of Applied Physiology* 66:161-6.

Kraemer, W. J., J. C. Volek, J. A. Bush, M. Putukian, and W. J. Sebastianelli. 1998. Hormonal responses to consecutive days of heavy-resistance exercise with or without nutritional supplementation. *Journal of Applied Physiology* 85:1544-55.

Krasnova, A. F., R. I. Lenkova, A. G. Leshkevitch, L. V. Maksimova, N. R. Tchagovets, and N. N. Yakovlev. 1972. Peculiarities of muscular energetic metabolism depending on the degree of adaptation. *Sechenov Physiological Journal of USSR* 58:114-21.

Kraus, H., and R. Kinne. 1970. Regulation der bei langandauerdem körperlichen Adaptation und Leistungssteingerung durch Thyroid-hormone. *Pflügers Archiv für gesante Physiologie* 321:332-45.

Kretzschmar, M., D. Müller, J. Hübscher, E. Marin, and W. Klinger. 1991. Influence of aging, training and acute physical exercise on plasma glutathione and lipid peroxides in men. *International Journal of Sports Medicine* 12:218-22.

Kreutz, L., R. Rose, and J. Jennings. 1972. Suppression of plasma testosterone level and psychological stress. *Archives of General Psychiatry* 26:479-82.

Krieder, R. B. 1998. Central fatigue hypothesis and overtraining. In *Overtraining in sport,* ed. Krieder, R. B., A. C. Fry, and M. L. O'Toole, 309-331. Champaign IL: Human Kinetics.

Krogh, A., and J. Lindhard. 1920. The relative value of fat and carbohydrate as sources of muscular energy. With appendices on the correlations between standard metabolism and the respiratory quotient during rest and work. *Biochemical Journal* 14:290-363.

Kuipers, H. 1998. Training and overtraining: an introduction. *Medicine and Science in Sports and Exercise* 30:1137-9.

Kuipers, H., E. J. Fransen, and H. A. Keizer. 1999. Preexercise ingestion of carbohydrate and transient hypoglycemia during exercise. *International Journal of Sport Medicine* 20(4):222-31.

Kuipers, H., and H. A. Keizer. 1988. Overtraining in elite athletes: review and directions for the future. *Sports Medicine* 6:79-92.

Kuoppasalmi, K. 1980. Plasma testosterone and sexhormone-binding globulin capacity in physical exercise. *Scandinavian Journal of Clinical and Laboratory Investigations* 40:411-8.

Kuoppasalmi, K., H. Näveri, M. Härkönen, and H. Adlercreutz. 1980. Plasma cortisol, androstenedione, testosterone and luteinizing hormone in running exercise of different intensities. *Scandinavian Journal of Clinical and Laboratory Investigations* 40:403-9.

Kurowski, T. T., R. T. Chatterton, and R. C. Hickson. 1984. Glucocorticoid-induced cardiac hypertrophy: Additive effects of exercise. *Journal of Applied Physiology* 57(2):514-9.

Kuznetsov, V. V. 1980. *Olympic Games and human capacities* [in Russian]. Moscow: Znanie.

Lacour, J. R., E. Borwat, and J. C. Barthélémy. 1990. Post-competition blood lactate concentrations as indicators of energy expenditure during 400-m and 800-m races. *European Journal of Applied Physiology* 61:172-6.

Ladu, M. J. 1991. Regulation of lipoprotein lipase in muscle and adipose tissue during exercise. *Journal of Applied Physiology* 71:404-9.

LaFontaine, T. P., B. R. Londeree, and W. K. Spath. 1981. The maximal steady state versus selected running events. *Medicine and Science in Sports and Exercise* 13:190-2.

Lamb, R. D. 1989. Anabolic steroids and athletic performance. In: *Hormones and Sport,* 257-273. New York: Raven Press.

Lamberts, S. W., H. A. T. Timmermans, M. Kramer-Blankenstijn, and J. C. Birkenhäger. 1975. The mechanism of the potentiating effect of glucocorticoids on catecholamine-induced lipolysis. *Metabolism* 24:681-9.

Landt, M., G. M. Lawson, J. M. Helgeson, V. G. Davila-Roman, J. H. Ladenson, A. S. Joffe, and R. C. Hickner. 1997. Prolonged exercise decreases serum leptin concentrations. *Metabolism* 46(10):1109-12.

Langfort, J., W. Pilis, R. Zarzeczny, K. Nazar, and H. Kaciuba-Uscilko. 1996. Effect of low-carbohydrate-ketogenic diet on metabolic and hormonal responses to graded exercise in men. *Journal of Physiology and Pharmacology* 47:361-71.

Larsson, L., and C. Skogsberg. 1988. Effects of the interval between removal and freezing of muscle biopsies on muscle fibre size. *Journal of Neurological Sciences* 85(1):27-38.

Lassarre, C., F. Girard, J. Durand, and J. Raynaud. 1974. Kinetics of human growth hormone during submaximal exercise. *Journal of Applied Physiology* 37:826-30.

Laurent, G. J., M. P. Sparrow, P. C. Bates, and D. J. Millward. 1978. Turnover of muscle protein in the fowl. Changes in rates of protein synthesis and breakdown during hypertrophy of the anterior and posterior latissimus dorsi muscles. *Biochemical Journal* 176:407-17.

Lavoie, J.-M., D. Consineau, F. Peronnet, and P. J. Provencher. 1983. Liver glycogen store and hypoglycemia during prolonged exercise in man. In *Biochemistry of exercise,* ed. Knuttgen, H. G., J. A. Vogel, and J. Poortmans, 297-301. Champaign, IL: Human Kinetics.

Lavoie, J.-M., D. Nicole, R. Helie, and G. M. Brison. 1987. Menstrual cycle phase dissociation of blood glucose homeostasis during exercise. *Journal of Applied Physiology* 62:1084-9.

Lehmann, M., P. Baumgarte, C. Wiesenack, A. Seidel, H. Baumann, S. Fisher, U. Spöri, G. Gendrish, R. Kaminski, and J. Keul. 1992c. Training-overtraining: influence of a defined increase in training volume vs. training intensity on performance catecholamines and some metabolic parameters in experienced middle- and long-distance runners. *European Journal of Applied Physiology* 64:169-77.

Lehmann, M., A. Berg, H. H. Dickhuth, E. Jacob, U. Korsten-Reck, W. Stockhausen, and J. Keul. 1989. Zur Bedeuteing von Katecholamin-und Adrenoreceptorverhalten für Leistungsdiagnostik und Trainingsbegleitung. *Leistungssport* 19(1):14-21.

Lehmann, M., A. Berg, R. Kapp, T. Wessingshage, and J. Keul. 1983. Correlation between laboratory testing and distance running performance in marathoners of similar performance ability. *International Journal of Sports Medicine* 4:226-30.

Lehmann, M., H. H. Dickhuth, G. Gendrisch, W. Lazar, M. Thum, R. Kaminski, J. F. Aramendi, E. Peterke, W. Wieland, and J. Keul. 1991. Training-overtraining. A prospective, experimental study with experienced middle- and long-distance runners. *International Journal of Sports Medicine* 12:444-52.

Lehmann, M., C. Foster, and J. Keul. 1993a. Overtraining in endurance athletes. A brief review. *Medicine and Science in Sports and Exercise* 25:854-62.

Lehmann, M., C. Foster, H. H. Dickhuth, and U. Gastmann. 1998b. Autonomic imbalance hypothesis and overtraining syndrome. *Medicine and Science in Sports and Exercise* 30:1140-5.

Lehmann, M., C. Foster, J. M. Steinacker, W. Lormes, A. Opitz-Gross, J. Keul, and U. Gastmann. 1997. Training and overtraining: overview and experimental results. *Journal of Sports Medicine and Physical Fitness* 37:7-17.

Lehmann, M., C. Foster, N. Netzer, W. Lormes, J. M. Steinacker, Y. Liu, A. Opitz-Gress, and U. Gastmann. 1998a. Physiological responses to short- and long-term overtraining in endurance athletes. In *Overtraining in sports,* ed. Krieder, R. B., A. C. Fry, and M. L. O'Toole, 19-46. Champaign IL: Human Kinetics.

Lehmann, M., C. Foster, U. Gastmann, H. Keizer, and J. M. Steinacker. 1999a. Definition, types, symptoms, findings, underlying mechanisms, and frequency of overtraining and overtraining syndrome. In *Overload, performance incompetence, and regression in sport,* ed. Lehmann, M., C. Foster, U. Gastmann, H. Keizer, and J.M. Steinacker. New York: Kluwer Academic/Plenum Publication.

Lehmann, M., U. Gastmann, K. G. Petersen, N. Bachl, A. Seidel, A. N. Khalaf, S. Fischer, and J. Keul. 1992a. Training-overtraining performance in experienced middle- and long-distance runners. *British Journal of Sports Medicine* 26:233-42.

Lehmann, M., U. Gastmann, S. Baur, Y. Liu, W. Lormes, A. Opiz-Gress, S. Reißnecker, C. Simisch, and J. M. Steinacker. 1999b. Selected parameters and mechanisms of peripheral and central fatigue and regeneration in overtrained athletes. In *Overload, performance and regeneration in sport,* ed. Lehmann, M., C. Foster, U. Gastmann, H. Keizer, and J. M. Steinacker, 7-25. New York: Kluwer Academic/Plenum Publication.

Lehmann, M., V. Gortmann, K. G. Peterson, W. Lormes, and J. M. Steinacker. 2000. Prolonged heavy training: possible influence of leptin and insulin B on gonadotropic axis function in male athletes. Fifth Congress of the European College of Sport Science. Jyväskyla, 438.

Lehmann, M., M. Huonker, F. Dimeo, N. Heinz, U. Gartmann, N. Treis, J. M. Steinacker, J. Keul, R. Kaewski, and D. Häussinger. 1995. Serum amino acid concentrations in nine athletes before and after the 1993 Colmar Ultra Triathlon. *International Journal of Sports Medicine* 16:155-9.

Lehmann, M., and J. Keul. 1985. Capillary-venous differences of free plasma catecholamines at rest and during graded exercise. *European Journal of Applied Physiology* 54:502-5.

Lehmann, M., J. Keul, G. Huber, and M. Prada. 1981. Plasma catecholamines in trained and untrained volunteers during graduated exercise. *International Journal of Sports Medicine* 2:143-7.

Lehmann, M., K. Knizia, U. Gastmann, K. G. Petersen, A. N. Khalaf, S. Bauer, L. Kerp, and J. Keul. 1993b. Influence of 6-week, 6 day per week, training on pituitary function in recreational athletes. *British Journal of Sports Medicine* 27:186-92.

Lehmann, M., H. Mann, U. Gastmann, J. Keul, D. Vetter, J. M. Steinacker, and D. Haussinger. 1996. Unaccustomed high-mileage vs intensity training-related changes in performance and serum amino acid levels. *International Journal of Sports Medicine* 17:187-92.

Lehmann, M., P. Schmid, and J. Keul. 1984. Age- and exercise-related sympathetic activity in untrained volunteers, trained athletes and patients with impaired left-ventricular contractility. *European Heart Journal,* Suppl. E:1-7.

Lehmann, M., P. Schmid, and J. Keul. 1985. Plasma catecholamine and blood lactate accumulation during incremental exhaustive exercise. *International Journal of Sports Medicine* 6:78-81.

Lehmann, M., W. Schnee, R. Scheu, W. Stockhausen, and N. Bachl. 1992b. Decreased nocturnal catecholamine excretion: parameter for an overtraining syndrome in athletes? *International Journal of Sports Medicine* 13:236-42.

Lemon, P. W. R., and J. P. Mullin. 1980. The effect of initial muscle glycogen levels on protein catabolism during exercise. *Journal of Applied Physiology* 48:624-9.

Lemon, P. W. R., and F. J. Nagle. 1981. Effects of exercise on protein and amino acid metabolism. *Medicine and Sciences of Exercise Sports* 13:141-9.

Lemon, P. W. R., F. J. Nagle, J. P. Mullin, and N. J. Benevenga. 1982. In vitro leucine oxidation at rest and during two intensities of exercise. *Journal of Applied Physiology* 53:947-54.

Levando, V. A., R. S. Suzdalnitski, and G. N. Kassil. 1990. Athletic activities and problems associated with related stress, adaptation and acutely developing pathological conditions. *Sports Training, Medicine, Rehabilitation* 1:305-15.

Levando, V. A., R. S. Suzdalnitski, B. B. Pershin, and M. P. Zykov. 1988. Study of secretory and antiviral immunity in sportsmen. *Sports Training, Medicine, Rehabilitation* 1:49-52.

Levine, S. A., B. Gordon, and C. L. Derick. 1924. Some changes in the chemical constituents of the blood following a marathon race. *Journal of American Medicine Association* 82:1778-9.

Lewicki, R., H. Tchorzewski, A. Denys, M. Kowalska, and A. Golinska. 1987. Effect of physical exercise on some parameters of immunity in conditioned sportsmen. *International Journal of Sports Medicine* 8:309-14.

Lewicki, R., H. Tchorzewski, E. Majewska, Z. Nowak, and Z. Baj. 1988. Effect of maximal physical exercise on T-lymphocyte subpopulations and on interleukin 1 (IL 1) and interleukin 2 (IL 2) production *in vitro. International Journal of Sports Medicine* 9:114-7.

Lexell, J., K. Henriksson-Larsen, B. Winblad, and M. Sjöström. 1983. Distribution of different fiber types in human skeletal muscles: Effects of aging studied in whole muscle cross-sections. *Muscle and Nerve* 6:588-95.

Liesen, H., B. Dufaux, and W. Hollmann. 1977. Modifications of serum glycoproteins the days following a prolonged physical exercise and the influence of physical training. *European Journal of Applied Physiology* 37:243-54.

Liesen, H. 1985. Trainingsteierung im Hochleistungssport einge Aspekte und Beispiele. *Deutsche Zeitschrift für Sportmedizin* 1:8-18.

Lijnen, P. J., A. K. Amery, R. H. Fagard, T. M. Reybrouck, E. J. Moerman, and A. F. de Schaepdryver. 1979. The effects of ß-adrenoreceptor blockade on renin, angiotensin, aldosterone and catecholamines at rest and during exercise. *British Journal of Clinical Pharmacology* 7:175-81.

Lijnen, P., P. Hespel, R. Fagard, R. Lysens, E. Van den Ende, M. Goris, W. Goossens, W. Lissens, and A. Amery. 1988. Indicators of cell breakdown in plasma of men during and after a marathon race. *International Journal of Sports Medicine* 9:108-13.

Lijnen, P., P. Hespel, J. R. M'Buyamba-Kabangu, M. L. G. Goris, E. Van den Ende, R. Fagard, and A. Amory. 1987. Plasma atrial natriuretic peptide and cycle nucleotide levels before and after a marathon. *Journal of Applied Physiology* 63:1180-4.

Lindena, J., and I. Trautschol. 1983. Enzyme in lymph: a review. *Journal of Clinical Chemistry and Biochemistry* 21:327-46.

Lindinger, M. J., and G. Sjøgaard. 1991. Potassium regulation during exercise and recovery. *Sports Medicine* 11:382-401.

Litvinova, L., and A. Viru. 1995a. Effect of exercise and adrenal insufficiency on urea production in rats. *European Journal of Applied Physiology* 70:536-40.

Litvinova, L., and A. Viru. 1997. Does the increased blood urea depend on lactate response to exercise. *Coaching and Sport Sciences Journal* 2(2):6-11.

Litvinova, L., A. Viru, and T. Smirnova. 1989. Renal urea clearance in normal and adrenalectomized rats after exercise. *Japanese Journal of Physiology* 39:713-23.

Liu, Y., S. Mayr, A. Opitz-Gress, C. Zeller, W. Lormes, S. Baur, M. Lehmann, and J. M. Steinacker. 1999.

Human skeletal muscle HSP70 response to training in highly trained rowers. *Journal of Applied Physiology* 86(1):101-4.

Ljunghall, S., H. Jaborn, L. E. Roxin, J. Rastad, L. Wide, and G. Anerston. 1986. Prolonged low-intensity exercises raises the serum parathyroid hormone level. *Clinical Endocrinology* 25:535-42.

Ljunghall, S., H. Jaborn, L. E. Roxin, E. T. Skarfon, L. E. Wide, and H. O. Lithall. 1988. Increase in serum parathyroid hormone levels after prolonged physical exercise. *Medicine and Science in Sports and Exercise* 20:122-5.

Lonnqvist, F., P. Arner, L. Nordfors, and M. Schalling. 1995. Overexpression of the obese (ob) gene in adipose tissue of human obese subjects. *Nature Medicine* 1(9):950-3.

Lorenz, R., and G. Gerber. 1979. Harnstoff bei körperlichen Belastungen: Veränderung der Synthese, der Blutkonzentration und der Ausscheidung. *Medicina und Sport* 19:240-52.

Lormes, W., M. Lehmann, and J. M. Steinacker. 1998. The problems to study plasma lactate. *International Journal of Sports Medicine* 19:223-5.

Lotzerich, H., H.-G. Fehr, and H.-J. Appell. 1990. Potentiation of cytostatic but not cytolytic activity of murine macrophages after running stress. *International Journal of Sports Medicine* 11:61-5.

Lu, S. C., C. Garcia-Ruiz, J. Kuhlenkamp, M. Ookhtens, M. Prato, and N. Kaplowicz. 1990. Hormonal regulation of glutathione efflux. *Journal of Biological Chemistry* 265:16088-95.

Lunde, P. K., E. Verburg, N. K. Vøllestad, and O. M. Sejersted. 1998. Skeletal muscle fatigue in normal subjects and heart failure patients. Is there a common mechanism. *Acta Physiologica Scandinavica* 162:215-28.

Lundswall, J., S. Mellander, H. Westling, and T. White. 1972. Fluid transfer between blood and tissues during exercise. *Acta Physiologica Scandinavica* 85:258-69.

Luyckx, A. S., A. Dresse, A. Cession-Fossion, and P. J. Lefebvre. 1975. Catecholamines and exercise-induced glucagon and fatty acid mobilization in the rat. *American Journal of Physiology* 229:376-83.

Luyckx, A. S., F. Pirnay, and P. J. Lefebvre. 1978. Effect of glucose on plasma glucagon and free fatty acids during prolonged exercise. *European Journal of Applied Physiology* 39:53-61.

Macdonald, A. M. (ed.). 1972. *Chambers twentieth century dictionary.* Edinburgh: Chambers, 849.

Macdonald, I. A., S. A. Wootton, B. Muñoz, P. H. Fentem, and C. Williams. 1983. Catecholamine response to maximal anaerobic exercise. In *Biochemistry of exercise,* ed. Knuttgen, H. G., J. A. Vogel, and J. Poortmans, 749-754. Champaign, IL: Human Kinetics.

MacDougall, J. D., G. R. Ward, D. G. Sale, and J. R. Sutton. 1977. Biochemical adaptation of human skeletal muscle to heavy resistance training and immobilization. *Journal of Applied Physiology* 43:700-3.

Mackinnon, L. T. 1992. *Exercise and immunology.* Champaign, IL: Human Kinetics.

Mackinnon, L. T., T. W. Chick, A. van As, and T. B. Tomasi. 1988. Effects of prolonged intense exercise on natural killer cell number and function. In *Exercise physiology: Current selected research,* ed, Dotson, C. O., and J. H. Humphrey, Vol. 3, 77-89. New York: AMS Press.

Mackinnon, L. T., T. W. Chick, A. van As, and T. B. Tomasi. 1989. Decreased secretory immunoglobulins following intense endurance exercise. *Sports Training, Medicine, and Rehabilitation* 1:209-18.

Mackinnon, L. T., E. Ginn, and G. Seymour. 1991. Temporal relationship between exercise-induced decreases in salivary IgA concentration and subsequent appearance of upper respiratory illness in elite athletes. *Medicine and Science in Sports and Exercise* 23:S45.

Mackinnon, L. T., and S. L. Hooper. 1994. Mucosal (secretory) immune system responses to exercise of varying intensity and during overtraining. *International Journal of Sports Medicine* 15: (Suppl. 3):S179-83.

Mackinnon, L. T., and S. L. Hooper. 1996. Plasma glutamine and upper respiratory tract infection during intensified training in swimmers. *Medicine and Science in Sports and Exercise* 28:285-90.

Mackinnon, L. T., and S. L. Hooper. 2000. Overtraining and overreaching: causes, effects, and prevention. In *Exercise and sport science,* ed. Garrett, W. E., and D. T. Kirkendall, 487-498. Philadelphia: Lippincott Williams & Wilkins.

Mackinnon, L. T., S. L. Hooper, S. Jones, R. D. Gordon, and A. W. Bachmann. 1997. Hormonal, immunological, and hematological responses to intensified training in elite swimmers. *Medicine and Science in Sports and Exercise* 29:1637-45.

MacLean, D. A., L. L. Spriet, E. Hultman, and T. E. Graham. 1991. Plasma and muscle amino acid and ammonia responses during prolonged exercise in humans. *Journal of Applied Physiology* 70:2095-103.

MacLennan, P. A., R. A. Brown, and M. J. Rennie. 1987. A positive relationship between protein synthesis rate and intracellular glutamine concentration in perfused rat skeletal muscle. *FEBS Letters* 215:187-91.

MacLennan, P. A., L. Smith, B. Veryk, and M. J. Rennie.

1988. Inhibition of protein breakdown by glutamine in perfused rat skeletal muscle. *FEBS Letters* 237:133-6.

Mader, A. 1988. A transcription-translation activation feedback circuit as a function of protein degradation, with the quality of protein mass adaptation related to the average functional load. *Journal of Theoretical Biology* 134:135-57.

Mader, A. 1991. Evaluation of the endurance performance of marathon runners and theoretical analysis of test results. *Journal of Sports Medicine and Physical Fitness* 31:1-19.

Mader, A., and H. Heck. 1986. A theory of the metabolic origin of "anaerobic threshold." *International Journal of Sports Medicine* 7(Suppl):45-65.

Mader, A., H. Liesen, H. Heck, H. Philipp, R. Rost, P. Schürch, and W. Hollmann. 1976. Zur Beurteilung der sportartspezifischen Ausdauerleistungsfähigkeit im Labor. *Sportarzt Sportmed* 27:80-88,109-12.

Maffei, M., J. Halaas, E. Ravussin, R. E. Pratley, G. H. Lee, Y. Zang, H. Fei, H. Kim, R. Lallone, S. Ranganathan, P. A. Herr, and J. M. Friedman. 1995. Leptin levels in human and rodent: measurement of plasma leptin and ob RNA in obese and weight-reduced subjects. *Nature Medicine* 1:1115-61.

Magnusson, B., L. Hallberg, L. Rossander, and B. Swolin. 1984. Iron metabolism and sports anemia. A study of several iron parameters in elite runners with differences in iron status. *Acta Medica Scandinavica* 216:149-55.

Maher, J. T., L. G. Jones, L. H. Hartley, G. H. Williams, and L. I. Rose. 1975. Aldosterone dynamics during graded exercise at see level and high altitude. *Journal of Applied Physiology* 39:18-22.

Makarova, A. F. 1958. Biochemical characteristics of strength exercises. *Ukrainskij Biokhimicheskij Zhurnal* 30:368-77.

Malig, H., D. Stern, P. Altland, B. Highman, and B. Brodie. 1966. The physiological role of the sympathetic nervous system in exercise. *Journal of Pharmacology and Experimental Therapy* 154:35-45.

Manganiello, V., and M. Vaughan. 1972. An effect of dexamethasone on adenosine 3',5' monophosphate content and 3',5'-monophosphate phosphodiesterase activity of cultural hepatoma cells. *Journal of Clinical Investigations* 51:2763-7.

Mannix, E. T., P. Palange, G. R. Aronoff, F. Manfredi, and M. O. Farber. 1990. Atrial natriuretic peptide and the renin-aldosterone axis during exercise in man. *Medicine and Science in Sports and Exercise* 22:785-9.

Margaria, R., P. Aghemo, and E. Rovelli. 1966. Measurement of muscular power (anaerobic) in man. *Journal of Physiology* 21:1662-4.

Margaria, R., P. Cerretelli, P. E. DiPrampero, C. Massari, and G. Torelli. 1963. Kinetics and mechanism of oxygen debt contraction in man. *Journal of Applied Physiology* 18:371-7.

Margaria, R., H. T. Edwards, and D. B. Dill. 1933. The possible mechanisms of contracting and paying the oxygen debt and the role of lactic acid in muscular contraction. *American Journal of Physiology.* 106:689-715.

Margaria, R., and A. Foa. 1929/1930. Der Einfluss von Muskelarbeit auf den Stickstoffweshel die Kreatin und Säuereausscheidung. *Arbeitsphysilogie* 395-400.

Marker, J. C., I. B. Hirch, L. J. Smith, C. A. Parvin, J. O. Holloszy, and P. E. Cryer. 1991. Catecholamines in prevention of hypoglycemia during exercise in humans. *American Journal of Physiology* 260:E705-12.

Markoff, P. A., P. Ryan, and T. Young. 1982. Endorphin and mood changes in long-distance runners. *Medicine and Science in Sports and Exercise* 14:11-5.

Maron, M. B., S. M. Horwarth, and J. E. Wilkerson. 1975. Acute blood biochemical alteration in response to marathon running. *European Journal of Applied Physiology* 34:173-81.

Marone, J. R., M. T. Falduto, D. A. Essig, and R. C. Hickson. 1994. Effect of glucocorticoids and endurance training on cytochrome oxidase expression in skeletal muscle. *Journal of Applied Physiology* 77:1685-90.

Marniemi, J., P. Peltonen, I. Vuori, and E. Hietanen. 1980. Lipoprotein-lipase of human post-heparin plasma and adipose-tissue in relation to physical training. *Acta Physiologica Scandinavica* 110:131-5.

Martin, W. H., G. P. Dalsky, B. F. Hurley, D. E. Matthews, D. M. Bier, J. M. Hagberg, M. A. Rogers, D. S. King, and J. O. Holloszy. 1993. Effect of endurance training on plasma free fatty acid turnover and oxidation during exercise. *American Journal of Physiology* 265:E708-14.

Mason, J. W., L. H. Hartley, T. A. Kotchan, E. H. Mougey, P. T. Ricketts, and L. G. Jones. 1973a. Plasma cortisol and norepinephrine responses in anticipation of muscular activity. *Psychosomatic Medicine* 35:406-14.

Mason, J. W., L. H. Hartley, T. A. Kotchen, F. E. Wherry, L. L. Pennington, and L. G. Jones. 1973b. Plasma thyroid-stimulating hormone response in anticipation of muscular exercise in the human. *Journal of Clinical Endocrinology* 37:403-6.

Matlina, E. 1984. Effects of physical activity and other types of stress on catecholamine metabolism in various animal species. *Journal of Neurology and Transmission* 60:11-8.

Matlina, E. S., Pukhova, G. S., S. D. Galimov, A. I. Galentchik, and S. N. Almaeva. 1976. The catecholamines metabolism during adaptation to muscular activity. *Sechenov Physiological Journal of the USSR* 62:431-6.

Matlina, E., G. Schreiberg, M. Voinova, and L. Dunaeva. 1978. The interrelationships between catecholamines and corticosteroids in the course of muscular fatigue. *Sechenov Physiological Journal of the USSR* 64:171-6.

Matsin, T., T. Mägi, M. Alaver, and A. Viru. 1997. Possibility of monitoring training and recovery in different conditions of endurance training. *Coaching Sport and Sciences Journal* 2(2):18-24.

Matveyev, L. P. 1964. *The problem of sports training periodicity* [in Russian]. Moscow: FiS.

Matveyev, L. P. 1980. A noteworthy experience [in Russian]. *Sports Science Messenger* (Moscow) 6:38-39.

Maughan, R. J. 1994. Fluid and electrolyte loss and replacement in exercise. In *Oxford textbook of sports medicine,* ed. Harris, M., C. Williams, W. D. Stanish, and L. J. Micheli, 82-93. New York: Oxford University Press.

Maughan, R. J., and M. Gleeson. 1988. Influence of 36 h fast followed by refeeding with glucose, glycerol or placebo or metabolism and performance during prolonged exercise in man. *European Journal of Applied Physiology* 57:570-6.

Maughan, R. T. 1998. Restoration of water and electrolyte balance after exercise. *International Journal of Sports Medicine* 19:S136-8.

Max, S. 1990. Glucocorticoid-mediated induction of glutamine synthetase in skeletal muscle. *Medicine and Science in Sports and Exercise* 22:325-30.

Mayer, M., and R. Rosen. 1977. Interaction of glucocortoids and androgens with skeletal muscle. *Metabolism* 96:937-62.

Mayer, S., and P. M. Clarkson. 1984. Serum creatine kinase levels following isometric exercise. *Research Quarterly for Exercise and Sport* 55:191-4.

McBride, J. M., W. J. Kraemer, T. Triplett-McBride, and W. Sebastianelli. 1998. Effect of resistance exercise on free radical production. *Medicine and Science in Sports and Exercise* 30:67-72.

McCall, G. E., W. C. Byrnes, A. Fleck, S. J. Dickinson, A. Dickinson, and W. J. Kraemer. 1999. Acute and chronic hormonal responses to resistance training designed to promote muscle hypertrophy. *Canadian Journal of Applied Physiology* 24:96-107.

McEven, B. C. 1979. Influences of adrenocortical hormones on pituitary and brain function. In *Glucocorticoid hormone action,* ed. by Baxter, J. D., and G. G. Rousseau, 467-82. New York: Springer-Verlag.

McKelvie, R. S., M. L. Lindger, N. L. Jones, and G. J. F. Heigenhauser. 1992. Erythrocyte ion regulation across inactive muscle during leg exercise. *Canadian Journal of Physiology and Pharmacology* 70:1625-33.

McKenna, M. J. 1992. The roles of ionic processes in muscular fatigue during intense exercise. *Sports Medicine* 13:134-45.

McKenna, M. J. 1999. Role of the skeletal Na^+-K^+-pump during exercise. In *Biochemistry of exercise X,* ed. Hargreaves, M., and M. Thompson, 71-97. Champaign IL: Human Kinetics.

McKenna, M. J., A. R. Harmer, S. F. Fraser, J. L. Li. 1996. Effects of training on potassium, calcium and hydrogen ion regulation in skeletal muscle and blood during exercise. *Acta Physiologica Scandinavica* 156:335-46.

McKenna, M. J., T. A. Schmidt, H. Hargreaves, L. Cameran, S. L. Skinner, and K. Kjeldsen. 1993. Sprint training increases human skeletal muscle Na^+-K^+-ATPase concentration and improves K^+ regulation. *Journal of Applied Physiology* 75:173-80.

McKenzie, D. C., W. S. Parkhouse, E. C. Rhodes, P. W. Hochochka, W. K. Ovalle, T. P. Mommsen, and S. L. Shinn. 1983. Skeletal muscle buffering capacity in elite athletes. In *Biochemistry of exercise,* ed. Knuttgen, H. G., J. A. Vogel, and J. Poortmans, 585-589. Champaign, IL: Human Kinetics.

McMurray, R. G., T. K. Eubank, and A. C. Hackney. 1995. Nocturnal hormonal responses to resistance exercise. *European Journal of Applied Physiology* 72:121-6.

Medbø, J. I., and S. Burgers. 1990. Effect of training on the anaerobic capacity. *Medicine and Science in Sports and Exercise* 22:501-7.

Medbø, J. I., A. C. Mohin, I. Tabata, R. Bahr, O. Vaage, and O. M. Sejersted. 1988. Anaerobic capacity determined by maximal accumulated O_2 deficit. *Journal of Applied Physiology* 64:50-60.

Medbø J. I., and O. M. Sejersted. 1990. Plasma potassium changes with high intensity exercise. *Journal of Physiology* 421:105-22.

Medbø, J. I., and O. M. Sejersted. 1985. Acid-base and electrolyte balance after exhausting exercise in endurance-trained and sprint-trained subjects. *Acta Physiologica Scandinavica* 125:97-109.

Meerson, F. Z. 1965. Intensity of function of structures of the differentiated cell as a determinant of activity of its genetic apparatus. *Nature* 206:483-4.

Meerson, F. Z. 1983. *The failing heart: Adaptation and deadaptation.* New York: Raven Press.

Meeusen, G., J. Roeykens, L. Magnus, H. Keizer, K. DeMeirleir. 1997. Endurance performance in humans: the effect of a dopamine precursor or a specific serotonin (5-$HT_{2A/2C}$) antagonist. *International Journal of Sports Medicine* 18:571-7.

Meeusen, R. 1999. Overtraining and the central nervous system. The missing link? In *Overload, performance incompetence, and regeneration in sport,* ed. Lehmann, M., C. Foster, U. Gastmann, H. Keizer, and J. M. Steinacker, 187-202. New York: Kluwer Academic/Plenum Publications.

Meeussen, R., and K. De Meirleir. 1995. Exercise and brain neurotransmission. *Sports Medicine* 20:160-88.

Meeusen, R., K. Thorré, F. Chaouloff, S. Sarre, K. DeMeirleir, G. Elinger, and Y. Michotte. 1996. Effects of tryptophan and/or acute running on extracellular 5-HT and 5-HIAA levels in the hippocampus of food-deprived rats. *Brain Research* 740:245-52.

Melamed, I., Y. Romen, G. Keren, Y. Epstein, and E. Dolev. 1982. March myoglobinemia: a hazard to renal function. *Archives of Internal Medicine* 142:1277-9.

Melin, B., J. P. Eclache, G. Geelen, G. Annat, A. M. Allevard, E. Janaillon, A. Zebidi, J. J. Legros, and C. Gharib. 1980. Plasma AVP, neurophysin, renin activity, and aldosterone during submaximal exercise performed until exhaustion in trained and untrained men. *European Journal of Applied Physiology* 44:141-51.

Mena, P., A. Maynar, J. Guitierrez, J. Maynar, J. Timon, and J. Campillo. 1991. Erythrocyte free radical scavenger enzymes in bicycle professional races. Adaptation to training. *International Journal of Sports Medicine* 12:563-6.

Meyer, R. A., G. A. Dudley, and R. L. Terjung. 1980. Ammonia and IMP in different skeletal muscle fibers after exercise in rat. *Journal of Applied Physiology* 49:1037-41.

Meyer, R. A., and R. L. Terjung. 1979. Differences in ammonia and adenylate metabolism in contracting fast and slow muscle. *American Journal of Physiology* 237:C111-8.

Meyerhof, O. 1930. *Chemische Vorgänge im Muskel.* Berlin: Springer.

Mikenes, K. J., B. Sonne, P. A. Farrell, B. Tronier, and H. Galbo. 1989. Effect of training on the dose-response relationship for insulin action in man. *Journal of Applied Physiology* 66:695-703.

Milasius, K. 1997. *Istverme lavinanciu sportininku organizmo adaptacja prie fiziniu kruviu* (Summary in English: Adaptation of the organism of endurance-training sportsmen to physical loads). Vilnius: Vilnius Pedagogius Universitatis.

Miller, B. J. 1990. Haemotological effects of running. A brief review. *Sports Medicine* 9:1-6.

Miller, B. J., R. R. Page, and W. Burgers. 1988. Foot impact and intravascular hemolysis during distance running. *International Journal Sports Medicine* 9:56-90.

Miller, N., C. Rice-Evans, M. Davies, V. Gopinathan, and A. Milner. 1993. A novel method for measuring antioxidant capacity and its application to monitoring the antioxidant status in premature neonates. *Clinical Sciences* 84:407-12.

Millward, D. J. 1980. Protein turnover in cardiac and skeletal muscle during normal growth and hypertrophy. In *Degenerative processes in heart and skeletal muscle,* ed. Wildenthal, E., 161-199. Amsterdam: Elsevier/North Holland.

Millward, D. J., and P. C. Bates. 1983. 3-methylhistidine turnover in the whole body and the contribution of skeletal muscles and intestine to urinary 3-methylhistidine excretion in the adult rat. *Biochemical Journal* 214:607-15.

Millward, D. J., P. C. Bates, G. K. Grimble, J. G. Brown, M. Nathan, and M. J. Rennie. 1980. Quantitative importance of non-skeletal-muscle sources of N tau-methylhistidine in urine. *Biochemical Journal* 190:225-8.

Millward, D. J., C. T. Davies, D. Halliday, S. L. Wolman, D. Mathews, and M. Rennie. 1982. Effect of exercise on protein metabolism in humans as explored with stable isotopes. *Federation Proceedings* 41:2686-91.

Millward, D. J., P. J. Garlick, W. P. T. James, D. O. Nganyelugo, and J. S. Ryatt. 1973. Relationship between protein synthesis and RNA content in skeletal muscles. *Nature* 241:204-5.

Mitolo, M. 1951. Allenamento all'esercizio fisico e "sindrome general of 'adappamento". *Studi di Medicine e Chirurgie dello Sport* 5:311-42.

Mittleman, K. D., M. R. Ricci, and S. P. Bailey. 1998. Branched-chain amino acids prolong exercise during heart stress in men and women. *Medicine and Science in Sports and Exercise* 30:83-91.

Moesch, H., and H. Howald. 1975. Heksokinase (HK), glyceraldehyde-3 P-dehydrogenase (GAPDH), succinate-dehydrogenase (SDH) and 3-hydroxyacyl-CoA-dehydrogenase (NAD) in skeletal muscle of trained and untrained men. In *Metabolic adaptation to prolonged physical exercise,* ed. Howald, H., and J. P. Poortmans, 463-465. Basel: Birkhäuser.

Mole, P. A., K. M. Baldwin, R. L. Terjung, and J. O. Holloszy. 1973. Enzymatic pathways of pyruvate metabolism in skeletal muscle: adaptations to exercise. *American Journal of Physiology* 244(1):50-4.

Montgomery, D. L. 2000. Physiology of ice hockey. In *Exercise and sport science,* ed. Garrett, W. E., and D. T. Kirkendall, 815-828. Philadelphia: Lippincott Williams & Wilkins.

Moorthy, A. V., and S. W. Zimmerman. 1978. Human leukocyte response to an endurance race. *European Journal of Applied Physiology* 38:271-6.

Morgan, T. E., F. A. Short, and L. A. Cobb. 1969. Effect of long-term exercise on skeletal muscle lipid composition. *American Journal of Physiology* 216:82-6.

Morrow, J. R., A. W. Jackson, J. G. Disch, and D. P. Mood (eds.). 1997. *Measurement and evaluation in human performance*. Champaign, IL: Human Kinetics.

Mujika, I., T. Busso, L. Lacoste, F. Barale, A. Geyssant, and J. C. Chatard. 1996a. Modeled responses to training and taper in competitive swimmers. *Medicine and Science in Sports and Exercise* 28:251-8.

Mujika, I., J. C. Chatard, S. Padilla, C. Y. Guezennec, and A. Geyssant. 1996b. Hormonal responses to training and its tapering off in competitive swimmers: relationships with performance. *European Journal of Applied Physiology* 74:361-6.

Munjal, D. D., J. A. McFadden, P. A. Matix, K. D. Coffman, and S. M. Cattaneo. 1983. Changes in serum myoglobin total creatine kinase, lactate dehydrogenase and creatine kinase MB levels in runners. *Clinical Biochemistry* 16:195-9.

Muns, G. 1994. Effect of long-distance running on polynuclear neutrophil phagocytotic function of the upper airways. *International Journal of Sports Medicine* 15:96-9.

Muns, G., H. Liesen, H. Riedel, and K.-Ch. Bergmann. 1989. Influence of long-distance running of IgA in nasal secretion and saliva. *Deutsche Zeitschrift für Sportmedizin* 40:63-5.

Nair, K. S., D. Halliday, and R. C. Griggs. 1988. Leucine incorporation into mixed skeletal muscle protein in humans. *American Journal of Physiology* 254:E208-13.

Näveri, H., K. Kuoppasalmi, and M. Härkönen. 1985a. Plasma glucagon and catecholamines during exhaustive short-term exercise. *European Journal of Applied Physiology* 93:308-11.

Näveri, H., K. Kuoppasalmi, and M. Härkönen. 1985b. Metabolic and hormonal changes in moderate and intense long-term running exercises. *International Journal of Sports Medicine* 6:276-81.

Nazar, K. 1981. Glucostatic control of hormonal responses to physical exercise in men. In *Biochemistry of exercise IV-A,* ed. Poortmans, J., and G. Niset, 188-195. Baltimore: University Park Press.

Nazar, K., D. Jezova, and E. Kowalik-Borowna. 1989. Plasma vasopressin, growth hormone and ACTH responses to static handgrip in healthy subjects. *European Journal of Applied Physiology* 58:400-4.

Nesher, R., I. E. Karl, and D. M. Kipnis. 1980. Epitrochlearis muscle. II. Metabolic effects of contraction and catecholamines. *American Journal of Physiology* 239:E461-7.

Neufer, P. D. 1989. The effect of detraining and reduced training on the physiological adaptation to aerobic exercise training. *Sports Medicine* 8:302-21.

Neufer, P. D., D. L. Costill, R. A. Fielding, M. G. Flynn, and J. P. Kirwan. 1987. Effect of reduced training on muscular strength and endurance in competitive swimmers. *Medicine and Science in Sports and Exercise* 19:486-90.

Neumann, G. 1992. Cycling. In *Endurance in sport,* ed. Shephard, R. J., and P.-O. Åstrand, 582-596. London: Blackwell Scientific.

Neumann, G., and K.-P. Schüler. 1989. *Sportmedizinische Funktionaldiagnostik*. Leipzig: J. A. Bash.

Neumann, G., H.-G. Scherster, and H. Buhl. 1980. Komplexe Stoffwechsel untersuchungen nach einer Marathonbelastung *Medicina und Sport* 20:12-7.

Nevill, M. E., L. H. Boobis, S. Brooks, and C. Williams. 1989. Effect of training on muscle metabolism during treadmill sprinting. *Journal of Applied Physiology* 67:2376-82.

Nevill, M. E., D. J. Holmyard, G. M. Hale, P. Allsop, A. von Oosterhout, J. M. Burrin, and A. M. Nevill. 1996. Growth hormone response to treadmill sprinting in sprint and endurance-trained athletes. *European Journal of Applied Physiology* 72:460-7.

Newsholme, E. A. 1979. The control of fuel utilization by muscle during exercise and starvation. *Diabetes* 28 (Suppl 1):1-7.

Newsholme, E. A. 1986. Application of principles of metabolic control to the problem of metabolic limitation in sprinting, middle-distance and marathon running. *International Journal of Sports Medicine* 7(Suppl.)66-70.

Newsholme, E. A. 1989. Metabolic causes of fatigue in track events and the marathon. In *Advances in myochemistry,* 263-271. London: John Libbey Eurotext Ltd.

Newsholme, E. A., E. Blomstrand, N. McAndrew, and M. Parry-Billings. 1992. Biochemical causes of fatigue and overtraining. In *Endurance in sport,* ed. Shephard, R. J., and P.-O. Åstrand, 351-364. London: Blackwell Scientific.

Newsholme, E. A., and A. R. Leich. 1983. *Biochemistry for the medical sciences*. Chichester: John Wiley.

Newshouse, J., and D. B. Clement. 1988. Iron status in athletes. An update. *Sports Medicine* 5:337-42.

Nichols, B., A. T. Miller, and E. P. Hiatt. 1951. Influence of muscular exercise on uric acid excretion in man. *Journal of Applied Physiology* 3:505-7.

Nielsen, B., R. Kubica, A. Bonnesen, I. B. Rasmussen, J. Stoklase, and B. Wilk. 1981. Physical work capacity after dehydration and hyperthermia. *Scandinavian Journal of Sports Sciences* 3:2-10.

Nieman, D. C. 1994a. Exercise, upper respiratory tract infections, and the immune system. *Medicine and Science in Sports and Exercise* 26:128-39.

Nieman, D. C. 1994b. Exercise, infections, and immunity. *International Journal of Sports Medicine* 15(Suppl. 3):S131-41.

Nieman, D. C. 1997. Exercise immunology: practical application. *International Journal of Sports Medicine* 18(Suppl. 1):S91-100.

Nieman, D. C. 2000. Exercise, the immune system and infectious diseases. In *Exercise and sport science,* ed. Garrett, W. E., and D. T. Kirkendall, 177-190. Philadelphia: Lippincott Williams & Wilkins.

Nieman, D. C., K. S. Buckley, D. A. Hekson, B. J. Warren, J. Suttles, J. C. Ahle, S. Simandle, O. R. Fagoaga, and S. L. Nehlsen-Canneerella. 1995. Immune function in marathon running versus sedentary controls. *Medicine and Science in Sports and Exercise* 27:986-92.

Nieman, D. C., L. S. Berk, M. Simpson-Westerberg, K. Arabatzis, S. Youngberg, S. A. Tan, J. W. Lee, and W. C. Eby. 1989b. Effects of long-endurance running on immune system parameters and lymphocyte function in experienced marathoners. *International Journal of Sports Medicine* 10:317-23.

Nieman, D. C., A. R. Millar, D. A. Henson, B. J. Warren, G. Gusewitch, R. L. Johnson, J. M. Davis, D. E. Butterworth, J. L. Herring, and S. L. Nehlsen-Cannarella. 1994. Effects of high versus moderate-intensity exercises on lymphocyte subpopulations and proliferative response. *International Journal of Sports Medicine* 15:199-206.

Nieman, D. C., and S. L. Nehlsen-Cannarella. 1991. The effect of acute and chronic exercise on immunoglobin. *Sports Medicine* 11:183-201.

Nieman, D. C., S. A. Tan, J. W. Lee, and L. S. Berk. 1989a. Complement and immunoglobulin levels in athletes and sedentary controls. *International Journal of Sports Medicine* 10:124-8.

Nikkilä, E. A., M. R. Taskinen, S. Rehunen, and M. Härkönen. 1978. Lipoprotein lipase activity in adipose tissue and skeletal muscle of runners: relation to serum lipoproteins. *Metabolism* 27:1662-7.

Noakes, T. D. 1998. Fluid and electrolyte disturbances in heat illness. *International Journal of Sports Medicine* 19(Suppl. 2):S146-9.

Noakes, T. D., and J. W. Carter. 1982. The responses of plasma biochemical parameters to a 56-km race in novice and experienced ultra-marathon runners. *European Journal of Applied Physiology* 49:179-86.

Nøgaard, A., K. Kjeldsen, O. Hansen, and T. Clausen. 1983. A simple and rapid method for the determination of the number of 3H-ouabain binding sites in biopsies of skeletal muscle. *Biochemistry and Biophysics Research Communications* 111(1):319-25.

Nosaka, K., and P. M. Clarkson. 1995. Muscle damage following repeated bouts of high force eccentric exercise. *Medicine and Science in Sports and Exercise* 27:1263-9.

Nummela, A., M. Alberts, R. P. Rijntes, P. Luhtanen, and P. Rusko. 1996a. Reliability and validity of the maximal anaerobic running test. *International Journal of Sport Medicine* 17(Suppl.2):S97-102.

Nummela, A., N. Andersson, K. Häkkinen, and H. Rusko. 1996b. Effect of inclination on the results of the maximal anaerobic running test. *International Journal of Sports Medicine* 17(Suppl. 2):S103-08.

Nummela, A., A. Mero, and H. Rusko. 1996c. Effects of sprint training on anaerobic performance characteristics determined by the MART. *International Journal of Sports Medicine* 17(Suppl.2):S114-9.

Nygaard, E. 1980. Number of fiber in skeletal muscle of man. *Muscle Nerve* 3:268.

Nygaard, E. 1981. Women and exercise—with special reference to muscle morphology and metabolism. In *Biochemistry of exercise IV-B,* International Series on Sports Sciences, ed. Poortmans J., and G. Niset, 161-175. Baltimore: University Park Press.

Nygaard, E. 1982. Skeletal muscle fibre characteristics in young women. *Acta Physiologica Scandinavica* 112:299-302.

Nygaard, F., and I. Sanchez. 1982. Intramuscular variation of fiber types in the brachial biceps and the lateral vastus muscles of elderly men. How representative is a small biopsy sample? *Anatomy Record* 203:541-9.

Nylin, G. 1947. The effect of heavy muscular work on the volume of circulating red corpuscles in man. *American Journal of Physiology* 149:180-4.

O'Neil, M. E., M. Wikinson, B. G. Robinson, D. B. McDowall, K. A. Cooper, A. S. Mihailidou, D. B. Frewin, P. Clifton-Bligh, and S. N. Huhyor. 1990. The effect of exercise on circulating immunoreactive calcitonin in men. *Hormone and Metabolic Research* 22:546-50.

Odagiri, Y., T. Shimomitsu, H. Ikiane, and T. Katsumura. 1996. Relationships between exhaustive mood state and changes in stress hormones following an ultraendurance race. *International Journal of Sports Medicine* 17:325-31.

Ogoltsov, I. G. 1968. Analysis of preparation of best cross-country skiers for X Winter Olympic Games [in Russian]. *Teoria i Praktika Fizicheskoi Kulturõ* 31(7):32-7.

Ohkuwa, T., H. Itoh, Y. Yamazaki, and Y. Sato. 1995. Salivary and blood lactate after supramaximal exercise in sprinters and long-distance runners. *Scandinavian Journal of Medicine and Sciences of Sports* 5:285-90.

Ohkuwa, T., M. Salto, and M. Mlymaura. 1984. Plasma LDH and CK activities after 400 m sprinting by well-trained sprint runners. *European Journal of Applied Physiology* 52:296-9.

Ohno, H., Y. Sato, K. Yamashita, R. Doi, K. Arai, T. Kono, and N. Taniguchi. 1986. The effect of brief physical exercise on free radical scavenger enzyme systems in human red blood cells. *Canadian Journal of Physiology and Pharmacology* 64:1263-5.

Olsson, K.-E., and B. Saltin. 1970. Variation in total body water with muscle glycogen changes in man. *Acta Physiologica Scandinavica* 80:11-8.

Olweus, D., A. Mattson, D. Schalling, and H. Low. 1980. Testosterone, aggression, physical and personality dimensions of normal adolescent males. *Psychosomatic Medicine* 42:253-69.

Ööpik, V., K. Alev, and V. Buchinkskayte. 1988. Dynamics of protein metabolism in skeletal muscle during daily repeated muscular work. *Acta et Commentationes Universitatis Tartuensis* 813:3-14.

Ööpik, V., K. Port, and A. Viru. 1991. Adrenocortical activity during daily repeated exercise. *Biology of Sport* 8:187-94.

Ööpik, V., and A. Viru. 1988. Specific nature of adaptive protein synthesis in systematic muscular activity. *Proceedings of the Academy of Sciences of the Estonia. Biology* 37:158-61.

Ööpik, V., and A. Viru. 1992. Changes of protein metabolism in skeletal muscle in response to endurance training. *Sports Medicine, Training and Rehabilitation* 3:55-64.

Ööpik, V., M. Viru, S. Timpmann, L. Medijainen, and A. Viru. 1993. Lack of stimulation of protein synthesis in skeletal muscles by creatine administration in rats. *Acta Commentationes Universitatis Tartuensis* 958:16-21.

Opstad, P. K., A. H. Haugen, O. M. Sejersted, R. Bahr, and K. V. Saredo. 1994. Atrial natriuretic peptide in plasma after prolonged physical strain, energy deficiency and sleep deprivation. *European Journal of Applied Physiology* 68:122-6.

Orlova, E. H., M. G. Pshennikova, A. D. Dmitriyev, and F. Z. Meerson. 1988. An increase of the content of immunoreactive opioid peptides in brain and adrenals of rats under the influence of adaptation to muscular activity. *Byulleten Eksperimentel'noi Biologij i Medicinõ* 105:145-8.

Oscai, L., B. T. Williams, and B. A. Hertig. 1968. Effect of exercise on blood volume. *Journal of Applied Physiology* 24:622-4.

Oshida, Y., K. Yamanouchi, S. Hayamizu, and Y. Sato. 1988. Effect of acute physical exercise on lymphocyte subpopulations in trained and untrained subjects. *International Journal of Sports Medicine* 9:137-40.

Östman, I., and N. O. Sjöstrand. 1971. Effect of prolonged physical training on the catecholamine levels of the heart and the adrenals of the rat. *Acta Physiologica Scandinavica* 82:202-8.

Owles, W. H. 1930. Alterations in the lactic acid contact of the blood as a result of light exercise and associated changes in the CO_2 combining power of the blood and in the alveolar CO_2 pressure. *Journal of Physiology* 69:214-37.

Ozolin, P. 1986. Blood flow in the extremities of athletes. *International Journal of Sports Medicine* 7:117-22.

Palmer, R. M., P. J. Reeds, T. Atkinson, and R. H. Smith. 1983. The influence of changes in tension on protein synthesis and prostaglandin release in isolated rabbit muscle. *Biochemical Journal* 214:1011-4.

Pardridge, W. M., L. Duducgian-Vartavarian, D. Casanello-Ertl, M. J. Jones, and J. D. Kopple. 1982. Arginase metabolism and urea synthesis in cultured rat skeletal muscle cells. *American Journal of Physiology* 242:E87-92.

Parizková, J., and R. Kvetnansky. 1980. Catecholamine metabolism and compositional growth in exercised and hypokinetic male rats. In *Catecholamines and stress: Recent advances,* ed. Usdin, E., R. Kvetnansky, and I. J. Kopin, 355-358. New York: Elsevier/North Holland.

Parkhouse, W. S., D. C. McKenzie, P. W. Hochachka, T. P. Mommsen, W. K. Ovalle, S. L. Shinn, and E. C. Rhodes. 1983. The relationship between carnosine levels, buffering capacity, fiber type and anaerobic capacity in elite athletes. In *Biochemistry of exercise,* ed. Knuttgen, H. G., J. A. Vogel, and J. Poortmans, 590-594. Champaign, IL: Human Kinetics.

Pärnat, J., A. Viru, T. Savi, F. Kudu, and F. Markusas. 1973. Untersuchunder der aeroben und anaeroben Leistungsfähigkeit von Zehnkümphern. *Medicine und Sport* 13:366-9.

Pärnat, J., T. Savi, and A. Viru. 1975b. Physical fitness of decathletes. *Teoria i Praktika Fizicheskoi Kulturõ* 11:30-1.

Pärnat, J., A. Viru, T. Savi, and A. Nurmekivi. 1975a. Indices of aerobic work capacity and cardiovascular response during exercise in athletes specializing in different events. *Journal of Sports Medicine and Physical Fitness* 15:100-5.

Parry-Billings, M., E. Blomstrand, N. McAndrew, and E. A. Newsholme. 1990. A communicational link between skeletal muscle, brain and cells of the immune system. *International Journal of Sports Medicine* 11(Suppl 2):S122-8.

Parry-Billings, M., R. Budgett, Y. Koutedakis, E. Blomstrand, S. Brooks, C. Williams, P. C. Calder, S. Pilling, R. Baigrie, and E. A. Newsholme. 1992. Plasma amino acid concentrations in the overtraining syndrome: possible effects on the immune system. *Medicine and Science of Sports Exercise* 24:1353-8.

Pedersen, B. K., and H. Brunsgaard. 1995. How physical exercise influences the establishment of infections. *Sports Medicine* 19:393-400.

Pedersen, B. K., M. Kappel, M. Klokker, H. B. Nielsen, and N. H. Secher. 1994. The immune system during exposure to extreme physiologic conditions. *International Journal of Sports Medicine* 15:S116-21.

Pedersen, B. K., N. Tvede, L. D. Christensen, K. Klarlund, S. Kragbak, and J. Halkjaer-Kristensen. 1989. Natural killer cell activity in peripheral blood of highly trained and untrained persons. *International Journal of Sports Medicine* 10:129-31.

Pedersen, B. K., N. Tvede, F. R. Hansen, V. Andersen, T. Bendix, G. Bendixen, K. Bendtzen, H. Galbo, P. M. Haahr, K. Klarlund, J. Sylvest, B. S. Thomsen, and J. Halkjaer-Kristensen. 1988. Modulation of natural killer cell activity in peripheral blood by physical exercise. *Scandinavian Journal of Immunology* 27:673-8.

Pedersen, B. K., N. Tvede, K. Klarlund, L. D. Christensen, F. R. Hansen, H. Galbo, and A. Kharazmi. 1990. Indomethacin *in vitro* and *in vivo* abolishes post-exercise suppression of natural killer cell activity in peripheral blood. *International Journal of Sports Medicine* 11:127-31.

Pedersen, O., and J. Bak. 1986. Effect of acute exercise and physical training on insulin receptor and insulin action. In *Biochemistry of exercise VI,* ed. Saltin, B., 87-94. Champaign, IL: Human Kinetics.

Pelayo, P., I. Mujika, M. Sidney, and J. C. Chatard. 1996. Blood lactate recovery measurements, training, and performance during a 23-week period of competitive swimming. *European Journal of Applied Physiology* 74:107-13.

Pellicia, A., and G. B. DiNucci. 1987. Anemia in swimmers: fact or fiction? Study of hemological and iron status in male and female top-level swimmers. *International Journal of Sports Medicine* 8:227-30.

Péquignot, J. M., L. Peyrin, R. Favier, and R. Flandros. 1979. Adrenergic response to emotivity and physical training. *European Journal of Applied Physiology* 40:117-35.

Péquignot, J. M., L. Peyrin, and G. Peres. 1980. Catecholamine-fuel interrelationships during exercise in fasting men. *Journal of Applied Physiology* 48:109-13.

Péronnet, F., P. Blier, G. Brisson, M. Ledoux, P. Diamond, M. Volle, and D. de Camfel. 1982. Relationship between trait-anxiety and plasma catecholamine concentration at rest and during exercise. *Medicine Sciences of Sports Exercise* 14:173-4.

Péronnet, F., J. Cléroux, H. Perrault, D. Cousineau, J. de Champlain, and R. Nadeau. 1981. Plasma norepinephrine response to exercise before and after training in humans. *Journal of Applied Physiology* 51:812-5.

Péronnet, F., and R. H. Morton. 1994. Plasma lactate concentration increases as a parabola with delay during ramp exercise. *European Journal of Applied Physiology* 68:228-33.

Péronnet, F., G. Thibault, H. Perrault, and D. Cousineau. 1986. Sympathetic response to maximal bicycle exercise before and after leg strength training. *European Journal of Applied Physiology* 55:1-4.

Perriello, G., R. Jorde, N. Nurjhan, N. Stumvoll, G. Dailey, T. G. Jenssen, D. M. Bier, and J. E. Gerich. 1995. Estimation of glucose-alanine-lactate-glutamine cycles in postabsorptive humans: role of skeletal muscle. *American Journal of Physiology* 269(3 Pt 1):E443-50.

Pershin, B. B., S. N. Kuzmin, R. S. Suzdalnitski, and V. A. Levando. 1988. Reserve potentials of immunity. *Sports Training, Medicine and Rehabilitation* 1:53-60.

Peterkofer, von M., and C. Voit. 1866. Untersuchung über den Stoffverbrauch des normalen Menschen. *Zeitschrift für Biologie* 2:489-573.

Petraglia, F., C. Barletta, F. Facchinetti, F. Spinazzola, A. Monzanni, D. Scavo, A. R. Genazzini. 1988. Response of circulating adrenocorticotropin, beta-endorphin, beta-lipotropin and cortisol to athletic competition. *Acta Endocrinologica* 188:332-6.

Pette, D., and G. Dölken. 1975. Some aspects of regulation of enzyme levels in muscle energy supplying metabolism. *Advances of Enzyme Regulation* 13:355-77.

Pfitzinger, P., and P. S. Freedson. 1998. The reliability of lactate measurements during exercise. *International Journal of Sports Medicine* 19:349-57.

Piehl, K., S. Adolfsson, and K. Nazar. 1974. Glycogen store and glycogen synthetase activity in trained and untrained muscle of man. *Acta Physiologica Scandinavica* 90:779-88.

Pills, W., J. Langfost, Pilsniak, M. Pyzik, and M. Btasiak. 1988. Plasma lactate dehydrogenase and creatine kinase after an aerobic exercise. *International Journal of Sports Medicine* 9:102-3.

Pivornik, J. M., J. F. Hickson, and I. Wolinsky. 1989. Urinary 3-methylhistidine excretion with repeated weight training exercise. *Medicine and Science in Sports and Exercise* 21:283-7.

Pizza, F. X., M. G. Flynn, J. B. Boone, J. R. Rodriguez-Zayas, and F. F. Andres. 1997. Serum haptoglobin and ferritin during a competitive running and swimming season. *International Journal of Sports Medicine* 18:233-7.

Platonov, V. N. 1986. *Training in qualified sportsmen* [in Russian]. Moscow: FiS.

Platonov, V. M., and S. M. Vaitsekhovski. 1985. *Training of high class swimmers* [in Russian]. Moscow: FiS.

Poland, J. L., and D. H. Blount. 1968. The effects of training on myocardial metabolism. *Proceedings for Society of Experimental Biology and Medicine* 129:171-4.

Poland, J. S., and D. A. Trauer. 1973. Adrenal influence on the supercompensation of cardiac glycogen following exercise. *American Journal of Physiology* 224:540-2.

Pollock, M. L., L. Garzarella, and J. E. Graves. 1995. The measurement of body compositors. In *Physiological assessment of human fitness,* ed., Maud, P. J., and C. Foster, 167-204. Champaign, IL: Human Kinetics.

Poortmans, J. R. 1970. Serum protein determination during short exhaustive physical activity. *Journal of Applied Physiology* 30:190-2.

Poortmans, J. R. 1975. Effects of long lasting physical exercises and training on protein metabolism. In *Metabolic adaptation to prolonged physical exercise,* ed. Howald, H., and J. R. Poortmans, 212-228. Basel: Birkhäuser.

Poortmans, J. R. 1984. Protein turnover and amino acid oxidation during and after exercise. *Medicine of Sports Sciences and Exercise* 17:130-47.

Poortmans, J. R. 1988. Protein metabolism. In *Principles of exercise biochemistry,* ed. Poortmans, J. R., 184-193. Basel: Karger.

Poortmans, J., K. H. Luke, A. Zipursky, and J. Bienenstock. 1971. Fibrinolytic activity and fibrinogen split products in exercise proteinuria. *Clinical Chimia Acta* 35:449-54.

Poortmans, J. R., G. Seast, M. M. Galteau, and D. Houst. 1974. Distribution of plasma amino acids in humans during submaximal prolonged exercise. *European Journal of Applied Physiology* 32:143-7.

Port, K. 1991. Serum and saliva cortisol responses and blood lactate accumulation during incremental exercise testing. *International Journal of Sports Medicine* 12:490-94.

Port, K., and A. Viru. 1987. Changes of cortisol concentration in blood of swimmers during improvement of performance level. In *Endocrine mechanisms of regulation of adaptation to muscular activity,* ed. Matsin, T., 65-67. Tartu: University of Tartu.

Powers, S. K., D. Criswell, J. Lawler, L. L. Ji, D. Martin, R. Herb, and G. Dudley. 1994. Influence of exercise and fiber type on antioxidant enzyme activities in rat skeletal muscle. *American Journal of Physiology* 266:R375-80.

Powers, S. K., L. L. Ji, and C. Leeuwenburgh. 1999. Exercise training-induced alterations in skeletal muscle antioxidant capacity: a brief review. *Medicine and Science in Sports and Exercise* 31:987-97.

Prokop, L. 1959. *Erfolg im sport.* Wien-München: Fürlinger.

Pruett, D. R. 1970a. Glucose and insulin during prolonged work stress in men living on different diets. *Journal of Applied Physiology* 28:199-208.

Pruett, E. D. R. 1970b. Plasma insulin concentrations during prolonged work at near maximal oxygen uptake. *Journal of Applied Physiology* 29:155-8.

Pruett, E. D. R. 1985. Insulin and exercise in non-diabetic and diabetic man. In *Exercise endocrinology,* ed. Fortherby, K., and S. B. Pal, 1-23. Berlin, New York: De Gruyter.

Pruett, E. D. R., and S. Oseid. 1970. Effect of exercise on glucose and insulin response to glucose infusion. *Scandinavian Journal of Clinical Physiology* 26:277-85.

Pugh, L. G. C. E. 1969. Blood volume changes in outdoor exercise of 8-10 hour duration. *Journal of Physiology* 200:345-51.

Pyne, D. B., M. S. Baker, P. A. Fricker, W. McDonald, R. Telford, and M. Weideman. 1995. Effects of an intensive 12-wk training program by elite swimmers on neutrophil oxidative activity. *Medicine and Science in Sports and Exercise* 27:536-42.

Raczek, J. 1989. Zur Optimierung der Trainingsbelastungen im Mittel- und Langstrechenlauf. *Leistunssport* 19(3)12-7.

Rahkila, P., E. Hakala, M. Alén, K. Salminen, and T. Laatikainen. 1988. ß-endorphin and corticotropin release is dependent on a threshold intensity of running exercise in male endurance athletes. *Life Sciences* 43:551-8.

Rakestraw, N. W. 1921. Chemical factors in fatigue. I. The effect of muscular exercise upon certain blood constituents. *Journal of Biological Chemistry* 47:565-91.

Raymond, L. W., J. Sode, and J. R. Tucci. 1972. Adrenocortical response to nonexhaustive muscular exercise. *Acta Endocrinologica* 70:73-80.

Raynaud, J., L. Droket, J. P. Martineaud, J. Bordachar, J. Coudert, and J. Durand. 1981. Time course of plasma growth hormone during exercise in humans at altitude. *Journal of Applied Physiology* 50:229-33.

Refsum, H. E., and S. B. Strömme. 1974. Urea and creatinine production and excretion in urine during and after prolonged heavy exercise. *Scandinavian Journal of Clinical and Laboratory Investigations* 33:247-54.

Rehunen, S. 1989. High-energy phosphates in human muscle. In *Paavo Nurmi Congress Book,* ed. Kvist, M., 40-42. Turku: Finnish Society of Sports Medicine.

Reid, M., K. Haack, K. Franchek, P. Valberg, L. Kobzik, and S. West. 1992a. Reactive oxygen in skeletal muscle. I. Intracellular oxidant kinetics and fatigue in vitro. *Journal of Applied Physiology* 73:1797-1804.

Reid, M., T. Shoji, M. Moody, and M. Entman. 1992b. Reactive oxygen in skeletal muscle. II. Extracellular release of radicals. *Journal of Applied Physiology* 73:1805-9.

Remes, K. 1979. Effect of long-term physical training on total red cell volume. *Scandinavian Journal of Clinical and Laboratory Investigation* 39:311-9.

Ren, J. M., and E. Hultman. 1988. Phosphorylase activity in needle biopsy samples-factors influencing transformation. *Acta Physiologica Scandinavica* 133:109-14.

Rennie, M. J., and R. H. Johnson. 1974. Alteration of metabolic and hormonal response to exercise by physical training. *European Journal of Applied Physiology* 33:215-26.

Rennie, M. J., R. H. T. Edwards, C. T. M. Davies, S. Krywawych, D. Halliday, J. C. Waterlow, and D. J. Millward. 1980. Protein and amino acid turnover during and after exercise. *Biochemical Societies Transactions* 6:499-501.

Rennie, M. J., R. H. T. Edwards, S. Krywawych, C. T. M. Davies, D. Halliday, J. C. Waterlow, and D. J. Millward. 1981. Effect of exercise on protein turnover in man. *Clinical Sciences* 61:627-39.

Resina, A., L. Gatteschi, M. A. Giamberardino, F. Imreh, L. Rubenni, and L. Vecchiet. 1991. Hematological comparison of iron status in trained top-level soccer players and control subjects. *International Journal of Sports Medicine* 12:453-6.

Richter, E. A. 1984. Influence of the sympatho-adrenal system on some metabolic and hormonal responses to exercise in the rat. With special reference to the effect on glycogenolysis in skeletal muscle. *Acta Physiologica Scandinavica* Suppl. 528.

Richter, E. A., H. Galbo, B. Sonne, J. J. Holst, and N. J. Christensen. 1980. Adrenal medullary control of muscular and hepatic glycogenolysis and of pancreatic hormonal secretion in exercising rats. *Acta Physiologica Scandinavica* 108:235-42.

Richter, E. A., H. Galbo, J. J. Holst, and B. Sonne. 1981. Significance of glucagon for insulin secretion and hepatic glycogenolysis during exercise in rats. *Hormone and Metabolic Research* 13:323-6.

Richter, E. A., N. B. Ruderman, and H. Galbo. 1983. Alpha and beta adrenergic effects on muscle metabolism on contracting, perfused muscle. In *Biochemistry of exercise,* ed. Knuttgen, H. G., J. A. Vogel, and J. Poortmans, 766-772. Champaign IL: Human Kinetics.

Richter, E. A., N. B. Ruderman, H. Gavras, E. R. Belur, and H. Galbo. 1982. Muscle glycogenolysis during exercise: dual control by epinephrine and contractions. *American Journal of Physiology* 242:625-32.

Rivoire, M., I. Rivoire, and M. Ponjol. 1953. La fatigue syndrome d'insuffisance surrenale fonctionelle. *Presse Medicale* 61:1431-3.

Robertson, J., R. Maughan, G. Duthie, and P. Morrice. 1991. Increased blood antioxidant systems of runners in response to training load. *Clinical Sciences* 80:611-8.

Roberts, A. C., R. D. Mcclure, R. I. Weiker, and G. A. Brooks. 1993. Overtraining affects male reproductive status. *Fertility and Sterility* 60:686-92.

Roberts, J. A. 1986. Viral illnesses and sport performance. *Sports Medicine* 3:296-303.

Robinson, S., and P. M. Harmon. 1941. The lactic acid mechanism and certain properties of the blood in relation to training. *American Journal of Physiology* 132:757-69.

Robinson, S., H. T. Edwards, and D. B. Dill. 1937. New records in human power. *Science* 85:409-10.

Röcker, L., K. Kirsch, U. Mund, and H. Stoboy. 1975. The role of plasma protein in the control of plasma volume during exercise and dehydration in long distance runners and cyclists. In *Metabolic adaptation to prolonged physical exercise,* ed. Howald, H., J. R. Poortmans, 238-244. Basel: Birkhäuser Verlag.

Rodemann, H. R., and A. L. Goldberg. 1972. Arachidonic acid, prostaglandin E_2 and F_2 influence rates of protein turnover in skeletal and cardiac muscle. *Journal Biological Chemistry* 257:1632-38.

Rogozkin, V. A. 1976. The role of low molecular weight compounds in the regulation of skeletal muscle genome activity during exercise. *Medicine and Science in Sports* 8:74-9.

Rogozkin, V. A. 1979. Metabolic effects of anabolic steroids on skeletal muscle. *Medicine and Science in Sports* 11:160-3.

Rogozkin, V. A., and B. I. Feldkoren. 1979. The effect of retabolil and training on activity of RNA polymerase in skeletal muscle. *Medicine and Science in Sports* 11:345-7.

Romijn, J. A., E. F. Coyle, L. S. Sidossis, X.-J. Zhang, and R. R. Wolfe. 1995. Relationship between fatty acid delivery and fatty acid oxidation during strenuous exercise. *Journal of Applied Physiology* 79:1939-45.

Rosdahl, H., U. Ungerstedt, L. Jorfeldt, and J. Hendriksson. 1993. Interstitial glucose and lactate balance in human skeletal muscle and adipose tissue studied by microdialysis. *Journal of Physiology* 471:637-57.

Rosing, D. R., P. Brakman, D. R. Redwood, R. E. Goldstein, G. D. Astrup, and S. E. Epstein. 1970. Blood fibrinolytic activity in man. Diurnal variation and the response to varying intensities of exercise. *Circulation Research* 27:171-84.

Rotstein, A., O. Bar-Or, and R. Dlin. 1982. Hemoglobin, hematocrit and calculated plasma volume changes induced by a short supramaximal task. *International Journal of Sports Medicine* 3:230-3.

Rougier, G., and J. P. Babin. 1975. A blood and urine study of heavy muscular work on ureic and uric metabolism in man. *Journal of Sports Medicine and Physical Fitness* 15:213-22.

Rowbottom, D. G. 2000. Periodization of training. In *Exercise and sport science,* ed. Garrott, W. E., and D. T. Kirkendall, 499-512. Philadelphia: Lippincott Williams & Wilkins.

Rowbottom, D. G., D. Keast, C. Goodman, and A. R. Morton. 1995. The hematological biochemical and immunological profile of athletes suffering from the overtraining syndrome. *European Journal of Applied Physiology* 70:502-9.

Rowbottom, D. G., D. Keast, and A. R. Morton. 1996. The emerging role of glutamine as an indicator of exercise stress and overtraining [review]. *Sports Medicine* 21(2):80-97.

Rowbottom, D. G., D. Keast, and A. R. Morton. 1998. Monitoring and prevention of overreaching and overtraining in endurance athletes. In *Overtraining in sport,* ed. Kreider, R. B., A. C. Fry, M. L. O'Toole, 47-66. Champaign IL: Human Kinetics.

Rowell, L. B. 1986. *Human circulation during physical stress.* New York: Oxford University Press.

Rowell, L. B. 1988. Muscle blood flow in humans: how high can it go? *Medicine and Science in Sports and Exercise* 20:S97-103.

Saari, M. 1979. *Juoksenisen salaisuudet.* Helsingissä Kustannusosakeyhtiö Otava.

Saheki, T., and N. Katunuma. 1975. Analysis of regulatory factors for urea synthesis by isolated perfused rat liver. Urea synthesis with ammonia and glutamine as nitrogen sources. *Journal of Biochemistry* (Tokyo) 77:659-69.

Sahlin, K. 1978. Intracellular pH and energy metabolism in skeletal muscle of man. *Acta Physiologica Scandinavica* Suppl. 455.

Sahlin, K., A. Alverstrand, R. Brandt, and E. Hultman. 1978. Intracellular pH and bicarbohydrate concentration in human muscle during recovery from exercise. *Journal of Applied Physiology* 45:474-80.

Sahlin, K., and J. Henriksson. 1984. Buffer capacity and lactate accumulation in skeletal muscle of trained and untrained men. *Acta Physiologica Scandinavica* 122:331-9.

Sahlin, K., A. Katz, and S. Broberg. 1990. Tricarboxylic acid cycle intermediates in human muscle during prolonged exercise. *American Journal of Physiology* 259:C834-41.

Sahlin, K., M. Tonokonogi, and K. Söderlund. 1999. Plasma hypoxanthine and ammonia in humans during prolonged exercise. *European Journal Applied Physiology* 80:417-22.

Salo, D. C., C. M. Donvan, and K. J. Davies. 1991. HSP70 and other possible heat shock or oxidative stress proteins are induced in skeletal muscle, heart, and liver during exercise. *Free Radical Biology and Medicine* 11(3):239-46.

Saltin, B. 1964a. Aerobic and anaerobic work capacity after dehydration. *Journal of Applied Physiology* 19:1114-8.

Saltin, B. 1964b. Circulatory response to submaximal and maximal exercise after thermal dehydration. *Journal of Applied Physiology* 19:1125-32.

Saltin, B. 1990a. Anaerobic capacity: part, present, and future. In *Biochemistry of exercise VII,* ed. Taylor, P. D. Gollnick, H. J. Green, C. D. Ianuzzo, E. G. Noble, G. Metiver, and J. R. Sutton, 387-412. Champaign. IL: Human Kinetics.

Saltin, B. 1990b. Maximal oxygen uptake: limitation and malleability. In *International perspectives in exercise physiology,* ed. Nazar, K., R. L. Terjung, H. Kaciuba-Usciko, and L. Budohoski, 26-40. Champaign, IL: Human Kinetics.

Saltin, B., and D. Costill. 1988. Fluid and electrolyte balance during prolonged exercise. In *Exercise, nutrition and energy metabolism,* ed. Horton, E. S., and R. L. Terjung, 150-158. New York: Macmillan.

Saltin, B., and P. D. Gollnick. 1983. Skeletal muscle adaptability. Significance for metabolism and performance. In *Handbook of physiology. Sect. 10. Skeletal muscle,* ed. Peachy R. H. Adrians, and S. R. Geiger, 555-631. Baltimore: Williams & Wilkins.

Saltin, B., K. Nazar, D. L. Costill, E. Stein, E. Jansson, B. Essén, and P. D. Gollnick. 1976. The nature of the training response, peripheral and central adaptation to one-legged exercise. *Acta Physiologica Scandinavica* 96:289-305.

Saltin, B., and L. B. Rowell. 1980. Functional adaptations to physical activity and inactivity. *Federation Proceedings* 39:1506-13.

Saltin, B., and J. Stenberg. 1964. Circulatory and respiratory adaptation to prolonged severe exercise. *Journal of Applied Physiology* 19:833-8.

Saltin, B., and S. Strange. 1992. Maximal oxygen uptake: "old" and "new" arguments for a cardiovascular limitation. *Medicine and Science in Sports and Exercise* 24:30-7.

Sanches, J., J. M. Pequignot, L. Peyrin, and H. Monod. 1980. Sex differences in the sympathoadrenal response to isometric exercise. *European Journal of Applied Physiology* 45:147-54.

Sato, Y., S. Hayamizu, C. Yamamoto, Y. Okhuwa, K. Yamanouchi, and N. Sakamoto. 1986. Improved insulin sensitivity in carbohydrate and lipid metabolism after physical training. *International Journal of Sports Medicine* 7:307-10.

Sato, Y., Y. Oshida, I. Ohsawa, N. Nakai, N. Ohsani, K. Yamanouchi, J. Sato, Y. Shimomura, and H. Ohno.

1996. The role of glucose transport in the regulation of glucose utilization by muscle. In *Biochemistry of exercise IX*, ed. Maughan, R. J., and S. M. Shirrefs, 37-50. Champaign, IL: Human Kinetics.

Savard, R., J. P. Despres, M. Marcotte, and C. Bouchard. 1985. Endurance training and glucose conversion into triglycerides in human fat cells. *Journal of Applied Physiology* 59:230-5.

Savi, T. and A. Viru. 1975. Functional activity of adrenal cortex during competition in decathletes. In *Endocrine mechanisms of regulation of adaptation of muscular activity* [in Russian]. Vol. 5., ed. Viru A., 107-115. Tartu: University of Tartu.

Sawka, M. N., V. A. Convertino, E. R. Eichner, S. M. Schnieder, and A. J. Young. 2000. Blood volume: importance and adaptation to exercise training, environmental stresses, and trauma/sickness [review]. *Medicine and Science in Sports and Exercise* 32:332-248.

Sawka, M. N., W. A. Latzka, R. P. Matott, and S. J. Montain. 1998. Hydration effects on temperature regulation. *International Journal of Sports Medicine* 19(Suppl. 2):S108-10.

Schaefer, R. M., K. Kokot, A. Heidland, and R. Plass. 1987. Joggers' leukocytes. *The New England Journal of Medicine* 316:223-4.

Schamadan, J. L., and W. D. Snilvely. 1967. Potassium depletion as a possible cause of heat stroke. *Indian Medicine & Surgery* 36:785-8.

Scheen, A. J., O. M. Buxton, M. Jison, O. Van Reeth, R. Leproult, M. L'Hermite-Balériaux, and E. Van Cauter. 1998. Effects of exercise on neuroendocrine secretions and glucose regulation at different times of day. *American Journal of Physiology* 274:E1040-9.

Schenk, P. 1930. *Die Ermüdung. Gesunds and kraunen Menschen*. Jena: Fischer Verlag.

Schenk, P., and K. Craemer. 1929. Der Einfluß der schwerer körplicher Arbeit auf den menschlichen Stoffwechsel. *Arbeitsphysiologie* 2:163-86.

Scheuer, J., L. Kapner, C. A. Stringfellow, C. L. Armstrong, and S. Penpargkul. 1970. Glycogen, lipid and high energy phosphate store in hearts from conditioned rats. *Journal of Laboratory and Clinical Medicine* 75:924-8.

Scheurink, A. J. W., A. B. Steffons, and R. P. A. Gaynema. 1990. Hypothalamic adrenoreceptors mediate sympathoadrenal activity in exercising rats. *American Journal of Physiology* 259:R470-7.

Schimke, R. T. 1962. Differential effects of fasting and protein-free diets on levels of urea cycle enzymes in rats liver. *Journal of Biology and Chemistry* 237:1921-4.

Schlicht, W., W. Naretz, D. Witt, and H. Rieckert. 1990. Ammonia and lactate: differential information on monitoring training load in sprint events. *International Journal of Sports Medicine* 11(Suppl. 2):S85-S90.

Schmid, P., H. H. Push, W. Wolf, E. Pilger, H. Pessenhofer, G. Schwaberger, H. Pristautz, and P. Pürstner. 1982. Serum FSH, LH, and testosterone in humans after physical exercise. *International Journal of Sports Medicine* 3:84-9.

Schmidt, W., G. Brabant, C. Kröger, S. Strauch, and A. Hilgendorf. 1990. Atrial natriuretic peptide during and after maximal and submaximal exercise under normoxic and hypoxic conditions. *European Journal of Applied Physiology* 61:398-407.

Schmidt, W., K. U. Eckardt, A. Hilgendorf, S. Strauch, and C. Bauer. 1991. Effects of maximal and submaximal exercises under normoxic and hypoxic conditions on serum erythropoietin level. *International Journal of Sports Medicine* 12:457-61.

Schmidt, W., N. Maasen, U. Tegtbur, and K. M. Braumann. 1989. Changes in plasma volume and red cell formation after a marathon competition. *European Journal of Applied Physiology* 58:453-8.

Schmidt, W., N. Maasen, F. Trost, and D. Böning. 1988. Training induced effects on blood volume, erythrocyte turnover and haemoglobin, oxygen binding properties. *European Journal of Applied Physiology* 57:490-98.

Schnabel, A., W. Kindermann, W. M. Schmitt, G. Biro, and H. Stegmann. 1982. Hormonal and metabolic consequences of prolonged running at the individual anaerobic threshold. *International Journal of Sports Medicine* 3:163-8.

Schwab, R., G. O. Johnson, T. J. Housh, J. E. Kinder, and J. P. Weir. 1993. Acute effects of different intensities of weight lifting in serum testosterone. *Medicine and Science in Sports and Exercise* 25:1381-5.

Schwandt, H. J., B. Heyduck, H. C. Gunga, and L. Röcker. 1991. Influence of prolonged physical exercise on the erythropoietin concentration in blood. *European Journal of Applied Physiology* 63:463-6.

Schwarz, L., and W. Kindermann. 1990. ß-endorphin, adrenocorticotropin hormone, cortisol and catecholamines during aerobic and anaerobic exercise. *European Journal of Applied Physiology* 61:165-71.

Scott, C. B., F. B. Rody, T. G. Lohman, and J. C. Bunt. 1991. The maximally accumulated oxygen deficit as an indicator of anaerobic capacity. *Medicine and Science in Sports and Exercise* 23:618-24.

Seene, T., and A. Viru. 1982. The catabolic effects of glucocorticoids on different types of skeletal muscle fibres and its dependence upon muscle activity and interaction with anabolic steroids. *Journal Steroid Biochemistry* 16:349-52.

Seene, T., R. Masso, M. Oks, A. Viru, and E. Seppet. 1978. Changes in the adrenal cortex during adaptation to various regimes of physical activity. *Sechenov Physiological Journal of USSR* 64:1444-50.

Sejersted, O. M. 1992. Electrolyte imbalance in body fluids as a mechanism of fatigue. In *Perspectives in exercise science and sports medicine. Energy metabolism in exercise and sport,* ed. Lamb, D. R., and C. V. Gisolfi, 149-207. Carmer: Brown & Benchmark.

Sejersted, O. M., N. K. Vøllenstad, and J. I. Medbø. 1986. Muscle fluid and electrolyte balance during and following exercise. *Acta Physiologica Scandinavica* 128(Suppl. 556):119-27.

Selby, G. B., and E. R. Eichner. 1986. Endurance swimming, intravascular hemolysis, anemia, and iron depletion. *American Journal of Medicine* 81:791-4.

Sellers, T. L., A. W. Jaussi, H. T. Yang, R. W. Heninger, and W. W. Winder. 1988. Effect of the exercise-induced increase in glucocorticoids on endurance in the rat. *Journal of Applied Physiology* 65:173-8.

Selye, H. 1950. *The physiology and pathology of exposure to stress.* Montreal: Med Publication.

Sen, C. 1995. Oxidants and antioxidants in exercise. *Journal of Applied Physiology* 79:675-86.

Sen, C. K., E. Marin, M. Kretzchmar, and O. Hänninen. 1992. Skeletal muscle and liver glutathione homeostasis in response to training, exercise, and immobilization. *Journal of Applied Physiology* 73:1265-72.

Sen, C., S. Rankinen, S. Vaisanen, and R. Rauramaa. 1994. Oxidative stress following human exercise. Effect of N-acetylcysteine supplementation. *Journal of Applied Physiology* 76:2570-7.

Senay, L. C. 1970. Movement of water, protein and crystalloids between vascular and extravascular compartments in heat-exposed men during dehydration and following limited relief of dehydration. *Journal of Physiology* 210:617-35.

Serresse, O., G. Lortie, C. Bouchard, and M. R. Boulay. 1988. Estimation of the contribution of various energy systems during maximal work of sport duration. *International Journal of Sports Medicine* 9:456-60.

Sforzo, G. A., T. F. Seeger, A. Pert, and C. O. Dotson. 1986. In vivo opioid receptor occupation in the rat brain following exercise. *Medicine and Science in Sports and Exercise* 18:380-4.

Sharp, R. L., L. E. Armstrong, D. S. King, and D. L. Costill. 1983. Buffer capacity of blood in trained and untrained males. In *Biochemistry of exercise,* ed. Knuttgen, H. G., J. A. Vogel, and J. Poortmans, 595-599. Champaign, IL: Human Kinetics.

Sharp, R. L., D. L. Costill, W. J. Fink, and D. S. King. 1986. Effects of eight weeks of bicycle ergometer sprint training on human muscle buffer capacity. *International Journal of Sports Medicine* 7:13-7.

Shaw, W. A .S., T. B. Issekutz, and B. Issekutz. 1975. Interrelationship of FFA and glycerol turnover in resting and exercising dogs. *Journal of Applied Physiology* 39:30-60.

Shephard, R. J. 1971. The oxygen conductance equation. In *Frontiers of fitness,* ed. Shephard, R. J., 129-154. Springfield, IL: Charles C Thomas.

Shephard, R. J. 1984. Tests of maximum oxygen intake. A critical review. *Sports Medicine* 1:99-124.

Shephard, R. J. 1986. Exercise and malignancy. *Sports Medicine* 3:235-41.

Shephard, R. J. 1992. Maximal oxygen uptake. In *Endurance in sport,* ed. Shephard, R. J., P.-O. Åstrand, 192-200. Oxford: Blackwell Scientific.

Shephard, R. J., and T. J. Kavanagh. 1975. Biochemical changes with marathon running. Observations on post-coronary patients. In *Metabolic adaptation to prolonged physical exercise,* ed. Howald, H., and J. R. Poortmans, 245-252. Basel: Birkhäuser.

Shephard, R. J., and P. N. Shek. 1998. Acute and chronic over-exertion: do depressed immune responses provide useful markers? *International Journal of Sports Medicine* 19:159-71.

Shimazu, T. 1987. Neural regulation of hepatic glucose metabolism in mammals. *Diabetes/Metabolism Reviews* 3:185-206.

Shinkai, S., S. Share, P. N. Shek, and R. J. Shephard. 1992. Acute exercise and immune function. *International Journal of Sports Medicine* 13:452-61.

Shirreffs, S. M., and R. J. Maughan. 1997. Whole body sweat collection in man: an improved method with some preliminary data on electrolyte composition. *Journal of Applied Physiology* 82:336-41.

Shyu, B.-C., S. A. Andersson, and P. Thoren. 1982. Endorphin mediated increase in pain threshold induced by long-lasting exercise in rats. *Life Sciences* 30:833-40.

Siggaard-Andersen, O. 1963. Blood acid-base alignment nomogram. *Scandinavian Journal of Clinical and Laboratory Investigation* 15:211-7.

Siggaard-Andersen, O., K. Engel, K. Jorgensen, and P. Åstrup. 1960. A micromethod for determination of pH, carbon dioxide tension base excess and standard bicarbonate in capillary blood. *Scandinavian Journal of Clinical and Laboratory Investigations* 12:172-6.

Simoneau, J. A., G. Lortie, M. R. Boulay, and C. Bouchard. 1983. Tests of anaerobic alactacid and lactacid capacities: description and reliability. *Canadian Journal of Applied Sports Science* 8:266-70.

Sjödin, B., and I. Jacobs. 1981. Onset of blood lactate accumulation and marathon running performance. *International Journal of Sports Medicine* 2:23-6.

Sjödin, B., I. Jacobs, and J. Karlsson. 1981. Onset of blood lactate accumulation and enzyme activities in vastus lateralis in man. *International Journal of Sports Medicine* 2:166-70.

Sjödin, B., I. Jacobs, and J. Svedenhag. 1982. Changes in onset of blood lactate accumulation (OBLA) and muscle enzymes after training at OBLA. *European Journal of Applied Physiology* 49:45-57.

Sjödin, B., and J. Svedenhag. 1985. Applied physiology of marathon running. *Sports Medicine* 2:83-99.

Sjödin, B., A. Thorstensson, K. Frith, and J. Karlsson. 1976. Effect of physical training on LDH activity and LDH isozyme pattern in human skeletal muscle. *Acta Physiologica Scandinavica* 97:150-7.

Sjøgaard, G. 1979. Water spaces and electrolyte concentrations in human skeletal muscle. Thesis. Copenhagen: University of Copenhagen.

Sjøgaard, G. 1990. Exercise-induced muscle fatigue: the significance of potassium. *Acta Physiologica Scandinavica* vol.140, Suppl. 593.

Sjøgaard, G. 1991. Role of exercise induced potassium fluxes underlying muscle fatigue: a brief review. *Canadian Journal of Physiology and Pharmacology* 69:238-45.

Sjøgaard, G., and B. Saltin. 1982. Extra and intracellular water spaces in muscle of man at rest and with dynamic exercises. *American Journal of Physiology* 243:R271-80.

Sjøgaard, G., R. P. Adams, and B. Saltin. 1985. Water and ion shifts in skeletal muscle of humans with intense dynamic knee extension. *American Journal of Physiology* 248:R190-6.

Skidmore, R., J. A. Gutierrez, V. Guerriero, and K. C. Kregel. 1995. HSP70 induction during exercise and heat stress in rats: Role of internal temperature. *American Journal of Physiology* 268(1 Pt. 2):R92-7.

Skinner, J. S., and T. M. McLellan. 1980. The transition from aerobic to anaerobic metabolism. *Research Quarterly for Exercise and Sport* 436:1-32.

Smalls, A. G. H., P. W. C. Kloppenborg, and T. J. Beuraad. 1976. Circannual cycle in plasma testosterone level in man. *Journal of Clinical Endocrinology* 42:979-82.

Smirnova, T., and A. Viru. 1977. Dependence of physical working capacity on the state of adrenocortical function and character of the tissue distribution of corticosterone. *Acta Commetationes Universitatis Tartuensis* 419:130-3.

Smith, B. W., R. G. McMurray, and J. D. Symanski. 1984. A comparison of the anaerobic threshold of sprint and endurance trained swimmers. *Journal of Sports Medicine and Physical Fitness* 24:94-9.

Smith, E. W., M. S. Skeleton, D. E. Kremer, D. D. Pascoe, and L. B. Gladden. 1997. Lactate distribution in the blood during progressive exercise. *Medicine and Science in Sports and Exercise* 29:654-60.

Smith, J. A., R. D. Telford, I. B. Mason, and M. J. Weidemann. 1990. Exercise, training and neutrophil microbicidal activity. *International Journal of Sports Medicine* 11:179-87.

Smith, K., and M. J. Rennie. 1990. Protein turnover and amino acid metabolism in humans skeletal muscle. *Clinical Endocrinology and Metabolism* 4:461-8.

Smith, R. H., R. M. Palmer, and P. J. Reeds. 1983. Protein synthesis in isolated rabbit forelimb muscle. *Biochemical Journal* 214:153-11.

Snegovskaya, V., and A. Viru. 1992. Growth hormone, cortisol and progesterone levels in rowers during a period of high intensity training. *Biology of Sport* 9:93-101.

Snegovskaya, V., and A. Viru. 1993b. Elevation of cortisol and growth hormone levels in the course of further improvement of performance capacity in trained rowers. *International Journal of Sports Medicine* 14:202-6.

Snegovskaya, V., and A. Viru. 1993a. Steroid and pituitary hormone responses to rowing exercises: Relative significance of exercise intensity and duration and performance. *European Journal of Applied Physiology* 67:59-65.

Snell, P. G., W. H. Martin, J. C. Buckey, and C. G. Blomquist. 1987. Maximal vascular leg conductance in trained and untrained men. *Journal of Applied Physiology* 62:606-10.

Snochowski, M., T. Saartok, E. Dahlberg, E. Eriksson, and J.-A. Gustafsson. 1981. Androgen and glucocorticoid receptors in human skeletal muscle cytosol. *Journal of Steroid Biochemistry*14:765-71.

Snyder, A. C. 1998. Overtraining and glycogen depletion hypothesis. *Medicine and Science in Sports and Exercise.* 30:1146-50.

Snyder, A. C., H. Kuipers, B. Cheng, R. Servais, and E. Fransen. 1995. Overtraining followed intensified training with normal muscle glycogen. *Medicine and Science in Sports and Exercise* 27:1063-70.

Sobel, B. E., and S. Kaufman. 1970. Enhanced RNA polymerase activity in skeletal muscle undergoing hypertrophy. *Archives of Biochemistry and Biophysics* 137:469-76.

Söderlund, K. 1991. *Energy metabolism in human skeletal muscle during intense contraction and recovery with reference to metabolic differences between type I and type II fibers.* Stockholm: Karolinska Institutet.

Söderlund, K., and E. Hultman. 1986. Effects of delayed freezing on content of phosphagens in human skeletal muscle biopsy samples. *Journal of Applied Physiology* 61(3):832-5.

Spikermann, M. 1989. Der Einsatz einer neuen Organisationsform der Belastung im Hochleistungssport. *Leistungssport* 19(1):33-5.

Spodaryk, K. 1993. Hematological and iron-related parameters of male and endurance and strength trained athletes. *European Journal of Applied Physiology* 67:66-70.

Spriet, L. L. 1995. Anaerobic metabolism during high-intensity exercise. In *Exercise metabolism,* ed. Hargreaves, M., 1-39. Champaign, IL: Human Kinetics.

Spriet, L. L., J. M. Ren, and E. Hultman. 1988. Epinephrine infusion enhances muscle glycogenolysis during prolonged electrical stimulation. *Journal of Applied Physiology* 64:1439-44.

Staehelin, D., A. Labhart, R. Froesch, and H. R. Kägi. 1955. The effect of muscular exercise and hypoglycemia on the plasma level of 17-hydroxysteroids in normal adults and in patients with the adrenogenital syndrome. *Acta Endocrinologica* 18:521-9.

Stansbie, D., J. P. Aston, N. S. Dallimore, M. S. Williams, and N. Willis. 1983. Effect of exercise on plasma pyruvate kinase and creatine kinase activity. *Clinical Chemistry Acta* 2588:127-32.

Steel, C. M., J. Evans, and M. A. Smith. 1974. Physiological variation in circulating B cell:T cell ratio in man. *Nature* 247:387-9.

Steffens, A. B., and J. H. Strubbe. 1983. CNS regulation of glucagon secretion. In *Advances in metabolic disorders*, ed. Szabo, A. J., 221-257. New York: Academic Press.

Stegmann, H., and W. Kindermann. 1982. Comparison of prolonged exercise test at the individual anaerobic threshold and the fixed anaerobic threshold of 4 mmol · l^{-1} lactate. *International Journal of Sports Medicine* 3:105-10.

Stegmann, H., W. Kindermann, and A. Schnabel. 1981. Lactate kinetics and individual anaerobic threshold. *International Journal of Sports Medicine* 2:160-5.

Stegmann, H., W. Kindermann, and A. Schnabel. 1982. Lactate kinetics and individual anaerobic threshold. *International Journal of Sports Medicine* 2:160-5.

Stegemann, J. 1981. *Exercise physiology. Physiological bases of work and sport.* Stuttgart, New York: G. Thieme Verlag.

Steinacker, J. M., M. Kellmann, B. O. Böhm, Y. Liu, A. Opitz-Gress, K. W. Kallus, M. Lehmann, D. Altenburg, and W. Lormes. 1999. Clinical findings and parameters of stress and regeneration in rowers before world championships. In *Overload, performance, incompetence, and regeneration in sport,* ed. Lehmann, M., C. Foster, U. Gastmann, H. Keizer, and J. M. Steinacker, 71-80. New York: Kluwer Academic/Plenum Publications.

Steinacker, J. M., R. Laske, W. D. Hetzel, W. Lormes, Y. Liu, and M. Stauch. 1993. Metabolic and hormonal reactions during training in junior oarsmen. *International Journal of Sports Medicine* 14(Suppl. 1):S24-8.

Steinhaus, A. H. 1933. Chronic effects of exercise. *Physiological Reviews* 13:103-47.

Stephenson, L. A., M. A. Kolka, R. Francesoni, and R. R. Gonzalez. 1989. Circadian variations in plasma renin activity, catecholamines and aldosterone in woman. *European Journal of Applied Physiology* 58:756-64.

Stock, M. J., C. Chapman, J. L. Stirling, and J. T. Cambell. 1978. Effect of exercise, altitude and food on blood hormone and metabolite levels. *Journal of Applied Physiology* 45:350-4.

Stokes, J. L., and M. Mancini. (eds.) 1988. *Hypercholesterolemia: clinical and therapeutic implications.* Atherosclerosis Reviews, Vol. 18. New York: Raven Press.

Stray-Gundersen, J., T. Videman, and P. G. Snell. 1986. Changes in selected objective parameters during overtraining (abstract). *Medicine and Science in Sports and Exercise* 18(Suppl.):S54-5.

Struck, P. J., and C. M. Tipton. 1974. Effect of acute exercise on glycogen levels in adrenalectomized rats. *Endocrinology* 95:1385-91.

Stupnicki, R., and Z. Obminski. 1992. Glucocorticoid response to exercise as measured by serum and salivary cortisol. *European Journal of Applied Physiology and Occupational Physiology* 65(6):546-9.

Sugden, P. H., and S. J. Fuller. 1991. Regulation of protein turnover in skeletal and cardiac muscle. *Biochemical Journal* 273:21-37.

Sundsfjord, J. A., S. B. Stromme, and A. Aakvaag. 1975. Plasma aldosterone (PA), plasma renin activity (PRA) and cortisol (F) during exercise. In *Metabolic adaptation to prolonged physical exercise,* ed. Howald, H., and J. R. Poortmans, 308-314. Basel: Birkhäuser Verlag.

Sutton, J. R. 1977. Effect of acute hypoxia on the hormonal response to exercise. *Journal of Applied Physiology* 42:587-92.

Sutton, J. R. 1978. Hormonal and metabolic responses to exercise in subjects of high and low work capacities. *Medicine and Science in Sports* 10:1-6

Sutton, J. R., and J. H. Casey. 1975. The adrenocortical response to competitive athletics in veteran athletes. *Journal of Clinical Endocrinology* 40:135-8.

Sutton, J. R., M. J. Coleman, J. Casey, L. Lazarus, J. B. Hickie, and J. Maksvytis. 1969. The hormonal response to physical exercise. *Australian Annals of Medicine* 18:84-90.

Sutton, J. R., M. J. Coleman, and J. H. Casey. 1978. Testosterone production rate during exercise. In *3rd International Symposium on Biochemistry of Exercise,* ed. Landry, F., and W. A. Orban, 227-234. Miami: Symposia Specialists.

Sutton, J. R., and L. Lazarus. 1976. Growth hormone in exercise: comparison of physiologic and pharmacologic stimuli. *Journal of Applied Physiology* 41:523-7.

Svedenhag, J., and B. Sjödin. 1984. Maximal and submaximal oxygen uptakes and blood lactate levels in elite male middle- and long-distance runners. *International Journal of Sports Medicine* 5:255-61.

Szczepanowska, E., T. Rychlewski, and A. Viru 1999. Effect of acute treadmill exercise on hormonal changes in 15-17 year-old female middle-distance runners. Significance of phase of the ovarian-menstrual cycle. *Medicina dello Sport* 52:41-7.

Szogy, A., and G. Cherebetiu. 1974. A one-minute bicycle ergometer test for determination of anaerobic capacity. *European Journal of Applied Physiology* 33:171-6.

Tabata, I., Y. Atomi, and M. Misyashita. 1989. Bi-phasic change of serum cortisol concentration in the marking during high-intensity physical training in man. *Hormone and Metabolic Research* 21:218-9.

Tabata, I., K. Nishimura, M. Kouzaki, Y. Hirai, F. Ogita, M. Miyachi, and K. Yamamoto. 1996. Effects of moderate-intensity endurance and high-intensity intermittent training on anaerobic capacity and $\dot{V}O_2max$. *Medicine and Science in Sports Exercise* 28:1327-30.

Tabor, C. W., and H. Tabor. 1976. 1.4-diaminobutene (putrescine), spermidine and spermine. *Annual Reviews of Biochemistry* 45:285-306.

Tanaka, H., K. A. West, G. E. Duncan, and D. R. Bassett. 1997. Changes in plasma tryptophan/branched chain amino acid ratio in response to training volume variation. *International Journal of Sports Medicine* 18:270-5.

Targan, S., L. Britvan, and F. Dorey. 1981. Activation of human NKCC by moderate exercise: increased frequency of NK cells with enhanced capability of effector-target lytic interactions. *Clinical Experimental Immunology* 45:352-60.

Taylor, A. W., M. A. Booth, and S. Rao. 1972. Human skeletal muscle phosphorylase activities with exercise and training. *Canadian Journal of Physiology and Pharmacology* 50:1038-42.

Tchagovets, N. R., L. V. Maksimova, R. I. Lenkova, and A. P. Kraskova. 1983. Peculiarities of metabolism in sportsmen during competition period [in Russian]. *Teoria i Praktika Fizicheskoi Kulturõ* (Moscow) 9:20-2.

Tchaikovsky, V. S., I. V. Astratenkova, and O. B. Bashirina. 1986. The effect of exercise on the content and receptor of the steroid hormones in rat skeletal muscle. *Journal of Steroid Biochemistry* 24:251-3.

Tchareyeva, A. A. 1986a. Biochemical criteria of fitness in high qualification ice-hockey players in various stages of training [in Russian]. In *Biochemical criteria of improvement of special fitness,* ed. Tchareyeva, A. A., 131-140. Moscow: Allunion Research Institute of Culture.

Tchareyeva, A. A. 1986b. Universal energy-transportic role of phosphocreatine mechanism and its significance for energetics of muscular activity [in Russian]. In *Biochemical criteria of improvement of special fitness,* ed. Tchareyeva, A. A., 4-14. Moscow: Allunion Research Institute of Physical Culture.

Tendzegolskis, Z., A. Viru, and E. Orlova. 1991. Exercise-induced changes of endorphin contents in hypothalamus, hypophysis adrenals and blood plasma. *International Journal of Sports Medicine* 12:495-7.

Tesch, P. A., and J. Karlsson. 1985. Muscle fiber types and size in trained and untrained muscles of elite athletes. *Journal of Applied Physiology* 59:1716-20.

Tharp, G. D., and M. W. Barnes. 1990. Reduction of saliva immunoglobulin levels by swim training. *European Journal of Applied Physiology* 60:61-4.

Thayer, R. 1983. Planning a training program. *Track Technique*, Annu 83:4-7.

Thomson, J. M., and K. J. Garvie. 1981. A laboratory method for determination of anaerobic energy expenditure during sprinting. *Canadian Journal of Applied Sport Sciences* 6:21- 6.

Thorn, G. W., D. Jenkins, J. C. Laidlaw, F. C. Goetz, and W. Reddy. 1953. Response of the adrenal cortex to stress in man. *Transactions of the Association of American Physicians* 66:48-64.

Thorstensson, A. 1976. Muscle strength, fiber types and enzyme activities in man. *Acta Physiologica Scandinavica* Suppl.443.

Thorstensson, A., L. Larsson, P. Tesch, and J. Karlsson. 1977. Muscle strength and fiber composition in athletes and sedentary men. *Medicine and Science in Sports* 9:26-30.

Tibes, U., B. Hemmer, U. Schweigart, D. Boning, and D. Fortescu. 1974. Exercise acidosis as cause of electrolyte change in femoral venous blood of trained and untrained men. *Pflügers Achic für die gesammte Physiologie* 341:145-58.

Tihanyi, J. 1989. Principen individualisierter Trainingsprotokolle auf der Basis der Muskel foserzusammensetzung und mechanischer Merkmale. *Leistungssport* 19(2):41-5.

Tihanyi, J. 1997. Principles of power training and control of dynamic muscle work. *Acta Academiae Olympiquae Estonia* 5:5-23.

Tihanyi, J., P. Apar, and G. Fekete. 1982. Force-velocity-power characteristics and fiber composition in human knee extensor muscles. *European Journal of Applied Physiology* 48:331-43.

Tiidus, P. M. 1998. Radical species in inflammation and overtraining. *Canadian Journal of Physiology and Pharmacology* 76:533-8.

Tiidus, P. M., and C. D. Ianuzzo. 1983. Effects of intensity and duration of muscular exercise on delayed soreness and serum enzyme activities. *Medicine and Science in Sports and Exercise* 15:461-5.

Tiidus, P. M., and M. E. Houston. 1995. Vitamin E status and responses to exercise training. *Sports Medicine* 20:12-23.

Tiidus, P. M., J. Pushkarenko, and M. E. Houston. 1996. Lack of antioxidant adaptation to short-term aerobic training in human muscle. *American Journal of Physiology* 271:R832-6.

Tipton, C. M., P. J. Struck, K. M. Baldwin, R. D. Matther, and R. T. Dowell. 1972. Response of adrenalectomized rats to chronic exercise. *Endocrinology* 91:573-99.

Tischler, M. E., M. Desautels, and A. L. Goldberg. 1982. Does leucine, leucyl-tRNA, or some metabolic of leucine regulate protein synthesis and degradation in skeletal and cardiac muscle? *Journal of Biological Chemistry* 257:1613-21.

Tomasi, T. B., F. B. Trudeau, D. Czerwinski, and S. Erredge. 1982. Immune parameters in athletes before and after strenuous exercise. *Journal of Clinical Immunology* 2:173-8.

Toode, K., T. Smirnova, Z. Tendzegolskis, and A. Viru. 1993. Growth hormone action on blood glucose, lipids, and insulin during exercise. *Biology of Sport* 10:99-106.

Torjman, M. C., A. Zafeiridis, A. M. Paolone, C. Wilkerson, and P. V. Considere. 1999. Serum leptin during recovery following maximal momental and prolonged exercise. *International Journal of Sports Medicine* 20(7):444-50.

Tran, Z. V., A. Weltman, G. V. Glass, and D. P. Mood. 1983. The effects of exercise on blood lipids and lipoproteins: a meta-analysis of studies. *Medicine and Science in Sports and Exercise* 15:393-402.

Tremblay, M. S., S. Y. Chu, and R. Mureika. 1995. Methodological and statistical considerations for exercise-related hormone evaluation. *Sports Medicine* 20:90-108.

Trump, M. E., G. J. E. Heigenhauser, C. T. Putman, and L. L. Spriet. 1996. Importance of muscle phosphocreatine during intermittent maximal cycling. *Journal of Applied Physiology* 80:1574-80.

Tsai, K. S., J. C. Lin, C. K. Chen, W. C. Cheng, and C. H. Yang. 1997. Effects of exercise and exogenous glucocorticoid on serum level of intact parathyroid hormone. *International Journal of Sports Medicine* 18:583-7.

Tsai, L., C. Johansson, Å. Pousette, R. Tegelman, K. Carlström, and P. Hemmingsson. 1991. Cortisol and androgen concentrations in female and male elite endurance athletes in relation to physical activity. *European Journal of Applied Physiology* 63:308-11.

Tsõbizov, G. G. 1978. Hormonal regulation of calcium and phosphorus homeostasis during physical excretion. *Sechenov Physiological Journal of the USSR* 65:1539-44.

Turcotte, L. P., E. A. Richter, and B. Kiens. 1995. Lipid metabolism during exercise. In *Exercise metabolism,* ed. by Hargreaves, M., 99-130. Champaign, IL: Human Kinetics.

Tvede, N., M. Kappel, J. Halkjaer-Christensen, H. Galbo, and B. K. Pederson. 1993. The effect of light, moderate and severe bicycle exercise on lymphocyte subsets, natural and lymphokine activated killer cells, lymphocyte proliferative response and interleukin 2 production. *International Journal of Sports Medicine* 14(5):275-82.

Tvede, N., B. K. Pedersen, F. R. Hansen, T. Bendix, L. D. Christensen, H. Galbo, and J. Halkjaer-Kristensen. 1989. Effect of physical exercise on blood mononuclear cell subpopulations and in vitro proliferative responses. *Scandinavian Journal of Immunology* 29:383-9.

Tvede, N., J. Steensberg, B. Baslund, J. Halkjaer-Kristensen, and B. K. Pedersen. 1991. Cellular immunity in highly-trained elite racing cyclists and controls during periods of training with high and low intensity. *Scandinavian Journal of Sports Medicine* 1:163-6.

Ullman, M., and A. Oldfors. 1986. Effects of growth hormone on skeletal muscle. I Studies on normal adult rats. *Acta Physiologica Scandinavica* 135:531-6.

Ungerstedt, U. 1991. Microdialysis—Principles and applications for studies in animals and man. *Journal of Internal Medicine* 230:365-73.

Urban, R. J., Y. H. Bodenburg, C. Gilkson, J. Foxwarth, A. R. Coggan, R. R. Wolfe, and A. Ferrando. 1995. Testosterone administration to elderly men increases skeletal muscle strength and protein synthesis. *American Journal of Physiology* 269:E820-6.

Urhausen, A., B. Coen, and W. Kindermann. 2000. Individual assessment of the aerobic-anaerobic transition by measurements of blood lactate. In *Exercise and sport science,* ed. Garrett, W. E., and D. T. Kirkendall, 267-275. Philadelphia: Lippincott, Williams & Wilkins.

Urhausen, A., H. Gabriel, and W. Kindermann. 1995. Blood hormones as markers of training stress and overtraining. *Sports Medicine* 20:351-76.

Urhausen, A., H. H. W. Gabriel, and W. Kindermann. 1998. Impaired pituitary hormonal responses to exhaustive exercise in overtrained athletes. *Medicine and Science in Sports and Exercise* 30:407-14.

Urhausen, A., and W. Kindermann. 1992a. Biochemical monitoring of training. *Clinical Journal of Sports Medicine* 2:52-61.

Urhausen, A., and W. Kindermann. 1992b. Blood ammonia and lactate concentrations during endurance exercise of differing intensities. *European Journal of Applied Physiology* 65:209-14.

Urhausen, A., and W. Kindermann. 1994. Monitoring of training by determination of hormone concentration in the blood. In *Regulations und repairmechanismen,* ed. Liesen, H., M. Weiss, and M. Baum, 551-554. Köln: Deutche Ärzte-Verlag.

Urhausen, A., and W. Kindermann. 1987. Behavior of testosterone, sex hormone binding globulin (SHBG) and cortisol before and after a triathlon competition. *International Journal of Sports Medicine* 8:305-8.

Urhausen, A., T. Kullmer, and W. Kindermann. 1987. A seven week follow-up study of the behaviour of testosterone and cortisol during the competition period of rowers. *European Journal of Applied Physiology* 56:528-33.

Urhausen, A., B. Coen, B. Weiler, and W. Kindermann. 1993. Individual anaerobic threshold and maximum lactate steady state. *International Journal of Sports Medicine* 14:134-9.

Urhausen, A., B. Weiler, B. Coen, and W. Kindermann. 1994. Plasma catecholamines during endurance exercise of different intensities as related to the individual anaerobic threshold. *European Journal of Applied Physiology* 69:16-20.

Uusitalo, A. L. T., P. Huttunen, Y. Hanin, A. J. Uusitalo, and H. K. Rusko. 1998. Hormonal responses to endurance training and overtraining in female athletes. *Clinical Journal of Sports Medicine* 8:178-86.

Vaernes, R., H. Ursin, and A. Darrogh. 1982. Endocrine response pattern and psychological correlates. *Journal of Psychosomatic Research* 26:123-31.

Van Acker, S. A. B. E., L. M. H. Koymans, and A. Bast. 1993. Molecular pharmacology of vitamin E: structural aspects of antioxidant activity. *Free Radical Biology and Medicine* 15:311-28.

Van Beaumont, W., J. E. Greenleaf, and L. Juhos. 1973. Disproportional changes in hematocrit, plasma volume and proteins during exercise and bed rest. *Journal of Applied Physiology* 34:102-6.

Van Hall, G., B. Saltin, G. J. Van der Vusse, K. Söderlund, and A. J. M. Wagenmakers. 1995. Deamination of amino acids as a source for ammonia production in human skeletal muscle during prolonged exercise. *Journal of Physiology* 489:251-61.

Van Handel, P. J., W. J. Fink, G. Branam, and D. L. Costill. 1980. Fate of ^{14}C glucose ingested during prolonged exercise. *International Journal of Sports Medicine* 1:127-31.

Van Slyke, D. D. 1922. On the measurement of buffer values and on the relationship of buffer values to concentration and reaction of the buffer solution. *Journal of Biological Chemistry* 52:525-70.

Vandenburgh, H. H., and S. Kaufman. 1981. Stretch-induced growth of skeletal myotubes correlates with activation of the sodium pump. *Journal of Cell Physiology* 109:205-14.

Vandewalle, H., G. Peres, and H. Monod. 1987. Standard anaerobic exercise tests. *Sports Medicine* 4:268-89.

Vanhelder, T. and M. W. Radomski. 1989. Sleep deprivation and the effect of exercise performance. *Sports Medicine* 7:235-47.

Vanhelder, W. P., K. Casey, and M. W. Radowski. 1987. Regulation of growth hormone during exercise by oxygen demand and availability. *European Journal of Applied Physiology* 56:628-32.

Vanhelder, W. P., R. C. Goode, and M. W. Radomski. 1984a. Effect of anaerobic and aerobic exercise of equal duration and work expenditure on plasma growth hormone levels. *European Journal of Applied Physiology* 52:255-7.

Vanhelder, W. P., M. W. Radomski, and R. C. Goode. 1984b. Growth hormone responses during intermittent weight lifting exercises in men. *European Journal of Applied Physiology* 53:31-4.

Vanhelder, W. P., M. W. Radomski, R. C. Goode, and K. Casey. 1985. Hormonal and metabolic responses to three types of exercise of equal duration and external work output. *European Journal of Applied Physiology* 54:337-42.

Varrik, E., and A. Viru. 1988. Excretion of 3-methylhistidine in exercising rats. *Biology of Sports* 5:195-204.

Varrik, E., A. Viru, V. Ööpik, and M. Viru. 1992. Exercise-induced catabolic responses to various muscle fibers. *Canadian Journal of Sport Sciences* 17:125-8.

Vasankari, T. J., U. M. Kujula, T. T. Viljanen, and I. T. Huhtaniemi. 1991. Carbohydrate ingestion during prolonged running exercise results in an increase of serum cortisol and decrease gonadotropins. *Acta Physiologica Scandinavica* 141:373-8.

Vasankari, T. J., H. Rusko, U. M. Kujala, and I. T. Huhtamiemi. 1993. The effect of ski training at altitude and racing on pituitary adrenal and testicular function in men. *European Journal of Applied Physiology* 66:221-5.

Vasiljeva, V. V., E. B. Kossovskaya, N. A. Stepochkina, and V. V. Trunin. 1972. Vascular reactions and aerobic working capacity in cyclists during various periods of training. *Teoria i Praktika Fizicheskoi Kulturõ* 35(6):28-30.

Vasiljeva, V. V., S. N. Popov, N. A. Stepochkina, Z. A. Teslenko, and V. V. Trunin. 1971. Experience of medical-biological observations in training process of long-distance runners. *Teoria i Praktika Fizicheskoi Kulturõ* 34(11):36-9.

Vaughan, M. H., and B. S. Hansen. 1973. Control of initiation of protein synthesis in human cells. Evidence for a role on uncharged transfer ribonucleic acid. *Journal Biological Chemistry* 248:7087-96.

Vedovato, M., E. De Paoli Vitali, C. Guglielmini, I. Casoni, G. Ricci, and M. Masotti. 1988. Erythropoietin in athletes of endurance events. *Nephron* 48:78-9.

Vendsalu, A. 1960. Studies on adrenaline and noradrenaline in human plasma. *Acta Physiologica Scandinavica* 49(Suppl 173).

Verde, T., S. Thomas, and R. J. Shephard. 1992. Potential markers of heavy training in highly trained distance runners. *British Journal of Sports Medicine* 26:167-75.

Verkhoshanski, Y., and A. Viru. 1990. Einige Geretzmäßigkeiten der langfristigen Adaptation des Organismus von Sportlern an körperliche Belastung. *Leistungsport* 3:10-3.

Verkhoshanski, Y. V. 1985. *Programming and organization of training process* [in Russian]. Moscow: FiS.

Vermulst, L. J. M., C. Vervoorn, A. M. Boelens-Quist, H. P. F. Koppeschaar, W. B. M. Erich, J. H. H. Thijssen, and W. R. de Vries. 1991. Analysis of seasonal training volume and working capacity in elite female rowers. *International Journal of Sports Medicine* 12:567-72.

Vervoorn, C., A. Quist, L. Vermulst, W. Erich, W. deVries, and J. Thjisen. 1991. The behaviour of the plasma free testosterone/cortisol ratio during a season of elite rowing training. *International Journal of Sports Medicine* 12:257-63.

Vervoorn, C., L. J. M. Vermulst, A. M. Boelens-Quist, H. P. F. Koppeschaar, W. B. M. Erich, J. H. H. Thijssen, and W. R. Vries. 1992. Seasonal changes in performance and free testosterone: cortisol ratio of elite female rowers. *European Journal of Applied Physiology* 64:14-21.

Vincent H. K., and K. R. Vincent. 1997. The effect of training status on the serum creatine kinase response, soreness and muscle function following resistance exercise. *International Journal of Sports Medicine* 18:431-7.

Vinnichuk, M., T. Smirnova, K. Karelson, and A. Viru. 1993. Effect of competition situation on catecholamine, cortisol, insulin and lactate responses to supramaximal exercises. *Acta et Commentationes Universitatis Tartuensis* 958:58-62.

Viru, A. 1964. Des changements dans le fonctionnement des surrénales avant la compétition. *Acta et Commentationes Universitatis Tartuensis* 154:70-7.

Viru, A. 1975a. Defense reaction theory of fatigue. *Schweizerische Zeitschrift für die Sportsmedizin* 4:171-87.

Viru, A. 1975b. Some methodological questions of the investigation of the endocrine regulation of metabolism during muscular work. *Acta et Commentationes Universitatis Tartuensis* 368:3-19.

Viru, A. 1976a. The physical working ability and the mouse organism's unspecific resistance in the course of training. *Sechenov Physiological Journal of the USSR* 62:636-9.

Viru, A. 1976b. The role of the adrenocortical reaction to exertion in the increase of body working capacity. *Bulleten Eksperimental'noi Biologij i Medicine (Moscow)* 82:774-6.

Viru, A. 1977. *Functions of adrenal cortex in muscular activity*. Moscow: Medicina.

Viru, A. 1983. Exercise metabolism and endocrine function. In *Biochemistry of exercise*, ed. Knuttgen, H. G., J. A. Vogel, and J. Poortmans, 76-86. Champaign, IL: Human Kinetics.

Viru, A. 1984. The mechanism of training effects: A hypothesis. *International Journal of Sports Medicine* 5:219-27.

Viru, A. 1985a. *Hormones in muscular activity*. Vol. I. Hormonal ensemble in exercise. Boca Raton, FL: CRC Press.

Viru, A. 1985b. *Hormones in muscular activity*. Vol. II. Adaptive effects of hormones in exercise. Boca Raton, FL: CRC Press.

Viru, A. 1987. Mobilization of structural protein during exercise. *Sports Medicine* 4:95-128.

Viru, A. 1991. Adaptive regulation of hormone interaction with receptors. *Experimental and Clinical Endocrinology* 97:13-28.

Viru, A. 1992. Mechanism of general adaptation. *Medical Hypothesis* 38:296-300.

Viru, A. 1993. Mobilization of the possibilities of the athlete's organism: a problem. *Journal of Sports Medicine and Physical Fitness* 33:413-25.

Viru, A. 1994a. Contemporary state and further perspectives on using muscle biopsy for metabolism studies on sportsmen. *Medicina dello Sport* 47:371-6.

Viru, A. 1994b. Molecular cellular mechanisms of training effects. *Journal of Sports Medicine and Physical Fitness* 34:309-22.

Viru, A. 1995. *Adaptation in sports training*. Boca Raton, Ann Arbor, London, Tokyo: CRC Press.

Viru, A. 1996. Postexercise recovery period: carbohydrate and protein metabolism. *Scandinavian Journal of Medicine Sciences and Sports* 6:2-14.

Viru, A., and H. Äkke. 1969. Effects of muscular work on cortisol and corticosterone content in the blood and adrenals of guinea pigs. *Acta Endocrinology* 62:385-90.

Viru, A., and A. Eller. 1976. Adrenocortical regulation of protein metabolism during prolonged physical exertion. *Bulleean Eksperimental'noi Biologij i Medicine (Moscow)* 82:1436-9.

Viru, A., J. Jürgenstein, and A. Pisuke. 1972. Influence of training methods on endurance. *Track Technique* 47:1494-6.

Viru, A., K. Karelson, and T. Smirnova. 1992a. Stability and variability in hormone responses to prolonged exercise. *International Journal of Sports Medicine* 13:230-5.

Viru, A., K. Karelson, T. Smirnova, and J. Ereline. 1995. Variability in blood glucose change during 2-h exercise. *Sports Medicine, Training and Rehabilitation* 6:127-37.

Viru, A., M. Viru, K. Karelson, and T. Janson. 1999. Hormones in biochemical monitoring of training (abstract). *Journal of Physiology and Pharmacology* 50(Suppl. 1):101.

Viru, A., and P. Kõrge. 1979. Role of anabolic steroids in the hormonal regulation of skeletal muscle adaptation. *Journal of Steroid Biochemistry* 11:931-2.

Viru, A., P. Kõrge, and E. Viru. 1973. Interrelations between glucocorticoid activity of adrenals, cardiovascular system and electrolyte metabolism during prolonged work. *Sechenov Physiological Journal of the USSR* 59:105-10.

Viru, A., L. Kostina, and L. Zhurkina. 1988. Dynamics of cortisol and somatotropin contents in blood of male and female sportsmen during their intensive training. *Fiziologicheskij Zhurnal (Kiev)* 34(4):61-6.

Viru, A., L. Laaneots, K. Karelson, T. Smirnova, and M. Viru. 1998. Exercise-induced hormone responses in girls at different stages of sexual maturation. *European Journal of Applied Physiology* 77:401-8.

Viru, A., L. Litvinova, M. Viru, and T. Smirnova. 1994. Glucocorticoids in metabolic control during exercise: alanine metabolism. *Journal of Applied Physiology* 76:801-5.

Viru, E., J. Pärnat, S. Täll, and A. Viru. 1979. Influence of frequent short-term training sessions on physical working capacity of female students. *Acta et Commentationes Universitatis Tartuensis* 497:12-8.

Viru, A., and V. Ööpik. 1989. Anabolic and catabolic responses to training. In *Paavo nurmi congress book*, ed. Kvist, M. 55-56. Turku: The Finnish Society of Sports Medicine.

Viru, A., and T. Seene. 1982. Peculiarities of adaptation to systematic muscular activity in adrenalectomized rats. *Endokrinologie* 80:235-7.

Viru, A., and T. Seene. 1985. Peculiarities of adjustments in the adrenal cortex to various training regimes. *Biology of Sport* 2:91-9.

Viru, A., and N. Seli. 1992. 3-methylhistidine excretion in training for improved power and strength. *Sports Medicine Training, Rehabilitation* 70:1624-8.

Viru, A., and T. Smirnova. 1982. Independence of physical working capacity from increased glucocorticoid level during short-term exercise. *International Journal of Sports Medicine* 3:80-3.

Viru, A., and T. Smirnova. 1985. Involvement of protein synthesis in action of glucocorticoids on the working capacity of adrenalectomized rats. *International Journal of Sports Medicine* 3:80-3.

Viru, A., and T. Smirnova. 1995. Health promotion and exercise training. *Sports Medicine* 19:123-36.

Viru, A., T. Smirnova, K. Karelson, V. Snegovskaya, and M. Viru. 1996. Determinants and inoculators of hormonal response to exercise. *Biology of Sport* 13:169-87.

Viru, A., T. Smirnova, K. Tomson, and T. Matsin. 1981. Dynamics of blood levels of pituitary trophic hormones during prolonged exercise. In *Biochemistry of exercise IVB*, ed. Poortmans, J., and G. Niset, 100-106. Baltimore: University Park Press.

Viru, A., and Z. Tendzegolskis. 1995. Plasma endorphin species during dynamic exercise in humans. *Clinical Physiology* 15:73-9.

Viru, A., Z. Tendzegolskis, and T. Smirnova. 1990. Changes of ß-endorphin levels in blood during prolonged exercise. *Endocrinologica Experimentalis* 24:63-8.

Viru, A., K. Toode, and A. Eller. 1992b. Adipocyte responses to adrenaline and insulin in active and former sportsmen. *European Journal of Applied Physiology* 64:345-9.

Viru, A., E. Välja, A. C. Hackney, M. Viru, K. Karelson, and T. Janson. 2000. Influence of prolonged aerobic exercise on hormonal responses to subsequent intensive exercise. *Submitted for publication.*

Viru, A., E. Varrik, V. Ööpik, and A. Pehme. 1984. Protein metabolism in muscles after their activity. *Sechenov Physiological Journal of USSR* 70:1624-28.

Viru, A., and M. Viru. 1997a. Adaptivity changes in athletes. *Coaching and Sport Science Journal* 2(2):26-35.

Viru, A., and M. Viru. 1997b. Adattabilità dell'-

organismo: the problema fondamentale in medicina dello sport. *Medicina dello Sport* 50:365-71.

Viru, A., and M. Viru. 2000. Nature of training effects. In *Exercise and sport science,* ed. Garrett, W. E., and D. T. Kirkendall, 67-95. Philadelphia: Lippincott Williams & Wilkins.

Viru, M. 1994. Differences in effects of various training regimes on metabolism of skeletal muscles. *Journal of Sports Medicine and Physical Fitness* 34:217-27.

Viru, M., E. Jansson, A. Viru, and C. J. Sundberg. 1998. Effect of restricted blood flow on exercise-induced hormone changes in healthy men. *European Journal of Applied Physiology* 77:517-22.

Viru, M., L. Litvinova, T. Smirnova, and A. Viru. 1994. Glucocorticoids in metabolic control during exercise: glycogen metabolism. *Journal of Sports Medicine and Physical Fitness* 34:377-82.

Viru, M., K. Lõhmus, T. Kiudma, K. Karelson, and A. Viru. 2000a. Hormonal monitoring of training in elite cross-country skiers. In preparation.

Viru, M., and C. J. Sundberg. 1994. Effects of exercise and training in ischaemic conditions on skeletal muscle metabolism and distribution of fibre types. *Medicina dello Sport* 47:385-90.

Viru, M., H. Valk, and J. Teppan. 2000b. Monitoring of development of aerobic working capacity and endurance performance in young skiers during a three year period. In preparation.

Viti, A., M. Muscettola, L. Paulesu, V. Bocci, and A. Almi. 1985. Effect of exercise on plasma interferon levels. *Journal of Applied Physiology* 59:426-8.

Volek, J. S., W. J. Kraemer, J. A. Bush, T. Incledon, and M. Boetes. 1997. Testosterone and cortisol in relation to dietary nutrients and resistance exercise. *Journal of Applied Physiology* 82:49-54.

Volkov, N. I. 1963. Oxygen consumption and lactic acid content of blood during strenuous muscular exercise. *Federation Proceedings* 22:118-26.

Volkov, N. I. 1974. The problems of biochemical assays in sports activities of man. In *Metabolism and biochemical evaluation of fitness of sportsmen,* ed. Yakovlev, N. N., 213-225. Leningrad: Leningrad Research Institute of Physical Culture.

Volkov, N. I. 1977. Biochemische Kontrolle in Sport. *Theorie und Praxis der Körperkultuur* 26:45-52.

Volkov, N. I. 1990. *Human bioenergetics in strenuous muscular activity and pathways for improved performance in sportsmen.* Moscow: Anokhin Research Institute of Normal Physiology [in Russian].

Vøllestad, N. K., J. Hallen, and O. M. Sejersted. 1995. Effect of exercise intensity on potassium balance in muscle and blood in man. *Journal of Physiology* 475:359-68.

Vora, N. M., S. C. Kukreja, P. A. J. York, E. N. Bowser, G. K. Hargis, and G. A. Williams. 1983. Effect of exercise on serum calcium and parathyroid hormones. *Journal of Clinical Endocrinology and Metabolism* 57:1067-9.

Voznesenskij, L., M. Zaleskij, G. Arzhanova, and V. Tõshkevich. 1979. Control by blood urea in cyclic sports events. *TeoriaiI Praktika Fizisheskoi Kulturõ* 10:21-3 [in Russian].

Vranic, M., R. Kawamori, S. Pek, N. Kovacevic, and G. Wrenshall. 1976. The essentiality of insulin and the role of glucagon in regulating glucose utilization and production during strenuous exercise in dogs. *Journal of Clinical Investigations* 57:245-56.

Vranic, M., R. Kawamori, and G. A. Wranshall. 1975. The role of insulin and glucagon in regulating glucose turnover in dogs during exercise. *Medicine and Science in Sports* 7:27-33.

Wade, C. E., and J. R. Claybaugh. 1980. Plasma renin activity, vasopressin concentration, and urinary excretory responses to exercise in men. *Journal of Applied Physiology* 49:930-6.

Wagenmakers, A. J. 1999. Skeletal muscle amino acid transport and metabolism. In *Biochemistry of exercise, X,* ed. by Hargreaves, M. and M. Thompson, 217-231. Champaign, IL: Human Kinetics.

Wagenmakers, A. J., H. J. Salden, and J. H. Veerkamp. 1985. The metabolic fate of branched-chain amino acids and 2-oxo acids in rat muscle homogenates and diaphragms. *International Journal of Biochemistry* 17(9):957-65.

Wahren, J. 1979. Metabolic adaptation to physical exercise in man. In *Endocrinology.* Vol. 3, ed. DeGroot, L. J., and G. F. Cahill et al., 1911-1926. San Francisco: Grune & Stratton.

Wahren, J., and O. Björkman. 1981. Hormones, exercise, and regulation of splanchnic glucose output in normal man. In *Biochemistry of exercise IV-A,* ed. Poortmans, J., and G. Niset, 149-160. Baltimore: University Park Press.

Wahren, J., P. Felig, L. Hagenfeldt, R. Hendler, and G. Ahlborg. 1975. Splanchnic and leg metabolism of glucose, free fatty acids and amino acids during prolonged exercise in man. In *Adaptation to prolonged physical exercise,* ed. Howald, M., and J. R. Poortmans, 144-153. Basel: Birkhäuser.

Wasserman, K. 1967. Lactate and related acid base and blood gas exchanges during constant load and graded exercise. *Canadian Medical Association Journal* 96:775-83.

Wasserman, K., and M. B. McIlroy. 1964. Detecting the threshold of anaerobic metabolism in cardiac patients during exercise. *American Journal of Cardiology* 14:844-52.

Waterlow, C. 1984. Protein turnover with special reference to man. *Quarterly Journal of Experimental Physiology* 69:409-38.

Waterlow, J. C., R. J. Neale, L. Rowe, and I. Palin. 1972. Effect of diet and infection on creatine turnover in the rat. *American Journal of Clinical Nutrition* 25:371-5.

Watson, R. D., W. A. Littler, and B. M. Eriksson. 1980. Changes in plasma noradrenaline and adrenaline during isometric exercise. *Clinical and Experimental Pharmacology and Physiology* 7(4):399-402.

Watt, E. W., M. L. Foss, and W. D. Block. 1972. Effects of training and detraining on the distribution of cholesterol, triglycerides and nitrogen in tissues of albino rats. *Circulation Research* 31:908-14.

Webb, M. L., J. P. Wallace, C. Hamill, J. L. Hodgson, and M. M. Mashaldi. 1984. Serum testosterone concentration during two hours of moderate intensity treadmill running in trained men and women. *Endocrinology Research* 10:27-38.

Wegmann, H. M., K. E. Klein, and H. Brüner. 1968. Submaximale Belastung und maximale Belastbarkeit. I Biochemische Untersuchung an Untrainierten under körperlicher Arbeit. *International Zeitschrift für die angewante Physiologie* 26:4-12.

Wegner, H., W. Helbig, and F. Reichel. 1965. Die Auswirkung der exogenen Beinflussung der Wasserhausaltes auf die Ausscheidung der Nebennierenrinde-Metabolite im Harn. *Acta Biologica et Medica Germanica* 15:222-8.

Weicker, H., H. Bert, A. Rettenmeier, U. Ottinger, H. Hägele, and U. Keilholz. 1983. Alanine formation during maximal short-term exercise. In *Biochemistry of exercise,* ed. Knuttgen, H. G., J. A. Vogel, and J. Poortmans, 385-394. Champaign, IL: Human Kinetics.

Weicker, H., A. Rettenmeier, F. Ritthaler, H. Frank, W. P. Bieger, and G. Klett. 1981. Influence of anabolic and catabolic hormones on substrate concentrations during various running distances. In *Biochemistry of exercise IV-A,* ed. Poortmans, J., and G. Niset, 208-218. Baltimore: University Park Press.

Weiss, L. W., K. J. Cureton, and F. N. Thomson. 1983. Comparison of serum testosterone and androstenedione responses to weight lifting in men and women. *European Journal of Applied Physiology* 50:413-9.

Weltman, A., D. Snead, P. Stein, R. Seip, R. Schurrer, R. Rutt, and J. Weltman. 1990. Reliability and validity of a continuous incremental treadmill protocol for the determination of lactate threshold, fixed blood lactate concentrations, and $\dot{V}O_2$max. *International Journal of Sports Medicine* 11:26-32.

Westin, A. R., K. H. Myburgh, F. H. Lindsay, S. C. Dennis, T. D. Noakes, and J. A. Hawley. 1997. Skeletal muscle buffering capacity and endurance performance after high intensity interval training by well-trained cyclists. *European Journal of Applied Physiology* 75:7-13.

White, J. A., A. H. Ismail, and G. D. Bottoms. 1976. Effect of physical fitness on the adrenocortical response to exercise stress. *Medicine and Science in Sports* 8:113-8.

Whittlesey, M. J., C. M. Marash, L. E. Armstrong, T. S. Morocco, D. R. Hannon, C. L. V. Gabaree, and J. R. Hoffman. 1996. Plasma volume responses to consecutive anaerobic exercise test. *International Journal of Sports Medicine* 17:268-71.

Wibom, R., and E. Hultman. 1990. ATP-production rate in mitochondria isolated from microsamples of human muscle. *American Journal of Physiology* 259:E204-9.

Wibom, R., K. Söderlund, A. Lundin, and E. Hultman. 1991. A luminometric method for the determination of ATP and phosphocreatine in single human skeletal muscle fibers. *Journal of Bioluminescence and Chemiluminescence* 6:123-9.

Wilkerson, J. E., S. M. Horwath, and G. Gutin. 1980. Plasma testosterone during treadmill exercise. *Journal of Applied Physiology* 49:249-53.

Williams, B. D., D. L. Chinkes, and R. R. Wolfe. 1998. Alanine and glutamine kinetics at rest and during exercise in humans. *Medicine and Science in Sports and Exercise* 30:1053-8.

Williams, M. H., and A. J. Ward. 1977. Hematological changes elicited by prolonged intermittent aerobic exercise. *Research Quarterly for Exercise and Sport* 48:606-16.

Wilmore, J. H., and D. L. Costill. 1994. *Physiology of sport and exercise.* Champaign, IL: Human Kinetics.

Wilson, J. D. 1988. Androgen abuse by athletes. *Endocrine Reviews* 9:181-99.

Wilson, W. M., and R. J. Maughan. 1992. Evidence for a possible role of 5-hydroxytryptamine in the genesis of fatigue in man: administration of paroxetine, a 5-HT re-uptake inhibitor, reduced the capacity to perform prolonged exercise. *Journal of Physiology* 77:921-4.

Winder, W. W., J. M. Hagberg, R. C. Hickson, A. A. Ehsahi, and J. A. McLane. 1978. Time course of sympathoadrenal adaptation to endurance exercise training in man. *Journal of Applied Physiology* 45:370-4.

Winder, W. W., R. C. Hickson, J. M. Hagberg, A. A. Ehsani, and J. A. McLane. 1979. Training-induced changes in hormonal and metabolic responses to submaximal exercise. *Journal of Applied Physiology* 46:766-71.

Winder, W. W., and H. T. Yang. 1987. Blood collection and processing for measurement of catecholamines in exercising rats. *Journal of Applied Physiology* 63:418-20.

Wit, B. 1984. Immunological responses of regularly trained athletes. *Biology of Sport* 1:221-35.

Withers, R. T., W. M. Sherman, D. G. Clark, P. C. Esselbach, S. R. Nolan, M. H. Mackay, and M. Brinkman. 1991. Muscle metabolism during 30, 60 and 90 s of maximal cycling on an air braked ergometer. *European Journal of Applied Physiology* 63:354-62.

Wittert, G. A., J. H. Livesy, E. A. Espiner, and R. A. Donald. 1996. Adaptation of the hypothalamopituitary-adrenal axis of chronic exercise stress in humans. *Medicine and Science in Sports and Exercise* 28:1015-9.

Wolfe, R. R., R. D. Goodenough, M. N. Wolfe, G. T. Royl, and E. K. Nadel. 1982. Isotopic analysis of leucine and urea metabolism in exercising humans. *Journal of Applied Physiology* 52:458-66.

Wolfe, R. R., S. Klein, F. Carraro, and J.-M. Weber. 1990. Role of triglyceride-fatty acid cycle in controlling fat metabolism in humans during and after exercise. *American Journal of Physiology* 258:E382-9.

Wolfe, R. R., E. R. Nadd, J. H. E. Shaw, L. A. Stephenson, and M. H. Wolfe. 1986. Role of changes in insulin and glucagon in glucose homeostasis in exercise. *Journal of Clinical Investigation* 77:900-7.

Wolfe, R. R., M. H. Wolfe, E. R. Nadel, and J. H. Shaw. 1984. Isotopic determination of amino acid-urea interaction in exercise in humans. *Journal of Applied Physiology* 56:221-9.

Wong, T. S., and F. W. Booth. 1990a. Protein metabolism in rat gastrocnemius muscle after stimulated chronic concentric exercise. *Journal of Applied Physiology* 69:1718-24.

Wong, T. S., and F. W. Booth. 1990b. Protein metabolism in rat tibialis anterior muscle after stimulated chronic eccentric exercise. *Journal of Applied Physiology* 69:1709-17.

Wool, I. G., and P. Cavicchi. 1966. Insulin regulation of protein synthesis by muscle ribosomes. *Proceeding of the National Academy of Sciences USA* 56:991-8.

Yakovlev, H. H., L. G. Leshkevitch, N. K. Popova, and L. K. Yampolskaya. 1954. Biochemical changes in blood of rowers in conditions of training and competition. *Teoria i Praktika Fizicheskoi Kulturõ* 17:576-82 [in Russian].

Yakovlev, N. N. 1955. *Survey on sports biochemistry.* Moscow: FiS [in Russian].

Yakovlev, N. N. 1962. Tasks of exercise biochemistry. *Teoria i Praktika Fizicheskoi Kulturõ* 25(4):6-9 [in Russian].

Yakovlev, N. N. 1970. Usage of biochemical criteria for evaluation of the functional state of the body in the sports training. *Teoria i Praktika Fizicheskoi Kulturõ* 33(7):28-30 [in Russian].

Yakovlev, N. N. 1972. Perspectives of development of exercise biochemistry and its significance for sports practice. *Teoria i Praktika Fizicheskoi Kulturõ* 3:21-3 [in Russian].

Yakovlev, N. N. 1975. The role of sympathetic nervous system in the adaptation of skeletal muscles to increased activity. In *Metabolic adaptation to physical exercise*, ed. Howald, H., and J. Poortmans, 293-300. Basel: Birkhäuser.

Yakovlev, N. N. 1977. *Sportbiochemie.* Leipzig: Barth.

Yakovlev, N. N. 1978. Biochemische und morphologische Veränderungen der Muskelfasern in Abhängigkeit von der Art des Training. *Medicine und Sport* 18:161-4.

Yakovlev, N. N. 1979. Ornithine metabolism and adaptation to increased muscular activity. *Sechenov Physiological Journal of the USSR* 65:979-84.

Yakovlev, N. N., and A. Viru. 1985. Adrenergic regulation of adaptation to muscular activity. *International Journal of Sports Medicine* 6:255-65.

Yakovlev, N. N., L. I. Yampolskaya, A. G. Leshkevitch, and N. K. Popova. 1952. Biochemical changes in blood of sportsmen during competition in sports games. *Sechenov Physiology Journal of USSR* 38:739-47.

Yakovlev, N. N., N. P. Yeremenko, A. G. Leshkevitch, A. F. Makarova, and N. K. Popova. 1959. Improvement in strength, speed and endurance gained by training for various kinds of sports. *Sechenov Physiology Journal of USSR* 45:1422-9.

Yarasheski, K. E., J. J. Zachwieja, and D. M. Bier. 1993. Acute effects of resistance exercise on muscle protein synthesis rate in young and elderly men and women. *American Journal of Physiology* 265:E210-4.

Yaresheski, K. E., J. J. Zachnieja, J. A. Cambell, and D. M. Bier. 1995. Effect of growth hormone and resistance exercise on muscle growth and strength in older men. *American Journal of Physiology* 268:E268-76.

Yoshida, T. 1989. Effect of dietary modification on lactate threshold and onset of blood lactate accumulation during incremental exercise. *European Journal of Applied Physiology* 53:200-5.

Young, D. A., H. Wallberg-Henriksson, J. Cranshaw, M. Chem, and J. O. Holloszy. 1985. Effect of catecholamines on glucose uptake and glycogenolysis in rat skeletal muscle. *American Journal of Physiology* 248:C406-9.

Young, V. R., and H. N. Munro. 1978. N^1-methylhistidine (3-methylhistidine) and muscle protein turnover: an overview. *Federation Proceedings* 37:2291-300.

Yu, B. 1994. Cellular defences against damage from reactive oxygen species. *Physiological Review* 74:139-62.

Zarzeczny, R., K. Madsen, W. Pilis, J. Langfort, and F. Kokot. 1999. Changes in plasma ions concentration in relation to lactate and ventilatory thresholds during graded exercise in men. *Biology of Sports* 16:245-56.

Zerbes, H., K. Kühne, and H.-C. Götte. 1983. Der Einfluss von körperlicher Belatung und Eiweißgehalt der Nahrung auf die Harkstoff-production, -exkretion und –retention. *Medizin und Sport* 23:299-301.

Zhimkin, N. V. 1968. Significance of increased muscular activity on improved function of human organism in contemporary society. In *Civilization, sport and heart,* ed. Geselevich, V. A., 5-11. Moscow: FiS.

Zhimkin, N. V. 1975. *Human physiology*, p. 361. Moscow: FiS [in Russian].

Zouhal, H., F. Rannou, A. Grates-Delamarche, M. Monnier, D. Bentué-Ferrer, and P. Delamarche. 1998. Adrenal medulla responsiveness to the sympathetic nervous activity in sprinters and untrained subjects during a supramaximal exercise. *International Journal of Sports Medicine* 19:172-6.

Zuliani, U., A. Novariani, A. Bonetti, M. Astorri, G. Mortani, I. Simoni, and A. Zappavigna. 1984. Endocrine modifications caused by sports activity. Effect in leisure-time cross-country skiers. *Journal of Sports Medicine and Physical Fitness* 24:263-9.

Zunts, N. 1901. Ueber die Bedeutung der verschiedeners Nährstoffe als Erzeuger der Muskelkraft. *Pflügers Archiv für die gesamte Physiologie* 83:557-71.

Index

Figures and tables are indicated with an italic *f* and *t*.

About the Authors

Atko-Meeme Viru, PhD, DSc, is a professor emeritus specializing in exercise physiology at the University of Tartu in Estonia. His investigations examine both the fundamental problems and applied questions related to the foundations of training monitoring. He earned a PhD from the University of Tartu in Estonia and a DSc from the Academy of Sciences of Estonia.

Mehis Viru, PhD, is a senior reseacher, head of the Laboratory of Sports Physiology, and the chair of coaching studies at the University of Tartu in Estonia. His main research areas are training monitoring, overtraining, and metabolic-hormonal adaptation to exercise and training. Dr. Viru has spent 15 years monitoring the training of Estonian elite athletes in different sport events. He studied and worked for four years at the Karolinska Institute in Stockholm, one of the world's leading medical institutes. Dr. Viru earned a PhD from the University of Tartu in Estonia.